INTERNATIONAL MATHEMATICAL SERIES

Series Editor: **Tamara Rozhkovskaya**
Novosibirsk, Russia

This series was founded in 2002 and is a joint publication of Springer and "Tamara Rozhkovskaya Publisher." Each volume presents contributions from the Volume Editors and Authors exclusively invited by the Series Editor Tamara Rozhkovskaya who also prepares the Camera Ready Manuscript. This volume is distributed by "Tamara Rozhkovskaya Publisher" (tamara@mathbooks.ru) in Russia and by Springer over all the world.

For other titles published in this series, go to
www.springer.com/series/6117

AROUND THE RESEARCH OF VLADIMIR MAZ'YA III

Analysis and Applications

Editor: **Ari Laptev**

Imperial College London, UK
Royal Institute of Technology, Sweden

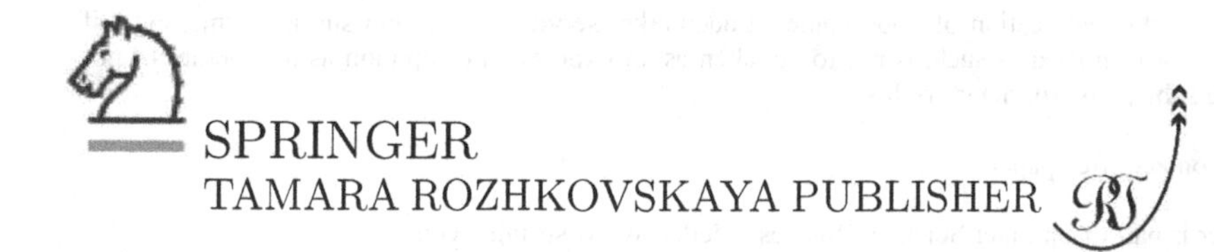

Editor
Ari Laptev
Department of Mathematics
Imperial College London
Huxley Building, 180 Queen's Gate
London SW7 2AZ
United Kingdom
a.laptev@imperial.ac.uk

ISSN 1571-5485 e-ISSN 1574-8944
ISBN 978-1-4614-2551-9 e-ISBN 978-1-4419-1345-6

DOI 10.1007/978-1-4419-1345-6
Springer New York Dordrecht Heidelberg London

Softcover reprint of the hardcover 1st edition 2010

Printed on acid-free paper

Springer is part of Springer Science+Business Media (www.springer.com)

V. Maz'ya. 1950

Vladimir Maz'ya was born on December 31, 1937, in Leningrad (present day, St. Petersburg) in the former USSR. His first mathematical article was published in Doklady Akad. Nauk SSSR when he was a fourth-year student of the Leningrad State University. From 1961 till 1986 V. Maz'ya held a senior research fellow position at the Research Institute of Mathematics and Mechanics of LSU, and then, during 4 years, he headed the Laboratory of Mathematical Models in Mechanics at the Institute of Engineering Studies of the Academy of Sciences of the USSR. Since 1990, V. Maz'ya lives in Sweden. At present, Vladimir Maz'ya is a Professor Emeritus at Linköping University and Professor at Liverpool University. He was elected a Member of Royal Swedish Academy of Sciences in 2002. The list of publications of V. Maz'ya contains 20 books and more than 450 research articles covering diverse areas in Analysis and containing numerous fundamental results and fruitful techniques. Research activities of Vladimir Maz'ya have strongly influenced the development of many branches in Analysis and Partial Differential Equations, which are clearly highlighted by the contributions to this collection of 3 volumes, where the world-recognized specialists present recent advantages in the areas I. *Function Spaces* (Sobolev type spaces, isoperimetric and capacitary inequalities in different contexts, sharp constants, traces, extnesion operators etc.) II. *Partial Differential Equations* (asymptotic analysis, homogenization, boundary value problems, boundary integral equations etc.) III. *Analysis and Applications* (various problems including the oblique derivative problem, ill-posed problems etc.)

The monograph *Theory of Sobolev Multipliers* by Vladimir Maz'ya and his wife Tatyana Shaposhnikova was published in 2009 (Springer). In 2003, the French Academy of Sciences awarded V. Maz'ya and T. Shaposhnikova the Verdaguer prize for the book *Jacques Hadamard, a universal mathematician.*

I. Function Spaces

Ari Laptev Ed.

II. Partial Differential Equations

Ari Laptev Ed.

III. Analysis and Applications

Ari Laptev Ed.

Contributors

Editor

Ari Laptev

President
The European Mathematical Society

Professor
Head of Department
Department of Mathematics
Imperial College London
Huxley Building, 180 Queen's Gate
London SW7 2AZ, UK
a.laptev@imperial.ac.uk

Professor
Department of Mathematics
Royal Institute of Technology
100 44 Stockholm, Sweden
laptev@math.kth.se

Ari Laptev is a world-recognized specialist in Spectral Theory of Differential Operators. He discovered a number of sharp spectral and functional inequalities. In particular, jointly with his former student T. Weidl, A. Laptev proved sharp Lieb–Thirring inequalities for the negative spectrum of multidimensional Schrödinger operators, a problem that was open for more than twenty five years.

A. Laptev was brought up in Leningrad (Russia). In 1971, he graduated from the Leningrad State University and was appointed as a researcher and then as an Assistant Professor at the Mathematics and Mechanics Department of LSU. In 1982, he was dismissed from his position at LSU due to his marriage to a British subject. Only after his emigration from the USSR in 1987 he was able to continue his career as a mathematician. Then A. Laptev was employed in Sweden, first as a lecturer at Linköping University and then from 1992 at the Royal Institute of Technology (KTH). In 1999, he became a professor at KTH and also Vice Chairman of its Department of Mathematics. From January 2007 he is employed by Imperial College London where from September 2008 he is the Head of Department of Mathematics.

A. Laptev was the Chairman of the Steering Committee of the five years long ESF Programme SPECT, the President of the Swedish Mathematical Society from 2001 to 2003, and the President of the Organizing Committee of the Fourth European Congress of Mathematics in Stockholm in 2004. He is now the President of the European Mathematical Society for the period January 2007–December 2010.

Authors

David R. Adams Vol. III
University of Kentucky
Lexington, KY 40506-0027
USA
dave@ms.uky.edu

Hiroaki Aikawa Vol. III
Hokkaido University
Sapporo 060-0810
JAPAN
aik@math.sci.hokudai.ac.jp

Farit Avkhadiev Vol. I
Kazan State University
420008 Kazan
RUSSIA
favhadiev@ksu.ru

Catherine Bandle Vol. II
Mathematisches Institut
Universität Basel
Rheinsprung 21, CH-4051 Basel
SWITZERLAND
catherine.bandle@unibas.ch

Gerassimos Barbatis Vol. II
University of Athens
157 84 Athens
GREECE
gbarbatis@math.uoa.gr

Sergey Bobkov Vol. I
University of Minnesota
Minneapolis, MN 55455
USA
bobkov@math.umn.edu

Victor I. Burenkov Vol. II
Università degli Studi di Padova
63 Via Trieste, 35121 Padova
ITALY
burenkov@math.unipd.it

Grigori Chechkin Vol. II
Lomonosov Moscow State University
Vorob'evy Gory, Moscow
RUSSIA
chechkin@mech.math.msu.su

Alberto Cialdea Vol. III
Università della Basilicata
Viale dell'Ateneo Lucano 10,
85100, Potenza
ITALY
cialdea@email.it

Andrea Cianchi Vol. I
Università di Firenze
Piazza Ghiberti 27, 50122 Firenze
ITALY
cianchi@unifi.it

Martin Costabel Vol. I
Université de Rennes 1
Campus de Beaulieu
35042 Rennes
FRANCE
martin.costabel@univ-rennes1.fr

Monique Dauge Vols. I, II
Université de Rennes 1
Campus de Beaulieu
35042 Rennes
FRANCE
monique.dauge@univ-rennes1.fr

Martin Dindoš Vol. II
Maxwell Institute of
Mathematics Sciences
University of Edinburgh
JCMB King's buildings Mayfield Rd
Edinbugh EH9 3JZ
UK
m.dindos@ed.ac.uk

András Domokos Vol. II
California State
University Sacramento
Sacramento 95819
USA
domokos@csus.edu

Yuri V. Egorov Vol. II
Université Paul Sabatier
118 route de Narbonne
31062 Toulouse Cedex 9
FRANCE
egorov@mip.ups-tlse.fr

Gregory Eskin Vol. III
University of California
Los Angeles, CA 90095-1555
USA
eskin@math.ucla.edu

Nicola Garofalo Vol. I
Purdue University
West Lafayette, IN 47906
USA
garofalo@math.purdue.edu
and

Università di Padova
35131 Padova
ITALY
garofalo@dmsa.unipd.it

Vladimir Gol'dshtein Vol. I
Ben Gurion University of the Negev
P.O.B. 653, Beer Sheva 84105
ISRAEL
vladimir@bgu.ac.il

Alexander Grigor'yan Vol. II
Bielefeld University
Bielefeld 33501
GERMANY
grigor@math.uni-bielefeld.de

Stathis Filippas Vol. I
University of Crete
71409 Heraklion
Institute of Applied and
Computational Mathematics
71110 Heraklion
GREECE
filippas@tem.uoc.gr

Rupert L. Frank Vol. I
Princeton University
Washington Road
Princeton, NJ 08544
USA
rlfrank@math.princeton.edu

Michael W. Frazier Vol. III
University of Tennessee
Knoxville, Tennessee 37922
USA
frazier@math.utk.edu

Bernard Helffer Vol. III
Université Paris-Sud
91 405 Orsay Cedex
FRANCE
Bernard.Helffer@math.u-psud.fr

Thomas Hoffmann-Ostenhof Vol. III
Institut für Theoretische Chemie
Universität Wien
Währinger Strasse 17, and
International Erwin
Schrödinger Institute
for Mathematical Physics
Boltzmanngasse 9
A-1090 Wien
AUSTRIA
thoffman@esi.ac.at

Volodymyr Hrynkiv Vol. III
University of Houston-Downtown
Houston, TX 77002-1014
USA
HrynkivV@uhd.edu

Niels Jacob Vol. I
Swansea University
Singleton Park
Swansea SA2 8PP
UK
n.jacob@swansea.ac.uk

Dmitry Khavinson Vol. II
University of South Florida
4202 E. Fowler Avenue, PHY114
Tampa, FL 33620-5700
USA
dkhavins@cas.usf.edu

Juha Kinnunen Vol. I
Institute of Mathematics
Helsinki University of Technology
P.O. Box 1100, FI-02015
FINLAND
juha.kinnunen@tkk.fi

Gerasim Kokarev Vol. II
University of Edinburgh
King's Buildings, Mayfield Road
Edinburgh EH9 3JZ
UK
G.Kokarev@ed.ac.uk

Vladimir A. Kondratiev Vol. II
Moscow State University
119992 Moscow
RUSSIA
vla-kondratiev@yandex.ru

Riikka Korte Vol. I
University of Helsinki
P.O. Box 68
Gustaf Hällströmin katu 2 b
FI-00014
FINLAND
riikka.korte@helsinki.fi

Pekka Koskela Vol. I
University of Jyväskylä
P.O. Box 35 (MaD), FIN-40014
FINLAND
pkoskela@maths.jyu.fi

Nikolay Kuznetsov Vol. II
Institute for Problems
in Mechanical Engineering
Russian Academy of Sciences
V.O., Bol'shoy pr. 61
199178 St. Petersburg
RUSSIA
nikolay.g.kuznetsov@gmail.com

Pier Domenico Lamberti Vol. II
Universitá degli Studi di Padova
63 Via Trieste, 35121 Padova
ITALY
lamberti@math.unipd.it

Ari Laptev Vol. I
Imperial College London
Huxley Building, 180 Queen's Gate
London SW7 2AZ
UK
a.laptev@imperial.ac.uk
and
Royal Institute of Technology
100 44 Stockholm
SWEDEN
laptev@math.kth.se

Suzanne Lenhart Vol. III
University of Tennessee
Knoxville, TN 37996-1300
USA
lenhart@math.utk.edu

Vitali Liskevich Vol. II
Swansea University
Singleton Park
Swansea SA2 8PP
UK
v.a.liskevich@swansea.ac.uk

Juan J. Manfredi Vol. II
University of Pittsburgh
Pittsburgh, PA 15260
USA
manfredi@pitt.edu

Moshe Marcus Vol. I
Israel Institute of Technology-Technion
33000 Haifa
ISRAEL
marcusm@math.technion.ac.il

Joaquim Martín Vol. I
Universitat Autònoma de Barcelona
Bellaterra, 08193 Barcelona
SPAIN
jmartin@mat.uab.cat

Eric Mbakop Vol. I
Worcester Polytechnic Institute
100 Institute Road
Worcester, MA 01609
USA
steve055@WPI.EDU

Nicolas Meunier Vol. II
Université Paris Descartes (Paris V)
45 Rue des Saints Pères
75006 Paris
FRANCE
nicolas.meunier@math-info.univ-paris5.fr

Emanuel Milman Vol. I
Institute for Advanced Study
Einstein Drive, Simonyi Hall
Princeton, NJ 08540
USA
emilman@math.ias.edu

Mario Milman Vol. I
Florida Atlantic University
Boca Raton, Fl. 33431
USA
extrapol@bellsouth.net

Michele Miranda Jr. Vol. I
University of Ferrara
via Machiavelli 35
44100, Ferrara
ITALY
michele.miranda@unife.it

Dorina Mitrea Vol. III
University of Missouri at Columbia
Columbia, MO 65211
USA
mitread@missouri.edu

Marius Mitrea Vol. III
University of Missouri
at Columbia
Columbia, MO 65211
USA
mitream@missouri.edu

Stanislav Molchanov Vol. III
University of North Carolina
at Charlotte
Charlotte, NC 28223
USA
smolchan@uncc.edu

Sylvie Monniaux Vol. III
Université Aix-Marseille 3
F-13397 Marseille Cédex 20
FRANCE
sylvie.monniaux@univ.u-3mrs.fr

Vitaly Moroz Vol. II
Swansea University
Singleton Park
Swansea SA2 8PP
UK
v.moroz@swansea.ac.uk

Umberto Mosco Vol. I
Worcester Polytechnic Institute
100 Institute Road
Worcester, MA 01609
USA
mosco@WPI.EDU

Oleg Motygin Vol. II
Institute for Problems
in Mechanical Engineering
Russian Academy of Sciences
V.O., Bol'shoy pr. 61
199178 St. Petersburg
RUSSIA
o.v.motygin@gmail.com

Nikolai Nadirashvili Vol. II
Centre de Mathématiques
et Informatique
Université de Provence
39 rue F. Joliot-Curie
13453 Marseille Cedex 13
FRANCE
nicolas@cmi.univ-mrs.fr

Yuri Netrusov Vol. III
University of Bristol
University Walk
Bristol BS8 1TW
UK
y.netrusov@bristol.ac.uk

Serge Nicaise Vol. I
Université Lille Nord de France
UVHC, 59313
Valenciennes Cedex 9
FRANCE
snicaise@univ-valenciennes.fr

Dian K. Palagachev Vol. III
Technical University of Bari
Via E. Orabona 4
70125 Bari
ITALY
palaga@poliba.it

Grigory P. Panasenko Vol. II
University Jean Monnet
23, rue Dr Paul Michelon
42023 Saint-Etienne
FRANCE
Grigory.Panasenko@univ-st-etienne.fr

Luboš Pick Vol. III
Charles University
Sokolovská 83, 186 75 Praha 8
CZECH REPUBLIC
pick@karlin.mff.cuni.cz

Sergei V. Poborchi Vol. II
St. Petersburg State University
28, Universitetskii pr., Petrodvorets
St. Petersburg 198504
RUSSIA
poborchi@mail.ru

Wolfgang Reichel Vol. II
Universität Karlsruhe (TH)
D-76128 Karlsruhe
GERMANY
wolfgang.reichel@math.uni-karlsruhe.de

Jürgen Roßmann Vol. II
University of Rostock
Institute of Mathematics
D-18051 Rostock
GERMANY
juergen.rossmann@uni-rostock.de

Grigori Rozenblum Vol. III
Chalmers University of Technology
University of Gothenburg
S-412 96, Gothenburg
SWEDEN
grigori@math.chalmers.se

Yuri Safarov Vol. III
King's College London
Strand, London WC2R 2LS
UK
yuri.safarov@kcl.ac.uk

Laurent Saloff-Coste Vol. I
Cornell University
Mallot Hall, Ithaca, NY 14853
USA
lsc@math.cornell.edu

Evariste Sanchez-Palencia Vol. II
Université Pierre et Marie Curie
4 place Jussieu
75252 Paris
FRANCE
sanchez@lmm.jussieu.fr

René L. Schilling Vol. I
Technische Universität Dresden
Institut für Stochastik
D-01062 Dresden
GERMANY
rene.schilling@tu-dresden.de

Gunther Schmidt Vol. II
Weierstrass Institute of
Applied Analysis and Stochastics
Mohrenstr. 39, 10117 Berlin
GERMANY
schmidt@wias-berlin.de

Robert Seiringer Vol. I
Princeton University
P. O. Box 708
Princeton, NJ 08544
USA
rseiring@princeton.edu

Christina Selby Vol. I
The Johns Hopkins University
11100 Johns Hopkins Road
Laurel, MD 20723.
USA
Christina.Selby@jhuapl.edu

Nageswari Shanmugalingam Vol. I
University of Cincinnati
Cincinnati, OH 45221-0025
USA
nages@math.uc.edu

Johannes Sjöstrand Vol. III
Université de Bourgogne
9, Av. A. Savary, BP 47870
FR-21078 Dijon Cédex
and UMR 5584, CNRS
FRANCE
johannes@u-bourgogne.fr

Igor I. Skrypnik Vol. II
Institute of Applied
Mathematics and Mechanics
Donetsk
UKRAINE
iskrypnik@iamm.donbass.com

Ruxandra Stavre Vol. II
Institute of Mathematics
"Simion Stoilow" Romanian Academy
P.O. Box 1-764
014700 Bucharest
ROMANIA
Ruxandra.Stavre@imar.ro

Nikos Stylianopoulos Vol. II
University of Cyprus
P.O. Box 20537
1678 Nicosia
CYPRUS
nikos@ucy.ac.cy

Susanna Terracini Vol. III
Università di Milano Bicocca
Via Cozzi, 53 20125 Milano
ITALY
susanna.terracini@unimib.it

Achilles Tertikas Vol. I
University of Crete and
Institute of Applied and
Computational Mathematics
71110 Heraklion
GREECE
tertikas@math.uoc.gr

Jesper Tidblom Vol. I
The Erwin Schrödinger Institute
Boltzmanngasse 9
A-1090 Vienna
AUSTRIA
Jesper.Tidblom@esi.ac.at

Sébastien Tordeux Vol. II
Institut de Mathématiques de Toulouse
31077 Toulouse
FRANCE
sebastien.tordeux@insa-toulouse.fr

Alexander Ukhlov Vol. I
Ben Gurion University of the Negev
P.O.B. 653, Beer Sheva 84105
ISRAEL
ukhlov@math.bgu.ac.il

Boris Vainberg Vol. III
University of North Carolina
at Charlotte
Charlotte, NC 28223
USA
brvainbe@uncc.edu

Igor E. Verbitsky Vol. III
University of Missouri
Columbia, Missouri 65211
USA
verbitskyi@missouri.edu

Gregory C. Verchota Vol. II
Syracuse University
215 Carnegie
Syracuse NY 13244
USA
gverchot@syr.edu

Laurent Véron Vol. I
Université de Tours
Parc de Grandmont
37200 Tours
FRANCE
veronl@univ-tours.fr

Grégory Vial Vol. II
IRMAR, ENS Cachan Bretagne
CNRS, UEB
35170 Bruz
FRANCE
gvial@bretagne.ens-cachan.fr

Jie Xiao Vol. I
Memorial University
of Newfoundland
St. John's, NL A1C 5S7
CANADA
jxiao@mun.ca

Boguslaw Zegarlinski Vol. I
Imperial College London
Huxley Building
180 Queen's Gate
London SW7 2AZ
UK
b.zegarlinski@imperial.ac.uk

Analysis and Applications

Contributions of Vladimir Maz'ya

The following should be also included into the long list of influential contributions of Vladimir Maz'ya.

— In 1989, V. Maz'ya and V. Kozlov developed new iterative procedures for solving ill-posed problems of mathematical physics. Unlike widely used numerical methods, where the equations are perturbed, they proposed to construct iteratively solutions to problems for the original equation and provide the convergence by an appropriate choice of the boundary conditions at each step. In 1994, V. Kozlov, V. Maz'ya, and A. Fomin established the uniqueness theorem for the inverse problem of coupled thermoelasticity. An inverse problem connected with a wave motion was also discussed in the paper by H. Hellsten, V. Maz'ya, and B. Vainberg (2003).

— Explicit criteria of the L^p-contractivity of semigroups generated by elliptic equations and systems were obtained by V. Maz'ya and A. Cialdea in 2005–2006.

— In 1972, V. Maz'ya made a breakthrough in the study of the oblique derivative problem (Poincaré problem). General degenerate elliptic pseudodifferential operators on a compact manifold were studied by V. Maz'ya and B. Paneyah (1974).

— V. Maz'ya obtained significant results in mathematical elasticity and fluid mechanics. In 1967, V. Maz'ya and S. Mikhlin published a paper concerning the Cosserat spectrum associated with the Lamé equations of elastostatics. A method of computation of stress intensity factors in fracture mechanics was developed by V. Maz'ya and B. Plamenevskii (1974).

— Several important results concerning Schrödinger and general second order differential operators were obtained by V. Maz'ya with coauthors. In particular, two papers with M. Shubin contain necessary and sufficient results on the discreteness of spectra for the Schrödinger operator.

— Necessary and sufficient conditions ensuring various estimates for general differential operators with pseudodifferential coefficients were found by I. Gelman and V. Maz'ya. These results were collected in their book of 1981.

— Among Maz'ya's contributions to the complex function theory are sharp approximation theorems and improvement of Carleson's uniqueness theorems obtained together with V. Khavin (1968-1969). A new unified approach to Hadamard type real part theorems for analytic functions was recently proposed by G. Kresin and V. Maz'ya.

— In the late 1980's, V. Maz'ya invented the so-called "approximate approximations," which deliver a numerical approximation within the desired accuracy even though the approximation process does not necessarily converge as the grid size tends to zero. On this subject, the book by V. Maz'ya and G. Schmidt was published in 2007.

Main Topics

In this volume, the following topics are discussed:

- Obstacle type variational inequality with biharmonic differential operator. /Adams–Hrynkiv–Lenhart/
- The Beurling minimum principle (studied by V. Maz'ya in 1972) is applied for treating the minimal thinness of nontangentially accessible domains. /Aikawa/
- The L^p-dissipativity of partial differential operators and the L^p-contractivity of the generated semigroups. Review of results by V. Maz'ya, G. Kresin, M. Langer, and A. Cialdea. /Cialdea/
- Uniqueness and nonuniqueness in inverse hyperbolic problems and the existence of black (white) holes. /Eskin/
- Global exponential bounds for Green's functions for a broad class of differential and integral equations with possibly singular coefficients, data, and boundaries of the domains. /Frazier–Verbitsky/
- Properties of spectral minimal partitions in the case of the sphere. /Helffer–T.Hoffmann-Ostenhof–Terracini/
- The boundedness of integral operators from Besov spaces on the boundary of a Lipschitz domain Ω into weighted Sobolev spaces of functions in Ω /D.Mitrea–M.Mitrea–Monniaux/
- The Cwikel–Lieb–Rozenblum and Lieb–Thirring inequalities for operators on functions in metric spaces. Applications to spectral problems with the Schrödinger operator. /Molchanov–Vainberg/
- The Weyl formula for the Laplace operator on a domain under minimal assumptions on the boundary. Estimates for the remainder. /Netrusov–Safarov/
- The $W^{2,p}$-theory of a degenerate oblique derivative problem for second order uniformly elliptic operators. /Palagachev/
- The Hardy integral operator. Weighted inequalities in the integral and supremum forms. /Pick/
- Finite rank Toeplitz operators and applications. /Rozenblum/
- Estimates for the resolvent of a non-selfadjoint pseudodifferential operator. An approach based on estimates of an associated semigroup. /Sjostrand/

Contents

Optimal Control of a Biharmonic Obstacle Problem

David R. Adams, Volodymyr Hrynkiv, and Suzanne Lenhart

Abstract We consider a variational inequality of the obstacle type where the underlying partial differential operator is biharmonic. We consider an optimal control problem where the state of the system is given by the solution of the variational inequality and the obstacle is taken to be a control. For a given target profile we want to find an obstacle such that the corresponding solution to the variational inequality is close the target profile while the norm of the obstacle does not get too large in the appropriate space. We prove the existence of an optimal control and derive the optimality system by using approximation techniques. Namely, the variational inequality and the objective functional are approximated by a semilinear partial differential equation and the corresponding approximating functional respectively.

1 Introduction

Let $\Omega \subset \mathbb{R}^n$ be a bounded domain with a smooth boundary $\partial\Omega$, and let $z \in L^2(\Omega)$ be a given target profile. Denote $V \stackrel{\text{def}}{=} H^3(\Omega) \cap H_0^2(\Omega)$. Given $\psi \in V$, define the closed convex set

David R. Adams
Department of Mathematics, University of Kentucky, Lexington, KY 40506-0027, USA
e-mail: dave@ms.uky.edu

Volodymyr Hrynkiv
Department of Computer and Mathematical Sciences, University of Houston-Downtown, Houston, TX 77002-1014, USA
e-mail: HrynkivV@uhd.edu

Suzanne Lenhart
Department of Mathematics, University of Tennessee, Knoxville, TN 37996-1300, USA
e-mail: lenhart@math.utk.edu

A. Laptev (ed.), *Around the Research of Vladimir Maz'ya III: Analysis and Applications*, International Mathematical Series 13, DOI 10.1007/978-1-4419-1345-6_1,

$$K(\psi) = \{v \in H_0^2(\Omega) : v \geqslant \psi \ \text{ a.e. in } \Omega\}$$

and consider the following obstacle problem for the biharmonic operator:

$$\text{find } u \in K(\psi) \text{ such that } \int_\Omega \Delta u\, \Delta(v-u)\, dx \geqslant 0 \ \ \forall v \in K(\psi). \tag{1.1}$$

The equivalent strong formulation of (1.1) is as follows:

$$\begin{aligned}
&\text{find } u \in H_0^2(\Omega) \text{ such that}\\
&u \geqslant \psi \text{ a.e. in } \Omega,\\
&\Delta^2 u \geqslant 0 \text{ a.e. in } \Omega,\\
&(\Delta^2 u)(u-\psi) = 0 \text{ a.e. in } \Omega.
\end{aligned}$$

The bilinear form

$$a(u,v) \stackrel{\text{def}}{=} \int_\Omega \Delta u\, \Delta(v-u)\, dx$$

can be shown to satisfy the coercivity condition and, consequently (cf. [19]), the problem 1.1 admits a unique solution $u \in H_0^2(\Omega)$ which is also the minimizer of the functional

$$v \mapsto \int_\Omega |\Delta v|^2\, dx$$

over the set $K(\psi)$. We will show in the next section that if u solves (1.1), then there exists a nonnegative Borel measure $\mu \in H^{-2}(\Omega)$ such that $\Delta^2 u = \mu$ in the sense of distributions.

We consider an optimal control problem for (1.1) in which we view ψ as the control and $u \stackrel{\text{def}}{=} T(\psi)$ as the corresponding state. We introduce an objective functional

$$J(\psi) = \frac{1}{2}\int_\Omega \{|T(\psi)-z|^2 + |\nabla\Delta\psi|^2\} dx \tag{1.2}$$

and formulate the following

Problem C. Find $\psi^* \in V$ such that $J(\psi^*) = \inf_{\psi\in V} J(\psi)$. (1.3)

In other words, for a given target profile $z \in L^2(\Omega)$ we want to find an obstacle $\psi^* \in V$ such that the corresponding solution $u^* = T(\psi^*)$ is close to z in the L^2 norm, while ψ^* is not too large in $H^3(\Omega)$.

Any ψ^* satisfying (1.3) is called an *optimal control*, the corresponding state $u^* = T(\psi^*)$ is called an *optimal state*, and (u^*, ψ^*) is referred to as an *optimal pair.*

Theoretical analysis of variational inequalities has received a significant amount of attention (cf. [6, 8, 13, 15, 19, 26, 27]).

The motivation for this paper is threefold.

- First, the obstacle problem (1.1) arises in elasticity theory when, for example, a two-dimensional plate is bent so that it must stay above the obstacle ψ. More specifically, it concerns the small transverse displacement of a plate when its boundary is fixed and the whole plate is simultaneously subject to a pressure to lie on one side of an obstacle.
- Second, use of the obstacle as the control is a novel feature for this operator.
- Third, a very "rough" obstacle in the framework of the given optimal control problem leads to the consideration of fine pointwise properties of functions from the Sobolev spaces, in particular, the need for "capacity of a set" [2].

The fundamental work on optimal control of variational inequalities was done by Barbu [8, 9]. Various coefficients and source terms were taken as the control, but not the obstacle. The first work to consider the obstacle as a control was by Adams, Lenhart, and Yong [3]. It was found there that the optimal obstacle is equal to the corresponding state, i.e., $\psi^* = T(\psi^*)$. The results from [3] were generalized in [4, 5, 11, 10, 12, 20, 21] to include source terms, bilateral problems as well as semilinear, quasilinear, and parabolic operators. Recent work on semilinear elliptic variational inequalities with control entering in lower order terms can be found in [10, 14]. A general framework for analysis of optimal control of elliptic variational inequalities using an augmented Lagrangian technique was developed by Ito and Kunisch [17, 18]. For results on other types of control of variational inequalities see [8]–[7].

To derive necessary conditions on an optimal control, we would like to differentiate the map $\psi \mapsto J(\psi)$, which would require differentiation of the map $\psi \mapsto T(\psi)$. Since the map $\psi \mapsto T(\psi)$ is not directly differentiable, an approximation problem with a semilinear PDE is introduced. The approximation map $\psi \mapsto T_\delta(\psi)$ will be then differentiable and approximate necessary conditions will be derived. This method of deriving necessary conditions through this approximation was first used by Barbu [8].

This paper is organized as follows. In Section 2, we review some results from the theory of Sobolev spaces and derive an estimate that shows explicitly the dependence of u on ψ in the H^2 norm. The existence of an optimal control is proved in Section 3. Section 4 introduces and deals with a family of approximation problems which is the main tool in the derivation of optimality system. These problems have the regularized state equations with the same cost functional (1.2). We prove that the optimal control for the approximation problems converges strongly in $H^2(\Omega)$ to the optimal control for Problem C. This allows us to obtain necessary conditions for the optimal control which is given in Section 5.

2 Preliminaries

For a given compact subset $e \subset \mathbb{R}^n$, define the closed convex set

$$K_e = \{v \in C_0^\infty(\Omega) : v \geqslant 1 \text{ on } e\}.$$

Then the $C_{2,2}$*-capacity* of e is defined by

$$C_{2,2}(e) = \inf_{v \in K_e} \|\Delta v\|^2_{L^2(\Omega)}.$$

For history and relevant information on capacity, the reader is referred to [2, 19, 25, 26]. We will need the following result concerning pointwise comparison of functions from Sobolev spaces.

Lemma 2.1. *Let $u, v \in H^2(\Omega)$. If $u \geqslant v$ a.e. (with respect to the Lebesgue measure), then $\widehat{u} \geqslant \widehat{v}$ $C_{2,2}$-a.e., where $\widehat{u}$ denotes a precise representative for u defined $C_{2,2}$-a.e. and*

$$\widehat{u}(x) = \lim_{R \to 0} \frac{1}{|B(x,R)|} \int_{B(x,R)} u \, dy, \tag{2.1}$$

where the limit exists $C_{2,2}$-a.e.

Proof. See [2]. □

In what follows, we will use the precise representatives and for the sake of notational convenience we drop the "hats" which denote precise representatives.

Lemma 2.2. *Let $u = T(\psi)$ solve (1.1). Then there exists a nonnegative Borel measure $\mu \in H^{-2}(\Omega)$ such that*

$$\Delta^2 u = \mu \text{ in } \Omega, \tag{2.2}$$

where u denotes a precise representative defined $C_{2,2}$-a.e. in Ω, and $\Delta^2 u = \mu$ is understood in the sense of distributions, i.e.,

$$\int_\Omega \Delta u \Delta \xi \, dx = \int_\Omega \xi \, d\mu \quad \forall \xi \in H_0^2(\Omega). \tag{2.3}$$

Proof. We follow [13, 19]. Let $\xi \in C_0^\infty(\Omega)$, $\xi \geqslant 0$, be arbitrary. It is clear that for an arbitrary $\varepsilon > 0$ we have $v = u + \varepsilon\xi \in K(\psi)$. Substituting this v into (1.1), we obtain

$$\int_\Omega \Delta u \Delta \xi \, dx \geqslant 0. \tag{2.4}$$

This implies that there exists a nonnegative Borel measure $\mu \in H^{-2}(\Omega)$ such that

$$\mu = \Delta^2 u,$$

where $\Delta^2 u$ is taken in the sense of distributions. By the density of $C_0^\infty(\Omega)$ in $H_0^2(\Omega)$, (2.3) also holds for any $\xi \in H_0^2(\Omega)$. □

Lemma 2.3. *Let $u = T(\psi)$ solve (1.1), and let $\mu \in H^{-2}(\Omega)$ be the Borel measure from Lemma 2.2. Then*

(i) *there exists a positive constant such that $\mu(e) \leqslant \text{const} \cdot C_{2,2}(e)$ for each set $e \subset \Omega$ whose distance from the boundary $\partial\Omega$ is positive, and*

(ii) *$u = \psi$ μ-a.e., where u and ψ are the representatives defined $C_{2,2}$-a.e. respectively.*

Proof. (i) We follow Kinderlehrer and Stampacchia [19, Theorem 6.11] with appropriate modifications. Take an arbitrary $\zeta \in C_0^\infty(\Omega)$ with $\zeta \geqslant 0$ on Ω and $\zeta \geqslant 1$ on e. Since $\Delta^2 u = \mu$ in Ω in the sense of distributions, we can write

$$\mu(e) = \int_\Omega \chi_e \, d\mu \leqslant \int_\Omega \zeta \, d\mu = \int_\Omega \Delta\zeta \Delta u \, dx \leqslant \|\Delta\zeta\|_{L^2(\Omega)} \|\Delta u\|_{L^2(\Omega)}.$$

Taking the infimum over all ζ, we obtain

$$\mu(e) \leqslant \text{const} \cdot \inf_{\zeta \in K_e} \|\Delta\zeta\|_{L^2(\Omega)} = \text{const} \cdot C_{2,2}(e)$$

with const $= \|\Delta u\|_{L^2(\Omega)} < \infty$.

(ii) Take $v = \psi$ in (1.1) to get

$$\int_\Omega \Delta u \, \Delta(\psi - u) \, dx \geqslant 0.$$

This implies

$$\int_\Omega (\psi - u) \, d\mu \geqslant 0. \tag{2.5}$$

From (2.5) it follows that

$$\mu\{x \in \Omega : u > \psi \text{ a.e.}\} = 0, \tag{2.6}$$

and we will show that

$$\mu\{x \in \Omega : u > \psi \ C_{2,2}\text{-a.e.}\} = 0. \tag{2.7}$$

Indeed, to show (2.7), we write

$$\{x \in \Omega : u > \psi \ C_{2,2}\text{-a.e.}\} = \{x \in \Omega : u > \psi \text{ a.e.}\} \cup Q,$$

where $C_{2,2}(Q) = 0$ and $Q \cap \{x \in \Omega : u > \psi \text{ a.e.}\} = \varnothing$. Therefore, taking into account (2.6) and part (i) of Lemma 2.3, we obtain

$$\mu\{x \in \Omega : u > \psi \ C_{2,2}\text{-a.e.}\} = \mu\{x \in \Omega : u > \psi \text{ a.e.}\} + \mu(Q) = 0.$$

This proves (2.7), which gives (ii). □

Remarks. 1. Let $N = \{x \in \Omega : u(x) > \psi(x) \text{ a.e.}\}$ and $I = \{x \in \Omega : u(x) = \psi(x) \text{ a.e.}\}$ denote the noncoincidence set and the coincidence set respectively. Then, by Lemma 2.1, we have $I = \{x \in \Omega : u(x) = \psi(x) \ C_{2,2}\text{-a.e.}\}$ and $C_{2,2}(\Omega \backslash (N \cup I)) = 0$.

2. Observe that $C_{2,2}\{x \in \Omega : u(x) = \psi(x) \ C_{2,2}\text{-a.e.}\} > 0$ if $\psi > 0$ on a set of positive capacity. Indeed, if $C_{2,2}\{x \in \Omega : u(x) = \psi(x) C_{2,2}\text{-a.e.}\} = 0$, then since $\operatorname{supp} \mu \subset \{u = \psi C_{2,2}\text{-a.e.}\}$ it follows that $C_{2,2}\{\operatorname{supp} \mu\} = 0$. Therefore, by part (i) of Lemma 2.3, we get $\mu\{\operatorname{supp} \mu\} = 0$. This would imply that $\mu \equiv 0$, which gives $\Delta^2 u = 0$ on Ω. Hence $u \equiv 0$, which implies $\psi \leqslant 0 \ C_{2,2}$-a.e. Thus, if $\psi > 0$ on a set of positive capacity, u is nonzero and $C_{2,2}\{x \in \Omega : u(x) = \psi(x) \ C_{2,2}\text{-a.e.}\} > 0$.

3. The class of measures in Lemma 2.3, appeared in Maz'ya's early works (with the first order derivatives in [22, 23] and the second order derivatives in [24]) and turned out to be very important in different areas of PDEs, analysis, geometry, probability theory etc.

The next lemma gives the key estimate which will be used in the proof of the existence of a solution to Problem C. Even though this estimate (for a smooth obstacle) can be found in [1], a more detailed proof is presented here. Also, in what follows, we denote by $\|\cdot\|_2$, the L^2 norm on Ω.

Lemma 2.4. *Let u solve* (1.1) *with obstacle ψ, and let u_k solve* (1.1) *with obstacle ψ_k. Then there exists $\widetilde{C} > 0$ such that*

$$\|u - u_k\|_{H^2(\Omega)} \leqslant \widetilde{C} \|\psi - \psi_k\|_{H^2(\Omega)}. \tag{2.8}$$

Proof. Since u and u_k are the solutions to (1.1) in $H_0^2(\Omega)$ with obstacles ψ and ψ_k respectively, it follows that there exist nonnegative measures $\mu, \mu_k \in H^{-2}(\Omega)$ such that

$$\begin{aligned} &\Delta^2 u = \mu \ \text{ in } \ \Omega, \quad u = \psi \ \mu\text{-a.e.}, \\ &\Delta^2 u_k = \mu_k \ \text{ in } \ \Omega, \quad u_k = \psi_k \ \mu_k\text{-a.e.} \end{aligned} \tag{2.9}$$

By Lemma 2.1, we have $u \geqslant \psi \ C_{2,2}$-a.e. Thus, $C_{2,2}\{x \in \Omega : u < \psi \text{ a.e.}\} = 0$. By part (i) of Lemma 2.3, we can write

$$\mu_k\{x \in \Omega : u < \psi \text{ a.e.}\} = 0. \tag{2.10}$$

Therefore,

$$-\int_\Omega u\, d\mu_k \leqslant -\int_\Omega \psi\, d\mu_k. \tag{2.11}$$

Similarly,

$$-\int_\Omega u_k\, d\mu \leqslant -\int_\Omega \psi_k\, d\mu. \tag{2.12}$$

Taking into account (2.9), (2.11), (2.12), and that functions involved are of class $H_0^2(\Omega)$, we have

$$\begin{aligned}
\|u-u_k\|^2_{H^2(\Omega)} &= \int_\Omega |\Delta(u-u_k)|^2 dx = \int_\Omega (u-u_k)\, d(\mu-\mu_k) \\
&= \int_\Omega u\, d\mu + \int_\Omega u_k\, d\mu_k - \int_\Omega u_k\, d\mu - \int_\Omega u\, d\mu_k \\
&= \int_\Omega \psi\, d\mu + \int_\Omega \psi_k\, d\mu_k - \int_\Omega u_k\, d\mu - \int_\Omega u\, d\mu_k \\
&\leqslant \int_\Omega \psi\, d\mu + \int_\Omega \psi_k\, d\mu_k - \int_\Omega \psi_k\, d\mu - \int_\Omega \psi\, d\mu_k \\
&= \int_\Omega (\psi-\psi_k)\, d(\mu-\mu_k) \\
&= \int_\Omega \Delta(\psi-\psi_k)\, \Delta(u-u_k)\, dx \\
&\leqslant \widetilde{C}\|\Delta\psi-\Delta\psi_k\|_2 \|\Delta u-\Delta u_k\|_2 \\
&\leqslant \widetilde{C}\|\psi-\psi_k\|_{H^2(\Omega)} \|u-u_k\|_{H^2(\Omega)}.
\end{aligned}$$

The lemma is proved. □

3 Existence of an Optimal Control

Now we are ready to prove the main theorem of the paper.

Theorem 3.1. *There exists a solution to Problem* C.

Proof. Since the objective functional $J(\psi) \geqslant 0$ for all $\psi \in V$, there exists a minimizing sequence $\{\psi_n\} \subset V$ such that

$$\lim_{n\to\infty} J(\psi_n) = \inf_{\psi\in V} J(\psi).$$

Let $u_n = T(\psi_n)$ be the corresponding solutions to the problem

$$\text{find } u_n \in K(\psi_n) \text{ such that } \int_\Omega \Delta u_n \, \Delta(v - u_n) \, dx \geqslant 0 \quad \forall v \in K(\psi_n). \quad (3.1)$$

We have (from (3.1) with $v = \psi_n$)

$$\|u_n\|_{H^2(\Omega)} \leqslant \|\psi_n\|_{H^2(\Omega)} \tag{3.2}$$

Since the sequence $\{\psi_n\}$ is bounded in $H^3(\Omega)$, from the format of $J(\psi)$, we can write

$$\|u_n\|_{H^2(\Omega)} \leqslant \|\psi_n\|_{H^2(\Omega)} \leqslant C_1 \|\psi_n\|_{H^3(\Omega)} \leqslant C_2 \quad \forall n. \tag{3.3}$$

The estimate in (3.3) implies that there exists $\psi^* \in V$ such that, on a subsequence, $\psi_n \xrightarrow{w} \psi^*$ in $H^3(\Omega)$. Since $H^3(\Omega) \subset\subset H^2(\Omega)$, it follows that $\psi_n \xrightarrow{s} \psi^*$ in $H^2(\Omega)$. By Lemma 2.4, for $u^* = T(\psi^*)$

$$\|u^* - u_n\|_{H^2(\Omega)} \leqslant \widetilde{C} \|\psi^* - \psi_n\|_{H^2(\Omega)}. \tag{3.4}$$

We have

$$\begin{aligned} &\psi_n \xrightarrow{s} \psi^* \text{ in } H^2(\Omega), \\ &u_n \xrightarrow{s} u^* \text{ in } H^2(\Omega). \end{aligned} \tag{3.5}$$

We show that ψ^* is optimal. Indeed, since $\psi_n \xrightarrow{w} \psi^*$ in $H^3(\Omega)$ and $H^3(\Omega) \subset\subset L^2(\Omega)$, it follows that

$$\varliminf_{n\to\infty} \int_\Omega |\nabla \Delta \psi_n|^2 \, dx \geqslant \int_\Omega |\nabla \Delta \psi^*|^2 \, dx.$$

Also, $T(\psi_n) = u_n \xrightarrow{s} u^* = T(\psi^*)$ in $H^2(\Omega)$ implies that

$$\lim_{n\to\infty} \int_\Omega |T(\psi_n) - z|^2 \, dx$$

exists and

$$\lim_{n\to\infty} \int_\Omega |T(\psi_n) - z|^2 \, dx = \int_\Omega |T(\psi^*) - z|^2 \, dx.$$

Since

$$\lim_{n\to\infty} \frac{1}{2} \left(\int_\Omega |T(\psi_n) - z|^2 \, dx + \int_\Omega |\nabla \Delta \psi_n|^2 \, dx \right)$$

exists and

$$\lim_{n\to\infty} \int_\Omega |T(\psi_n) - z|^2 \, dx$$

exists, it follows that

$$\lim_{n\to\infty} \int_\Omega |\nabla \Delta \psi_n|^2 \, dx$$

exists. Therefore,

$$\begin{aligned}
\inf_{\psi\in V} J(\psi) &= \lim_{n\to\infty} J(\psi_n) \\
&= \lim_{n\to\infty} \frac{1}{2}\int_\Omega |T(\psi_n)-z|^2\,dx + \lim_{n\to\infty}\frac{1}{2}\int_\Omega |\nabla\Delta\psi_n|^2\,dx \\
&\geqslant \lim_{n\to\infty} \frac{1}{2}\int_\Omega |T(\psi_n)-z|^2\,dx + \varliminf_{n\to\infty}\frac{1}{2}\int_\Omega |\nabla\Delta\psi_n|^2\,dx \\
&\geqslant \frac{1}{2}\int_\Omega |T(\psi^*)-z|^2\,dx + \frac{1}{2}\int_\Omega |\nabla\Delta\psi^*|^2\,dx = J(\psi^*).
\end{aligned}$$

The theorem is proved. □

4 Approximation Problems

To derive necessary conditions for the optimal control, we need to differentiate the map $\psi \longmapsto J(\psi)$ which depends on $T(\psi)$. The map $\psi \longmapsto u = T(\psi)$ cannot be differentiated directly thereby necessitating introduction of a family of approximation problems. Appropriate necessary conditions will be derived as in [8]. We follow the basic structure of the derivation of the necessary conditions in [3, 13] with appropriate modifications. We define a nonnegative function γ by the formula

$$\gamma(r) = \begin{cases} 0, & r\in[0,\infty), \\ -\frac{1}{3}r^3, & r\in[-\frac{1}{2},0], \\ \frac{1}{2}r^2+\frac{1}{4}r+\frac{1}{24}, & r\in(-\infty,-\frac{1}{2}), \end{cases} \tag{4.1}$$

and its derivative β (which is nonpositive)

$$\beta(r) \stackrel{\text{def}}{=} \gamma'(r) = \begin{cases} 0, & r\in[0,\infty), \\ -r^2, & r\in[-\frac{1}{2},0], \\ r+\frac{1}{4}, & r\in(-\infty,-\frac{1}{2}). \end{cases} \tag{4.2}$$

For an arbitrary $\delta>0$ the problem

$$\text{minimize}_{v\in H_0^2(\Omega)} \int_\Omega \left[\frac{1}{2}|\Delta v|^2 + \frac{1}{\delta}\gamma(v-\psi)\right] dx \tag{4.3}$$

has a unique solution u^δ. It can be shown that u^δ solves the problem

$$\begin{aligned}&\Delta^2 u^\delta + \frac{1}{\delta}\beta(u^\delta - \psi) = 0 \quad \text{in } \Omega,\\ &u^\delta = 0 \quad \text{on } \partial\Omega,\\ &\frac{\partial u^\delta}{\partial n} = 0 \quad \text{on } \partial\Omega.\end{aligned} \tag{4.4}$$

Taking into account that $\beta(u^\delta - \psi) \in H_0^1(\Omega)$, by the standard elliptic theory for this semilinear case (cf. [16]), it follows that there exists a solution $u^\delta \in H_0^2(\Omega) \cap H^4(\Omega)$ to (4.4). We set $T_\delta(\psi) = u^\delta$ and define the following approximate objective functional:

$$J_\delta(\psi) := \frac{1}{2}\int_\Omega \{|T_\delta(\psi) - z|^2 + |\nabla\Delta\psi|^2\}dx.$$

Now we can state our approximation problem

Problem C$_\delta$. Find $\psi^\delta \in V$ such that $J_\delta(\psi^\delta) = \inf_{\psi\in V} J_\delta(\psi)$.

With the help of this approximation, we can derive necessary conditions for the corresponding optimal controls. As was already mentioned in Section 1, the existence and uniqueness of a solution to the variational inequality (1.1) follows from the coercivity of the bilinear form

$$a(u,v) = \int_\Omega \Delta u\,\Delta(v-u)\,dx$$

(cf. [19]). For the sake of completeness, we give a more constructive proof of the existence of a solution to (1.1) by incorporating approximation techniques. Also, it is used in the proof of Theorem 4.2.

Theorem 4.1. *Let $\psi \in H^3(\Omega) \cap H_0^2(\Omega)$ be arbitrary. If $u^\delta = T_\delta(\psi)$ is a solution to (4.4), then there exists $u \in H_0^2(\Omega)$ such that*

$$T_\delta(\psi) \stackrel{w}{\rightharpoonup} u \text{ in } H_0^2(\Omega)$$

as $\delta \to 0^+$, where $u = T(\psi)$. Furthermore, the following estimate holds:

$$\|\Delta T_\delta(\psi)\|_2 \leqslant \|\Delta\psi\|_2. \tag{4.5}$$

Proof. First, we derive (4.5). We know that for a given $\psi \in V$ there exists a solution $T_\delta(\psi) = u^\delta \in H_0^2(\Omega) \cap H^4(\Omega)$ to (4.4). Let $v \in K(\psi)$ be arbitrary. Multiplying (4.4) by $v - u^\delta$, integrating by parts, and using properties of β, we get

$$\int_\Omega \Delta u^\delta \Delta(v - u^\delta)\,dx = -\frac{1}{\delta}\int_\Omega \beta(u^\delta - \psi)(v - u^\delta)\,dx \geqslant 0.$$

Thus, we can write

$$\int_{\Omega} \Delta u^{\delta} \Delta(v - u^{\delta})\, dx \geqslant 0. \tag{4.6}$$

Now, if we take $v = \psi$ in (4.6), we obtain

$$\|\Delta u^{\delta}\|_2^2 = \int_{\Omega} (\Delta u^{\delta})^2\, dx \leqslant \int_{\Omega} \Delta u^{\delta} \Delta \psi\, dx \leqslant \|\Delta u^{\delta}\|_2 \|\Delta \psi\|_2.$$

Hence we obtain (4.5).

By the *a priori* estimate (4.5), there exists $u \in H_0^2(\Omega)$ such that, on a subsequence,

$$u^{\delta} \xrightarrow{w} u \text{ in } H_0^2(\Omega) \tag{4.7}$$

as $\delta \to 0^+$. For an arbitrary $\varphi \in H_0^2(\Omega)$ we obtain from (4.4)

$$\begin{aligned} -\frac{1}{\delta} \int_{\Omega} \beta(u^{\delta} - \psi)\varphi\, dx &= \int_{\Omega} \Delta u^{\delta} \Delta \varphi\, dx \\ &\leqslant \|\Delta u^{\delta}\|_2 \|\Delta \varphi\|_2 \leqslant C \|\Delta \psi\|_2 \|\Delta \varphi\|_2. \end{aligned} \tag{4.8}$$

First, we show that $u \geqslant \psi$ a.e. in Ω. Take an arbitrary $\varphi \in H_0^2(\Omega)$, $\varphi \geqslant 0$, a.e. in Ω. Rewrite (4.8) as

$$0 \leqslant -\int_{\Omega} \beta(u^{\delta} - \psi)\varphi\, dx \leqslant C\delta \|\Delta \psi\|_2 \|\Delta \varphi\|_2. \tag{4.9}$$

Letting $\delta \to 0^+$ in (4.9), we obtain

$$\int_{\Omega} \beta(u - \psi)\varphi\, dx = 0 \quad \forall \varphi \in H_0^2(\Omega), \varphi \geqslant 0 \text{ a.e. in } \Omega.$$

This implies $u \geqslant \psi$ a.e. in Ω. Hence $u \in K(\psi)$. We show that u satisfies

$$\int_{\Omega} \Delta u\, \Delta(v - u)\, dx \geqslant 0 \quad \forall\, v \in K(\psi), \tag{4.10}$$

i.e., we need to show that u minimizes

$$\int_{\Omega} |\Delta v|^2\, dx, \quad v \in K(\psi).$$

Indeed, let $v \in K(\psi)$ be arbitrary. Taking into account (4.3) and the definition of γ, we can write

$$\int_{\Omega} \frac{1}{2} |\Delta v|^2\, dx = \int_{\Omega} \left[\frac{1}{2} |\Delta v|^2 + \frac{1}{\delta} \gamma(v - \psi) \right] dx$$

$$\geqslant \int_\Omega \left[\frac{1}{2}|\Delta u^\delta|^2 + \frac{1}{\delta}\gamma(u^\delta - \psi) \right] dx \geqslant \int_\Omega \frac{1}{2}|\Delta u^\delta|^2\, dx. \tag{4.11}$$

Using (4.7), (4.11), and the lower semicontinuity of weak convergence, we get

$$\int_\Omega |\Delta v|^2\, dx \geqslant \varliminf_{\delta\to 0+} \int_\Omega |\Delta u^\delta|^2\, dx \geqslant \int_\Omega |\Delta u|^2\, dx. \tag{4.12}$$

Hence u satisfies (4.10). Thus, we have shown that $u = T(\psi)$, i.e., u is a solution to (1.1) associated with the obstacle ψ. □

Now we prove the following assertion.

Proposition 4.1. *Problem* C_δ *admits an optimal pair* $(\overline{u}^\delta, \psi^\delta)$, *where* $\overline{u}^\delta = T_\delta(\psi^\delta)$. *Moreover,*

$$\|\Delta \overline{u}^\delta\|_2^2 \leqslant C \int_\Omega |\nabla \Delta \psi^\delta|^2\, dx \leqslant C\|z\|_2^2. \tag{4.13}$$

Proof. Let $\{\psi_k\}_{k=1}^\infty \subset V$ be a minimizing sequence for J_δ. Since

$$\int_\Omega |\nabla \Delta \psi_k|^2\, dx$$

is bounded (because $\{\psi_k\}$ is a minimizing sequence),

$$\|\psi_k\|_{H^3(\Omega)} \leqslant C.$$

This implies that there exists $\psi^\delta \in V$ such that, on a subsequence,

$$\psi_k \xrightarrow{w} \psi^\delta \text{ in } H^3(\Omega).$$

Let $u_k^\delta = T_\delta(\psi_k)$ be the corresponding solution to (4.4) (with $\psi = \psi_k$). By (4.5),

$$\|\Delta T_\delta(\psi_k)\|_2 \leqslant \|\Delta \psi_k\|_2 \leqslant C_4. \tag{4.14}$$

Then there exists $\overline{u}^\delta \in H_0^2(\Omega)$ such that, on a subsequence, $u_k^\delta \xrightarrow{w} \overline{u}^\delta$ in $H_0^2(\Omega)$ as $k \to \infty$. Thus, on a subsequence,

$$\psi_k \xrightarrow{w} \psi^\delta \text{ in } H^3(\Omega), \quad u_k^\delta \xrightarrow{w} \overline{u}^\delta \text{ in } H_0^2(\Omega) \quad \text{as } k \to \infty. \tag{4.15}$$

Note that the convergences in (4.15) imply that, on a subsequence,

$$\psi_k \xrightarrow{s} \psi^\delta \text{ in } H^1(\Omega), \quad u_k^\delta \xrightarrow{s} \overline{u}^\delta \text{ in } H_0^1(\Omega) \quad \text{as } k \to \infty. \tag{4.16}$$

Multiplying (4.4) by a test function $v \in H_0^2(\Omega)$ and integrating by parts, we get

$$\int_\Omega \Delta u_k^\delta \Delta v\,dx = -\frac{1}{\delta}\int_\Omega \beta(u_k^\delta - \psi_k)v\,dx. \tag{4.17}$$

Taking into account (4.16) and the fact that β is continuous, we have

$$\int_\Omega \beta(u_k^\delta - \psi_k)v\,dx \to \int_\Omega \beta(\overline{u}^\delta - \psi^\delta)v\,dx$$

as $k \to \infty$. Now, letting $k \to \infty$ in (4.17), we obtain

$$\int_\Omega \Delta\overline{u}^\delta \Delta v\,dx = -\frac{1}{\delta}\int_\Omega \beta(\overline{u}^\delta - \psi^\delta)v\,dx. \tag{4.18}$$

This implies that $\overline{u}^\delta$ is the solution associated with ψ^δ, i.e., $\overline{u}^\delta = T_\delta(\psi^\delta)$. We show that ψ^δ is optimal for J_δ. Indeed, we have

$$\begin{aligned}\inf_{\psi\in V} J_\delta(\psi) &= \lim_{k\to\infty} J_\delta(\psi_k)\\ &\geqslant \frac{1}{2}\int_\Omega \bigl(T_\delta(\psi^\delta) - z\bigr)^2\,dx + \frac{1}{2}\varliminf_{k\to\infty}\int_\Omega |\nabla\Delta\psi_k|^2\,dx\\ &\geqslant \frac{1}{2}\int_\Omega \bigl(T_\delta(\psi^\delta) - z\bigr)^2\,dx + \frac{1}{2}\int_\Omega |\nabla\Delta\psi^\delta|^2\,dx = J_\delta(\psi^\delta).\end{aligned}$$

This proves that ψ^δ is optimal. Now we prove (4.13). Denoting

$$d\mu^\delta \overset{\text{def}}{=} -\frac{1}{\delta}\beta(\overline{u}^\delta - \psi^\delta)\,dx,$$

we estimate

$$\begin{aligned}\int_\Omega |\Delta\overline{u}^\delta|^2\,dx &= \int_\Omega \Delta\overline{u}^\delta\Delta\overline{u}^\delta\,dx = -\int_\Omega \overline{u}^\delta\frac{1}{\delta}\beta(\overline{u}^\delta - \psi^\delta)\,dx\\ &= \int_\Omega \overline{u}^\delta\,d\mu^\delta \leqslant \int_\Omega \psi^\delta\,d\mu^\delta \qquad (4.19)\\ &= \int_\Omega \Delta\psi^\delta\Delta\overline{u}^\delta\,dx = -\int_\Omega \nabla\overline{u}^\delta\cdot\nabla\Delta\psi^\delta\,dx\\ &\leqslant \left|\int_\Omega \nabla\overline{u}^\delta\cdot\nabla\Delta\psi^\delta\,dx\right| \leqslant \|\nabla\overline{u}^\delta\|_2\cdot\|\nabla\Delta\psi^\delta\|_2 \qquad (4.20)\\ &\leqslant C_5\|\Delta\overline{u}^\delta\|_2\cdot\|\nabla\Delta\psi^\delta\|_2\end{aligned}$$

where we used properties of β in (4.19) and the Poincaré inequality in (4.20). Hence

$$\|\Delta\overline{u}^\delta\|_2^2 \leqslant C\|\nabla\Delta\psi^\delta\|_2^2 = C\int_\Omega |\nabla\Delta\psi^\delta|^2\,dx$$
$$\leqslant CJ_\delta(\psi^\delta) \leqslant CJ_\delta(0) = C\|z\|_2^2.$$

The proposition is proved. □

Now we prove the main result of this section; namely, the convergence of the optimal controls for the approximation problems to the minimizer for the original problem.

Theorem 4.2. *Let $(\overline{u}^\delta, \psi^\delta)$ be an optimal pair for Problem C_δ. Then there exist $\overline{u} \in H_0^2(\Omega)$ and $\overline{\psi} \in V$ such that, on a subsequence,*

$$\psi^\delta \xrightarrow{s} \overline{\psi} \text{ in } H^2(\Omega),$$
$$\overline{u}^\delta \xrightarrow{s} \overline{u} \text{ in } H_0^2(\Omega),$$

where $\overline{u} = T(\overline{\psi})$ and $(\overline{u}, \overline{\psi})$ is an optimal pair for Problem C.

Proof. By the *a priori* estimates (4.13), there exist $\overline{u} \in H_0^2(\Omega)$ and $\overline{\psi} \in V$ such that, on a subsequence,

$$\begin{aligned} \psi^\delta &\xrightarrow{w} \overline{\psi} \quad \text{in } H^3(\Omega), \\ \overline{u}^\delta &\xrightarrow{w} \overline{u} \quad \text{in } H_0^2(\Omega) \end{aligned} \tag{4.21}$$

as $\delta \to 0^+$. Since $H^3(\Omega) \subset\subset H^2(\Omega)$, it follows that, on a subsequence,

$$\psi^\delta \xrightarrow{s} \overline{\psi} \quad \text{in } H^2(\Omega). \tag{4.22}$$

For an arbitrary $\varphi \in H_0^2(\Omega)$ we obtain from (4.4) (with $\psi = \psi^\delta$)

$$\begin{aligned} -\frac{1}{\delta}\int_\Omega \beta(\overline{u}^\delta - \psi^\delta)\varphi\,dx &= \int_\Omega \Delta\overline{u}^\delta\Delta\varphi\,dx \\ &\leqslant \|\Delta\overline{u}^\delta\|_2\|\Delta\varphi\|_2 \leqslant C\|z\|_2\|\Delta\varphi\|_2. \end{aligned} \tag{4.23}$$

This implies that there exists $\overline{\mu} \in H^{-2}(\Omega)$ such that

$$\mu^\delta = \Delta^2\overline{u}^\delta \xrightarrow{*} \overline{\mu} \quad \text{in } H^{-2}(\Omega). \tag{4.24}$$

The convergence in (4.24) means that for an arbitrary function $f \in H_0^2(\Omega)$

$$\int_\Omega f\,d\mu^\delta \to \int_\Omega f\,d\overline{\mu}. \tag{4.25}$$

First, we show that $\overline{u} \geqslant \overline{\psi}$ a.e. in Ω. Take an arbitrary $\varphi \in H_0^2(\Omega)$, $\varphi \geqslant 0$, a.e. in Ω. We can rewrite (4.23) as follows:

$$0 \leqslant -\int_{\Omega} \beta(\overline{u}^{\delta} - \psi^{\delta})\varphi \, dx \leqslant C\delta \|z\|_2 \|\Delta\varphi\|_2. \tag{4.26}$$

Letting $\delta \to 0^+$ in (4.26) we obtain

$$\int_{\Omega} \beta(\overline{u} - \overline{\psi})\varphi \, dx = 0 \quad \forall \varphi \in H_0^2(\Omega), \varphi \geqslant 0 \text{ a.e. in } \Omega.$$

This implies $\overline{u} \geqslant \overline{\psi}$ a.e. in Ω. Hence $\overline{u} \in K(\overline{\psi})$.

We show that $\overline{u}$ satisfies

$$\int_{\Omega} \Delta\overline{u} \, \Delta(v - \overline{u}) \, dx \geqslant 0 \quad \forall v \in K(\overline{\psi}), \tag{4.27}$$

i.e., we need to show that $\overline{u}$ minimizes

$$\int_{\Omega} |\Delta v|^2 \, dx, \quad v \in K(\overline{\psi}).$$

Indeed, let $v \in K(\overline{\psi})$ be arbitrary. Taking into account (4.3) and the definition of γ, we can write

$$\begin{aligned} \int_{\Omega} \frac{1}{2}|\Delta v|^2 \, dx &= \int_{\Omega} \left[\frac{1}{2}|\Delta v|^2 + \frac{1}{\delta}\gamma(v - \overline{\psi})\right] dx \\ &\geqslant \int_{\Omega} \left[\frac{1}{2}|\Delta\overline{u}^{\delta}|^2 + \frac{1}{\delta}\gamma(\overline{u}^{\delta} - \overline{\psi})\right] dx \\ &\geqslant \int_{\Omega} \frac{1}{2}|\Delta\overline{u}^{\delta}|^2 \, dx. \end{aligned} \tag{4.28}$$

Using (4.21), (4.28), and the lower semicontinuity of weak convergence, we get

$$\int_{\Omega} |\Delta v|^2 \, dx \geqslant \varliminf_{\delta \to 0+} \int_{\Omega} |\Delta\overline{u}^{\delta}|^2 \, dx \geqslant \int_{\Omega} |\Delta\overline{u}|^2 \, dx. \tag{4.29}$$

Hence $\overline{u}$ satisfies (4.27). Thus, we have shown that $\overline{u} = T(\overline{\psi})$. Let us show that $\overline{\psi}$ minimizes $J(\psi)$. For this purpose, we denote $T_{\delta}(\psi^*) := u_*^{\delta}$, where ψ^* is the optimal control from Theorem 3.1 (cf. formula (3.5)). By Theorem 4.1,

$$u_*^{\delta} \xrightarrow{w} u^* = T(\psi^*) \;\text{ in } H_0^2(\Omega) \text{ as } \delta \to 0^+, \tag{4.30}$$

and the optimality of ψ^{δ} implies

$$J_{\delta}(\psi^{\delta}) \leqslant J_{\delta}(\psi^*). \tag{4.31}$$

Taking into account (4.30), the lower semicontinuity of weak convergence, Theorem 4.1, and (4.31), we can write

$$
\begin{aligned}
J(\overline{\psi}) &\leqslant \varliminf_{\delta\to 0+} J_\delta(\psi^\delta) \leqslant \varlimsup_{\delta\to 0+} J_\delta(\psi^\delta) \leqslant \varlimsup_{\delta\to 0+} J_\delta(\psi^*) \\
&= \frac{1}{2} \varlimsup_{\delta\to 0+} \int_\Omega \{|T_\delta(\psi^*) - z|^2 + |\nabla\Delta\psi^*|^2\}\, dx \\
&= \frac{1}{2} \int_\Omega \{|T(\psi^*) - z|^2 + |\nabla\Delta\psi^*|^2\}\, dx = J(\psi^*) = \inf_{\psi\in V} J(\psi).
\end{aligned}
$$

Thus,

$$
J(\overline{\psi}) = \inf_{\psi\in V} J(\psi).
$$

This proves that $\overline{\psi}$ minimizes $J(\psi)$.

Recall that we obtained (cf. formula (4.24))

$$
\mu^\delta \overset{*}{\rightharpoonup} \overline{\mu} \ \text{ in } H^{-2}(\Omega).
$$

Taking $f \in H_0^2(\Omega)$ in (4.25), we get

$$
\int_\Omega f\, d\overline{\mu} = \lim_{\delta\to 0+} \int_\Omega f\, d\mu^\delta = \lim_{\delta\to 0+} \int_\Omega \Delta f \cdot \Delta\overline{u}^\delta\, dx = \int_\Omega \Delta f \cdot \Delta\overline{u}\, dx.
$$

This implies

$$
\Delta^2 \overline{u} = \overline{\mu} \ \text{ in } \Omega. \tag{4.32}
$$

Since μ^δ is a nonnegative Borel measure for each $\delta > 0$ and the set of all nonnegative Borel measures supported in Ω is convex and weak-$*$ closed in $H^{-2}(\Omega)$, it follows that $\overline{\mu} \geqslant 0$ (as a weak-$*$ limit of (μ^δ)'s).

Before we proceed to the proof of the strong convergence of $\overline{u}^\delta$ to $\overline{u}$ in $H_0^2(\Omega)$, we make the following observation. Since $\overline{u} = T(\overline{\psi})$ solves the variational inequality (1.1), there exists a nonnegative measure $\mu^\dagger \in H^{-2}(\Omega)$ such that

$$
\begin{aligned}
\Delta^2 \overline{u} &= \mu^\dagger \ \text{ in } \Omega, \\
\overline{u} &= \overline{\psi} \quad \mu^\dagger\text{-a.e.},
\end{aligned}
$$

i.e.,

$$
\int_\Omega \xi\, d\mu^\dagger = \int_\Omega \Delta\overline{u}\Delta\xi\, dx \quad \forall\, \xi \in H_0^2(\Omega).
$$

On the other hand, by (4.32), we have

$$
\int_\Omega \xi\, d\overline{\mu} = \int_\Omega \Delta\overline{u}\Delta\xi\, dx \quad \forall\, \xi \in H_0^2(\Omega).
$$

Hence

$$
\int_\Omega \xi\, d\overline{\mu} = \int_\Omega \xi\, d\mu^\dagger \quad \forall \xi \in H_0^2(\Omega).
$$

Therefore, we can write

$$\|\mu^\dagger - \overline{\mu}\|_{H^{-2}(\Omega)} = \sup |(\mu^\dagger - \overline{\mu})(\xi)| = 0,$$

where the supremum is taken over all $\xi \in H_0^2(\Omega)$ with $\|\xi\|_{H_0^2(\Omega)} = 1$. Thus, $\mu^\dagger = \overline{\mu}$.

Now we are ready to prove that $\overline{u}^\delta$ converges strongly to $\overline{u}$ in $H_0^2(\Omega)$ as $\delta \to 0^+$. Indeed,

$$\begin{aligned}
&\int_\Omega (\Delta\overline{u}^\delta - \Delta\overline{u})^2 dx \\
&= \int_\Omega \Delta(\overline{u}^\delta - \overline{u})\Delta(\overline{u}^\delta - \overline{u})\, dx = \int_\Omega (\overline{u}^\delta - \overline{u})\, d(\mu^\delta - \overline{\mu}) \\
&= \int_\Omega \overline{u}^\delta\, d\mu^\delta + \int_\Omega \overline{u}\, d\overline{\mu} - \int_\Omega \overline{u}\, d\mu^\delta - \int_\Omega \overline{u}^\delta\, d\overline{\mu} \\
&= \int_\Omega (\overline{u}^\delta - \psi^\delta)\, d\mu^\delta + \int_\Omega \psi^\delta\, d\mu^\delta + \int_\Omega \overline{u}\, d\overline{\mu} - \int_\Omega \overline{u}\, d\mu^\delta - \int_\Omega \overline{u}^\delta\, d\overline{\mu} \\
&\leqslant \int_\Omega \psi^\delta\, d\mu^\delta + \int_\Omega \overline{\psi}\, d\overline{\mu} - \int_\Omega \overline{u}\, d\mu^\delta - \int_\Omega \overline{u}^\delta\, d\overline{\mu}
\end{aligned} \tag{4.33}$$

where we take into account that

$$\int_\Omega (\overline{u}^\delta - \psi^\delta)\, d\mu^\delta \leqslant 0$$

(by properties of β) and $\overline{u} = \overline{\psi}$ $\overline{\mu}$-a.e. Letting $\delta \to 0^+$ in (4.33) and taking into account convergences in (4.22) and (4.24), we get

$$\begin{aligned}
\lim_{\delta\to 0} \|\overline{u}^\delta - \overline{u}\|^2_{H^2(\Omega)} &\leqslant \int_\Omega \overline{\psi}\, d\overline{\mu} + \int_\Omega \overline{\psi}\, d\overline{\mu} - \int_\Omega \overline{u}\, d\overline{\mu} - \int_\Omega \overline{u}\, d\overline{\mu} \\
&\leqslant \int_\Omega \overline{u}\, d\overline{\mu} + \int_\Omega \overline{u}\, d\overline{\mu} - \int_\Omega \overline{u}\, d\overline{\mu} - \int_\Omega \overline{u}\, d\overline{\mu} = 0.
\end{aligned}$$

This shows that $\overline{u}^\delta \xrightarrow{s} \overline{u}$ in $H_0^2(\Omega)$. The theorem is proved. □

We use this convergence in the next section (actually, we need only the strong L^2 convergence of $\overline{u}^\delta$ to $\overline{u}$).

5 Characterization of the Optimal Control

In order to characterize an optimal control, we need to derive necessary conditions which include the original state system coupled with an adjoint system. We must first obtain necessary conditions for the approximate problems. For this purpose, we differentiate the objective functional $J_\delta(\psi)$ with respect to the control. Since the objective functional depends on u^δ, we need to differentiate u^δ with respect to control ψ. We use the following assertion.

Lemma 5.1 (sensitivity). *The map $\psi \longmapsto u^\delta \equiv u^\delta(\psi)$ is differentiable in the following sense: given $\psi, \ell \in V$, there exists $\xi^\delta \in H_0^2(\Omega)$ such that*

$$\frac{u^\delta(\psi + \varepsilon\ell) - u^\delta(\psi)}{\varepsilon} \xrightarrow{w} \xi^\delta \quad \text{in } H_0^2(\Omega) \text{ as } \varepsilon \to 0. \tag{5.1}$$

Moreover, ξ^δ solves the problem

$$\begin{aligned} &\Delta^2\xi^\delta + \frac{1}{\delta}\beta'(u^\delta - \psi)(\xi^\delta - \ell) = 0 \quad \text{in } \Omega, \\ &\xi^\delta = 0 \quad \text{on } \partial\Omega, \\ &\frac{\partial\xi^\delta}{\partial n} = 0 \quad \text{on } \partial\Omega. \end{aligned} \tag{5.2}$$

Proof. The proof is similar to that of Lemma 5.1 in [3]. Subtracting $u^\delta(\psi+\varepsilon\ell)$ PDE from $u^\delta(\psi)$ PDE gives

$$\begin{aligned} &\Delta^2(u^\delta(\psi + \varepsilon\ell) - u^\delta(\psi)) \\ &\qquad = -\frac{1}{\delta}\beta(u^\delta(\psi + \varepsilon\ell) - \psi - \varepsilon\ell) + \frac{1}{\delta}\beta(u^\delta(\psi) - \psi) \quad \text{in } \Omega, \\ &u^\delta(\psi + \varepsilon\ell) - u^\delta(\psi) = 0 \quad \text{on } \partial\Omega, \\ &\frac{\partial(u^\delta(\psi + \varepsilon\ell) - u^\delta(x\psi))}{\partial n} = 0 \quad \text{on } \partial\Omega. \end{aligned} \tag{5.3}$$

Multiplying (5.3) by the test function $\phi = u^\delta(\psi + \varepsilon\ell) - u^\delta(\psi)$, integrating over Ω, and integrating by parts twice, we get

$$\begin{aligned} &\int_\Omega \left(\Delta[u^\delta(\psi + \varepsilon\ell) - u^\delta(\psi)]\right)^2 dx \\ &\quad = -\frac{1}{\delta}\int_\Omega \int_0^1 \beta'[\theta(u^\delta(\psi + \varepsilon\ell) - \psi - \varepsilon\ell) + (1 - \theta)(u^\delta(\psi) - \psi)]\, d\theta \\ &\qquad \times (u^\delta(\psi + \varepsilon\ell) - u^\delta(\psi) - \varepsilon\ell)(u^\delta(\psi + \varepsilon\ell) - u^\delta(\psi))\, dx \\ &\quad \leqslant -\frac{1}{\delta}\int_\Omega \int_0^1 \beta'(\cdot)\, d\theta (u^\delta(\psi + \varepsilon\ell) - u^\delta(\psi))^2\, dx \\ &\qquad + \frac{\varepsilon}{\delta}\int_\Omega \ell \int_0^1 \beta'(\cdot)\, d\theta\, [u^\delta(\psi + \varepsilon\ell) - u^\delta(\psi)]\, dx \end{aligned}$$

$$
\begin{aligned}
&\leqslant \frac{\varepsilon}{\delta}\|\ell\|_2\|u^\delta(\psi+\varepsilon\ell)-u^\delta(\psi)\|_2 \\
&\leqslant \frac{C'\varepsilon}{\delta}\|\ell\|_2\|\nabla(u^\delta(\psi+\varepsilon\ell)-u^\delta(\psi))\|_2 \\
&\leqslant \frac{C\varepsilon}{\delta}\|\ell\|_2\|\Delta(u^\delta(\psi+\varepsilon\ell)-u^\delta(\psi))\|_2,
\end{aligned}
$$

where we used the Poincaré inequality twice. Thus,

$$
\left\|\Delta\left(\frac{u^\delta(\psi+\varepsilon\ell)-u^\delta(\psi)}{\varepsilon}\right)\right\|_2 \leqslant \frac{C}{\delta}\|\ell\|_2,
$$

which gives the desired *a priori* estimate of the difference quotients, and therefore justifies the weak convergence on a subsequence to a limit which we denote ξ^δ. To see that ξ^δ satisfies (5.2), we take an arbitrary $\varphi \in H_0^2(\Omega)$ and let $\varepsilon \to 0$ in the equation below

$$
\begin{aligned}
&\int_\Omega \Delta\Big(\frac{u^\delta(\psi+\varepsilon\ell)-u^\delta(\psi)}{\varepsilon}\Big)\Delta\varphi\, dx \\
&\qquad = -\frac{1}{\delta}\int_\Omega\int_0^1 \beta'(\cdot)\, d\theta\Big(\frac{u^\delta(\psi+\varepsilon\ell)-u^\delta(\psi)}{\varepsilon}-\ell\Big)\varphi\, dx.
\end{aligned}
$$

The lemma is proved. □

We prove an assertion which gives necessary conditions for optimal pairs of Problem C_δ.

Theorem 5.1. *Let $(\overline{u}^\delta,\psi^\delta)$ be an optimal pair for problem C_δ. Then there exists an adjoint function $p^\delta \in H_0^2(\Omega)$ such that the triple $(\overline{u}^\delta,p^\delta,\psi^\delta)$ satisfies the following system of equations:*

$$
\begin{aligned}
&\Delta^2\overline{u}^\delta + \frac{1}{\delta}\beta(\overline{u}^\delta-\psi^\delta) = 0 \quad \text{in } \Omega, \\
&\Delta^2 p^\delta + \frac{1}{\delta}\beta'(\overline{u}^\delta-\psi^\delta)p^\delta = \overline{u}^\delta - z \quad \text{in } \Omega, \\
&-\Delta^3\psi^\delta + \frac{1}{\delta}\beta'(\overline{u}^\delta-\psi^\delta)p^\delta = 0 \quad \text{in } \Omega, \\
&\overline{u}^\delta = \frac{\partial\overline{u}^\delta}{\partial n} = 0 \quad \text{on } \partial\Omega, \\
&p^\delta = \frac{\partial p^\delta}{\partial n} = 0 \quad \text{on } \partial\Omega, \\
&\psi^\delta = \frac{\partial\psi^\delta}{\partial n} = \frac{\partial(\Delta\psi^\delta)}{\partial n} = 0 \quad \text{on } \partial\Omega.
\end{aligned}
\tag{5.4}
$$

Proof. Let $(\overline{u}^\delta, \psi^\delta)$ be an optimal pair for problem C_δ. Take an arbitrary $\ell \in V$. First, we derive a system which is adjoint to (5.2). Rewrite (5.2) as follows (with ψ replaced by ψ^δ):

$$\mathcal{L}\xi^\delta = \frac{1}{\delta}\beta'(\overline{u}^\delta - \psi^\delta)\ell, \tag{5.5}$$

where

$$\begin{aligned} &\mathcal{L}\xi^\delta \stackrel{\text{def}}{=} \Delta^2\xi^\delta + \frac{1}{\delta}\beta'(\overline{u}^\delta - \psi^\delta)\xi^\delta, \\ &\xi^\delta = 0 \quad \text{on } \partial\Omega, \\ &\frac{\partial \xi^\delta}{\partial n} = 0 \quad \text{on } \partial\Omega. \end{aligned} \tag{5.6}$$

Note that the boundary conditions in (5.6) are part of the definition of operator $\mathcal{L}$.

Using the adjoint of the operator $\mathcal{L}$, we see that the adjoint function p^δ satisfies

$$\begin{aligned} &\Delta^2 p^\delta + \frac{1}{\delta}\beta'(\overline{u}^\delta - \psi^\delta)p^\delta = \overline{u}^\delta - z \quad \text{in } \Omega, \\ &p^\delta = \frac{\partial p^\delta}{\partial n} = 0 \quad \text{on } \partial\Omega, \end{aligned} \tag{5.7}$$

where the right-hand side of (5.7) contains the derivative of the integrand of the objective functional $J_\delta(\psi^\delta)$ with respect to the state $\overline{u}^\delta$. By the standard elliptic theory (cf., for example, [16]), there exists a solution $p^\delta \in H_0^2(\Omega) \cap H^4(\Omega)$ satisfying (5.7). We have

$$J_\delta(\psi^\delta + \varepsilon\ell) = \frac{1}{2}\int_\Omega \left\{ |\overline{u}^\delta(\psi^\delta + \varepsilon\ell) - z|^2 + |\nabla\Delta(\psi^\delta + \varepsilon\ell)|^2 \right\} dx,$$

$$J_\delta(\psi^\delta) = \frac{1}{2}\int_\Omega \left\{ |\overline{u}^\delta(\psi^\delta) - z|^2 + |\nabla\Delta(\psi^\delta)|^2 \right\} dx.$$

Hence

$$\begin{aligned} 0 \leqslant \varliminf_{\varepsilon \to 0+} & \frac{J_\delta(\psi^\delta + \varepsilon\ell) - J_\delta(\psi^\delta)}{\varepsilon} \\ = \varliminf_{\varepsilon \to 0+} & \frac{1}{2}\int_\Omega \left\{ \left(\frac{\overline{u}^\delta(\psi^\delta + \varepsilon\ell) - \overline{u}^\delta(\psi^\delta)}{\varepsilon} \right) \left(\overline{u}^\delta(\psi^\delta + \varepsilon\ell) + \overline{u}^\delta(\psi^\delta) \right) \right. \\ & \left. - 2\left(\frac{\overline{u}^\delta(\psi^\delta + \varepsilon\ell) - \overline{u}^\delta(\psi^\delta)}{\varepsilon} \right) z + 2\nabla\Delta\psi^\delta \cdot \nabla\Delta\ell + \varepsilon|\nabla\Delta\ell|^2 \right\} dx \end{aligned}$$

$$
\begin{aligned}
&= \int_\Omega \Big\{ (\overline{u}^\delta - z)\xi^\delta + \nabla \Delta \psi^\delta \cdot \nabla \Delta \ell \Big\}\, dx \\
&= \int_\Omega \Big\{ \Big(\Delta^2 p^\delta + \frac{1}{\delta}\beta'(\overline{u}^\delta - \psi^\delta) p^\delta \Big) \xi^\delta + \nabla \Delta \psi^\delta \cdot \nabla \Delta \ell \Big\}\, dx \\
&= \int_\Omega \Big\{ \Delta p^\delta \Delta \xi^\delta + \frac{1}{\delta}\beta'(\overline{u}^\delta - \psi^\delta) p^\delta \xi^\delta + \nabla \Delta \psi^\delta \cdot \nabla \Delta \ell \Big\}\, dx \\
&= \int_\Omega \Big\{ \frac{1}{\delta}\beta'(\overline{u}^\delta - \psi^\delta)\, \ell p^\delta + \nabla \Delta \psi^\delta \cdot \nabla \Delta \ell \Big\}\, dx
\end{aligned}
$$

where we take into account (5.5) and (5.7). Since $\ell \in V$ is arbitrary, taking $\partial(\Delta\psi^\delta)/\partial n = 0$ on $\partial\Omega$ and using $\ell = \frac{\partial \ell}{\partial n} = 0$ on $\partial\Omega$, we obtain in the sense of distributions

$$
\begin{aligned}
\int_\Omega \nabla \Delta \psi^\delta \cdot \nabla \Delta \ell\, dx &= \int_\Omega \sum_{i=1}^n (\Delta \psi^\delta)_{x_i} (\Delta \ell)_{x_i}\, dx \\
&= -\int_\Omega \sum_{i=1}^n (\Delta \psi^\delta)_{x_i x_i} \Delta \ell\, dx + \int_{\partial\Omega} \sum_{i=1}^n (\Delta \psi^\delta)_{x_i} (\Delta \ell) \eta_i\, ds \\
&= -\int_\Omega (\Delta^2 \psi^\delta) \Delta \ell\, dx + \int_{\partial\Omega} \frac{\partial(\Delta \psi^\delta)}{\partial n} \Delta \ell\, ds \\
&= -\int_\Omega (\Delta^2 \psi^\delta) \Delta \ell\, dx \\
&= \int_\Omega \sum_{i=1}^n (\Delta^2 \psi^\delta)_{x_i} \ell_{x_i}\, dx - \int_{\partial\Omega} \sum_{i=1}^n (\Delta^2 \psi^\delta) \ell_{x_i} \eta_i\, ds \\
&= \int_\Omega \sum_{i=1}^n (\Delta^2 \psi^\delta)_{x_i} \ell_{x_i}\, dx - \int_{\partial\Omega} (\Delta^2 \psi^\delta) \frac{\partial \ell}{\partial n}\, ds \\
&= \int_\Omega \sum_{i=1}^n (\Delta^2 \psi^\delta)_{x_i} \ell_{x_i}\, dx \\
&= -\int_\Omega \sum_{i=1}^n (\Delta^2 \psi^\delta)_{x_i x_i} \ell\, dx + \int_{\partial\Omega} \sum_{i=1}^n (\Delta^2 \psi^\delta)_{x_i} \ell \eta_i\, ds \\
&= -\int_\Omega (\Delta^3 \psi^\delta) \ell\, dx.
\end{aligned}
$$

Therefore,

$$\begin{aligned} -\Delta^3\psi^\delta + \frac{1}{\delta}\beta'(\overline{u}^\delta - \psi^\delta)p^\delta = 0 \ \text{ in } \Omega, \\ \psi^\delta = \frac{\partial \psi^\delta}{\partial n} = \frac{\partial(\Delta\psi^\delta)}{\partial n} = 0 \text{ on } \partial\Omega. \end{aligned} \tag{5.8}$$

This completes the proof of the theorem. □

The following assertion gives necessary conditions for an optimal control for Problem C.

Theorem 5.2. *Let $(\overline{u}, \overline{\psi}) \in H_0^2(\Omega) \times V$ be an optimal pair for Problem* C. *Then there exist $\overline{p} \in H_0^2(\Omega)$ and $\overline{q} \in H^{-2}(\Omega)$ such that*

$$\begin{aligned} &\overline{u} = T(\overline{\psi}), \\ &\Delta^2\overline{p} + \overline{q} = \overline{u} - z \quad \text{in } \Omega, \\ &\Delta^3\overline{\psi} + \overline{q} = 0 \quad \text{in } \Omega, \\ &\overline{\psi} = \frac{\partial\overline{\psi}}{\partial n} = \frac{\partial(\Delta\overline{\psi})}{\partial n} = 0 \quad \text{on } \partial\Omega, \\ &\overline{u} = \frac{\partial\overline{u}}{\partial n} = 0 \quad \text{on } \partial\Omega, \\ &\overline{p} = \frac{\partial\overline{p}}{\partial n} = 0 \quad \text{on } \partial\Omega, \\ &\langle \overline{q}, \overline{p} \rangle_{H^{-2}, H_0^2} \geqslant 0. \end{aligned} \tag{5.9}$$

Proof. In order to pass to a limit in (5.4), we need to show an *a priori* estimate for p^δ first. Multiplying (5.7) by a test function $v = p^\delta$, integrating over Ω, and integrating by parts twice, we get

$$\int_\Omega (\Delta p^\delta)^2\,dx = -\frac{1}{\delta}\int_\Omega \beta'(\overline{u}^\delta - \psi^\delta)(p^\delta)^2\,dx + \int_\Omega (\overline{u}^\delta - z)p^\delta\,dx.$$

Since $\beta'(\overline{u}^\delta - \psi^\delta) \geqslant 0$, we can write

$$\begin{aligned} \int_\Omega (\Delta p^\delta)^2\,dx &= -\frac{1}{\delta}\int_\Omega \beta'(\overline{u}^\delta - \psi^\delta)(p^\delta)^2\,dx + \int_\Omega (\overline{u}^\delta - z)p^\delta\,dx \\ &\leqslant \int_\Omega (\overline{u}^\delta - z)p^\delta\,dx. \end{aligned} \tag{5.10}$$

Therefore,

$$\|\Delta p^\delta\|_2^2 = \int_\Omega (\Delta p^\delta)^2\,dx \leqslant \|\overline{u}^\delta - z\|_2 \cdot \|p^\delta\|_2. \tag{5.11}$$

The fact that $p^\delta \in H_0^2(\Omega)$ and (4.13) give us the desired H^2 estimate on p^δ. By (5.11), there exists $\overline{p} \in H_0^2(\Omega)$ such that, on a subsequence,

$$p^\delta \stackrel{w}{\rightharpoonup} \overline{p} \ \text{ in } H_0^2(\Omega).$$

Denote $q^\delta := \frac{1}{\delta}\beta'(\overline{u}^\delta - \psi^\delta)p^\delta$. Let $\varphi \in H_0^2(\Omega)$ be arbitrary. From (5.7) we find

$$\begin{aligned}
|q^\delta(\varphi)| &= \left|\frac{1}{\delta}\int_\Omega \beta'(\overline{u}^\delta - \psi^\delta)p^\delta \varphi\, dx\right| \\
&= \left| - \int_\Omega (\Delta^2 p^\delta)\varphi\, dx + \int_\Omega (\overline{u}^\delta - z)\varphi\, dx\right| \\
&\leqslant \left| - \int_\Omega \Delta p^\delta \Delta\varphi\, dx + \int_\Omega (\overline{u}^\delta - z)\varphi\, dx\right| \\
&\leqslant \left|\int_\Omega \Delta p^\delta \Delta\varphi\, dx\right| + \left|\int_\Omega (\overline{u}^\delta - z)\varphi\, dx\right| \\
&\leqslant \|\Delta p^\delta\|_2 \cdot \|\Delta\varphi\|_2 + \|\overline{u}^\delta - z\|_2 \cdot \|\varphi\|_2 \\
&\leqslant C\|\overline{u}^\delta - z\|_2 \cdot \|\varphi\|_{H^2(\Omega)}.
\end{aligned}$$

Hence

$$\|q^\delta\|_{H^{-2}(\Omega)} = \sup |q^\delta(\varphi)| \leqslant C\|z\|_2,$$

where the supremum is taken over all $\varphi \in H_0^2(\Omega)$ such that $\|\varphi\|_{H^2(\Omega)} = 1$. This implies that there exists $\overline{q} \in H^{-2}(\Omega)$ such that

$$q^\delta \stackrel{w*}{\rightharpoonup} \overline{q} \ \text{ in } H^{-2}(\Omega).$$

By Theorem 4.2, $(\overline{u}^\delta, \psi^\delta)$ converges to an optimal pair $(\overline{u}, \overline{\psi})$ as $\delta \to 0^+$. Letting $\delta \to 0^+$ in (5.4), we obtain the first six lines in (5.9). To show that $\langle \overline{q}, \overline{p} \rangle_{H^{-2},\, H_0^2} \geqslant 0$, we note the following. By the lower semicontinuity of the L^2-norm with respect to weak convergence, from (5.10) we obtain

$$\int_\Omega (\Delta\overline{p})^2\, dx \leqslant \int_\Omega (\overline{u} - z)\overline{p}\, dx. \tag{5.12}$$

Using $\overline{p}$ as a test function in the weak formulation of the second equation in (5.9), we find

$$\int_\Omega (\Delta\overline{p})^2\, dx + \langle \overline{q}, \overline{p} \rangle_{H^{-2},\, H_0^2} = \int_\Omega (\overline{u} - z)\overline{p}\, dx. \tag{5.13}$$

Combining (5.12) and (5.13), we obtain $\langle \overline{q}, \overline{p} \rangle_{H^{-2},\, H_0^2} \geqslant 0$. □

References

1. Adams, D.R.: The biharmonic obstacle problem with varying obstacles and a related maximal operator. Operator Theory: Advances and Applications **110**, 1–12 (1999)
2. Adams, D.R., Hedberg, L.I.: Function Spaces and Potential Theory. Springer (1999)
3. Adams, D.R., Lenhart, S.M., Yong, J.: Optimal control of the obstacle for an elliptic variational inequality. Appl. Math. Optim. **38**, 121–140 (1998)
4. Adams, D.R., Lenhart, S.M.: An obstacle control problem with a source term. Appl. Math. Optim. **47**, 79–95 (2002)
5. Adams, D.R., Lenhart, S.M.: Optimal control of the obstacle for a parabolic variational inequality. J. Math. Anal. Appl. **268**, 602–614 (2002)
6. Baiocchi, C., Capelo, A.: Variational and Quasivariational Inequalities. Wiley, New York (1984)
7. Barbu, V.: Necessary conditions for distributed control problems governed by parabolic variational inequalities. SIAM J. Control Optim. **19**, 64–86 (1981)
8. Barbu, V.: Optimal Control of Variational Inequalities. Pitman, London (1984)
9. Barbu, V.: Analysis and Control of Nonlinear Infinite Dimensional Systems. Academic Press, New York (1993)
10. Bergounioux, M.: Optimal control of semilinear elliptic obstacle problem. J. Nonlin. Convex Anal. **3**, 25–39 (2002)
11. Bergounioux, M., Lenhart, S.: Optimal control of the obstacle in semilinear variational inequalities. Positivity **8**, 229–242 (2004)
12. Bergounioux, M., Lenhart, S.: Optimal control of bilateral obstacle problems. SIAM J. Control Optim. **43**, 240–255 (2004)
13. Caffarelli, L.A., Friedman, A.: The obstacle problem for the biharmonic operator. Ann. Sco. Norm. Sup. Pisa **6**, 151–184 (1979)
14. Chen, Q.: Optimal control for semilinear evolutionary variational bilateral problems. J. Math. Anal. Appl. **277**, 303–323 (2003)
15. Friedman, A.: Variational Principles and Free Boundary Problems. Wiley, New York (1982)
16. Gilbarg, D., Trudinger, N.S.: Elliptic Partial Differential Equations of Second Order. 2nd ed. Springer, Berlin (1983)
17. Ito, K., Kunisch, K.: An augmented lagrangian technique for variational inequalities. Appl. Math. Optim. **21**, 223–241 (1990)
18. Ito, K., Kunisch, K.: Optimal control of elliptic variational inequalities. Appl. Math. Optim. **41**, 343–364 (2000)
19. Kinderlehrer, D., Stampacchia, G.: An Introduction to Variational Inequalities and Their Applications. Academic Press, New York (1980)
20. Lou, H.: On the regularity of an obstacle control problem. J. Math. Anal. Appl. **258**, 32–51 (2001)
21. Lou, H.: An Optimal Control Problem Governed by Quasilinear Variational Inequalities. Preprint.
22. Maz'ya, V.G.: The negative spectrum of the higher-dimensional Schrödinger operator (Russian). Dokl. Akad. Nauk SSSR **144**, 721–722 (1962); English transl.: Sov. Math. Dokl. **3**, 808–810 (1962)
23. Maz'ya, V.G.: On the theory of the higher-dimensional Schrödinger operator (Russian). Izv. Akad. Nauk SSSR. Ser. Mat. **28**, 1145–1172 (1964)
24. Maz'ya, V.G.: Certain integral inequalities for functions of several variables (Russian). Probl. Mat. Anal. **3**, 33–68 (1972)
25. Maz'ya, V.G.: Sobolev Spaces. Springer, Berlin etc. (1985)
26. Rodrigues, J.: Obstacle Problems in Mathematical Physics. North Holland, Amsterdam (1987)
27. Troianiello, G.M.: Elliptic Differential Equations and Obstacle Problems. Plenum, New York (1987)

Minimal Thinness and the Beurling Minimum Principle

Hiroaki Aikawa

Abstract Every domain with the Green function has the Martin boundary and every positive harmonic function on the domain is represented as the integral over the Martin boundary. The classical Fatou theorem concerning nontangential limits of harmonic functions is extended to this general context by using the notion of the minimal thinnes. The identification of the Martin boundaries for specific domains is an interesting problem and has attracted many mathematicians. From smooth domains to nonsmooth domains the study of the Martin boundary has been expanded. It is now known that the Martin boundary of a uniform domain is homeomorphic to the Euclidean boundary. On the other hand, the study of the minimal thinnes has been exploited comparatively a little. The minimal thinnes of an NTA domain was studied by the author with the aid of the quasiadditivity of capacity, the Hardy inequality, and the Beurling minimum principle. Maz'ya was one of the first mathematicians who recognized the significance of the Beurling minimum principle. This article illustrates the backgrounds of the characterization of the minimal thinness and gives a characterization of the minimal thinness of a uniform domain.

1 Introduction

Every domain with the Green function has the Martin boundary and every positive harmonic function on the domain is represented as the integral over the Martin boundary [26]. The classical Fatou theorem concerning nontangential limits of harmonic functions [17] is extended to this general context by using the notion of the minimal thinnes [30]. The identification of the

Hiroaki Aikawa
Hokkaido University, Sapporo 060-0810, Japan
e-mail: aik@math.sci.hokudai.ac.jp

A. Laptev (ed.), *Around the Research of Vladimir Maz'ya III: Analysis and Applications*, International Mathematical Series 13, DOI 10.1007/978-1-4419-1345-6_2,

Martin boundaries for specific domains is an interesting problem and has attracted many mathematicians. From smooth domains to nonsmooth domains the study of the Martin boundary has been developed and a number of papers have been published. It is now known that the Martin boundary of a uniform domain is homeomorphic to the Euclidean boundary [4].

On the other hand, the study of the minimal thinnes has been exploited comparatively a little (cf., for example, [1, 2, 3, 5, 6, 14, 15, 19, 25, 28, 29, 35]). The minimal thinnes of an NTA domain was studied by the author [2, 3]. The quasiadditivity of capacity, the Hardy inequality and the Beurling minimum principle played important roles [2, 3, 7, 11, 32]

Maz'ya [27] *was one of the first mathematicians who recognized the significance of the Beurling minimum principle.* The term "Beurling theorem" was introduced by Maz'ya. Unfortunately, Maz'ya's work [27] was buried in the literature; Dahlberg [13] independently proved the same results four years later.

The author is grateful to Professor Vladimir Maz'ya for drawing his attention to the paper [27] on the Beurling minimum principle. Professor Maz'ya informed the author for this paper on the occasion of the international conference on potential theory in Amersfort (ICPT91), and this information was very fruitful for the author.

Starting with a heuristic introduction of the Martin boundary, this article illustrates the backgrounds of the minimal thinness and gives some new results, i.e., characterizations of the minimal thinness in a uniform domain.

2 Martin Boundary

Throughout the paper, D is a proper domain in $\mathbb{R}^n$ with $n \geqslant 2$. Let $\delta_D(x) = \operatorname{dist}(x, \partial D)$. We write $B(x,r)$ and $S(x,r)$ for the open ball and sphere of center at x and radius r respectively. We denote by M an absolute positive constant which value is unimportant and may change. Sometimes, we write $M_0, M_1, \dots$ to specify such constants. If two positive quantities f and g satisfies the inequality $M^{-1} \leqslant f/g \leqslant M$ with some constant $M \geqslant 1$, we say f and g are *comparable* and write $f \approx g$.

2.1 General Martin boundary theory

We begin with a heuristic introduction of the Martin boundary [26]. Let $G_D(x,y)$ be the normalized Green function for D, i.e.,

1) For each $x \in D$ fixed, $G_D(x, \cdot) = 0$ on ∂D except for a polar set.

2) For each $y \in D$ fixed, $\Delta G_D(\cdot, y) = -\delta_y$, where Δ and δ_y stand for the distributional Laplacian and Dirac measure at y respectively.

For the sake of simplicity, we write $G(x, y)$ for $G_D(x, y)$.

If D is sufficiently smooth, we have the Poisson integral formula: every harmonic function h on D with continuous boundary values is represented as

$$h(x) = \int_{\partial D} \frac{\partial G(x, y)}{\partial n_y} h(y) d\sigma(y), \tag{2.1}$$

where n_y and σ stand for the inward normal and surface measure on ∂D. The kernel $\dfrac{\partial G(x, y)}{\partial n_y}$ is called the *Poisson kernel*.

Let us try to generalize (2.1) as far as possible. First, let us extend the representation to an arbitrary positive harmonic function, so that the Poisson kernel itself can be represented. For this purpose, we extend the measure $h(y)d\sigma(y)$ to a general measure on ∂D. We have the following *Herglotz theorem*: every nonnegative harmonic function h on D has a measure μ_h on ∂D such that

$$h(x) = \int_{\partial D} \frac{\partial G(x, y)}{\partial n_y} d\mu_h(y) \quad \text{for } x \in D. \tag{2.2}$$

Next, we extend (2.2) to more general domains which may not have normals. For this purpose, we need to understand the normal derivative $\partial G(x, y)/\partial n_y$ in an appropriate way. Recall that $\partial/\partial n_y$ is decomposed into two operations:

1) divide by the distance $\delta_D(z)$ for $z \in D$,
2) take the limit as $z \to y$, i.e.,

$$\lim_{z \to y} \frac{G(x, z)}{\delta_D(z)}.$$

We fix $x_0 \in D$. It is known that

$$G(x_0, z) \approx \delta_D(z) \quad \text{for } y \in D \text{ close to the boundary } \partial D,$$

provided that D is a smooth domain, say a $C^{1,\alpha}$-domain $(0 < \alpha < 1)$ or a Lyapunov–Dini domain [34].

Thus, the limit of the ratio of the Green functions

$$\lim_{z \to y} \frac{G(x, z)}{G(x_0, z)}$$

would be a substitute of the normal derivative. This limit can be considered if the domain has the Green function, whereas the regularity of the boundary has no influence. This limit is called the *Martin kernel* $K(x, y)$. Formula (2.2) will be generalized to the Martin integral representation:

$$h(x) = \int K(x,y)d\mu_h(y) \quad \text{for } x \in D.$$

More exactly, let us consider the ratio of the Green functions

$$K(x,y) = K_y(x) = \frac{G(x,y)}{G(x_0,y)} \quad \text{for } y \in D \setminus \{x_0\}.$$

Note that K_y is a positive harmonic function on $D \setminus \{y\}$. Suppose that $\{y_j\}_j \subset D$ is a sequence without accumulation points in D. On any relatively compact subdomain, $\{K_{y_j}\}_j$ becomes a sequence of positive harmonic functions with $K_{y_j}(x_0) = 1$ for large j. Hence the Harnack principle implies that they are locally uniformly bounded, so that there is a subsequence converging to a positive harmonic function h^*. For the sake of simplicity, we assume that K_{y_j} itself converges to h^*. We regard that $\{y_j\}_j$ defines the boundary point ξ and write

$$h^* = K(\cdot, \xi).$$

Considering all sequences $\{y_j\}_j$, we obtain a compactification $\widehat{D}$ of D, called the *Martin compactification* of D. The boundary $\widehat{D} \setminus D = \Delta$ is called the *Martin boundary.* Note that $K(x,\cdot)$ is continuous on $\widehat{D} \setminus \{x_0\}$ for each $x \in D$ and $\{K(x,\cdot) : x \in D\}$ separates the boundary point in Δ. These two properties characterize the Martin compactification, known as *Q-compactification* [12].

For a nonnegative superharmonic function u on D and $E \subset D$ we denote by $\widehat{R}_u^E$ the *regularized reduced function* of u on E (cf., for example, [22]). Note that if E is a compact subset of D, then $\widehat{R}_u^E = u$ in the interior of E and $v = \widehat{R}_u^E$ is the Dirichlet solution to the equation $\Delta v = 0$ on $D \setminus E$ with boundary data $v = u$ on ∂E and $v = 0$ on ∂D.

The Martin compactification is closely related to the integral representation of positive harmonic functions h on D. Using the exhaustion of D and regularized reduced functions, we easily find that there is a measure μ on Δ such that (cf. [22, Chapter 12])

$$h(x) = \int_\Delta K(x,\xi)d\mu(\xi) \quad \text{for } x \in D.$$

Unfortunately, the measure μ is not necessarily unique: if there is a point $\xi_0 \in \Delta$ such that

$$K(x,\xi_0) = \sum \alpha_j K(x,\xi_j),$$

where $\xi_j \in \Delta \setminus \{\xi_0\}$ and positive numbers α_j are such that $\sum \alpha_j = 1$, then the measures μ and

$$\mu + \mu(\{\xi_0\})(\delta_{\xi_0} - \sum \alpha_j \delta_{\xi_j})$$

define the same harmonic function h, but these measures do not coincide unless $\mu(\{\xi_0\}) = 0$. Thus, we eliminate such points ξ_0.

A positive harmonic function h on D is said to be *minimal* if any other positive harmonic function less than h coincides with a constant multiple of h. Let

$$\Delta_1 = \{y \in \Delta : K(\cdot, \xi) \text{ is a minimal harmonic function}\}.$$

The set Δ_1 is called the *minimal Martin boundary*. The complement $\Delta \setminus \Delta_1$ is the *nonminimal Martin boundary* and is denoted by Δ_0. We state the Martin representation theorem.

Theorem 2.1 (Martin representation theorem). *For a positive harmonic function h there exists a unique measure μ_h on Δ_1 such that*

$$h(x) = \int_{\Delta_1} K(x, \xi) d\mu_h(\xi) \quad \text{for } x \in D.$$

Combining with the Riesz decomposition theorem, we obtain the following assertion.

Theorem 2.2 (Riesz–Martin representation theorem). *For a nonnegative superharmonic function u on D there exists a unique measure μ_u on $D \cup \Delta_1$ such that*

$$u(x) = \int_{D \cup \Delta_1} K(x, \xi) d\mu_u(\xi) \quad \text{for } x \in D.$$

For the sake of simplicity, we write $K\mu$ for

$$\int_{D \cup \Delta_1} K(\cdot, y) d\mu(y)$$

if μ is a measure on $D \cup \Delta_1$.

2.2 Identification of the Martin boundary

We present some facts of the abstract Martin theory. Identification of the Martin boundaries of specific domains is an interesting independent problem attracting the attention of many mathematicians. As one may expect, if D possesses geometrical smoothness, then $\Delta = \Delta_1 = \partial D$.

Definition 2.1 (cf. [23]). A domain is called an NTA *domain* if there exist positive constants M and r_1 such that

(i) *Corkscrew condition.* For any $\xi \in \partial D$ and small positive r there exists a point $A_\xi(r) \in D \cap S(\xi, r)$ such that $\delta_D(A_\xi(r)) \approx r$. A point $A_\xi(r)$ is referred to as a *nontangential point* at ξ.
(ii) The complement of D satisfies the corkscrew condition.

(iii) *Harnack chain condition.* If $\varepsilon > 0$ and $x, y \in D$, $\delta_D(x) \geqslant \varepsilon$, $\delta_D(y) \geqslant \varepsilon$, and $|x - y| \leqslant M\varepsilon$, then there exists a Harnack chain from x and y whose length is independent of ε.

We introduce a larger class of uniform domains. Roughly speaking, D is a uniform domain if D satisfies (i) and (iii).

Definition 2.2 (cf. [20]). A domain D is said to be *uniform* if there exist constants M and M' such that each pair of points $x_1, x_2 \in D$ can be joined by a rectifiable curve $\gamma \subset D$ such that

$$\ell(\gamma) \leqslant M|x_1 - x_2|,$$
$$\min\{\ell(\gamma(x_1, y)), \ell(\gamma(x_2, y))\} \leqslant M'\delta_D(y) \quad \text{for all } y \in \gamma.$$

Here, $\ell(\gamma)$ and $\gamma(x_j, y)$ denote the length of γ and the subarc of γ connecting x_j and y respectively.

Remark 2.1. Let D be a uniform domain, and let $\xi \in \partial D$. For small $r > 0$ we have a nontangential point $A_\xi(r) \in D \cap S(\xi, r)$ at ξ.

Note that a uniform domain need not satisfy the exterior condition (ii), hence it may be irregular for the Dirichlet problem. However, even in this nasty situation, the following assertion holds.

Theorem 2.3 (cf. [4]). *Let D be a uniform domain. Then the Martin boundary of D is homeomorphic to its Euclidean boundary ∂D; and each boundary point is minimal, which is symbolically written as*

$$\Delta = \Delta_1 = \partial D.$$

3 Minimal Thinness

3.1 Minimal thinness and the Fatou–Näim–Doob theorem

We consider the boundary behavior of nonnegative superharmonic functions. For this purpose, the notion of the minimal thinness is important.

Definition 3.1. Let $\xi \in \Delta_1$ and $E \subset D$. We say that E is minimally thin at ξ if

$$\widehat{R}_{K_\xi}^{E} \neq K_\xi.$$

This is equivalent to the fact that $\widehat{R}_{K_\xi}^{E}$ is a Green potential. We say that a function f on D has the *minimal fine limit* l at ξ if there is a set E minimally thin at ξ such that

$$\lim_{x \to \xi, x \in D \setminus E} \frac{u(x)}{K_\xi(x)} = l.$$

We write

$$\mathrm{mf}\lim_{x \to \xi} \frac{u(x)}{K_\xi(x)} = l$$

if f has the minimal fine limit l at ξ.

Using the notion of minimal thinness, we extend the classical Fatou theorem. The extension is referred to as the *Fatou–Naïm–Doob theorem* (cf. [30]).

Theorem 3.1 (Fatou–Naïm–Doob theorem). *Let u and v be positive harmonic functions on D, and let $v = K\mu_v$ with the Martin representation measure μ_v. Then the ratio u/v has the minimal fine limit at μ_v-a.e. boundary points in Δ_1.*

This theorem is rather abstract. The identification of the minimal thinness for specific domains is of interest, but there are only a few papers dealing with this problem.

3.2 Wiener type criteria for the minimal thinness

For the upper half-space a Wiener type criterion for the minimal thinness can be found in [25]. In the rest of this section, for D we consider the upper half-space $\mathbb{R}^n_+ = \{x : x_n > 0\}$. Following [16, Definition 2.2], we define the Green energy. For the sake of simplicity, let $n \geqslant 3$. Observe that the Martin kernel at $y \in \partial D$ has the form

$$K(x,y) = \begin{cases} \dfrac{x_n}{|x-y|^n} & \text{if } y \subset \partial D \\ x_n & \text{if } y = \infty \end{cases}$$

up to a positive continuous function of y. Introduce the Green energy, a kind of capacity, as follows. Let E be a compact set. Consider the regularized reduced function $\widehat{R}^E_{x_n}$ which can be represented as the Green potential $G\mu_E$ of a measure μ_E on E, i.e.,

$$\widehat{R}^E_{x_n} = \int_E G(x,y)\, d\mu_E(y).$$

By the *Green energy* of E we mean

$$\gamma(E) = \iint G(x,y)\, d\mu_E(x)\, d\mu_E(y).$$

In a standard way, $\gamma(E)$ is extended to open sets, and then to general sets (cf. Definition 3.2 below). We write $\gamma(E)$ for the extension as well. Observe that

$\gamma(E)$ is homogeneous of degree n, i.e., $\gamma(rE) = r^n\gamma(E)$ for $r > 0$ because the kernel $\dfrac{G(x,y)}{x_n y_n}$ is homogeneous of degree $-n$. Taking into account of the homogeneity, we set $I_i = \{x : 2^{-i-1} \leqslant |x| < 2^{-i}\}$ and consider the series

$$\sum_{i=1}^{\infty} 2^{in}\gamma(E \cap I_i).$$

Then we obtain the following characterization.

Theorem 3.2 (cf. [25]). *Let $E \subset \mathbb{R}^n_+$. Then the following assertions are equivalent:*

(i) *E is minimally thin at 0,*

(ii) $\sum_{i=1}^{\infty} 2^{in}\gamma(E \cap I_i) < \infty$.

This theorem provides us with a complete characterization of minimal thinness for the upper half-space. It can be generalized to uniform domains. For such a generalization, we need to generalize the notion of Green energy.

Definition 3.2. Let u be a nonnegative superharmonic function on D. For a compact subset K of D denote by $\widehat{R}^K_u$ the regularized reduced function of u with respect to K. Observe that $\widehat{R}^K_u$ is a Green potential of a measure λ^K_u on K. The energy

$$\gamma_u(K) = \iint G(x,y)d\lambda^K_u(x)d\lambda^K_u(y)$$

is called the *Green energy* of K relative to u. We set

$$\gamma_u(V) = \sup\{\gamma_u(K) : K \text{ is compact, } K \subset V\}$$

for an open subset V of D and

$$\gamma_u(E) = \inf\{\gamma_u(V) : V \text{ is open, } E \subset V\}$$

for a general subset E of D. The quantity $\gamma_u(E)$ is also called the *Green energy* relative to u.

Remark 3.1. If $u \equiv 1$, then $\gamma_u(E)$ is the usual Green capacity $C_G(E)$ (cf. [24, pp.174–177]). If $D = \{x = (x_1, \ldots, x_n) : x_n > 0\}$ and $u(x) = x_n$, then $\gamma_u(E)$ is the Green energy in the sense of Essén and Jackson [16, Definition 2.2].

Let $u(x) = g(x) = \min\{G(x,x_0), 1\}$ in Definition 3.2. Then

$$\widehat{R}^E_g(x_0) = \gamma_g(E) \quad \text{for } E \subset \{x \in D : g(x) < 1\}. \tag{3.1}$$

Theorem 3.3. *Let D be a uniform domain. Suppose that $E \subset D$ and $\xi \in \partial D$. Let $A_\xi(r) \in D \cap S(\xi, r)$ be a nontangential point for small $r > 0$ (cf. Remark 2.1). Then the following assertions are equivalent:*

(i) *E is minimally thin at ξ,*

(ii) *for some $i_0 \geqslant 1$*

$$\sum_{i=i_0}^{\infty} 2^{i(n-2)} g(A_\xi(2^{-i}))^{-2} \gamma_g(E \cap B(\xi, 2^{1-i}) \backslash B(\xi, 2^{-i})) < \infty. \tag{3.2}$$

Since $g(x) \approx \delta_D(x)$ for a smooth domain D, say a $C^{1,\alpha}$-domain $(0 < \alpha < 1)$ or a Lyapunov–Dini domain [34], this theorem generalizes Theorem 3.2.

3.3 Proof of Theorem 3.3

Let $\Theta(x, y) = G(x, y)/(g(x)g(y))$. In view of Theorem 2.3, $\Theta(x, y)$ admits a continuous extension to $\overline{D} \times \overline{D}$. We preserve the same notation for the continuous extension. The kernel Θ is referred to as the *Naïm's kernel* for D (cf. [30]). By definition, Θ is symmetric.

Lemma 3.1. *Let $\xi \in \partial D$. For small $r > 0$ we set $\theta_\xi(r) = \Theta(A_\xi(r), \xi)$, where $A_\xi(r) \in D \cap S(\xi, r)$ (cf. Remark 2.1). Then*

$$\theta_\xi(r) \approx g(A_\xi(r))^{-2} r^{2-n}. \tag{3.3}$$

Moreover, if $x \in S(\xi, r) \cap D$, then $\Theta(x, \xi) \approx \theta_\xi(r)$. Furthermore, if $r \approx R$, then $\theta_\xi(r) \approx \theta_\xi(R)$.

Proof. This assertion is an easy consequence of the scale-invariant boundary Harnack principle. More directly, Lemma 3 in [4] yields the following estimates (note that the Green function for $D \cap B(\xi, Mr)$ is comparable with the original Green function for D near ξ provided that D satisfies the capacity density condition):

$$\frac{G(x, y)}{G(x', y)} \approx \frac{G(x, y')}{G(x', y')} \quad \text{for } x, x' \in D \backslash B(\xi, r) \text{ and } y, y' \in D \cap B(\xi, r/6).$$

An elementary calculation shows that

$$\Theta(x, y) \frac{g(x)g(y)}{G(x', y)G(x, y')} \approx \frac{1}{G(x', y')},$$

where $\Theta(x, y)$ is the continuous extension of $G(x, y)/(g(x)g(y))$. Setting $x' = A_\xi(r)$ and $y' = A_\xi(r/6)$, we find that $G(x', y') \approx r^{2-n}$, so that

$$\Theta(x,y)\frac{g(x)g(y)}{G(A_\xi(r),y)G(x,A_\xi(r/6))} \approx r^{n-2}.$$

Hence

$$\Theta(x,y) \approx \frac{r^{n-2}G(A_\xi(r),y)G(x,A_\xi(r/6))}{g(x)g(y)}.$$

Setting $x = A_\xi(r)$, we obtain

$$\Theta(A_\xi(r),y) \approx \frac{r^{n-2}G(A_\xi(r),y)G(A_\xi(r),A_\xi(r/6))}{g(A_\xi(r))g(y)} \approx \frac{G(A_\xi(r),y)}{g(A_\xi(r))g(y)}$$
$$= g(A_\xi(r))^{-1}\frac{G(A_\xi(r),y)}{g(y)}.$$

The last term tends to

$$g(A_\xi(r))^{-1}K(A_\xi(r),\xi) \approx g(A_\xi(r))^{-2}r^{2-n}$$

as $y \to \xi$. Thus, (3.3) follows. The remaining can be easily proved by the Harnack inequality. □

Proof of Theorem 3.3. As in [1], we can show that E is minimally thin at ξ if and only if

$$\sum_{i=i_0}^{\infty} \widehat{R}_{K_\xi}^{E_i}(x_0) < \infty \quad \text{for some } i_0 \geqslant 1, \tag{3.4}$$

where $E_i = \{x \in E : 2^{-i} \leqslant |x-\xi| < 2^{1-i}\}$. By Lemma 3.1, $K_\xi \approx \theta_\xi(2^{-i})g$ on $\{x \in D : 2^{-i} \leqslant |x-\xi| < 2^{1-i}\}$. Hence (3.4) is equivalent to

$$\sum_i \theta_\xi(2^{-i})\gamma_g(E_i) < \infty. \tag{3.5}$$

The proof is complete. □

3.4 Whitney decomposition and the minimal thinness

The Green energy in Theorem 3.3 involves the Green function, which means that the above Wiener type criterion is not so clear as the classical Wiener criterion.

Is it possible to characterize the minimal thinness by a usual (or Newtonian) capacity? This question was posed by Hayman. Several results were obtained by Aikawa, Essén, Jackson, Rippon, Miyamoto, Yanagishita, Yoshida, and others. It turned out that the key role is played by the Whitney decomposition, which is finer than the usual dyadic spherical decomposition.

A family of closed cubes $\{Q_j\}$ with sides parallel to the coordinate axes is called the *Whitney decomposition* of an open set D with $\partial D \neq \varnothing$ if (cf., for example, [33, Chapter VI] for details)

(i) $\bigcup_j Q_j = D$,

(ii) the interiors of Q_j are mutually disjoint,

(iii) for each Q_j

$$\operatorname{diam}(Q_j) \leqslant \operatorname{dist}(Q_j, \partial D) \leqslant 4 \operatorname{diam}(Q_j).$$

The necessity of considering the Whitney decomposition can be justified as follows. In a Whitney cube Q_j, the Green function and Newtonian kernel are comparable for $n \geqslant 3$, i.e.,

$$G(x,y) \approx \frac{|x-y|^{2-n}}{\operatorname{diam}(Q_j)^2} \quad \text{for } x, y \in Q_j,$$

and the Green energy is estimated as

$$\gamma(E) \approx \operatorname{diam}(Q_j)^2 \operatorname{Cap}(E) \quad \text{for } E \subset Q_j,$$

where Cap stands for the Newtonian capacity. If $n = 2$, then $\mathbb{R}^n$ is recurrent and the fundamental harmonic function $-\log|x-y|$ takes both signs. Hence the logarithmic capacity, the counterpart of the Newtonian capacity, is defined in a special fashion (cf. [10, p. 150] for the logarithmic capacity). This fact illustrates the difference between the cases $n = 2$ and $n \geqslant 3$. Using the same symbol Cap for the logarithmic capacity, we summarize: if $E \subset Q_j$, then

$$\gamma(E) \approx \begin{cases} \dfrac{\operatorname{diam}(Q_j)^2}{\log(4 \operatorname{diam}(Q_j)/\operatorname{Cap}(E))} & \text{for } n = 2, \\ \operatorname{diam}(Q_j)^2 \operatorname{Cap}(E) & \text{for } n \geqslant 3 \end{cases}$$

provided that D is sufficiently smooth.

We return to this question after introducing the Beurling minimum principle. It will turn out that the Beurling minimum principle and the minimal thinness are inextricably related.

4 The Beurling Minimum Principle

Beurling [11] proved the following theorem. Assume, for a moment, that D is a simply connected planar domain with Green function G_D and $\xi \in \partial D$. Let $K(\cdot, \xi)$ be the Martin kernel at ξ in D. Suppose that $S = \{z_j\}_1^\infty \subset D$ is a sequence of points in D converging to ξ as $j \to \infty$.

Definition 4.1. We say that S is an *equivalence sequence* for ξ if

$$h(z_j) \geqslant \lambda K(z_j, \xi) \text{ for } j = 1, 2, \ldots, \implies h(z) \geqslant \lambda K(z, \xi) \text{ for all } z \in D,$$

whenever $\lambda > 0$ and h is a positive harmonic function on D.

Theorem 4.1 (Beurling minimum principle). *Let D, ξ, and S be as above. Then S is an equivalence sequence for ξ if and only if it contains a subsequence $\{z_{j_k}\}_k^\infty$ with the following two properties:*

(i) $\sup_k G_D(z_{j_k}, z_{j_l}) < \infty$ *if* $k \neq l$,

(ii) $\sum_{k=1}^\infty G_D(z, z_{j_k}) K(z_{j_k}, \xi) = \infty$ *for* $z \in D$.

This beautiful theorem has attracted many mathematicians (cf. [2, 3, 9, 15, 19, 27, 28, 31, 32]).

Recall that if D is the unit disk with center 0, the Green function is

$$G_D(z, w) = \log \left| \frac{1 - \overline{z}w}{z - w} \right| \quad \text{for } z, w \in D,$$

where $\overline{z}$ is the complex conjugate of z. This explicit form gives

$$G_D(z, w) \approx \frac{\delta_D(z)\delta_D(w)}{|z - w|^2}$$

whenever

$$|z - w| \geqslant \frac{1}{2} \min\{\delta_D(z), \delta_D(w)\}. \tag{4.1}$$

In fact, for $z, w \in D$

$$G_D(z, w) = \log \left(1 + \frac{(2\delta_D(z) - \delta_D(z)^2)(2\delta_D(w) - \delta_D(w)^2)}{|z - w|^2}\right) \leqslant \frac{4\delta_D(z)\delta_D(w)}{|z - w|^2},$$

where the last inequality follows from the elementary inequality $\log(1+t) \leqslant t$ for $t > 0$. If, in addition, z and w satisfy (4.1), then

$$\frac{4\delta_D(z)\delta_D(w)}{|z - w|^2} \leqslant 32,$$

so that the concavity of $\log(1 + t)$ yields

$$G_D(z, w) \geqslant \frac{\log 33}{32} \cdot \frac{4\delta_D(z)\delta_D(w)}{|z - w|^2}.$$

Similar estimates hold in the higher dimensional case. Let D be a $C^{1,\alpha}$-domain or a Lyapunov–Dini domain in $\mathbb{R}^n$ with $n \geqslant 2$. Then

$$G(x, y) \approx \frac{\delta_D(x)\delta_D(y)}{|x - y|^n}$$

provided that $|x-y| \geqslant \frac{1}{2}\min\{\delta_D(x), \delta_D(y)\}$ (cf. [34]). Here, the constant $\frac{1}{2}$ is not important, we can take any positive number less than 1. It is easy to see that $G(x,y) \leqslant c$ if and only if $|x-y| \geqslant M\min\{\delta_D(x), \delta_D(y)\}$, where M depends only on c. So, we give the following definition.

Definition 4.2. A sequence $\{x_j\} \subset D$ is said to be *separated* if $|x_j - x_k| \geqslant M\delta_D(x_j)$ for $j \neq k$ with some positive constant M independent of j and k.

It is easy to see that the Poisson kernel satisfies

$$P(x,\xi) \approx \frac{\delta_D(x)}{|x-\xi|^n} \quad \text{for } \xi \in \partial D \text{ and } x \in D$$

if D is a $C^{1,\alpha}$-domain or a Lyapunov-Dini domain. Since the Martin kernel is the multiple of the Poisson kernel and a positive continuous function of $\xi \in \partial D$, we can generalize Theorem 4.1 as follows.

Theorem 4.2 (cf. [27, 13]). *Let D be a $C^{1,\alpha}$-domain or a Lyapunov–Dini domain in $\mathbb{R}^n$ with $n \geqslant 2$. Suppose that $\xi \in \partial D$ and a sequence S converges to ξ. Then S is an equivalence sequence for ξ if and only if it contains a separated subsequence $\{x_j\}_j$ such that*

$$\sum_j \left(\frac{\delta_D(x_j)}{|x_j - \xi|}\right)^n = \infty.$$

This theorem was proved by hard calculus with the help of estimates for the Green function and its derivatives [34]. The author [2, 3] proved the theorem by using the *quasiadditivity of capacity*. This approach will be illustrated in the following sections.

5 Quasiadditivity of Capacity (General Theory)

Following [3], we state the quasiadditivity of capacity in a general setting. The quasiadditivity will play a crucial role for refined Wiener type criteria.

Let X be a locally compact Hausdorff space, and let k be a nonnegative lower semicontinuous function on $X \times X$. For a measure μ on X we write

$$k(x,\mu) = \int_X k(x,y)d\mu(y) \quad \text{for } x \in X.$$

Define the capacity C_k with respect to k by

$$C_k(E) = \inf\{\|\mu\| : k(\cdot,\mu) \geqslant 1 \text{ on } E\},$$

where $\|\mu\|$ stands for the total mass of μ. It is well known that C_k is countably subadditive, i.e.,

$$C_k(E) \leqslant \sum_j C_k(E_j) \quad \text{if } E = \bigcup_j E_j.$$

This section is devoted to a reverse inequality, up to a multiplicative constant, for some decomposition of E.

Definition 5.1. Let $\{Q_j\}$ and $\{Q_j^*\}$ be families of Borel subsets of X such that

(i) $Q_j \subset Q_j^*$,

(ii) $X = \bigcup_j Q_j$,

(iii) Q_j^* do not overlap so often, i.e., $\sum \chi_{Q_j^*} \leqslant N$.

Then $\{Q_j, Q_j^*\}$ is called a *quasidisjoint decomposition* of X. If Q_j^* is clear from the context, we write simply $\{Q_j\}$.

We note that Q_j and Q_j^* need not be disjoint.

Definition 5.2. A (Borel) measure σ on X is said to be *comparable* to C_k with respect to $\{Q_j\}$ if

$$\sigma(Q_j) \approx C_k(Q_j) \text{ for every } Q_j, \tag{5.1}$$

$$\sigma(E) \leqslant MC_k(E) \text{ for every Borel set } E \subset X. \tag{5.2}$$

Definition 5.3. We say that a kernel k has the *Harnack property* with respect to $\{Q_j, Q_j^*\}$ if

$$k(x,y) \approx k(x',y) \quad \text{for } x, x' \in Q_j \text{ and } y \in X \setminus Q_j^*$$

with the comparison constant independent of Q_j.

Now, we formulate and prove Theorem 1 in [3].

Theorem 5.1. *Let $\{Q_j, Q_j^*\}$ be a quasidisjoint decomposition of X. Suppose that the kernel k has the Harnack property with respect to $\{Q_j, Q_j^*\}$. If there is a measure σ comparable to C_k with respect to $\{Q_j, Q_j^*\}$, then for every $E \subset X$*

$$C_k(E) \approx \sum_j C_k(E \cap Q_j). \tag{5.3}$$

We say that C_k is quasiadditive with respect to $\{Q_j, Q_j^*\}$ if (5.3) holds.

Proof. Let $E \subset X$. We can assume that $C_k(E) < \infty$. By definition, we can find a measure μ such that $k(\cdot, \mu) \geqslant 1$ on E and $\|\mu\| \leqslant 2C_k(E)$. For each Q_j we have the following two cases:

(i) $k(x, \mu|_{Q_j^*}) \geqslant \frac{1}{2}$ for all $x \in E \cap Q_j$,

(ii) $k(x, \mu|_{Q_j^*}) \geqslant \frac{1}{2}$ for some $x \in E \cap Q_j$.

If (i) holds, then $C_k(E \cap Q_j) \leqslant 2\|\mu|_{Q_j^*}\|$ by definition. Since Q_j^* do not overlap so often, we obtain

$$\sum{}' C_k(E \cap Q_j) \leqslant 2 \sum{}' \|\mu_j\| \leqslant M\|\mu\| \leqslant MC_k(E), \tag{5.4}$$

where $\sum'$ denotes the summation over all Q_j for which (i) holds.

If (ii) holds, then the Harnack property of k yields $k(\cdot, \mu) \geqslant k(\cdot, \mu|_{X \setminus Q_j^*}) \geqslant M$ on Q_j, so that

$$k(\cdot, \mu) \geqslant M \text{ on } \bigcup{}'' Q_j,$$

where $\bigcup''$ denotes the union over all Q_j for which (ii) holds. Hence

$$C_k\Big(\bigcup{}'' Q_j\Big) \leqslant M\|\mu\|.$$

Since σ is comparable to C_k, from the countable additivity of σ it follows that

$$\begin{aligned}\sum{}'' C_k(E \cap Q_j) &\leqslant \sum{}'' C_k(Q_j) \leqslant M \sum{}'' \sigma(Q_j)\\ &\leqslant M\sigma\Big(\bigcup{}'' Q_j\Big) \leqslant MC_k\Big(\bigcup{}'' Q_j\Big) \leqslant M\|\mu\| \leqslant MC_k(E).\end{aligned}$$

This, together with (5.4), completes the proof. □

6 Quasiadditivity of the Green Energy

In general, it is difficult to find a measure comparable to a given capacity. In [2], we found such a measure for the Riesz capacity by using a certain weighted norm inequality. Following [3], we show that the Hardy inequality yields a measure comparable to the Green energy.

We begin with the capacity density condition. Recall that Cap stands for the logarithmic capacity if $n = 2$ and the Newtonian capacity if $n \geqslant 3$.

Definition 6.1. We say that D satisfies the *capacity density condition* if there exist constants $M > 1$ and $r_0 > 0$ such that

$$\operatorname{Cap}(B(\xi, r) \setminus D) \geqslant \begin{cases} M^{-1} r & \text{if } n = 2,\\ M^{-1} r^{n-2} & \text{if } n \geqslant 3,\end{cases}$$

whenever $\xi \in \partial D$ and $0 < r < r_0$.

Ancona [7] proved the most general Hardy inequality for domains possessing the capacity density condition. We note that a domain satisfies the capacity density condition if and only if it is uniformly Δ-regular in the sense of [7].

Lemma 6.1 (Hardy inequality). *Assume that D satisfies the capacity density condition. Then there is a positive constant M depending only on D such that*

$$\int_D \left| \frac{\psi(x)}{\delta_D(x)} \right|^2 dx \leqslant M \int_D |\nabla \psi(x)|^2 dx \quad \textit{for all } \psi \in W_0^{1,2}(D),$$

where $W_0^{1,2}(D)$ stands for the usual Sobolev space, namely the completion of $C_0^\infty(D)$ with the norm

$$\left(\int_D (|\psi|^2 + |\nabla \psi|^2) dx \right)^{1/2}.$$

Let u be a positive superharmonic function on D. Define the measure σ_u on D by

$$\sigma_u(E) = \int_E \left(\frac{u(x)}{\delta_D(x)} \right)^2 dx \quad \text{for } E \subset D.$$

Let $k(x,y) = G(x,y)/(u(x)u(y))$. It is easy to see that $\gamma_u(E) = C_k(E)$ (cf. [18]). Let $\{Q_j\}$ be the Whitney decomposition of D. For each Whitney cube Q_j we denote by x_j and Q_j^* the center and double of Q_j respectively. Suppose that u satisfies the Harnack property with respect to $\{Q_j, Q_j^*\}$. It is easy to see that for $x, y \in Q_j$

$$k(x,y) \approx \begin{cases} u(x_j)^{-2} \log \dfrac{4\operatorname{diam}(Q_j)}{|x-y|} & \text{if } n = 2, \\ u(x_j)^{-2}|x-y|^{2-n} & \text{if } n \geqslant 3. \end{cases}$$

Hence the following assertion holds.

Lemma 6.2. *Let $\{Q_j\}$ be the Whitney decomposition of D with doubles $\{Q_j^*\}$. Suppose that u is a positive superharmonic function on D satisfying the Harnack property with respect to $\{Q_j, Q_j^*\}$. If $E \subset Q_j$, then*

$$\gamma_u(E) \approx \begin{cases} \dfrac{u(x_j)^2}{\log(4\operatorname{diam}(Q_j)/\operatorname{Cap}(E))} & \textit{for } n = 2, \\ u(x_j)^2 \operatorname{Cap}(E) & \textit{for } n \geqslant 3. \end{cases}$$

In particular,

$$\gamma_u(Q_j) \approx \sigma_u(Q_j) \approx u(x_j)^2 \operatorname{diam}(Q_j)^{n-2} \quad \textit{for } n \geqslant 2.$$

The Hardy inequality leads to the following assertion.

Lemma 6.3. *Let D satisfy the capacity density condition, and let $\{Q_j\}$ be the Whitney decomposition of D with doubles $\{Q_j^*\}$. Suppose that u is a positive superharmonic function on D satisfying the Harnack property with respect to $\{Q_j, Q_j^*\}$. Then*

$$\sigma_u(E) \leqslant M\gamma_u(E) \quad \text{for Borel sets } E \subset D.$$

Proof. For a compact subset K of E we write $v_K = \widehat{R}_u^K = G\lambda_u^K$. Then

$$\gamma_u(K) = \iint G(x,y)d\lambda_u^K(x)d\lambda_u^K(y) = \int_D |\nabla v_K|^2 dx$$

(cf., for example, [12, Section 7.2]). Since $v_K = u$ on K except for a polar set, the same equality holds a.e. on K. By Lemma 6.1, we have

$$\begin{aligned}\gamma_u(E) \geqslant \gamma_u(K) &= \int_D |\nabla v_K|^2 dx \\ &\geqslant M \int_D \left(\frac{v_K(x)}{\delta_D(x)}\right)^2 dx \geqslant M \int_K \left(\frac{u(x)}{\delta_D(x)}\right)^2 dx = M\sigma_u(K).\end{aligned}$$

Since $K \subset E$ is an arbitrary compact subset of E, we obtain the required inequality. □

Theorem 6.1. *Let D satisfy the capacity density condition, and let $\{Q_j\}$ be the Whitney decomposition of D with doubles Q_j^*. Suppose that a positive superharmonic function u satisfies*

$$\sup_{Q_j^*} u \leqslant M_1 \inf_{Q_j^*} u \tag{6.1}$$

with M_1 independent of Q_j. Then σ_u is comparable to γ_u with respect to $\{Q_j\}$, and hence γ_u is quasiadditive with respect to $\{Q_j\}$, i.e., $\gamma_u(E) \approx \sum_j \gamma_u(E \cap Q_j)$ for $E \subset D$.

Proof. Lemmas 6.2 and 6.3 show (5.1) and (5.2) respectively. By Theorem 5.1, we obtain the required assertion. □

Let h be a positive harmonic function. Then the Harnack inequality shows that $u(x) = \min\{h(x)^a, b\}$, where $0 < a \leqslant 1$ and $b > 0$, satisfies (6.1). Hence the following assertion holds.

Corollary 6.1. *Let D and $\{Q_j\}$ be as in Theorem 6.1. Let $x_0 \in D$, and let $g(x) = \min\{G(x,x_0), 1\}$. Then γ_g is quasiadditive with respect to $\{Q_j\}$, i.e.,*

$$\gamma_g(E) \approx \sum_j \gamma_g(E \cap Q_j) \approx \begin{cases} \sum_j \dfrac{g(x_j)^2}{\log(4\,\mathrm{diam}(Q_j)/\mathrm{Cap}(E \cap Q_j))} & \text{if } n = 2, \\ \sum_j g(x_j)^2 \mathrm{Cap}(E \cap Q_j) & \text{if } n \geqslant 3. \end{cases}$$

7 Refined Wiener Type Criteria for the Minimal Thinness

In this section, D is a uniform domain possessing the capacity density condition. Let $\{Q_j\}_j$ be the Whitney decomposition of D. For each Whitney cube Q_j we denote by x_j the center. Recall that $A_\xi(r) \in S(\xi, r) \cap D$ is a nontangential point for $\xi \in \partial D$ (cf. Remark 2.1). The quasiadditivity of the Green energy leads to the following Wiener type criteria which is finer than Theorem 3.3.

Theorem 7.1. *Let D be a uniform domain possessing the capacity density condition. Suppose that $\xi \in \partial D$ and $E \subset D$. Then E is minimally thin at ξ if and only if*

$$\sum_j \left(\frac{g(x_j)}{g(A_\xi(|x_j - \xi|))} \right)^2 \frac{1}{\log(4 \operatorname{diam}(Q_j)/\operatorname{Cap}(E \cap Q_j))} < \infty \quad \text{if } n = 2,$$

$$\sum_j \left(\frac{g(x_j)}{g(A_\xi(|x_j - \xi|))} \right)^2 |x_j - \xi|^{2-n} \operatorname{Cap}(E \cap Q_j) < \infty \quad \text{if } n \geqslant 3.$$

Proof. By Corollary 6.1 we can write (3.2) as

$$\sum_j \theta_\xi(|x_j - \xi|) \gamma_g(E \cap Q_j) < \infty,$$

where x_j is the center of a Whitney cube Q_j. Thus, the Wiener type condition (3.5) is refined as in Theorem 7.1. □

Corollary 7.1. *Let D be a uniform domain possessing the capacity density condition. Suppose that $\xi \in \partial D$ and $E \subset D$. If E is measurable and minimally thin at ξ, then*

$$\int_E \left(\frac{g(x)}{g(A_\xi(|x - \xi|))} \right)^2 \frac{|x - \xi|^{2-n}}{\delta_D(x)^2} dx < \infty. \tag{7.1}$$

Proof. The refined Wiener criterion and Lemma 6.2 imply that

$$\sum_j g(A_\xi(|x_j - \xi|))^{-2} |x_j - \xi|^{2-n} \frac{g(x_j)^2}{\operatorname{diam}(Q_j)^2} \int_{E \cap Q_j} dx < \infty. \qquad \square$$

Remark 7.1. Essén [14] introduced first the refined Wiener criterion for the half-space. He used the weak L^1-estimate due to Sjögren [32]. We note that the weak L^1-estimate need not hold for a uniform domain.

Definition 7.1. Suppose that $\xi \in \partial D$ and $E \subset D$. We say that E is *minimally thin* at ξ for harmonic functions if there is a finite measure μ concentrated on ∂D such that $\mu(\{\xi\}) = 0$ and $K_\xi \leqslant K\mu$ on E. We say that E *determines the point measure* at $\xi \in \partial D$ if

$$K\mu \geqslant K_\xi \quad \text{on} \quad E \implies \mu(\{\xi\}) > 0$$

for every finite measure μ concentrated on ∂D (cf. [11, 13, 27, 32]).

Remark 7.2. We note that E is minimally thin at ξ for harmonic functions if and only if E does not determine the point measure at ξ. A sequence S is equivalent for ξ in the sense of Beurling (cf. Definition 4.1) if and only if S is not minimally thin at ξ for harmonic functions. Obviously, if E is minimally thin at ξ for harmonic functions, then it is minimally thin; but the converse is not neccssarily true.

Remark 7.3. Hayman and Kennedy [21, p. 481 and Theorem 7.37] defined sets "rarefied for harmonic functions." Note that the term "rarefied sets" was introduced by Essén and Jackson [16]. A set is "rarefied" in the terminology of Lelong–Ferrand [25] if and only if it is "semirarefied" in the terminology of Essén and Jackson. A set rarefied for harmonic functions corresponds to a rarefied set defined by Essén and Jackson (cf. [15] for further details).

Theorem 7.2. *Let D be a uniform domain possessing the capacity density condition. Suppose that $\xi \in \partial D$ and $E \subset D$. Then E is minimally thin at ξ for harmonic functions if and only if*

$$\sum_{E\cap Q_j\neq\varnothing} \left(\frac{g(x_j)}{g(A_\xi(|x_j-\xi|))}\right)^2 \left(\frac{\operatorname{diam}(Q_j)}{|x_j-\xi|}\right)^{n-2} < \infty.$$

Proof. It is not difficult to see that E is minimally thin at ξ for harmonic functions if and only if $\widetilde{E} = \cup_{E\cap Q_j\neq\varnothing} Q_j$ is minimally thin at ξ. Hence Lemma 3.1 and Theorem 7.2 readily imply the required assertion. □

Remark 7.4. It is easy to see that $\{x_i\}$ is separated if and only if the number of points x_i included in a Whitney cube Q_j is bounded by a positive constant independent of Q_j. Thus, Theorem 7.2 coincides with the Ancona criterion [8, Theorem 7.4] for a sequence to determine a point measure.

Corollary 7.2. *Let D be a uniform domain possessing the capacity density condition. Suppose that $\xi \in \partial D$ and $0 < \rho < 1$. If E is minimally thin at ξ for harmonic functions, then $E_\rho = \bigcup_{x\in E} B(x, \rho\delta_D(x))$ satisfies*

$$\int_{E_\rho} \left(\frac{g(x)}{g(A_\xi(|x-\xi|))}\right)^2 \frac{|x-\xi|^{2-n}}{\delta_D(x)^2} dx < \infty.$$

If D is a $C^{1,\alpha}$-domain or a Lyapunov–Dini domain, then $g(x) \approx \delta_D(x)$ [34], so that the above theorems and corollaries are generalizations of the results of Beurling [11], Dahlberg [13], Essén [15, Section 2], Maz'ya [27], and Sjögren [32]. For the sake of completeness, we formulate the results in the form of corollaries.

Corollary 7.3. *Let D be a $C^{1,\alpha}$-domain or a Lyapunov–Dini domain. Suppose that $\xi \in \partial D$ and $E \subset D$. Then E is minimally thin at ξ if and only if*

$$\sum_j \frac{\operatorname{diam}(Q_j)^2}{|x_j - \xi|^2 \log(4\operatorname{diam}(Q_j)/\operatorname{Cap}(E \cap Q_j))} < \infty \quad \text{if } n = 2,$$

$$\sum_j \frac{\operatorname{diam}(Q_j)^2 \operatorname{Cap}(E \cap Q_j)}{|x_j - \xi|^n} < \infty \qquad \text{if } n \geqslant 3.$$

Corollary 7.4. *Let D be a $C^{1,\alpha}$-domain or a Lyapunov–Dini domain. Suppose that $\xi \in \partial D$ and $E \subset D$. If E is measurable and minimally thin at ξ, then*

$$\int_E \frac{dx}{|x - \xi|^n} < \infty.$$

Corollary 7.5. *Let D be a $C^{1,\alpha}$-domain or a Lyapunov–Dini domain. Suppose that $\xi \in \partial D$ and $E \subset D$. Then E is minimally thin at ξ for harmonic functions if and only if*

$$\sum_{E \cap Q_j \neq \varnothing} \left(\frac{\operatorname{diam}(Q_j)}{|x_j - \xi|} \right)^n < \infty.$$

Corollary 7.6. *Let D be a $C^{1,\alpha}$-domain or a Lyapunov–Dini domain. Let $\xi \in \partial D$, and let $0 < \rho < 1$. If E is minimally thin at ξ for harmonic functions, then $E_\rho = \bigcup_{x \in E} B(x, \rho\delta_D(x))$ satisfies*

$$\int_{E_\rho} \frac{dx}{|x - \xi|^n} < \infty.$$

8 Further Remarks

We conclude the paper by raising some open problems:

1. In our argument, we used the geometric assumption of the domain to characterize the minimal thinness. However, there might be a direct extension

of the Beurling minimum principle to the higher dimensional case; such an extension allows us to give complete understanding of the minimal thinnes within the framework of the general Martin boundary theory.

2. The boundary Harnack principle holds for a uniform domain without any exterior condition [4]. Is the capacity density condition necessary for the refined Wiener criterion (cf. Theorem 7.1)?

3. Essén and Jackson [16] gave the covering properties of minimally thin sets in a half-space. Such covering properties of minimally thin sets in a uniform domain remain open.

Acknowledgment. This work was supported in part by Kakenhi No. 19654023 and No. 20244007.

References

1. Aikawa, H.: On the minimal thinness in a Lipschitz domain. Analysis **5**, no. 4, 347–382 (1985)
2. Aikawa, H.: Quasiadditivity of Riesz capacity. Math. Scand. **69**, no. 1, 15–30 (1991)
3. Aikawa, H.: Quasiadditivity of capacity and minimal thinness. Ann. Acad. Sci. Fenn. Ser. A I Math. **18**, no. 1, 65–75 (1993)
4. Aikawa, H.: Boundary Harnack principle and Martin boundary for a uniform domain. J. Math. Soc. Japan **53**. no. 1, 119–145 (2001)
5. Ancona, A.: Démonstration d'une conjecture sur la capacité et l'effilement. C. R. Acad. Sci. Paris Sér. I Math. **297**, no. 7, 393–395 (1983)
6. Ancona, A.: Sur une conjecture concernant la capacité et l'effilement. In: Seminar on Harmonic Analysis, 1983–1984, Publ. Math. Orsay Vol. 85, pp. 56–91. Univ. Paris XI, Orsay (1985)
7. Ancona, A.: On strong barriers and an inequality of Hardy for domains in $\mathbf{R}^n$. J. London Math. Soc. (2) **34**, no. 2, 274–290 (1986)
8. Ancona, A.: Positive harmonic functions and hyperbolicity. In: Potential Theory — Surveys and Problems (Prague, 1987). Lect. Notes Math. **1344**, 1–23 (1988)
9. Armitagem, D.H., Kuran, Ü.: On positive harmonic majorization of y in $R^n \times (0, +\infty)$. J. London Math. Soc. (2) **3**, 733–741 (1971)
10. Armitagem, D.H., Gardiner, S.J.: Classical Potential Theory. Springer, London (2001)
11. Beurling, A.: A minimum principle for positive harmonic functions. Ann. Acad. Sci. Fenn. Ser. A I **372**, 7 (1965)
12. Constantinescu, C., Cornea, A.: Ideale Ränder Riemannscher Flächen. Springer, Berlin (1963)
13. Dahlberg, B.: A minimum principle for positive harmonic functions. Proc. London Math. Soc. (3) **33**, no. 2, 238–250 (1976)
14. Essén, M.: On Wiener conditions for minimally thin and rarefied sets. In: Complex Analysis. pp. 41–50. Birkhäuser, Basel (1988)

15. Essén, M.: On minimal thinness, reduced functions and Green potentials. Proc. Edinburgh Math. Soc. (2) **36**, no. 1, 87–106 (1993)
16. Essén, M., Jackson, H. L.: On the covering properties of certain exceptional sets in a half-space. Hiroshima Math. J. **10**, no. 2, 233–262 (1980)
17. Fatou, P.: Séries trigonométriques et séries de Taylor. Acta. Math. **30**, 335–400 (1906)
18. Fuglede, B.: Le théorème du minimax et la théorie fine du potentiel. Ann. Inst. Fourier (Grenoble) **15**, no. 1, 65–88 (1965)
19. Gardiner, S.J.: Sets of determination for harmonic functions. Trans. Am. Math. Soc. **338**, no. 1, 233–243 (1993)
20. Gehring, F.W., Osgood, B.G.: Uniform domains and the quasihyperbolic metric. J. Analyse Math. **36**, 50–74 (1979)
21. Hayman, W.K., Kennedy, P.B.: Subharmonic Functions I. Academic Press, London (1976)
22. Helms, L.L.: Introduction to Potential Theory. Robert E. Krieger Publishing Co., Huntington, New York (1975)
23. Jerison, D.S., Kenig, C.E.: Boundary behavior of harmonic functions in non-tangentially accessible domains. Adv. Math. **46**, no. 1, 80–147 (1982)
24. Landkof, N.S.: Foundations of Modern Potential Theory. Springer, New York (1972)
25. Lelong-Ferrand, J.: Étude au voisinage de la frontière des fonctions subharmoniques positives dans un demi-espace. Ann. Sci. École Norm. Sup. (3) **66**, 125–159 (1949)
26. Martin, R.S.: Minimal positive harmonic functions. Trans. Am. Math. Soc. **49**, 137–172 (1941)
27. Maz'ya, V.G.: On Beurling's theorem on the minimum principle for positive harmonic functions (Russian). Zap. Nauch. Semin. LOMI **30**, 76–90 (1972)
28. Miyamoto, I., Yanagishita, M., Yoshida, H.: On harmonic majorization of the Martin function at infinity in a cone. Czechoslovak Math. J. **55(130)**, no. 4, 1041–1054 (2005)
29. Miyamoto, I., Yanagishita, M.: Some characterizations of minimally thin sets in a cylinder and Beurling–Dahlberg–Sjögren type theorems. Proc. Am. Math. Soc. **133**, no. 5, 1391–1400 (2005) (electronic)
30. Naïm, L.: Sur le rôle de la frontière de R. S. Martin dans la théorie du potentiel. Ann. Inst. Fourier, Grenoble **7**, 183–281 (1957)
31. Sjögren, P.: Une propriété des fonctions harmoniques positives, d'après Dahlberg. Séminaire de Théorie du Potentiel de Paris, Lect. Notes Math. **563**, 275–283 (1976)
32. Sjögren, P.: Weak L_1 characterizations of Poisson integrals, Green potentials and H^p spaces. Trans. Am. Math. Soc. **233**, 179–196 (1977)
33. Stein, E.M.: Singular Integrals and Differentiability Properties of Functions. Princeton Univ. Press, Princeton, NJ (1970)
34. Widman, K.-O.: Inequalities for the Green function and boundary continuity of the gradient of solutions of elliptic differential equations. Math. Scand. **21**, 17–37 (1968)
35. Zhang, Y.: Comparaison entre l'effilement interne et l'effilement minimal. C. R. Acad. Sci. Paris Sér. I Math. **304**, no. 1, 5–8 (1987)

Progress in the Problem of the L^p-Contractivity of Semigroups for Partial Differential Operators

Alberto Cialdea

Abstract This is a survey of results concerning the L^p-dissipativity of partial differential operators and the L^p-contractivity of the generated semigroups, mostly obtained by V. Maz'ya, G. Kresin, M. Langer, and A. Cialdea. Necessary and sufficient conditions are discussed for scalar second order operators with complex coefficients and some systems, including the two-dimensional elasticity. The case of higher order operators is reviewed as well.

1 Introduction

The classical theorems of Hille–Yosida (1948) and Lumer–Phillips (1962) play a central role in the theory of semigroups of linear operators. They give two different characterizations of generators of a strongly continuous semigroup of contractions.

Maz'ya and Sobolevskii [26] obtained independently of Lumer and Phillips the same result under the assumption that the norm of the Banach space is Gâteaux-differentiable. Applications to second order elliptic operators were also given in [26]. It is interesting to note that the paper [26] was sent to the journal in 1960, before the paper [22] by Lumer and Phillips appeared.

During the last half a century, various aspects of the L^p-theory of semigroups generated by linear differential operators were studied (cf., for example, [3, 7, 1, 34, 8, 14, 32, 16, 9, 10, 19, 20, 17, 18, 2, 6, 13, 21, 33, 29, 4, 27, 5]). An account of the subject can be found in the book [30] which contains also an extensive bibliography.

Alberto Cialdea
Dipartimento di Matematica e Informatica, Università della Basilicata, Viale dell'Ateneo Lucano 10, 85100, Potenza, Italy
cialdea@email.it

A. Laptev (ed.), *Around the Research of Vladimir Maz'ya III: Analysis and Applications*, International Mathematical Series 13, DOI 10.1007/978-1-4419-1345-6_3,

In the present paper, we survey results mostly obtained by Vladimir Maz'ya with his coauthors G. Kresin, M. Langer, and A. Cialdea about the L^p-dissipativity of partial differential operators and the related contractivity of the generated semigroup. Special attention is paid to the possibility of giving algebraic necessary and sufficient conditions.

The paper is organized as follows. In Section 2, we consider necessary and sufficient conditions for the L^p-dissipativity of the Dirichlet problem for second order scalar operators with complex coefficients.

Section 3 is devoted to weakly coupled systems.

Systems which are uniformly parabolic in the sense of Petrovskii are the topic of Section 4. We discuss necessary and sufficient conditions for the validity of maximum principles with respect to different norms.

The goal of Section 5 is to give necessary and sufficient conditions for the L^p-dissipativity of the two-dimensional elasticity system in terms of the Poisson ratio.

In the first part of Section 6, some auxiliary results for ordinary differential systems are considered. In the second part, these results are applied to obtain necessary and sufficient conditions for the L^p-dissipativity of the Dirichlet problem for a class of systems of partial differential equations.

Section 7 is concerned with higher order partial differential operators.

2 Scalar Second Order Operators with Complex Coefficients

A linear operator $A : D(A) \subset X = L^p(\Omega) \to X = L^p(\Omega)$ (Ω is a domain of $\mathbb{R}^n$, $1 < p < \infty$) is said to be *L^p-dissipative* if

$$\operatorname{Re} \int_\Omega \langle Au, u\rangle |u|^{p-2} dx \leqslant 0 \tag{2.1}$$

for any $u \in D(A)$, where $\langle \cdot, \cdot \rangle$ denotes the scalar product in $\mathbb{C}$. Hereinafter, the integrand is extended by zero on the set where u vanishes.

The case of a scalar second order operator in divergence form with real coefficients is well known (cf., for example, [31, p. 215]). Here, we have the L^p-dissipativity for any p. If $2 \leqslant p < \infty$, this can be deduced easily by integration by parts. If $A = \partial_i(a_{ij}\partial_j)$ ($a_{ji} = a_{ij} \in C^1(\overline{\Omega})$), we can write

$$\int_\Omega \langle Au, u\rangle |u|^{p-2} dx = -\int_\Omega a_{ij}\partial_j u\, \partial_i(\overline{u}\, |u|^{p-2})\, dx\, .$$

If we suppose that $a_{ij}\xi_i\xi_j \geqslant 0$ for any $\xi \in \mathbb{R}^n$, an easy calculation shows that

$$\operatorname{Re} \int_\Omega a_{ij}\partial_j u\, \partial_i(\overline{u}\, |u|^{p-2})\, dx \geqslant 0,$$

and the L^p-dissipativity of A follows. Some extra arguments are necessary in the case $1 \leqslant p < 2$.

Necessary and sufficient conditions for the L^∞-contractivity for general second order strongly elliptic systems with smooth coefficients were given in [16] (cf. also Subsection 4.1), where scalar second order elliptic operators with complex coefficients were handled as a particular case. Such operators generating L^∞-contractive semigroups were later characterized in [2] under the assumption that the coefficients are measurable and bounded.

These results imply the L^p-dissipativity for any $p \in [1, +\infty)$, but recently Maz'ya and the author of the present paper characterized the L^p-dissipativity individually, for each p. In the next subsections, we describe these results which are contained in [4] and are related to the L^p-dissipativity of a scalar second order operator with complex coefficients.

2.1 Generalities

Let Ω be a domain in $\mathbb{R}^n$, and let A be the operator

$$Au = \nabla^t(\mathscr{A}\nabla u) + \mathbf{b}\nabla u + \nabla^t(\mathbf{c}u) + au$$

where ∇^t is the divergence operator and the coefficients satisfy the following very general assumptions:

— $\mathscr{A}$ is an $n \times n$ matrix whose entries are complex-valued measures a^{hk} belonging to $(C_0(\Omega))^*$. This is the dual space of $C_0(\Omega)$, the space of complex-valued continuous functions with compact support contained in Ω;

— $\mathbf{b} = (b_1, \dots, b_n)$ and $\mathbf{c} = (c_1, \dots, c_n)$ are complex-valued vectors with $b_j, c_j \in (C_0(\Omega))^*$;

— a is a complex-valued scalar distribution in $({C_0}^1(\Omega))^*$, where ${C_0}^1(\Omega) = C^1(\Omega) \cap C_0(\Omega)$.

Consider the related sesquilinear form $\mathscr{L}(u, v)$

$$\mathscr{L}(u,v) = \int_\Omega (\langle \mathscr{A}\nabla u, \nabla v\rangle - \langle \mathbf{b}\nabla u, v\rangle + \langle u, \overline{\mathbf{c}}\nabla v\rangle - a\langle u, v\rangle)$$

on ${C_0}^1(\Omega) \times {C_0}^1(\Omega)$.

The operator A acts from ${C_0}^1(\Omega)$ to $({C_0}^1(\Omega))^*$ through the relation

$$\mathscr{L}(u,v) = -\int_\Omega \langle Au, v\rangle$$

for any $u, v \in {C_0}^1(\Omega)$. The integration is understood in the sense of distributions.

The following definition was given in [4]. Let $1 < p < \infty$. A form $\mathscr{L}$ is called *L^p-dissipative* if for all $u \in {C_0}^1(\Omega)$

$$\operatorname{Re}\mathscr{L}(u, |u|^{p-2}u) \geqslant 0 \quad \text{if} \ \ p \geqslant 2,$$
$$\operatorname{Re}\mathscr{L}(|u|^{p'-2}u, u) \geqslant 0 \quad \text{if} \ \ 1 < p < 2,$$

where $p' = p/(p-1)$ (we use here that $|u|^{q-2}u \in {C_0}^1(\Omega)$ for $q \geqslant 2$ and $u \in {C_0}^1(\Omega)$).

The following lemma is basic and provides a necessary and sufficient condition for the L^p-dissipativity of the form $\mathscr{L}$.

Lemma 2.1 ([4]). *A form $\mathscr{L}$ is L^p-dissipative if and only if for all $w \in {C_0}^1(\Omega)$*

$$\operatorname{Re}\int_\Omega \Big[\langle \mathscr{A}\nabla w, \nabla w\rangle - (1-2/p)\langle(\mathscr{A}-\mathscr{A}^*)\nabla(|w|), |w|^{-1}\overline{w}\nabla w\rangle$$
$$- (1-2/p)^2\langle \mathscr{A}\nabla(|w|), \nabla(|w|)\rangle\Big] + \int_\Omega \langle \operatorname{Im}(\mathbf{b}+\mathbf{c}), \operatorname{Im}(\overline{w}\nabla w)\rangle$$
$$+ \int_\Omega \operatorname{Re}(\nabla^t(\mathbf{b}/p - \mathbf{c}/p') - a)|w|^2 \geqslant 0.$$

From this lemma we obtain the following assertion.

Corollary 2.1 ([4]). *If a form $\mathscr{L}$ is L^p-dissipative, then*

$$\langle \operatorname{Re}\mathscr{A}\xi, \xi\rangle \geqslant 0 \quad \forall \xi \in \mathbb{R}^n.$$

This condition is not sufficient for the L^p-dissipativity if $p \neq 2$. Lemma 2.1 implies the following sufficient condition.

Corollary 2.2 ([4]). *Let α and β be two real constants. If*

$$\begin{aligned}\frac{4}{pp'}\langle \operatorname{Re}\mathscr{A}\xi, \xi\rangle + \langle \operatorname{Re}\mathscr{A}\eta, \eta\rangle + 2\langle(p^{-1}\operatorname{Im}\mathscr{A} + p'^{-1}\operatorname{Im}\mathscr{A}^*)\xi, \eta\rangle \\ + \langle \operatorname{Im}(\mathbf{b}+\mathbf{c}), \eta\rangle - 2\langle \operatorname{Re}(\alpha\mathbf{b}/p - \beta\mathbf{c}/p'), \xi\rangle \\ + \operatorname{Re}\left[\nabla^t\left((1-\alpha)\mathbf{b}/p - (1-\beta)\mathbf{c}/p'\right) - a\right] \geqslant 0\end{aligned} \tag{2.2}$$

for any $\xi, \eta \in \mathbb{R}^n$, then the form $\mathscr{L}$ is L^p-dissipative.

Putting $\alpha = \beta = 0$ in (2.2), we find that if

$$\begin{aligned}\frac{4}{pp'}\langle \operatorname{Re}\mathscr{A}\xi, \xi\rangle + \langle \operatorname{Re}\mathscr{A}\eta, \eta\rangle + 2\langle(p^{-1}\operatorname{Im}\mathscr{A} + p'^{-1}\operatorname{Im}\mathscr{A}^*)\xi, \eta\rangle \\ + \langle \operatorname{Im}(\mathbf{b}+\mathbf{c}), \eta\rangle + \operatorname{Re}\left[\nabla^t\left(\mathbf{b}/p - \mathbf{c}/p'\right) - a\right] \geqslant 0\end{aligned} \tag{2.3}$$

for any $\xi, \eta \in \mathbb{R}^n$, then the form $\mathscr{L}$ is L^p-dissipative.

Generally speaking, this condition (and the more general condition in Corollary 2.2) is not necessary.

Example 2.1. Let $n = 2$ and

$$\mathscr{A} = \begin{pmatrix} 1 & i\gamma \\ -i\gamma & 1 \end{pmatrix},$$

where γ is a real constant, $\mathbf{b} = \mathbf{c} = a = 0$. In this case, the polynomial (2.3) is given by

$$(\eta_1 + \gamma\xi_2)^2 + (\eta_2 - \gamma\xi_1)^2 - (\gamma^2 - 4/(pp'))|\xi|^2.$$

For $\gamma^2 > 4/(pp')$ the condition (2.2) is not satisfied, whereas the L^p-dissipativity holds because the corresponding operator A is the Laplacian.

Note that the matrix $\operatorname{Im}\mathscr{A}$ is not symmetric. Below (after Corollary 2.6), we give another example showing that the condition (2.3) is not necessary for the L^p-dissipativity even for symmetric matrices $\operatorname{Im}\mathscr{A}$.

Corollary 2.3 ([4]). *If a form $\mathscr{L}$ is simultaneously L^p- and $L^{p'}$-dissipative, it is also L^r-dissipative for any r between p and p', i.e., for any r such that*

$$1/r = t/p + (1-t)/p' \qquad (0 \leqslant t \leqslant 1). \tag{2.4}$$

Corollary 2.4 ([4]). *Suppose that either*

$$\operatorname{Im}\mathscr{A} = 0, \quad \operatorname{Re}\nabla^t\mathbf{b} = \operatorname{Re}\nabla^t\mathbf{c} = 0$$

or

$$\operatorname{Im}\mathscr{A} = \operatorname{Im}\mathscr{A}^t, \ \operatorname{Im}(\mathbf{b}+\mathbf{c}) = 0, \ \operatorname{Re}\nabla^t\mathbf{b} = \operatorname{Re}\nabla^t\mathbf{c} = 0.$$

If $\mathscr{L}$ is L^p-dissipative, it is also L^r-dissipative for any r satisfying (2.4).

2.2 The operator $\nabla^t(\mathscr{A}\nabla u)$. The main result

In this subsection, the operator A is supposed to be without lower order terms, i.e.,

$$Au = \nabla^t(\mathscr{A}\nabla u). \tag{2.5}$$

The following assertion gives a necessary and sufficient condition for the L^p-dissipativity of the operator (2.5).

The coefficients a^{hk} belong to $(C_0(\Omega))^*$, as in the previous subsection.

Theorem 2.1 ([4]). *Let $\operatorname{Im}\mathscr{A}$ be symmetric, i.e., $\operatorname{Im}\mathscr{A}^t = \operatorname{Im}\mathscr{A}$. The form*

$$\mathscr{L}(u,v) = \int_\Omega \langle \mathscr{A}\nabla u, \nabla v\rangle$$

is L^p-dissipative if and only if

$$|p-2|\,|\langle \operatorname{Im} \mathscr{A}\xi,\xi\rangle| \leqslant 2\sqrt{p-1}\,\langle \operatorname{Re} \mathscr{A}\xi,\xi\rangle \tag{2.6}$$

for any $\xi \in \mathbb{R}^n$, where $|\cdot|$ denotes the total variation.

The condition (2.6) is understood in the sense of comparison of measures.

Note that from Theorem 2.1 we immediately derive the following well known results.

Corollary 2.5. *Let A be such that $\langle \operatorname{Re} \mathscr{A}\xi,\xi\rangle \geqslant 0$ for any $\xi \in \mathbb{R}^n$. Then*

1) A is L^2-dissipative,

2) if A is an operator with real coefficients, then A is L^p-dissipative for any p.

The condition (2.6) is equivalent to the positivity of some polynomial in ξ and η. More exactly, (2.6) is equivalent to the following condition:

$$\frac{4}{pp'}\langle \operatorname{Re} \mathscr{A}\xi,\xi\rangle + \langle \operatorname{Re} \mathscr{A}\eta,\eta\rangle - 2(1-2/p)\langle \operatorname{Im} \mathscr{A}\xi,\eta\rangle \geqslant 0 \tag{2.7}$$

for any $\xi,\eta \in \mathbb{R}^n$.

Let us assume that either A has lower order terms or it has no lower order terms and $\operatorname{Im} \mathscr{A}$ is not symmetric. Then the (2.6) is still necessary for the L^p-dissipativity of A, but not sufficient, which will be shown in Example 2.2 (cf. also Theorem 2.2 below for the case of constant coefficients). In other words, for such general operators the algebraic condition (2.7) is necessary but not sufficient, whereas the condition (2.3) is sufficient, but not necessary.

Example 2.2. Let $n = 2$, and let Ω be a bounded domain. Denote by σ a real function of class $\mathring{C}^2(\Omega)$ which does not vanish identically. Let $\lambda \in \mathbb{R}$. Consider the operator (2.5) with

$$\mathscr{A} = \begin{pmatrix} 1 & i\lambda\partial_1(\sigma^2) \\ -i\lambda\partial_1(\sigma^2) & 1 \end{pmatrix},$$

i.e.,

$$Au = \partial_1(\partial_1 u + i\lambda\partial_1(\sigma^2)\,\partial_2 u) + \partial_2(-i\lambda\partial_1(\sigma^2)\,\partial_1 u + \partial_2 u),$$

where $\partial_i = \partial/\partial x_i$ $(i = 1,2)$. By definition, we have L^2-dissipativity if and only if

$$\operatorname{Re}\int_\Omega ((\partial_1 u + i\lambda\partial_1(\sigma^2)\,\partial_2 u)\partial_1\overline{u} + (-i\lambda\partial_1(\sigma^2)\,\partial_1 u + \partial_2 u)\partial_2\overline{u})\,dx \geqslant 0$$

for any $u \in \mathring{C}^1(\Omega)$, i.e., if and only if

$$\int_\Omega |\nabla u|^2 dx - 2\lambda\int_\Omega \partial_1(\sigma^2)\operatorname{Im}(\partial_1\overline{u}\,\partial_2 u)\,dx \geqslant 0$$

for any $u \in C_0{}^1(\Omega)$. Taking $u = \sigma \exp(itx_2)$ $(t \in \mathbb{R})$, we obtain, in particular,

$$t^2 \int_\Omega \sigma^2 dx - t\lambda \int_\Omega (\partial_1(\sigma^2))^2 dx + \int_\Omega |\nabla \sigma|^2 dx \geqslant 0. \tag{2.8}$$

Since

$$\int_\Omega (\partial_1(\sigma^2))^2 dx > 0,$$

we can choose $\lambda \in \mathbb{R}$ so that (2.8) is impossible for all $t \in \mathbb{R}$. Thus, A is not L^2-dissipative, although (2.6) is satisfied. Since A can be written as

$$Au = \Delta u - i\lambda(\partial_{21}(\sigma^2)\, \partial_1 u - \partial_{11}(\sigma^2)\, \partial_2 u),$$

this example shows that (2.6) is not sufficient for the L^2-dissipativity of an operator with lower order terms, even if $\operatorname{Im} \mathscr{A}$ is symmetric.

2.3 General equation with constant coefficients

Generally speaking, it is impossible to obtain an algebraic characterization for an operator with lower order terms. Indeed, let us consider, for example, the operator

$$Au = \Delta u + a(x)u$$

in a bounded domain $\Omega \subset \mathbb{R}^n$ with zero Dirichlet boundary data. Denote by λ_1 the first eigenvalue of the Dirichlet problem for the Laplace equation in Ω. A sufficient condition for the L^2-dissipativity of A has the form $\operatorname{Re} a \leqslant \lambda_1$, and we cannot give an algebraic characterization of λ_1.

Consider, as another example, the operator

$$A = \Delta + \mu$$

where μ is a nonnegative Radon measure on Ω. The operator A is L^p-dissipative if and only if

$$\int_\Omega |w|^2 d\mu \leqslant \frac{4}{pp'} \int_\Omega |\nabla w|^2 dx \tag{2.9}$$

for any $w \in C_0{}^\infty(\Omega)$ (cf. Lemma 2.1). Maz'ya [23, 24, 25] proved that the following condition is sufficient for (2.9):

$$\frac{\mu(F)}{\operatorname{cap}_\Omega(F)} \leqslant \frac{1}{pp'} \tag{2.10}$$

for all compact set $F \subset \Omega$ and the following condition is necessary:

$$\frac{\mu(F)}{\mathrm{cap}_\Omega(F)} \leqslant \frac{4}{pp'} \tag{2.11}$$

for all compact set $F \subset \Omega$. Here, $\mathrm{cap}_\Omega(F)$ is the capacity of F relative to Ω, i.e.,

$$\mathrm{cap}_\Omega(F) = \inf\Big\{ \int_\Omega |\nabla u|^2 dx \ : \ u \in C_0^\infty(\Omega), \ u \geqslant 1 \text{ on } F \Big\}.$$

The condition (2.10) is not necessary and the condition (2.11) is not sufficient.

However. it is possible to find necessary and sufficient conditions in the case of constant coefficients. In this subsection, we consider such equations.

Consider the operator

$$Au = \nabla^t(\mathscr{A}\nabla u) + \mathbf{b}\nabla u + au \tag{2.12}$$

with constant complex coefficients. Without loss of generality, we can assume that the matrix $\mathscr{A}$ is symmetric.

The following assertion provides a necessary and sufficient condition for the L^p-dissipativity of the operator A.

Theorem 2.2 ([4]). *Suppose that Ω is an open set in $\mathbb{R}^n$ which contains balls of arbitrarily large radius. The operator* (2.12) *is L^p-dissipative if and only if there exists a real constant vector V such that*

$$\begin{aligned} &2\,\mathrm{Re}\,\mathscr{A}V + \mathrm{Im}\,\mathbf{b} = 0, \\ &\mathrm{Re}\,a + \langle\, \mathrm{Re}\,\mathscr{A}V, V\rangle \leqslant 0 \end{aligned}$$

and for any $\xi \in \mathbb{R}^n$

$$|p-2|\,|\langle\, \mathrm{Im}\,\mathscr{A}\xi,\xi\rangle| \leqslant 2\sqrt{p-1}\,\langle\, \mathrm{Re}\,\mathscr{A}\xi,\xi\rangle. \tag{2.13}$$

If the matrix $\mathrm{Re}\,\mathscr{A}$ is not singular, the following assertion holds.

Corollary 2.6 ([4]; cf. also [16]). *Suppose that Ω is an open set in $\mathbb{R}^n$ which contains balls of arbitrarily large radius. Assume that the matrix $\mathrm{Re}\,\mathscr{A}$ is not singular. The operator A is L^p-dissipative if and only if* (2.13) *holds and*

$$4\,\mathrm{Re}\,a \leqslant -\langle (\mathrm{Re}\,\mathscr{A})^{-1}\,\mathrm{Im}\,\mathbf{b}, \mathrm{Im}\,\mathbf{b}\rangle. \tag{2.14}$$

Now, we can show that the condition (2.2) is not necessary for the L^p-dissipativity, even if the matrix $\mathrm{Im}\,\mathscr{A}$ is symmetric.

Example 2.3. Let $n = 1$, and let $\Omega = \mathbb{R}^1$. Consider the operator

$$\left(1 + 2\,\frac{\sqrt{p-1}}{p-2}\,i\right) u'' + 2iu' - u,$$

where $p \neq 2$ is fixed. The conditions (2.13) and (2.14) are satisfied, and this operator is L^p-dissipative in view of Corollary 2.6.

On the other hand, the polynomial in (2.3) has the form

$$\left(2\,\frac{\sqrt{p-1}}{p}\,\xi-\eta\right)^2+2\eta+1,$$

i.e., it is not nonnegative for any $\xi,\eta\in\mathbb{R}$.

2.4 Smooth coefficients

In this subsection, we consider the operator

$$Au=\nabla^t(\mathscr{A}\nabla u)+\mathbf{b}\nabla u+a\,u \tag{2.15}$$

with coefficients $a^{hk}, b^h \in C^1(\overline{\Omega})$, $a\in C^0(\overline{\Omega})$. Here, Ω is a bounded domain in $\mathbb{R}^n$ with boundary of class $C^{2,\alpha}$ for some $\alpha\in[0,1)$. Let A be defined on the set

$$D(A)=W^{2,p}(\Omega)\cap {W_0}^{1,p}(\Omega).$$

We can compare the classical concept of the L^p-dissipativity of the operator A (cf. (2.1)) with the L^p-dissipativity of the corresponding form $\mathscr{L}$.

Theorem 2.3 ([4]). *The operator A is L^p-dissipative if and only if the form $\mathscr{L}$ is L^p-dissipative.*

If the operator A has smooth coefficients and no lower order terms, one can determine the best interval of p's where the operator A is L^p-dissipative.

Define

$$\lambda=\inf_{(\xi,x)\in\mathscr{M}}\frac{\langle\,\mathrm{Re}\,\mathscr{A}(x)\xi,\xi\rangle}{|\langle\,\mathrm{Im}\,\mathscr{A}(x)\xi,\xi\rangle|}$$

where $\mathscr{M}$ is the set of (ξ,x) with $\xi\in\mathbb{R}^n$, $x\in\Omega$ such that $\langle\,\mathrm{Im}\,\mathscr{A}(x)\xi,\xi\rangle\neq 0$.

Theorem 2.4 ([4]). *Let A be an operator of the form*

$$Au=\nabla^t(\mathscr{A}\nabla u). \tag{2.16}$$

Suppose that the matrix $\mathrm{Im}\,\mathscr{A}$ is symmetric and

$$\langle\,\mathrm{Re}\,\mathscr{A}(x)\xi,\xi\rangle\geqslant 0$$

for any $x\in\Omega$, $\xi\in\mathbb{R}^n$. If $\mathrm{Im}\,\mathscr{A}(x)=0$ for any $x\in\Omega$, then A is L^p-dissipative for any $p>1$. If $\mathrm{Im}\,\mathscr{A}$ does not vanish identically in Ω, then A is L^p-dissipative if and only if

$$2+2\lambda(\lambda-\sqrt{\lambda^2+1})\leqslant p\leqslant 2+2\lambda(\lambda+\sqrt{\lambda^2+1}).$$

This theorem leads to a characterization of operators that are L^p-dissipative only for $p=2$.

Corollary 2.7 ([4]). *Let A be the same as in Theorem 2.4. The operator A is L^p-dissipative only for $p = 2$ if and only if* $\operatorname{Im}\mathscr{A}$ *does not vanish identically and $\lambda = 0$.*

We know that the condition (2.6) is necessary and sufficient for the L^p-dissipativity of the operator (2.16) provided that $\operatorname{Im}\mathscr{A}$ is symmetric and lower order terms are absent, but this is not true for a more general operator (2.15). The next result shows that the condition (2.6) is necessary and sufficient for the operator (2.15) to be L^p-quasidissipative, i.e., there exists $\omega \geqslant 0$ such the operator $A - \omega I$ is L^p-dissipative:

$$\operatorname{Re} \int_\Omega \langle Au, u\rangle |u|^{p-2} dx \leqslant \omega \, \|u\|_p^p$$

for any $u \in D(A)$.

In the next theorem, the operator A is assumed to be strongly elliptic, i.e.,

$$\langle \operatorname{Re}\mathscr{A}(x)\xi, \xi\rangle > 0$$

for any $x \in \overline{\Omega}$, $\xi \in \mathbb{R}^n \setminus \{0\}$.

Theorem 2.5 ([4]). *A strongly elliptic operator* (2.15) *is L^p-quasidissipative if and only if*

$$|p-2| \, |\langle \operatorname{Im}\mathscr{A}(x)\xi, \xi\rangle| \leqslant 2\sqrt{p-1}\, \langle \operatorname{Re}\mathscr{A}(x)\xi, \xi\rangle \tag{2.17}$$

for any $x \in \Omega$, $\xi \in \mathbb{R}^n$.

We emphasize that the symmetry of $\operatorname{Im}\mathscr{A}$ is not required here.

2.5 Contractivity and quasicontractivity of the generated semigroup

From the results of the previous subsection we can obtain some information about the semigroup generated by A.

Theorem 2.6 ([4]). *Let A be a strongly elliptic operator of the form* (2.16) *with* $\operatorname{Im}\mathscr{A} = \operatorname{Im}\mathscr{A}^t$. *The operator A generates a contraction semigroup on L^p if and only if*

$$|p-2| \, |\langle \operatorname{Im}\mathscr{A}(x)\xi, \xi\rangle| \leqslant 2\sqrt{p-1}\, \langle \operatorname{Re}\mathscr{A}(x)\xi, \xi\rangle$$

for any $x \in \Omega$, $\xi \in \mathbb{R}^n$.

This theorem holds if the operator A has no lower order terms and $\operatorname{Im}\mathscr{A} = \operatorname{Im}\mathscr{A}^t$. For a more general operator (2.15) we have the following criterion, where the notion of quasicontraction semigroup is used. We say that an operator A generates a quasicontraction semigroup if there exists $\omega \geqslant 0$ such that $A - \omega I$ generates a contraction semigroup.

Theorem 2.7 ([4]). *Let A be a strongly elliptic operator of the form* (2.15). *The operator A generates a quasicontraction semigroup on L^p if and only if* (2.17) *holds for any $x \in \Omega$, $\xi \in \mathbb{R}^n$.*

2.6 Dissipativity angle

Consider the operator

$$A = \nabla^t(\mathscr{A}(x)\nabla),$$

where $\mathscr{A}(x) = \{a_{ij}(x)\}$ $(i, j = 1, \ldots, n)$ is a matrix with complex locally integrable entries defined in a domain $\Omega \subset \mathbb{R}^n$. In this subsection, we consider the problem of determining the angle of dissipativity of A, i.e., necessary and sufficient conditions for the L^p-dissipativity of the Dirichlet problem for the differential operator zA, where $z \in \mathbb{C}$. If $\mathscr{A}$ is a real matrix, it is well known (cf., for example, [11, 12, 28]) that the dissipativity angle is independent of the operator and is given by

$$|\arg z| \leqslant \arctan\left(\frac{2\sqrt{p-1}}{|p-2|}\right). \tag{2.18}$$

If the entries of the matrix $\mathscr{A}$ are complex, the situation is different because the dissipativity angle depends on the operator, as the next theorem shows.

Theorem 2.8 ([5]). *Let a matrix $\mathscr{A}$ be symmetric. Suppose that the operator A is L^p-dissipative. Let*

$$\Lambda_1 = \operatorname*{ess\,inf}_{(x,\xi)\in\Xi} \frac{\langle \operatorname{Im} \mathscr{A}(x)\xi, \xi\rangle}{\langle \operatorname{Re} \mathscr{A}(x)\xi, \xi\rangle}, \qquad \Lambda_2 = \operatorname*{ess\,sup}_{(x,\xi)\in\Xi} \frac{\langle \operatorname{Im} \mathscr{A}(x)\xi, \xi\rangle}{\langle \operatorname{Re} \mathscr{A}(x)\xi, \xi\rangle},$$

where

$$\Xi = \{(x, \xi) \in \Omega \times \mathbb{R}^n \mid \langle \operatorname{Re} \mathscr{A}(x)\xi, \xi\rangle > 0\}.$$

The operator zA is L^p-dissipative if and only if

$$\vartheta_- \leqslant \arg z \leqslant \vartheta_+,$$

where [1]

[1] Here, $0 < \operatorname{arccot} y < \pi$, $\operatorname{arccot}(+\infty) = 0$, $\operatorname{arccot}(-\infty) = \pi$, and

$$\operatorname*{ess\,inf}_{(x,\xi)\in\Xi} \frac{\langle \operatorname{Im} \mathscr{A}(x)\xi, \xi\rangle}{\langle \operatorname{Re} \mathscr{A}(x)\xi, \xi\rangle} = +\infty, \qquad \operatorname*{ess\,sup}_{(x,\xi)\in\Xi} \frac{\langle \operatorname{Im} \mathscr{A}(x)\xi, \xi\rangle}{\langle \operatorname{Re} \mathscr{A}(x)\xi, \xi\rangle} = -\infty$$

if Ξ has zero measure.

$$\vartheta_- = \begin{cases} \operatorname{arccot}\left(\frac{2\sqrt{p-1}}{|p-2|} - \frac{p^2}{|p-2|}\frac{1}{2\sqrt{p-1}+|p-2|\varLambda_1}\right) - \pi & \text{if } p \neq 2, \\ \operatorname{arccot}(\varLambda_1) - \pi & \text{if } p = 2, \end{cases}$$

$$\vartheta_+ = \begin{cases} \operatorname{arccot}\left(-\frac{2\sqrt{p-1}}{|p-2|} + \frac{p^2}{|p-2|}\frac{1}{2\sqrt{p-1}-|p-2|\varLambda_2}\right) & \text{if } p \neq 2 \\ \operatorname{arccot}(\varLambda_2) & \text{if } p = 2. \end{cases}$$

Note that for a real matrix $\mathscr{A}$ we have $\varLambda_1 = \varLambda_2 = 0$ and, consequently,

$$\frac{2\sqrt{p-1}}{|p-2|} - \frac{p^2}{2\sqrt{p-1}|p-2|} = -\frac{|p-2|}{2\sqrt{p-1}}.$$

Theorem 2.8 asserts that zA is dissipative if and only if

$$\operatorname{arccot}\left(-\frac{|p-2|}{2\sqrt{p-1}}\right) - \pi \leqslant \arg z \leqslant \operatorname{arccot}\left(\frac{|p-2|}{2\sqrt{p-1}}\right),$$

i.e., if and only if (2.18) holds.

3 L^p-Contractivity for Weakly Coupled Systems

In this section, we consider the operator

$$\mathcal{A}_p u = \partial_i(a_{ij}\partial_j u) + a_i\partial_i u + Au, \quad u \in (W^{2,p}(\Omega) \cap W_0^{1,p}(\Omega))^N, \tag{3.1}$$

where a_{ij} and a_i are real $C^1(\overline{\Omega})$ functions and A is an $N \times N$-matrix with complex $C^0(\overline{\Omega})$ entries. We assume that the matrix $\{a_{ij}\}$ is pointwise symmetric.

With the operator (3.1) we associate the operator

$$\mathcal{A}u = \frac{4}{pp'}\partial_i(a_{ij}\partial_j u) + \frac{1}{2p}(p(A + A^*) - 2\partial_i a_i I)u, \tag{3.2}$$

where $u \in (H^2(\Omega) \cap H_0^1(\Omega))^N$.

The following result due to Langer and Maz'ya [17] shows that the L^p-contractivity of the operator $\mathcal{A}_p$ is connected with the L^2-contractivity of the operator $\mathcal{A}$.

Theorem 3.1 ([17]). *Let $\Omega \subset \mathbb{R}^n$ be a bounded domain with $C^{2,\alpha}$ boundary $(0 < \alpha \leqslant 1)$, and let $\mathcal{A}_p$ be an elliptic operator of the form* (3.1)*. If the operator* (3.2) *generates a contraction semigroup on $(L^2(\Omega))^N$, then $\mathcal{A}_p$ generates a contraction semigroup on $(L^p(\Omega))^N$. Conversely, if there exists a basis of constant eigenvectors of $A+A^*$, then $\mathcal{A}$ generates a contraction semigroup on $(L^2(\Omega))^N$ provided that $\mathcal{A}_p$ generates a contraction semigroup on $(L^p(\Omega))^N$. In particular, the converse assertion holds in the scalar case.*

To prove this theorem, the following functionals were introduced:

$$J(w) = \int_\Omega \Big(\frac{4}{pp'} a_{ij} \langle \partial_i w, \partial_j w\rangle + \mathrm{Re}\big\langle (p^{-1}\partial_i a_i I - A)w, w\big\rangle\Big)\, dx,$$

$$J_p(w) = \int_\Omega \Big(a_{ij} \langle \partial_i w, \partial_j w\rangle + \mathrm{Re}\big\langle (p^{-1}\partial_i a_i I - A)w, w\big\rangle\Big)\, dx$$
$$- \frac{(p-2)^2}{p^2} \int_{\{w\neq 0\}} a_{ij}\, \mathrm{Re}\,\langle \partial_i w, w\rangle\, \mathrm{Re}\,\langle \partial_j w, w\rangle |w|^{-2}\, dx,$$

and the related constants

$$\mu = \inf\{J(w) : w \in (H_0^1(\Omega))^N, \|w\|_2 = 1\},$$
$$\mu_p = \inf\{J_p(w) : w \in (H_0^1(\Omega))^N, \|w\|_2 = 1\}.$$

Lemma 3.1 ([17]). *An operator $\mathcal{A}_p$ is dissipative in $(L^p(\Omega))^N$ if and only if $\mu_p \geqslant 0$.*

Lemma 3.2 ([17]). *Let $1 < p < \infty$. Assume that the principal part of $\mathcal{A}_p$ is positive. Then $\mathcal{A}$ and $\mathcal{A}_p$ generate the semigroups T on $(L^2(\Omega))^N$ and T_p on $(L^p(\Omega))^N$ respectively, and the following inequalities hold:*

$$\|T(t)\| \leqslant e^{-\mu t}, \quad \|T_p(t)\| \leqslant e^{-\mu_p t}, \quad t \geqslant 0,$$

where μ and μ_p are the best possible constants.

By Lemma 3.2, in the case $\mu = \mu_p$, the operator $\mathcal{A}_p$ generates a contraction semigroup on $(L^p)^N$ if and only if the operator $\mathcal{A}$ generates a contraction semigroup on $(L^2)^N$. Therefore, it is of interest to clarify relations between μ and μ_p.

Lemma 3.3 ([17]). *Let $1 < p < \infty$. Assume that the principal part of $\mathcal{A}_p$ is positive. Then $\mu = \mu_p$ if and only if at least one of nonzero generalized solutions of the equation*

$$-\frac{4}{pp'} \partial_i (a_{ij}\partial_j w) + \frac{1}{2p}(2\partial_i a_i I - p(A + A^*))w = \mu w, \quad w \in (H_0^1(\Omega))^N$$

has the form

$$w = fc,$$

where f is a real-valued scalar function and $c \in \mathbb{C}^N$. Moreover, μ is the least eigenvalue of the left-hand side of the equation.

Corollary 3.1 ([17]). *Suppose that the principal part of $\mathcal{A}_p$ is positive. Then $\mu \leqslant \mu_p$. If $\mu = \mu_p$, then there is a constant eigenvector of $A + A^*$ in $\overline{\Omega}$.*

Based on these results, Langer and Maz'ya proved Theorem 3.1 above.

The following example concerns the equality $\mu = \mu_p$ which is always satisfied for $N = 1$ in view of Lemma 3.3.

Example 3.1. Let A be the matrix

$$A(x) = \begin{pmatrix} 1 & |x| \\ |x| & -1 \end{pmatrix}, \quad x \in \overline{\Omega}.$$

Since A has no constant eigenvectors, $\mu < \mu_p$ (cf. Corollary 3.1). Therefore, it is possible to choose a suitable constant c such that $\mathcal{A}_p + cI$ generates a contraction semigroup on $(L^p(\Omega))^2$, whereas $\mathcal{A} + cI$ does not generate a contraction semigroup on $(L^2(\Omega))^2$.

4 Parabolic Systems

4.1 The maximum modulus principle for a parabolic system

In this subsection, we discuss the L^∞ case. This case was investigated by Kresin and Maz'ya (cf. [15] for a survey of their results). In particular, they proved that for a uniformly parabolic system in the sense of Petrovskii with coefficients independent of t the maximum modulus principle holds if and only if the principal part of the system is scalar and the coefficients of the system satisfy a certain algebraic inequality (cf. Theorem 4.1 below). Kresin and Maz'ya also considered the case where the coefficients depend on t and found necessary and, separately, sufficient conditions for the validity of the maximum modulus principle. They studied also maximum principles where the norm is understood in a generalized sense, i.e., as the Minkowski functional of a compact convex body in $\mathbb{R}^n$ containing the origin. In this general case, they obtained necessary and (separately, if the coefficients depend on t) sufficient conditions for the validity of the maximum norm principle.

Let $\Omega \subset \mathbb{R}^n$ be a bounded domain with $C^{2,\alpha}$ boundary $(0 < \alpha \leqslant 1)$. Denote by Q_T the cylinder $\Omega \times (0, T)$. Consider the differential operator $\mathcal{A}$

$$\mathcal{A}u = \partial_i(A_{ij}\partial_j u) + A_i \partial_i u + Au,$$

where A_{ij}, A_i, and A are $N \times N$ matrices with complex-valued entries. The entries of A_{ij}, A_i, and A belong to $C^{2,\alpha}(\overline{\Omega})$, $C^{1,\alpha}(\overline{\Omega})$, and $C^{0,\alpha}(\overline{\Omega})$ respectively. Moreover, we assume that $A_{ij} = A_{ji}$ and there exists $\delta > 0$ such that for any $x \in \overline{\Omega}$ and $\xi = (\xi_1, \ldots, \xi_N) \in \mathbb{R}^N$ the zeros of the polynomial

$$\lambda \mapsto \det(\xi_i \xi_j A_{ij} + \lambda I)$$

satisfy the inequality

$$\operatorname{Re} \lambda \leqslant -\delta |\xi|^2.$$

Consider the problem

$$\begin{aligned} &\partial_t u - \mathcal{A}u = 0 \quad \text{in } \ Q_T, \\ &u(\cdot, 0) = \varphi \quad \text{on } \ \Omega, \\ &u|_{\partial\Omega \times [0,T]} = 0, \end{aligned} \tag{4.1}$$

where $\varphi \in (C^{2,\alpha}(\overline{\Omega}))^N$ vanishes on $\partial\Omega$.

Theorem 4.1 ([16]). *Let u be the solution to the problem* (4.1). *The solution u satisfies the estimate*

$$\|u(\cdot, t)\|_\infty \leqslant \|\varphi\|_\infty, \quad \forall t \in [0, T],$$

for all $\varphi \in (C^{2,\alpha}(\overline{\Omega}))^N$ vanishing on $\partial\Omega$ if and only if

(a) *there are real-valued scalar functions a_{ij} in $\overline{\Omega}$ such that $A_{ij} = a_{ij} I$ for every i, j, and the $n \times n$ matrix $\{a_{ij}\}$ is positive definite,*

(b) *for all $\eta_i, \zeta \in \mathbb{C}^N$, $i = 1, \ldots, n$, such that* $\operatorname{Re}\langle \eta_i, \zeta\rangle = 0$

$$\operatorname{Re}\big\{a_{ij}\langle \eta_i, \eta_j\rangle - \langle A_i \eta_i, \zeta\rangle - \langle A\zeta, \zeta\rangle\big\} \geqslant 0 \quad \textit{on } \Omega.$$

In the scalar case $n = 1$, condition (b) is reduced to the inequality

$$-4 \operatorname{Re} A \geqslant b_{ij} \operatorname{Im} A_i \operatorname{Im} A_j \quad \text{on } \Omega,$$

where $\{b_{ij}\} = \{a_{ij}\}^{-1}$ (cf. (2.14)).

Now, consider the problem

$$\begin{aligned} &\partial_t w + \mathcal{A}^* w = 0 \quad \text{in } \ Q_T, \\ &w(\cdot, T) = \psi \quad \text{on } \ \Omega, \\ &w|_{\partial\Omega \times [0,T]} = 0, \end{aligned} \tag{4.2}$$

where $\mathcal{A}^*$ is the formally adjoint operator of $\mathcal{A}$:

$$\mathcal{A}^* w = \partial_i(A^*_{ij}\partial_j w) - A^*_i \partial_i w + (A^* - \partial_i A^*_i) w.$$

Theorem 4.1 implies the following assertion.

Corollary 4.1. *Let w be the solution to the problem* (4.2). *The solution u satisfies the inequality*

$$\|w(\cdot, t)\|_\infty \leqslant \|\psi\|_\infty \quad \forall t \in [0, T],$$

for all $\psi \in (C^{2,\alpha}(\overline{\Omega}))^N$ vanishing on $\partial\Omega$ if and only if

(a) *there are real-valued scalar functions a_{ij} on $\overline{\Omega}$ such that $A_{ij} = a_{ij} I$ for every i, j and the $n \times n$ matrix $\{a_{ij}\}$ is positive definite,*

(b) *for all $\eta_i, \zeta \in \mathbb{C}^N$, $i = 1, \ldots, n$, such that* $\operatorname{Re}\langle \eta_i, \zeta\rangle = 0$

$$\operatorname{Re}\big\{a_{ij}\langle \eta_i, \eta_j\rangle + \langle A_i \zeta, \eta_i\rangle - \langle (A - \partial_i A_i)\zeta, \zeta\rangle\big\} \geqslant 0 \quad \textit{on } \Omega.$$

Based on Theorem 4.1 and Corollary 4.1 and using interpolation, it is possible to obtain necessary and sufficient conditions for the validity of the L^p maximum principle for all $p \in [1, \infty]$ simultaneously.

Corollary 4.2 ([17]). *Let u be the solution to the problem* (4.1). *The solution u satisfies the inequality*

$$\|u(\cdot, t)\|_p \leqslant \|\varphi\|_p, \quad \forall t \in [0, T],$$

for all $\varphi \in (C^{2,\alpha}(\overline{\Omega}))^N$ vanishing on $\partial\Omega$ and for all $p \in [1, \infty]$ if and only if

(a) *there are real-valued scalar functions a_{ij} on $\overline{\Omega}$ such that $A_{ij} = a_{ij} I$ for every i, j and the $n \times n$ matrix $\{a_{ij}\}$ is positive definite,*

(b) *for all $\eta_i, \zeta \in \mathbb{C}^N$, $i = 1, \ldots, n$, such that* $\operatorname{Re}\langle \eta_i, \zeta\rangle = 0$ *the following inequalities hold on Ω:*

$$\operatorname{Re}\{a_{ij}\langle \eta_i, \eta_j\rangle - \langle A_i \eta_i, \zeta\rangle - \langle A\zeta, \zeta\rangle\} \geqslant 0,$$
$$\operatorname{Re}\{a_{ij}\langle \eta_i, \eta_j\rangle + \langle A_i \zeta, \eta_i\rangle - \langle (A - \partial_i A_i)\zeta, \zeta\rangle\} \geqslant 0.$$

4.2 Applications to semigroup theory

From Theorem 4.2 we obtain the following contractivity property of the semigroup generated by $\mathcal{A}_p$, where $\mathcal{A}_p$ is the extension of $\mathcal{A}$ to the space

$$D(\mathcal{A}_p) = (W^{2,p}(\Omega) \cap \overset{\circ}{W}{}^{1,p}_0(\Omega))^N, \quad 1 < p < \infty$$

and $\mathcal{A}_1$ is the $(L^1(\Omega))^N$ closure of the operator $\mathcal{A}$ defined on the set of functions in $(C^{2,\alpha}(\overline{\Omega}))^N$ vanishing on $\partial\Omega$.

Theorem 4.2 ([17]). *Operators $\mathcal{A}_p$ generate contraction semigroups on $(L^p(\Omega))^N$ for all $p \in [1, \infty)$ and on $(C_0(\overline{\Omega}))^N$ for $p = \infty$ simultaneously if and only if*

(a) *there are real-valued scalar functions a_{ij} on $\overline{\Omega}$ such that $A_{ij} = a_{ij} I$ for every i, j and the $n \times n$ matrix $\{a_{ij}\}$ is positive definite,*

(b) *for all $\eta_i, \zeta \in \mathbb{C}^N$, $i = 1, \ldots, n$, such that* $\operatorname{Re}\langle \eta_i, \zeta\rangle = 0$ *the following inequalities hold on Ω :*

$$\operatorname{Re}\{a_{ij}\langle \eta_i, \eta_j\rangle - \langle A_i \eta_i, \zeta\rangle - \langle A\zeta, \zeta\rangle\} \geqslant 0,$$
$$\operatorname{Re}\{a_{ij}\langle \eta_i, \eta_j\rangle + \langle A_i \zeta, \eta_i\rangle - \langle (A - \partial_i A_i)\zeta, \zeta\rangle\} \geqslant 0.$$

Example 4.1. Consider the Schrödinger operator with magnetic field

$$-(i\nabla + m)^t (i\nabla + m) - V,$$

where m is an $\mathbb{R}^n$-valued function on Ω and V is complex-valued. Theorem 4.2 shows that this operator generates contraction semigroups on $L^p(\Omega)$ for all $p \in [1, \infty]$ simultaneously if and only if

$$-4 \operatorname{Re} A \geqslant \sum_{j=1}^{n} (\operatorname{Im} A_j)^2, \qquad -4 \operatorname{Re}(\overline{A - \partial_j A_j}) \geqslant \sum_{j=1}^{n} (-\operatorname{Im} \overline{A_j})^2,$$

which is equivalent to the condition $\operatorname{Re} V \geqslant 0$ on Ω.

5 Two-Dimensional Elasticity System

Consider the classical operator of two-dimensional elasticity

$$Eu = \Delta u + (1 - 2\nu)^{-1} \nabla \nabla^t u, \tag{5.1}$$

where ν is the Poisson ratio. As is known, E is strongly elliptic if and only if either $\nu > 1$ or $\nu < 1/2$. To obtain a necessary and sufficient condition for the L^p-dissipativity of this elasticity system, we formulate some facts about systems of partial differential equations of the form

$$A = \partial_h (\mathscr{A}^{hk}(x) \partial_k), \tag{5.2}$$

where $\mathscr{A}^{hk}(x) = \{a_{ij}^{hk}(x)\}$ are $m \times m$ matrices whose entries are complex locally integrable functions defined in an arbitrary domain Ω of $\mathbb{R}^n$ $(1 \leqslant i, j \leqslant m,\ 1 \leqslant h, k \leqslant n)$.

Lemma 5.1 ([5]). *An operator A of the form (5.2) is L^p-dissipative in $\Omega \subset \mathbb{R}^n$ if and only if*

$$\begin{aligned} \int_\Omega \Big(&\operatorname{Re} \langle \mathscr{A}^{hk} \partial_k w, \partial_h w \rangle \\ &- (1 - 2/p)^2 |w|^{-4} \operatorname{Re} \langle \mathscr{A}^{hk} w, w \rangle \operatorname{Re} \langle w, \partial_k w \rangle \operatorname{Re} \langle w, \partial_h w \rangle \\ &- (1 - 2/p) |w|^{-2} \operatorname{Re} \big(\langle \mathscr{A}^{hk} w, \partial_h w \rangle \operatorname{Re} \langle w, \partial_k w \rangle \\ &- \langle \mathscr{A}^{hk} \partial_k w, w \rangle \operatorname{Re} \langle w, \partial_h w \rangle \big) \Big) dx \geqslant 0 \end{aligned}$$

for any $w \in ({C_0}^1(\Omega))^m$.

In the case $n = 2$, Lemma 5.1 yields a necessary algebraic condition.

Theorem 5.1 ([5]). *Let Ω be a domain of $\mathbb{R}^2$. If an operator A of the form (5.2) is L^p-dissipative, then*

$$\operatorname{Re}\langle(\mathscr{A}^{hk}(x)\xi_h\xi_k)\lambda,\lambda\rangle-(1-2/p)^2\operatorname{Re}\langle(\mathscr{A}^{hk}(x)\xi_h\xi_k)\omega,\omega\rangle(\operatorname{Re}\langle\lambda,\omega\rangle)^2$$
$$-(1-2/p)\operatorname{Re}(\langle(\mathscr{A}^{hk}(x)\xi_h\xi_k)\omega,\lambda\rangle-\langle(\mathscr{A}^{hk}(x)\xi_h\xi_k)\lambda,\omega\rangle)\operatorname{Re}\langle\lambda,\omega\rangle\geqslant 0$$

for almost every $x\in\Omega$ and for any $\xi\in\mathbb{R}^2$, λ, $\omega\in\mathbb{C}^m$, $|\omega|=1$.

Based on Lemma 5.1 and Theorem 5.1, it is possible to obtain the following criterion for the L^p-dissipativity of the two-dimensional elasticity system.

Theorem 5.2 ([5]). *The operator* (5.1) *is L^p-dissipative if and only if*

$$\left(\frac{1}{2}-\frac{1}{p}\right)^2\leqslant\frac{2(\nu-1)(2\nu-1)}{(3-4\nu)^2}.$$

By Theorems 5.1 and 5.2, it is easy to compare E and Δ from the point of view of L^p-dissipativity.

Corollary 5.1 ([5]). *There exists $k>0$ such that $E-k\Delta$ is L^p-dissipative if and only if*

$$\left(\frac{1}{2}-\frac{1}{p}\right)^2<\frac{2(\nu-1)(2\nu-1)}{(3-4\nu)^2}.$$

There exists $k<2$ such that $k\Delta-E$ is L^p-dissipative if and only if

$$\left(\frac{1}{2}-\frac{1}{p}\right)^2<\frac{2\nu(2\nu-1)}{(1-4\nu)^2}.$$

6 A Class of Systems of Partial Differential Operators

In this section, we consider systems of partial differential operators of the form

$$Au=\partial_h(\mathscr{A}^h(x)\partial_h u),\tag{6.1}$$

where $\mathscr{A}^h(x)=\{a^h_{ij}(x)\}$ $(i,j=1,\dots,m)$ are matrices with complex locally integrable entries defined in a domain $\Omega\subset\mathbb{R}^n$ $(h=1,\dots,n)$. Note that the elasticity system is not a system of this kind.

To characterize the L^p-dissipativity of such operators, one can reduce the consideration to the one-dimensional case. Auxiliary facts are given in the following two subsections.

6.1 Dissipativity of systems of ordinary differential equations

In this subsection, we consider the operator

$$Au = (\mathscr{A}(x)u')', \tag{6.2}$$

where $\mathscr{A}(x) = \{a_{ij}(x)\}$ $(i,j = 1,\dots,m)$ is a matrix with complex locally integrable entries defined in a bounded or unbounded interval (a,b). The corresponding sesquilinear form $\mathscr{L}(u,w)$ takes the form

$$\mathscr{L}(u,w) = \int_a^b \langle \mathscr{A}u', w'\rangle\, dx.$$

Theorem 6.1 ([5]). *The operator A is L^p-dissipative if and only if*

$$\begin{aligned}&\operatorname{Re}\langle \mathscr{A}(x)\lambda,\lambda\rangle - (1-2/p)^2 \operatorname{Re}\langle \mathscr{A}(x)\omega,\omega\rangle (\operatorname{Re}\langle\lambda,\omega\rangle)^2\\ &- (1-2/p)\operatorname{Re}(\langle \mathscr{A}(x)\omega,\lambda\rangle - \langle \mathscr{A}(x)\lambda,\omega\rangle)\operatorname{Re}\langle\lambda,\omega\rangle \geqslant 0\end{aligned}$$

for almost every $x \in (a,b)$ and for any $\lambda,\omega \in \mathbb{C}^m$, $|\omega| = 1$.

This theorem implies the following assertion.

Corollary 6.1 ([5]). *If the operator A is L^p-dissipative, then*

$$\operatorname{Re}\langle \mathscr{A}(x)\lambda,\lambda\rangle \geqslant 0$$

for almost every $x \in (a,b)$ and for any $\lambda \in \mathbb{C}^m$.

We can precisely determine the angle of dissipativity of the matrix ordinary differential operator (6.2) with complex coefficients.

Theorem 6.2 ([5]). *Let the operator* (6.2) *be L^p-dissipative. The operator zA is L^p-dissipative if and only if*

$$\vartheta_- \leqslant \arg z \leqslant \vartheta_+$$

where

$$\vartheta_- = \operatorname{arccot}\left(\operatorname*{ess\,inf}_{(x,\lambda,\omega)\in\Xi} (Q(x,\lambda,\omega)/P(x,\lambda,\omega))\right) - \pi,$$

$$\vartheta_+ = \operatorname{arccot}\left(\operatorname*{ess\,sup}_{(x,\lambda,\omega)\in\Xi} (Q(x,\lambda,\omega)/P(x,\lambda,\omega))\right),$$

$$\begin{aligned}P(x,\lambda,\omega) = {}&\operatorname{Re}\langle \mathscr{A}(x)\lambda,\lambda\rangle - (1-2/p)^2 \operatorname{Re}\langle \mathscr{A}(x)\omega,\omega\rangle (\operatorname{Re}\langle\lambda,\omega\rangle)^2\\ &- (1-2/p)\operatorname{Re}(\langle \mathscr{A}(x)\omega,\lambda\rangle - \langle \mathscr{A}(x)\lambda,\omega\rangle)\operatorname{Re}\langle\lambda,\omega\rangle,\end{aligned}$$

$$\begin{aligned}Q(x,\lambda,\omega) = {}&\operatorname{Im}\langle \mathscr{A}(x)\lambda,\lambda\rangle - (1-2/p)^2 \operatorname{Im}\langle \mathscr{A}(x)\omega,\omega\rangle (\operatorname{Re}\langle\lambda,\omega\rangle)^2\\ &- (1-2/p)\operatorname{Im}(\langle \mathscr{A}(x)\omega,\lambda\rangle - \langle \mathscr{A}(x)\lambda,\omega\rangle)\operatorname{Re}\langle\lambda,\omega\rangle\end{aligned}$$

and Ξ *is the set*

$$\Xi = \{(x, \lambda, \omega) \in (a, b) \times \mathbb{C}^m \times \mathbb{C}^m \mid |\omega| = 1,\ P^2(x, \lambda, \omega) + Q^2(x, \lambda, \omega) > 0\}.$$

Another consequence of Theorem 6.1 is the possibility to compare the operators A and $I(d^2/dx^2)$.

Corollary 6.2 ([5]). *There exists* $k > 0$ *such that* $A - kI(d^2/dx^2)$ *is* L^p*-dissipative if and only if*

$$\operatorname*{ess\,inf}_{\substack{(x,\lambda,\omega)\in(a,b)\times\mathbb{C}^m\times\mathbb{C}^m \\ |\lambda|=|\omega|=1}} P(x, \lambda, \omega) > 0.$$

There exists $k > 0$ *such that* $kI(d^2/dx^2) - A$ *is* L^p*-dissipative if and only if*

$$\operatorname*{ess\,sup}_{\substack{(x,\lambda,\omega)\in(a,b)\times\mathbb{C}^m\times\mathbb{C}^m \\ |\lambda|=|\omega|=1}} P(x, \lambda, \omega) < \infty.$$

There exists $k \in \mathbb{R}$ *such that* $A - kI(d^2/dx^2)$ *is* L^p*-dissipative if and only if*

$$\operatorname*{ess\,inf}_{\substack{(x,\lambda,\omega)\in(a,b)\times\mathbb{C}^m\times\mathbb{C}^m \\ |\lambda|=|\omega|=1}} P(x, \lambda, \omega) > -\infty.$$

6.2 Criteria in terms of eigenvalues of $\mathscr{A}(x)$

If the coefficients a_{ij} of the operator (6.2) are real, it is possible to give a necessary and sufficient condition for the L^p-dissipativity of A in terms of eigenvalues of the matrix $\mathscr{A}$.

Theorem 6.3 ([5]). *Let* $\mathscr{A}$ *be a real matrix* $\{a_{hk}\}$ *with* $h, k = 1, \ldots, m$. *Suppose that* $\mathscr{A} = \mathscr{A}^t$ *and* $\mathscr{A} \geqslant 0$ *(in the sense that* $\langle \mathscr{A}(x)\xi, \xi\rangle \geqslant 0$ *for almost every* $x \in (a, b)$ *and for any* $\xi \in \mathbb{R}^m$*). The operator* A *is* L^p*-dissipative if and only if*

$$\left(\frac{1}{2} - \frac{1}{p}\right)^2 (\mu_1(x) + \mu_m(x))^2 \leqslant \mu_1(x)\mu_m(x)$$

almost everywhere, where $\mu_1(x)$ *and* $\mu_m(x)$ *are the smallest and largest eigenvalues of the matrix* $\mathscr{A}(x)$ *respectively. In the particular case* $m = 2$, *this condition is equivalent to*

$$\left(\frac{1}{2} - \frac{1}{p}\right)^2 (\mathrm{tr}\mathscr{A}(x))^2 \leqslant \det \mathscr{A}(x)$$

almost everywhere.

Corollary 6.3 ([5]). *Let $\mathscr{A}$ be a real symmetric matrix. Let $\mu_1(x)$ and $\mu_m(x)$ be the smallest and largest eigenvalues of $\mathscr{A}(x)$ respectively. There exists $k > 0$ such that $A - kI(d^2/dx^2)$ is L^p-dissipative if and only if*

$$\operatorname*{ess\,inf}_{x\in(a,b)} \left[(1+\sqrt{pp'}/2)\,\mu_1(x) + (1-\sqrt{pp'}/2)\,\mu_m(x)\right] > 0. \tag{6.3}$$

In the particular case $m = 2$, the condition (6.3) is equivalent to

$$\operatorname*{ess\,inf}_{x\in(a,b)} \left[\operatorname{tr}\mathscr{A}(x) - \frac{\sqrt{pp'}}{2}\sqrt{(\operatorname{tr}\mathscr{A}(x))^2 - 4\det\mathscr{A}(x)}\right] > 0.$$

Under an extra condition on the matrix $\mathscr{A}$, the following assertion holds.

Corollary 6.4 ([5]). *Let $\mathscr{A}$ be a real symmetric matrix. Suppose that $\mathscr{A} \geqslant 0$ almost everywhere. Denote by $\mu_1(x)$ and $\mu_m(x)$ the smallest and largest eigenvalues of $\mathscr{A}(x)$ respectively. If there exists $k > 0$ such that $A - kI(d^2/dx^2)$ is L^p-dissipative, then*

$$\operatorname*{ess\,inf}_{x\in(a,b)} \left[\mu_1(x)\mu_m(x) - \left(\frac{1}{2}-\frac{1}{p}\right)^2 (\mu_1(x)+\mu_m(x))^2\right] > 0. \tag{6.4}$$

If, in addition, there exists C such that

$$\langle \mathscr{A}(x)\xi, \xi\rangle \leqslant C|\xi|^2 \tag{6.5}$$

for almost every $x \in (a,b)$ and for any $\xi \in \mathbb{R}^m$, the converse assertion is also true. In the particular case $m = 2$, the condition (6.4) is equivalent to

$$\operatorname*{ess\,inf}_{x\in(a,b)} \left[\det\mathscr{A}(x) - \left(\frac{1}{2}-\frac{1}{p}\right)^2 (\operatorname{tr}\mathscr{A}(x))^2\right] > 0.$$

Generally speaking, the assumption (6.5) cannot be removed even if $\mathscr{A} \geqslant 0$.

Example 6.1. Consider $(a,b) = (1,\infty)$, $m = 2$, $\mathscr{A}(x) = \{a_{ij}(x)\}$, where

$$a_{11}(x) = (1 - 2/\sqrt{pp'})x + x^{-1}, \quad a_{12}(x) = a_{21}(x) = 0,$$
$$a_{22}(x) = (1 + 2/\sqrt{pp'})x + x^{-1}.$$

Then

$$\mu_1(x)\mu_2(x) - \left(\frac{1}{2}-\frac{1}{p}\right)^2 (\mu_1(x)+\mu_2(x))^2 = (8+4x^{-2})/(pp')$$

and (6.4) holds. But (6.3) is not satisfied because

$$(1+\sqrt{pp'}/2)\,\mu_1(x)+(1-\sqrt{pp'}/2)\,\mu_2(x)=2x^{-1}.$$

Corollary 6.5 ([5]). *Let $\mathscr{A}$ be a real symmetric matrix. Let $\mu_1(x)$ and $\mu_m(x)$ be the smallest and largest eigenvalues of $\mathscr{A}(x)$ respectively. There exists $k>0$ such that $kI(d^2/dx^2)-A$ is L^p-dissipative if and only if*

$$\operatorname*{ess\,sup}_{x\in(a,b)}\Big[(1-\sqrt{pp'}/2)\,\mu_1(x)+(1+\sqrt{pp'}/2)\,\mu_m(x)\Big]<\infty. \tag{6.6}$$

In the particular case $m=2$, the condition (6.6) is equivalent to

$$\operatorname*{ess\,sup}_{x\in(a,b)}\Big[\operatorname{tr}\mathscr{A}(x)+\frac{\sqrt{pp'}}{2}\sqrt{(\operatorname{tr}\mathscr{A}(x))^2-4\det\mathscr{A}(x)}\Big]<\infty.$$

If $\mathscr{A}$ is positive, the following assertion holds.

Corollary 6.6 ([5]). *Let $\mathscr{A}$ be a real symmetric matrix. Suppose that $\mathscr{A}\geqslant 0$ almost everywhere. Let $\mu_1(x)$ and $\mu_m(x)$ be the smallest and largest eigenvalues of $\mathscr{A}(x)$ respectively. There exists $k>0$ such that $kI(d^2/dx^2)-A$ is L^p-dissipative if and only if*

$$\operatorname*{ess\,sup}_{x\in(a,b)}\mu_m(x)<\infty.$$

6.3 L^p-dissipativity of the operator (6.1)

We represent necessary and sufficient conditions for the L^p-dissipativity of the system (6.1), obtained in [5].

Denote by y_h the $(n-1)$-dimensional vector $(x_1,\dots,x_{h-1},x_{h+1},\dots,x_n)$ and set $\omega(y_h)=\{x_h\in\mathbb{R}\mid x\in\Omega\}$.

Lemma 6.1 ([5]). *The operator (6.1) is L^p-dissipative if and only if the ordinary differential operators*

$$A(y_h)[u(x_h)]=d(\mathscr{A}^h(x)du/dx_h)/dx_h$$

are L^p-dissipative in $\omega(y_h)$ for almost every $y_h\in\mathbb{R}^{n-1}$ $(h=1,\dots,n)$. This condition is void if $\omega(y_h)=\varnothing$.

Theorem 6.4 ([5]). *The operator (6.1) is L^p-dissipative if and only if*

$$\begin{aligned}&\operatorname{Re}\langle\mathscr{A}^h(x_0)\lambda,\lambda\rangle-(1-2/p)^2\operatorname{Re}\langle\mathscr{A}^h(x_0)\omega,\omega\rangle(\operatorname{Re}\langle\lambda,\omega\rangle)^2\\&\quad-(1-2/p)\operatorname{Re}(\langle\mathscr{A}^h(x_0)\omega,\lambda\rangle-\langle\mathscr{A}^h(x_0)\lambda,\omega\rangle)\operatorname{Re}\langle\lambda,\omega\rangle\geqslant 0\end{aligned} \tag{6.7}$$

for almost every $x_0\in\Omega$ and for any $\lambda,\omega\in\mathbb{C}^m$, $|\omega|=1$, $h=1,\dots,n$.

In the scalar case ($m = 1$), the operator (6.1) can be considered as an operator from Section 2.

In fact, if $Au = \sum_{h=1}^{n} \partial_h(a^h \partial_h u)$, a^h is a scalar function, then A can be written in the form (2.5) with $\mathscr{A} = \{c_{hk}\}$, $c_{hh} = a^h$, $c_{hk} = 0$ if $h \neq k$. The conditions obtained in Section 2 can be directly compared with (6.7). We know that the operator A is L^p-dissipative if and only if (2.7) holds. In this particular case, it is clear that (2.7) is equivalent to the following n conditions:

$$\frac{4}{pp'}(\operatorname{Re} a^h)\,\xi^2 + (\operatorname{Re} a^h)\,\eta^2 - 2(1-2/p)(\operatorname{Im} a^h)\,\xi\eta \geqslant 0 \tag{6.8}$$

almost everywhere and for any $\xi, \eta \in \mathbb{R}$, $h = 1, \ldots, n$. On the other hand, in this case, (6.7) reads as

$$\begin{aligned}(\operatorname{Re} a^h)|\lambda|^2 - (1-2/p)^2(\operatorname{Re} a^h)(\operatorname{Re}(\lambda\overline{\omega})^2 \\ - 2(1-2/p)(\operatorname{Im} a^h)\operatorname{Re}(\lambda\overline{\omega})\operatorname{Im}(\lambda\overline{\omega}) \geqslant 0\end{aligned} \tag{6.9}$$

almost everywhere and for any $\lambda, \omega \in \mathbb{C}$, $|\omega| = 1$, $h = 1, \ldots, n$. Setting $\xi + i\eta = \lambda\overline{\omega}$ and observing that $|\lambda|^2 = |\lambda\overline{\omega}|^2 = (\operatorname{Re}(\lambda\overline{\omega}))^2 + (\operatorname{Im}(\lambda\overline{\omega}))^2$, we see that the conditions (6.8) (hence (2.7)) are equivalent to (6.9).

Theorem 6.4 allows us to determine the angle of dissipativity of the operator (6.1).

Theorem 6.5 ([5]). *Let A be L^p-dissipative. The operator zA is L^p-dissipative if and only if*

$$\vartheta_- \leqslant \arg z \leqslant \vartheta_+,$$

where

$$\vartheta_- = \max_{h=1,\ldots,n} \operatorname{arccot}\left(\operatorname*{ess\,inf}_{(x,\lambda,\omega)\in\Xi_h} (Q_h(x,\lambda,\omega)/P_h(x,\lambda,\omega))\right) - \pi,$$

$$\vartheta_+ = \min_{h=1,\ldots,n} \operatorname{arccot}\left(\operatorname*{ess\,sup}_{(x,\lambda,\omega)\in\Xi_h} (Q_h(x,\lambda,\omega)/P_h(x,\lambda,\omega))\right)$$

and

$$\begin{aligned}P_h(x,\lambda,\omega) = \operatorname{Re}\langle \mathscr{A}^h(x)\lambda,\lambda\rangle - (1-2/p)^2 \operatorname{Re}\langle \mathscr{A}^h(x)\omega,\omega\rangle(\operatorname{Re}\langle\lambda,\omega\rangle)^2 \\ - (1-2/p)\operatorname{Re}(\langle \mathscr{A}^h(x)\omega,\lambda\rangle - \langle \mathscr{A}^h(x)\lambda,\omega\rangle)\operatorname{Re}\langle\lambda,\omega\rangle,\end{aligned}$$

$$\begin{aligned}Q_h(x,\lambda,\omega) = \operatorname{Im}\langle \mathscr{A}^h(x)\lambda,\lambda\rangle - (1-2/p)^2 \operatorname{Im}\langle \mathscr{A}^h(x)\omega,\omega\rangle(\operatorname{Re}\langle\lambda,\omega\rangle)^2 \\ - (1-2/p)\operatorname{Im}(\langle \mathscr{A}^h(x)\omega,\lambda\rangle - \langle \mathscr{A}^h(x)\lambda,\omega\rangle)\operatorname{Re}\langle\lambda,\omega\rangle,\end{aligned}$$

$$\Xi_h = \{(x, \lambda, \omega) \in \Omega \times \mathbb{C}^m \times \mathbb{C}^m \mid |\omega| = 1,\ P_h^2(x, \lambda, \omega) + Q_h^2(x, \lambda, \omega) > 0\}.$$

If A has real coefficients, we can characterize the L^p-dissipativity in terms of the eigenvalues of the matrices $\mathscr{A}^h(x)$.

Theorem 6.6 ([5]). *Let A be an operator of the form (6.1), where $\mathscr{A}^h$ are real matrices $\{a_{ij}^h\}$ with $i, j = 1, \ldots, m$. Suppose that $\mathscr{A}^h = (\mathscr{A}^h)^t$ and $\mathscr{A}^h \geqslant 0$ $(h = 1, \ldots, n)$. The operator A is L^p-dissipative if and only if*

$$\left(\frac{1}{2} - \frac{1}{p}\right)^2 (\mu_1^h(x) + \mu_m^h(x))^2 \leqslant \mu_1^h(x)\, \mu_m^h(x)$$

for almost every $x \in \Omega$, $h = 1, \ldots, n$, where $\mu_1^h(x)$ and $\mu_m^h(x)$ are the smallest and largest eigenvalues of the matrix $\mathscr{A}^h(x)$ respectively. In the particular case $m = 2$, this condition is equivalent to

$$\left(\frac{1}{2} - \frac{1}{p}\right)^2 (\operatorname{tr} \mathscr{A}^h(x))^2 \leqslant \det \mathscr{A}^h(x)$$

for almost every $x \in \Omega$, $h = 1, \ldots, n$.

7 Higher Order Differential Operators

The contractivity of semigroups generated by second order partial differential operators, scalar or vector, was studied in many papers by different authors. The case of higher order operators is quite different because of peculiarities explained below. Apparently, only the paper [18] by Langer and Maz'ya dealt with similar questions for higher order differential operators.

7.1 Noncontractivity of higher order operators

The following simple example shows that it is not reasonable to expect the L^1-contractivity for higher order operators even in the one-dimensional case. Consider the problem

$$\frac{\partial}{\partial t} u(x, t) + (-1)^m \frac{\partial^{2m}}{\partial x^{2m}} u(x, t) = 0 \qquad x \in \mathbb{R}, \quad t \geqslant 0.$$

The solution u is expressed as

$$u(x, t) = \int_{\mathbb{R}} K_t(x - y)\, u(y, 0)\, dy$$

with the kernel K_t such that

$$\widehat{K}_t(\xi) = e^{-\xi^{2m}t}, \quad \xi \in \mathbb{R}, t \geqslant 0,$$

where $\widehat{K}_t$ is the Fourier transform of K_t. Since for $m > 1$

$$1 = \widehat{K}_t(0) = \int_{\mathbb{R}} K_t(x)\, dx, \qquad 0 = \widehat{K}_t''(0) = -\int_{\mathbb{R}} x^2 K_t(x)\, dx,$$

we have $\|K_t\|_{L^1} > 1$. Consequently, the semigroup generated by the operator

$$(-1)^{m+1} \frac{d^{2m}}{dx^{2m}}$$

cannot be contractive.

Maz'ya and Langer [18] considered multidimensional operators with locally integrable coefficients. In the case $1 \leqslant p < \infty$, $p \neq 2$, they proved that, in the class of linear partial differential operators of order higher than two, with the domain containing $({C_0}^\infty(\Omega))^N$, there are no generators of a contraction semigroup on $(L^p(\Omega))^N$. In the one-dimensional case, they found necessary and sufficient conditions.

Theorem 7.1 ([18]). *Let $k \in \mathbb{N}$ and $p \in [1, \infty)$. The integral*

$$\int w^{(k)} |w|^{p-1} \mathrm{sgn} w\, dx$$

preserves the sign as w runs over the real-valued functions in ${C_0}^\infty(\Omega)$ if and only if $p = 2$ or $k \in \{0, 1, 2\}$.

Consider a linear partial differential operator A of the form

$$A = \sum_{|\alpha| \leqslant k} a_\alpha D^\alpha \tag{7.1}$$

where a_α belong to $L^1_{\mathrm{loc}}(\Omega)$, and Ω is a domain in $\mathbb{R}^n$.

Theorem 7.2 ([18]). *Let $p \in [1, \infty)$, $p \neq 2$. If*

$$\mathrm{Re} \int_\Omega \langle Au, u \rangle\, |u|^{p-2}\, dx$$

preserves the sign as u runs over ${C_0}^\infty(\Omega)$, then A is of order 0, 1, or 2.

If u runs over not ${C_0}^\infty(\Omega)$, but only $({C_0}^\infty(\Omega))^+$ (i.e., the class of nonnegative functions of ${C_0}^\infty(\Omega)$), then the result for operators with real coefficients becomes quite different.

Theorem 7.3 ([18]). *Let $p \in (1, \infty)$, $p \neq 2$. Let A be a linear partial differential operator with real-valued coefficients. Assume that*

$$\int_{\Omega} (Au)u^{p-1}\, dx$$

preserves the sign as u *runs over* $(C_0^{\infty}(\Omega))^+$. *Then either* A *is of order* 0, 1, *or* 2 *or* A *is of order* 4 *and* $\frac{3}{2} \leqslant p \leqslant 3$.

Theorem 7.2 means the nondissipativity of higher order operators of the form (7.1), where a_α are $N \times N$ matrices with entries in $L^1_{\text{loc}}(\Omega)$ (N is a positive integer). For such operators the following assertion holds.

Theorem 7.4 ([18]). *If* $1 \leqslant p < \infty$, $p \neq 2$, *then there are no linear partial differential operator of order higher than two, with the domain containing* $(C_0^{\infty}(\Omega))^N$, *that generates a contraction semigroup on* $(L^p(\Omega))^N$.

7.2 The cone of nonnegative functions

In some cases, it is known that a solution to the Cauchy problem

$$\begin{aligned} s'(t) &= A[s(t)] \\ s(0) &= s_0 \end{aligned} \tag{7.2}$$

is nonnegative on some interval. Hence the question about the contractivity on the cone of nonnegative functions arises.

As is known, the norms of the solutions to the Cauchy problem decrease if the corresponding semigroup is contractive. In the case of a cone of nonnegative functions, when the semigroup theory cannot be applied, the following lemma can be regarded as a parallel result.

Note that the spaces L^p are assumed to be real in this subsection.

Lemma 7.1 ([18]). *Suppose that the Cauchy problem* (7.2) *has a unique solution of class* $C^1(\mathbb{R}^+, L^p)$ *for every* $s_0 \in D(A)$.

In the case $1 < p < \infty$,

$$\frac{d}{dt}\|s(t)\|_p \Big|_{t=0^+} \leqslant 0$$

for every $s(0) \in (D(A))^+$ *if and only if*

$$\int_{\Omega} (Au)u^{p-1}\, dx \leqslant 0 \tag{7.3}$$

for every $u \in (D(A))^+$.

In the case $p = 1$, *the inequality* (7.3) *holds for every* $u \in (D(A))^+$ *if*

$$\liminf_{t \to 0^+} t^{-1}(\|s(t)\|_1 - \|s(0)\|_1) \leqslant 0 \tag{7.4}$$

for every $s(0) \in (D(A))^+$.

Lemma 7.1 and Theorem 7.3 lead to the following assertion.

Theorem 7.5 ([18]). *Let* $1 < p < \infty$, $p \neq 2$. *Consider a linera partial differential operator* A *such that the domain* $D(A)$ *of* A *contains* $C_0^\infty(\Omega)$, *the coefficients of* A *belong to* $L^1_{\text{loc}}(\Omega)$, *and the Cauchy problem* (7.2) *has a unique solution for every nonnegative initial data in* $D(A)$. *If*

$$\frac{d}{dt}\|s(t)\|_p\bigg|_{t=0+} \leqslant 0$$

for every $s(0) \in (D(A))^+$, *then either* A *is of order* 0, 1, *or* 2 *or* A *is of order* 4 *and* $\frac{3}{2} \leqslant p \leqslant 3$.

This theorem does not cover the case $p = 1$. In this special case, it is possible to show that under the assumptions of Theorem 7.5 and the condition (7.4), the distribution

$$-\sum_{|\alpha|\leqslant k} (-1)^{|\alpha|}\partial^\alpha a_\alpha$$

is a positive measure.

The following theorem is related to the case where the operator has constant coefficients and the domain Ω has smooth boundary.

Theorem 7.6 ([18]). *Let* $1 < p < \infty$. *Suppose that* $\Omega \subset \mathbb{R}^n$ *is an open bounded domain with* C^∞*-boundary. Suppose that* $\{a_{ijk\ell}\}$ *are real constants such that*

$$a_{ijk\ell} = a_{jki\ell} = a_{j\ell ik}$$

for all i, j, k, ℓ *and*

$$\sum_{1\leqslant i,j,k,\ell\leqslant n} a_{ijk\ell}\xi_{ij}\xi_{k\ell} \geqslant 0$$

for all real symmetric $n \times n$ *matrices* $\xi = \{\xi_{ij}\}$. *Then*

$$\int_\Omega (a_{ijk\ell}\partial_i\partial_j\partial_k\partial_\ell u)u^{p-1}\,dx \geqslant 0$$

for all nonnegative functions $u \in W^{4,p}(\Omega) \cap W_0^{2,p}(\Omega)$ *if and only if* $\frac{3}{2} \leqslant p \leqslant 3$.

Corollary 7.1 ([18]). *Let* $\frac{3}{2} \leqslant p \leqslant 3$. *Suppose that* Ω *and the coefficients of an operator*

$$A = -a_{ijk\ell}\partial_{ijk\ell},$$

with the domain $W^{4,p}(\Omega) \cap W_0^{2,p}(\Omega)$, *satisfy the assumptions of Theorem* 7.6. *Then any differentiable solution* s *to the Cauchy problem* (7.2) *with nonnegative initial value* $s(0) \in D(A)$ *satisfies the estimate*

$$\frac{d}{dt}\|s(t)\|_p\bigg|_{t=0+} \leqslant 0.$$

In particular, $A = -\Delta^2$ satisfies the assumptions of Corollary 7.1.

In conclusion, we mention two open questions. The first one concerns necessary and sufficient conditions for the L^p-dissipativity of the operator (2.5) when the matrix $\operatorname{Im} \mathscr{A}$ is not necessarily symmetric. The second question is to find exact conditions for an L^p-dissipative operator to generate a contraction semigroup in L^p, without the smoothness assumptions of Sections 2.4 and 2.5.

References

1. Amann, H.: Dual semigroups and second order elliptic boundary value problems. Israel J. Math. **45**, 225–254 (1983)
2. Auscher, P., Barthélemy, L., Bénilan, P., Ouhabaz, El M.: Absence de la L^∞-contractivité pour les semi-groupes associés auz opérateurs elliptiques complexes sous forme divergence. Poten. Anal. **12**, 169–189 (2000)
3. Brezis, H., Strauss, W.A.: Semi-linear second order elliptic equations in L^1. J. Math. Soc. Japan **25**, 565–590 (1973)
4. Cialdea, A., Maz'ya, V.: Criterion for the L^p-dissipativity of second order differential operators with complex coefficients. J. Math. Pures Appl. **84**, 1067–1100 (2005)
5. Cialdea, A., Maz'ya, V.: Criteria for the L^p-dissipativity of systems of second order differential equations. Ric. Mat. **55**, 233–265 (2006)
6. Daners, D.: Heat kernel estimates for operators with boundary conditions. Math. Nachr. **217**, 13–41 (2000)
7. Davies, E.B.: One-Parameter Semigroups. Academic Press, London etc. (1980)
8. Davies, E.B.: Heat Kernels and Spectral Theory. Cambridge Univ. Press, Cambridge (1989)
9. Davies, E.B.: L^p spectral independence and L^1 analyticity. J. London Math. Soc. (2) **52**, 177–184 (1995)
10. Davies, E.B.: Uniformly elliptic operators with measurable coefficients. J. Funct. Anal. **132**, 141–169 (1995)
11. Fattorini, H.O.: The Cauchy Problem. In: Encyclopedia Math. Appl. **18**, Addison-Wesley, Reading, Mass. (1983)
12. Fattorini, H.O.: On the angle of dissipativity of ordinary and partial differential operators. Functional Analysis, Holomorphy and Approximation Theory II. Math. Stud. **86**, 85–111 (1984)
13. Karrmann, S.: Gaussian estimates for second order operators with unbounded coefficients. J. Math. Anal. Appl. **258**, 320–348 (2001)
14. Kovalenko, V., Semenov, Y.: C_0-semigroups in $L^p(\mathbb{R}^d)$ and $\widehat{C}(\mathbb{R}^d)$ spaces generated by the differential expression $\mathrm{d}+b\cdot\nabla$. Theor. Probab. Appl. **35**, 443–453 (1990)
15. Kresin, G.: Sharp constants and maximum principles for elliptic and parabolic systems with continuous boundary data. In: The Maz'ya Anniversary Collection 1m pp. 249–306. Birkhäuser, Basel (1999)
16. Kresin, G., Maz'ya, V.: Criteria for validity of the maximum modulus principle for solutions of linear parabolic systems. Ark. Mat. **32**, 121–155 (1994)

17. Langer, M.: L^p-contractivity of semigroups generated by parabolic matrix differential operators. In: The Maz'ya Anniversary Collection, 1, pp. 307–330. Birkhäuser, Basel (1999)
18. Langer, M., Maz'ya, V.: On L^p-contractivity of semigroups generated by linear partial differential operators. J. Funct. Anal. **164**, 73–109 (1999)
19. Liskevich, V.: On C_0-semigroups generated by elliptic second order differential expressions on L^p-spaces. Differ. Integral Equ. **9**, 811–826 (1996)
20. Liskevich, V.A., Semenov, Yu.A.: Some problems on Markov semigroups, In: Schrödinger Operators, Markov Semigroups, Wavelet Analysis, Operator Algebras. Math. Top. **11**, pp. 163–217. Akademie-Verlag, Berlin (1996)
21. Liskevich, V., Sobol, Z., Vogt, H.: On the L_p-theory of C_0 semigroups associated with second order elliptic operators. II. J. Funct. Anal. **193**, 55–76 (2002)
22. Lumer, G., Phillips, R.S.: Dissipative operators in a Banach space. Pacific J. Math. **11**, 679–698 (1961)
23. Maz'ya, V.G.: The negative spectrum of the higher-dimensional Schrödinger operator (Russian). Dokl. Akad. Nauk SSSR **144**, 721–722 (1962); English transl.: Sov. Math. Dokl. **3**, 808–810 (1962)
24. Maz'ya, V.: On the theory of the higher-dimensional Schrödinger operator (Russian). Izv. Akad. Nauk SSSR Ser. Mat. **28**, 1145–1172 (1964)
25. Maz'ya, V.: Analytic criteria in the qualitative spectral analysis of the Schrödinger operator. Proc. Sympos. Pure Math. **76**, no. 1, 257–288 (2007)
26. Maz'ya, V., Sobolevskii. P.: On the generating operators of semigroups (Russian). Usp. Mat. Nauk **17**, 151–154 (1962)
27. Metafune, G., Pallara, D., Prüss, J., Schnaubelt, R.: L^p-theory for elliptic operators on $\mathbb{R}^d$ with singular coefficients. Z. Anal. Anwe. **24**, 497–521 (2005)
28. Okazawa, N.: Sectorialness of second order elliptic operators in divergence form. Proc. Am. Math. Soc. **113**, 701–706 (1991)
29. Ouhabaz, El M.: Gaussian upper bounds for heat kernels of second order elliptic operators with complex coefficients on arbitrary domains. J. Operator Theory **51**, 335–360 (2004)
30. Ouhabaz, El M.: Analysis of Heat Equations on Domains. Princeton Univ. Press, Princeton, NJ (2005)
31. Pazy, A.: Semigroups of Linear Operators and Applications to Partial Differential Equations. Springer, New York (1983)
32. Robinson, D.W.: Elliptic Operators on Lie Groups. Oxford Univ. Press, Oxford (1991)
33. Sobol, Z., Vogt, H.: On the L_p-theory of C_0 semigroups associated with second order elliptic operators. I. J. Funct. Anal. **193**, 24–54 (2002)
34. Strichartz: R.S.: L^p contractive projections and the heat semigroup for differential forms. J. Funct. Anal. **65**, 348–357 (1986)

Uniqueness and Nonuniqueness in Inverse Hyperbolic Problems and the Black Hole Phenomenon

Gregory Eskin

Abstract We review recent results on inverse problems for the wave equation in an $(n+1)$-dimensional space equipped with a pseudo-Riemannian metric with Lorentz signature and discuss conditions for the existence of black (white) holes for these wave equations. We prove energy type estimates on a finite time interval in the presence of black or white holes. These estimates are used to prove the nonuniqueness in inverse problems.

1 Introduction

Inverse problems form an important field of mathematical research. The paper [14] by V. Kozlov, V. Maz'ya, and A. Fomin is one of the excellent contributionts to the subject. An important development in the theory of inverse problems occured on more than twenty years ago, when M. Belishev suggested a powerful method, called the Boundary Control method, for solving the inverse hyperbolic problem for equations

$$\frac{\partial^2 u}{\partial t^2} + Au = 0,$$

where A is the Laplace–Beltrami operator with time independent coefficients. The method was further developed by different authors (cf., for example, [1, 13] and the references therein). An important role in the solution of the hyperbolic inverse problem is played by the unique continuation theorem due to Tataru [16].

Gregory Eskin
Department of Mathematics, University of California, Los Angeles, CA 90095-1555, USA
e-mail: eskin@math.ucla.edu

A. Laptev (ed.), *Around the Research of Vladimir Maz'ya III: Analysis and Applications*,
International Mathematical Series 13, DOI 10.1007/978-1-4419-1345-6_4,

In [3, 4], the author, based on some ideas of the Boundary Control method, proposed a new approach to the study of inverse hyperbolic problems. This approach allowed one to solve some inverse hyperbolic problems that were not covered by the Boundary Control method. In particular, the case of hyperbolic equations with time-dependent coefficients was considered in [5], and the case of the hyperbolic equation with general pseudo-Riemannian time-independent metric was treated in [6]. We describe the main results of [6, 7] in the following sections.

An interesting phenomenon discussed in [7] is the appearance of black holes, called *artificial* (acoustic or optical) black holes in order to distinguish them from black holes in the general relativity theory. Artificial black holes attracted a great interest of physicists (cf. [15, 17] and the references therein) because the physisists believe that the study of artificial black holes will be useful for understanding the black holes in the universe.

In the last two sections, we prove energy type estimates on a finite time interval in the presence of black or white holes. These estimates are used to prove nonuniqueness in inverse problems.

2 Inverse Hyperbolic Problems

Let Ω be a bounded domain in $\mathbf{R}^n$ with smooth boundary $\partial\Omega$, and let $\Gamma \subset \partial\Omega$ be an open subset of $\partial\Omega$. Consider the following hyperbolic equation in the cylinder $\Omega \times \mathbf{R}$:

$$\sum_{j,k=0}^{n} \frac{1}{\sqrt{(-1)^n g(x)}} \frac{\partial}{\partial x_j} \left(\sqrt{(-1)^n g(x)} g^{jk}(x) \frac{\partial u(x_0, x)}{\partial x_k} \right) = 0, \tag{2.1}$$

where $x = (x_1, \dots, x_n) \in \overline{\Omega}$, $x_0 \in \mathbf{R}$ is the time variable, the coefficients in (2.1) are smooth and independent of x_0, $[g_{jk}(x)]_{j,k=0}^n = ([g^{jk}(x)]_{j,k=0}^n)^{-1}$ is a pseudo-Riemannian metric with Lorentz signature, i.e., the quadratic form $\sum_{j,k=0}^n g^{jk}(x)\xi_j\xi_k$ has signature $(1, -1, -1, \dots, -1)$ for all $x \in \overline{\Omega}$. Here, $g(x) = (\det[g^{jk}(x)]_{j,k=0}^n)^{-1}$. Note that $(-1)^n g(x) > 0$ for all $x \in \overline{\Omega}$.

We assume that

$$g^{00}(x) > 0, \quad x \in \overline{\Omega}, \tag{2.2}$$

i.e., $(1, 0, \dots, 0)$ is not a characteristic direction, and that

$$\sum_{j,k=1}^{n} g^{jk}(x)\xi_j\xi_k < 0 \quad \forall(\xi_1, \dots, \xi_n) \neq (0, \dots, 0), \forall x \in \overline{\Omega}, \tag{2.3}$$

i.e., the quadratic form (2.3) is negative definite. Note that (2.3) is equivalent to the condition

$$g_{00}(x) > 0, \quad x \in \overline{\Omega}, \tag{2.4}$$

i.e., $(1,0,\dots,0)$ is a time-like direction.

We consider the initial-boundary value problem for Equation (2.1) in the cylinder $\Omega \times \mathbf{R}$:

$$u(x_0,x) = 0, \quad x \in \Omega,\ x_0 \ll 0, \tag{2.5}$$

$$u(x_0,x)\,|_{\partial\Omega\times\mathbf{R}} = f(x_0,x'), \quad x' \in \partial\Omega, \tag{2.6}$$

where $f(x_0,x')$ has compact support in $\partial\Omega \times \mathbf{R}$.

Let Λf be the Dirichlet-to-Neumann (DN) operator, i.e.,

$$\Lambda f = \sum_{j,k=0}^{n} g^{jk}(x)\frac{\partial u}{\partial x_j}\nu_k(x)\Big(-\sum_{p,r=1}^{n} g^{jk}(x)\nu_p\nu_r\Big)^{-\frac{1}{2}}\Big|_{\partial\Omega\times\mathbf{R}}, \tag{2.7}$$

where $\nu_0 = 0, (\nu_1,\dots,\nu_n)$ is the outward unit normal vector to $\partial\Omega \subset \mathbf{R}^n$, $u(x_0,x)$ is a solution of the problem (2.1), (2.5), (2.6).

Consider a smooth change of variables of the form:

$$\begin{aligned} \widehat{x}_0 &= x_0 + a(x), \\ \widehat{x} &= \varphi(x), \end{aligned} \tag{2.8}$$

where $\varphi(x)$ is a diffeomorphism from $\overline{\Omega}$ onto some domain $\overline{\widehat{\Omega}}$ such that $\overline{\Gamma} \subset \partial\widehat{\Omega}$, $\varphi(x) = x$ on $\overline{\Gamma}$, and $a(x) = 0$ on $\overline{\Gamma}$. Note that (2.8) is the identity map on $\overline{\Gamma} \times \mathbf{R}$. Note also that the map (2.8) transforms (2.1) to an equation of the same form in $\widehat{\Omega} \times \mathbf{R}$.

Theorem 2.1 (cf. [6]). *Let L and $\widehat{L}$ be two operators of the form (2.1) in $\Omega \times \mathbf{R}$ and $\widehat{\Omega} \times \mathbf{R}$ respectively. Consider initial-boundary value problems of the form (2.5), (2.6) for L and $\widehat{L}$. Suppose that $\Lambda f = \widehat{\Lambda} f$ on $\Gamma \times \mathbf{R}$ for all $f \in C_0^\infty(\Gamma \times \mathbf{R})$, where Λ and $\widehat{\Lambda}$ are DN operators for L and $\widehat{L}$ respectively. Suppose that the conditions (2.2) and (2.3) hold for L and $\widehat{L}$. Then there exists a map of the form (2.8) such that*

$$[\widehat{g}^{jk}(\widehat{x})]_{j,k=0}^{n} = J^T(x)[g^{jk}(x)]_{j,k=0}^{n}J(x), \tag{2.9}$$

where $([\widehat{g}^{jk}(\widehat{x})]_{j,k=0}^{n})^{-1}$ is the metric tensor for $\widehat{L}$ and $J(x)$ is the Jacobi matrix of (2.8).

Remark 2.1. It is enough to know the DN operator on $\Gamma \times (0,T_0)$ for some $T_0 > 0$ instead of $\Gamma \times \mathbf{R}$. More precisely, let T_+ be the smallest number such that

$$D_+(\overline{\Gamma} \times \{x_0 = 0\}) \supset \overline{\Omega} \times \{x_0 = T_+\},$$

where $D_+(\overline{\Gamma} \times \{x_0 = 0\})$ is the forward domain of influence of $\overline{\Gamma} \times \{x_0 = 0\}$ corresponding to (2.1). Similarly, let T_- be the smallest number such that

$$D_-(\overline{\Gamma} \times \{x_0 = T_-\}) \supset \overline{\Omega} \times \{x_0 = 0\},$$

where $D_-(\overline{\Gamma} \times \{x_0 = T_-\})$ is the backward domain of influence of $\overline{\Gamma} \times \{x_0 = T_-\}$. If $T_0 > T_- + T_+$, then $\Lambda = \widehat{\Lambda}$ on $\Gamma \times (0, T_0)$ implies (2.9), i.e., the metrics $[g_{jk}(x)]$ and $[\widehat{g}_{jk}(\widehat{x})]$ are isometric.

3 Equation of Light Propagation in a Moving Dielectric Medium

In this section, we apply Theorem 2.1 to the equation of light propagation in a moving medium.

It was discovered by Gordon [10] that the light propagation in a moving medium is described by a hyperbolic equation of the form (2.1) with the metric tensor

$$g^{jk}(x) = \eta^{jk} + (n^2(x) - 1)v^j(x)v^k(x), \tag{3.1}$$

$0 \leqslant j, k \leqslant n$, $n = 3$, where $[\eta_{jk}] = [\eta^{jk}]^{-1}$ is the Lorentz metric tensor $\eta^{jk} = 0$ if $j \neq k$, $\eta^{00} = 1$, $\eta^{jj} = -1$, $1 \leqslant j \leqslant n$, $x_0 = t$ is the time, $n(x) = \sqrt{\varepsilon(x)\mu(x)}$ is the refraction index, $w(x) = (w_1(x), w_2(x), w_3(x))$ is the velocity of flow,

$$v^{(0)} = \left(1 - \frac{|w|^2}{c^2}\right)^{-\frac{1}{2}}, \quad v^{(j)} = \left(1 - \frac{|w|^2}{c^2}\right)^{-\frac{1}{2}} \frac{w_j(x)}{c}, \quad 1 \leqslant j \leqslant 3,$$

is the four-velocity field of the flow, c is the speed of light in the vacuum. Equation (2.1) with metric (3.1) is called the *Gordon equation.*

Let Ω be a smooth domain in $\mathbf{R}^n$ of the form $\Omega = \Omega_0 \setminus \bigcup_{j=1}^m \overline{\Omega}_j$, where Ω_0 is simply connected, Ω_j, $1 \leqslant j \leqslant m$, are smooth domains called *obstacles*, $\overline{\Omega}_j \subset \Omega_0$, $1 \leqslant j \leqslant m$, $\overline{\Omega}_j \bigcap \overline{\Omega}_k = \varnothing$ if $j \neq k$.

For the Gordon equation we consider the initial-boundary value problem

$$\begin{aligned} &u(x_0, x) = 0, \quad x_0 \ll 0,\ x \in \Omega, \\ &u(x_0, x)|_{\partial\Omega_j \times \mathbf{R}} = 0, \quad 1 \leqslant j \leqslant m, \\ &u(x_0, x)|_{\partial\Omega_0 \times \mathbf{R}} = f(x_0, x), \end{aligned}$$

i.e., $\partial\Omega_0 = \Gamma$ in the notation of Theorem 2.1.

Note that the condition (2.2) is always satisfied since

$$g^{00} = 1 + (n^2 - 1)(v^0)^2 > 0.$$

The condition that any direction $(0, \xi_1, \ldots, \xi_n)$ is not characteristic (cf. (2.3)) holds if

$$|w(x)|^2 < \frac{c^2}{n^2(x)}. \tag{3.2}$$

We impose some restrictions on the flow $w(x)$. Let $x = x(s)$ be the trajectory of the flow, i.e.,

$$\frac{dx}{ds} = w(x(s)), \quad 0 \leqslant s \leqslant 1,$$

where $w(x(s)) \neq 0$ for $0 \leqslant s \leqslant 1$. We assume the following condition:

(A) The trajectories that start and end on $\partial\Omega_0$ or are closed curves in Ω are dense in $\overline{\Omega}$.

Theorem 3.1 (cf. [7]). *Let $[g_{jk}(x)]_{j,k=0}^n$ and $[\widehat{g}_{jk}(y)]_{j,k=0}^n$ be two Gordon metrics in domains Ω and $\widehat{\Omega}$ respectively. Consider two initial-boundary value problems of the form (3.2) in $\Omega \times \mathbf{R}$ and $\widehat{\Omega} \times \mathbf{R}$ respectively, where $\Omega = \Omega_0 \setminus \bigcup_{j=1}^m \overline{\Omega}_j$ and $\widehat{\Omega} = \Omega_0 \setminus \bigcup_{j=1}^{\widehat{m}} \overline{\widehat{\Omega}}_j$. Assume that the refraction indexes n and $\widehat{n}$ are constant and the flow $w(x)$ satisfies condition (A). Assume also that (3.2) holds for both metrics. Then $\Lambda = \widehat{\Lambda}$ on $\partial\Omega_0 \times \mathbf{R}$ implies that $\widehat{n} = n$, $\widehat{\Omega} = \Omega$, and the flows $\widehat{w}(x)$ and $w(x)$ are equal.*

4 Light Propagation in a Slowly Moving Medium

In this case, one drops the terms of order $\frac{|w|^2}{c^2}$. Then the metric tensor has the form:

$$g^{jk}(x) = \eta^{jk}, \quad 1 \leqslant j,k \leqslant n, \ n = 3, \tag{4.1}$$
$$g^{00}(x) = n^2(x), \quad g^{0j}(x) = g^{j0}(x) = (n^2(x) - 1)\frac{w_j(x)}{c}, \quad 1 \leqslant j \leqslant n.$$

The wave equation with metric (4.1) describes the light propagation in a slowly moving medium. We shall see that the inverse problem for such an equation exhibits some nonuniqueness.

Denote $v_j(x) = g^{0j} = g^{j0}$. We say that a flow $v = (v_1, \dots, v_n)$ is a *gradient flow* if $v(x) = \frac{\partial b(x)}{\partial x}$, where $b(x) \in C^\infty(\overline{\Omega})$, $b(x) = 0$ on $\partial\Omega_0$.

Theorem 4.1 (cf. [6]). *Consider two initial-boundary value problems in domains $\Omega \times \mathbf{R}$ and $\widehat{\Omega} \times \mathbf{R}$ for operators of the form (2.1) with metrics $[g^{jk}(x)]$ and $[\widehat{g}^{jk}(\widehat{x})]$ of the form (4.1). Assume that* DN *operators Λ and $\widehat{\Lambda}$ are equal on $\partial\Omega_0 \times \mathbf{R}$. Assume that there exists an open connected dense set $O \subset \Omega$ such that $v(x)$ does not vanish on O. Then $\widehat{\Omega} = \Omega$, $\widehat{n}(x) = n(x)$, and $\widehat{v}(x) = v(x)$ if $v(x)$ is not a gradient flow. Otherwise, there are two solutions of the inverse problem:*

$$\widehat{v}(x) = v(x), \quad \widehat{v}(x) = -v(x).$$

Remark 4.1 (cf. [6]). Suppose that an open set O, where $v(x) \neq 0$, consists of several open components $O_1, \dots, O_r$. Suppose that there exists $b_j(x) \in C^\infty(\overline{\Omega})$ such that $b_j(x) = 0$ on $\partial\Omega_0$, $\dfrac{\partial b_j}{\partial x} = v(x)$ on O_j, $b_j = 0$ in $\overline{\Omega} \setminus O_j$,

$j = 1, 2, \ldots, r$. Then we have 2^r solutions of the inverse problem; moreover, each of these solutions is equal to either $\frac{\partial b_j}{\partial x}$ or $-\frac{\partial b_j}{\partial x}$ on O_j.

5 Artificial Black Holes

Let $S(x) = 0$ be a smooth closed surface in $\mathbf{R}^n$ such that the surface $S \times \mathbf{R} \subset \mathbf{R}^{n+1}$ is a characteristic surface for Equation (2.1), i.e.,

$$\sum_{j,k=1}^{n} g^{jk}(x) S_{x_j}(x) S_{x_k}(x) = 0 \quad \text{if } S(x) = 0. \tag{5.1}$$

Let Ω_{int} and Ω_{ext} be the interior and exterior of S respectively. A domain $\Omega_{\text{int}} \times \mathbf{R}$ is called an *artificial black hole* if no signal emanating from it can reach $\Omega_{\text{ext}} \times \mathbf{R}$. Similarly, $\Omega_{\text{int}} \times \mathbf{R}$ is an *artificial white hole* if no signal from $\Omega_{\text{ext}} \times \mathbf{R}$ can penetrate the interior of $S \times \mathbf{R}$.

Let y be any point of S, i.e., $S(y) = 0$.

Lemma 5.1. *If $S \times \mathbf{R}$ is a characteristic surface, then*

$$\sum_{j=1}^{n} g^{j0}(y) S_{x_j}(y) \neq 0.$$

Proof. Since (2.1) is hyperbolic, the equation

$$\sum_{j,k=0}^{n} g^{jk}(y) \xi_j \xi_k = 0$$

has two distinct real roots $\xi_0^{(1)}(\xi)$ and $\xi_0^{(2)}(\xi)$ for any $\xi = (\xi_1, \ldots, \xi_n) \neq 0$. Taking $\xi = S_x(y)$ and using (5.1), we get

$$g^{00}(y)\xi_0^2 + 2\sum_{j=1}^{n} g^{0j}(y)\xi_0 S_{x_j}(y) = 0.$$

Therefore,

$$\xi_0^{(1)} = 0, \xi_0^{(2)} = -2(g^{00}(y))^{-1} \sum_{j=1}^{n} g^{jk}(y) S_{x_j}(y) \neq 0. \qquad \square$$

From Lemma 5.1 it follows that either

$$\sum_{j=1}^{n} g^{j0}(y) S_{x_j}(y) > 0, \quad S(y) = 0 \tag{5.2}$$

or

$$\sum_{j=1}^{n} g^{j0}(y) S_{x_j}(y) < 0, \quad S(y) = 0. \tag{5.3}$$

Denote by $K^+(y) \subset \mathbf{R}^{n+1}$ the half-cone

$$K^+(y) = \Big\{(\xi_0, \xi_1, \ldots, \xi_n) : \sum_{j,k=0}^{n} g^{jk}(y)\xi_j\xi_k > 0\Big\} \tag{5.4}$$

containing $(1, 0, \ldots, 0)$ and by $K_+(y)$ the dual half-cone

$$K_+(y) = \Big\{(\dot{x}_0, \dot{x}_1, \ldots, \dot{x}_n) \in \mathbf{R}^{n+1} : \sum_{j,k=0}^{n} g_{jk}(y)\dot{x}_j\dot{x}_k > 0, \dot{x}_0 > 0\Big\}. \tag{5.5}$$

Since $K^+(y)$ and $K_+(y)$ are dual, we have

$$\sum_{j=0}^{n} \dot{x}_j \xi_j > 0 \tag{5.6}$$

for any $(\dot{x}_0, \ldots, \dot{x}_n) \in K_+(y)$ and any $(\xi_0, \ldots, \xi_n) \in K^+(y)$. We choose $S_x(y)$ to be the outward normal to S. Assuming (5.2), we have $(\varepsilon, S_x(y)) \in K^+(y)$ for any $\varepsilon > 0$. Using (5.6) and taking the limit as $\varepsilon \to 0$, we find that $\sum_{j=1}^{n} \dot{x}_j S_{x_j}(y) \geqslant 0$ for all $(\dot{x}_0, \ldots, \dot{x}_n) \in K_+(y)$, i.e., $\overline{K}_+(y)$ is contained in the half-space

$$\overline{P}_+(y) = \Big\{(\alpha_0, \alpha_1, \ldots, \alpha_n) : \sum_{j=1}^{n} \alpha_j S_{x_j}(y) \geqslant 0\Big\}.$$

In particular, $K_+(y)$ is contained in the open half-space P_+.

A ray $x_0 = x_0(s)$, $x = x(s)$, $s \geqslant 0$, is called a *forward time-like* if

$$\Big(\frac{dx_0(s)}{ds}, \frac{dx(s)}{ds}\Big) \in K_+(x(s))$$

for all s. As is known [2], the domain of influence of a point (y_0, y) is the closure of all forward time-like rays starting at (y_0, y). Therefore, since $K_+(y)$ is contained in the open half-space $P_+(y)$ for all $(y_0, y) \in S \times \mathbf{R}$, the domain of influence of $\Omega_{\text{ext}} \times \mathbf{R}$ is contained in $\overline{\Omega}_{\text{ext}} \times \mathbf{R}$, i.e., $\Omega_{\text{int}} \times \mathbf{R}$ is a white hole because no signal from $\Omega_{\text{ext}} \times \mathbf{R}$ may reach $\Omega_{\text{int}} \times \mathbf{R}$.

Consider the case where (5.3) holds. Then $(\varepsilon, -S_x(y)) \in K^+(y)$ for any $\varepsilon > 0$, $y \in S$. Therefore, passing to the limit as $\varepsilon \to 0$, we see that $K_+(y)$ is contained in the half-space

$$P_-(y) = \Big\{(\alpha_0, \alpha_1, \dots, \alpha_n) : \sum_{j=1}^{n} \alpha_j S_{x_j}(y) < 0\Big\}.$$

Since $S_x(y)$ is the outward normal to S, $y \in S$, the domain of influence of $\Omega_{\text{int}} \times \mathbf{R}$ is contained in $\overline{\Omega}_{\text{int}} \times \mathbf{R}$, i.e., $\Omega_{\text{int}} \times \mathbf{R}$ is a black hole. We have proved the following theorem.

Theorem 5.1 (cf. [7]). *Let $S \times \mathbf{R}$ be a characteristic surface for* (2.1). *Then $\Omega_{\text{int}} \times \mathbf{R}$ is a white hole if* (5.2) *holds and a black hole if* (5.3) *holds.*

In Sections 8 and 9, we give another proof of this theorem.

Let $\Delta(x) = \det[g^{jk}(x)]_{j,k=1}^n$. We assume that $S_\Delta = \{x : \Delta(x) = 0\}$ is a smooth closed surface. Let Ω_{int} be the interior of S_Δ and Ω_{ext} the exterior of S_Δ. We assume that $\Delta(x) > 0$ in $\overline{\Omega} \bigcap \Omega_{\text{ext}}$ and $\Delta(x) < 0$ in $\overline{\Omega} \bigcap \Omega_{\text{int}}$. Borrowing the terminology from the general relativity, we call S_Δ an *ergosphere*. If $S_\Delta \times \mathbf{R}$ is a characteristic surface for (2.1), then $\Omega_{\text{int}} \times \mathbf{R}$ is a black hole if (5.3) holds or a white hole if (5.2) holds. In the case of the Gordon equation, the ergosphere has the form

$$S_\Delta = \Big\{x : |w(x)|^2 = \frac{c^2}{n^2}\Big\},$$

and

$$g^{0j}(x) = \frac{(n^2(x) - 1)cw_j(x)}{c^2 - |w|^2}.$$

If $S_\Delta \times \mathbf{R}$ is a characteristic surface, then the normal to S_Δ is collinear to $w(y)$ and $\Omega_{\text{int}} \times \mathbf{R}$ is a black hole if $w(y)$ is pointed inside Ω_{int} and a white hole if $w(y)$ is pointed inside Ω_{ext}.

Note that the black or white holes with boundary $S_\Delta \times \mathbf{R}$ are not stable: If we perturb slightly the metric $[g_{jk}(x)]_{j,k=0}^n$, then the ergosphere changes slightly. However, it is not necessary remain a characteristic surface, and the black or the white hole will disappear (cf. [8]). In the next section, we find stable black and white holes.

6 Stable Black and White Holes

Consider the case $n = 2$, i.e., the case of two spatial variables $x = (x_1, x_2)$. Let S_Δ be an ergosphere, i.e.,

$$\Delta(x) = g^{11}(x)g^{22}(x) - (g^{12}(x))^2 = 0$$

on S_Δ. Suppose that S_Δ is a closed smooth curve. Let S_1 be a smooth closed curve inside S_Δ. Denote by Ω_e the domain between S_Δ and S_1 and assume that $\Omega_e \subset \Omega$. We call Ω_e the *ergoregion*. We assume that $\Delta(x) < 0$ on $\overline{\Omega}_e \backslash S_\Delta$

and S_Δ is not characteristic at any $y \in S_\Delta$, i.e.,

$$\sum_{j,k=1}^{2} g^{jk}(y)\nu_j(y)\nu_k(y) \neq 0 \quad \forall y \in S_\Delta, \tag{6.1}$$

where $(\nu_1(y), \nu_2(y))$ is the normal to S_Δ. Since $\Delta(x) < 0$ in Ω_e, we can define (locally) two families of characteristic curves $S^\pm(x) = \text{const}$ satisfying

$$\sum_{j,k=1}^{2} g^{jk}(x)S^\pm_{x_j}(x)S^\pm_{x_k}(x) = 0, \quad x \in \Omega_e. \tag{6.2}$$

As is shown in [7], there are two families $f^\pm(x)$ of vector fields such that $f^\pm(x) \neq (0,0)$ for all $x \in \overline{\Omega}_e$, $f^+(x) \neq f^-(x)$ for $x \in \overline{\Omega}_e \setminus S_\Delta$, $f^+(y) - f^-(y)$ on S_Δ, and $f^\pm(x)$ are tangent to $S^\pm(x) = \text{const}$.

Consider two systems of differential equations

$$\frac{d\widehat{x}^+(\sigma)}{d\sigma} = f^+(\widehat{x}^+(\sigma)), \quad \sigma \geqslant 0, \; \widehat{x}^+(0) = y \in S_\Delta, \tag{6.3}$$

$$\frac{d\widehat{x}^-(\sigma)}{d\sigma} = f^-(\widehat{x}^-(\sigma)), \quad \sigma \geqslant 0, \; \widehat{x}^+(0) = y \in S_\Delta. \tag{6.4}$$

Note that $x = \widehat{x}^\pm(\sigma) = y$, $\sigma \geqslant 0$, are parametric equations of characteristics (6.2). From (6.1) it follows that $f^+(y) = f^-(y)$ is not tangent to S_Δ for all $y \in S_\Delta$. Since the rank of $[g^{jk}(y)]_{j,k=1}^2$ on S_Δ is 1, one can choose a smooth vector $b(y)$, $y \in S_\Delta$, such that

$$\sum_{k=1}^{2} g^{jk}(y)b_k(y) = 0, \quad j = 1, 2. \tag{6.5}$$

Note that $f^\pm(y) \cdot b(y) = 0$. We choose $f^\pm(y)$ to be pointed inside S_Δ.

Consider the equations for the null-bicharacteristics:

$$\frac{dx_j(s)}{ds} = 2\sum_{k=0}^{2} g^{jk}(x(s))\xi_k(s), \quad x_j(0) = y_j, \quad 0 \leqslant j \leqslant 2, \tag{6.6}$$

$$\frac{d\xi_p(s)}{ds} = -\sum_{j,k=0}^{2} g^{jk}_{x_p}((s))\xi_j(s)\xi_k(s), \quad \xi_p(0) = \eta_p, \quad 0 \leqslant p \leqslant 2. \tag{6.7}$$

Here, $x(s) = (x_1(s), x_2(s))$. Since $g^{jk}(x)$ are independent of x_0, we have $\xi_0(s) = \eta_0$ for all s and choose $\eta_0 = 0$.

The bicharacteristic (6.6), (6.7) is a null-bicharacteristic if

$$\sum_{j,k=0}^{2} g^{jk}(y)\eta_j\eta_k = 0.$$

Choosing $\eta_j = \pm b(y)$, $1 \leqslant j \leqslant 2$, $\eta_0 = 0$, we get two null-bicharacteristics $x_0 = x_0^{\pm}(s)$, $x = x^{\pm}(s)$, $\xi_0 = 0$, $\xi = \xi^{\pm}(s)$ such that the projection of these null-bicharacteristics on the (x_1, x_2)-plane coincide with solutions $x = \widehat{x}^{\pm}(\sigma)$ of the systems (6.3), (6.4), i.e., $x = \widehat{x}^{\pm}(\sigma), \sigma \geqslant 0$ and $x = x^{\pm}(s)$ are equal after a re-parametrization $\sigma = \sigma^{\pm}(s)$, $\dfrac{d\sigma^{\pm}(s)}{ds} > 0$ for $s > 0$. We consider forward null-bicharacteristics, i.e., $\dfrac{dx_0^{\pm}(s)}{ds} > 0$ for all s. Therefore, one can take the time variable x_0 as a parameter on $x = x^{\pm}(\sigma)$.

The key observation in [7] is that for one of $x = \widehat{x}^{\pm}(\sigma)$, say for $x = \widehat{x}^{+}(\sigma)$, $\sigma = \sigma^{+}(s^{+}(x_0))$ increases when x_0 increases, and for $x = \widehat{x}^{-}(\sigma)$, $\sigma = \sigma^{-}(s^{-}(x_0))$ decreases when x_0 increases.

Now we impose conditions on S_1 that will guarantee the existence of black and white holes in Ω_e. We assume that S_1 is not characteristic.

Let $N(y)$ be the outward unit normal to S_1, $y \in S_1$. Suppose that either

(a) $\overline{K}_+(y)$ is contained in the open half-space

$$Q_+ = \{(\alpha_0, \alpha_1, \alpha_2) : (\alpha_0, \alpha_1, \alpha_2) \cdot (0, N(y)) > 0$$

or

(b) $\overline{K}_+(y)$ is contained in the open half-space

$$Q_- = \{(\alpha_0, \alpha_1, \alpha_2) : (\alpha_0, \alpha_1, \alpha_2) \cdot (0, N(y)) < 0.$$

Remark 6.1. There are equivalent forms of conditions (a) and (b). Since

$$\sum_{j,k=0}^{n} g_{jk}(x(s))\frac{dx_j(s)}{ds}\frac{dx_k(s)}{ds} = 0$$

for null-bicharacteristics, we have

$$\left(\frac{dx_0(s)}{ds}, \frac{dx_1(s)}{ds}, \frac{dx_2(s)}{ds}\right) \in \overline{K}_+(y)$$

for forward null-bicharacteristic if $x(s_1) = y \in S_1$. Therefore, condition (a) is equivalent to the following condition.

(a_1) The projection on (x_1, x_2)-plane of all forward null-characteristics passing through $y \in S_1$ leave Ω_e when x_0 increases.

Further, (a_1) is equivalent to the following simpler condition.

Let $x = x^{\pm}(s)$ be the projection onto the (x_1, x_2)-plane of two forward null-bicharacteristics such that $x = x^{\pm}(s)$ are the parametric equations of the characteristics $S^{\pm}(x) = \text{const}$, i.e., $x = x^{\pm}(s)$ are solutions of the differential equations (6.3), (6.4) after re-parametrization. Assume that

$$(a_2) \quad \frac{dx^{\pm}(s_1)}{ds} \cdot N(y) > 0 \text{ if } x^{\pm}(s_1) = y.$$

Condition (a_2) follows from (a_1). The converse implication is also true since the set of directions of the projections of all forward null-bicharacteristics passing through y is bounded by $\frac{dx^{+}(s_1)}{ds}$ and $\frac{dx^{-}(s_1)}{ds}$.

Conditions (b_1) and (b_2) are similar to (a_1) and (a_2) when the sign of the inner product in (a) is negative.

Theorem 6.1 (cf. [7]). *Let $\partial\Omega_e = S_\Delta \cup S_1$, where S_Δ is an ergosphere, i.e., $\Delta(y) = 0$ on S_Δ. Suppose that* (6.1) *holds on S_Δ and either* (a) *or* (b) *holds on S_1. Then there exists a closed Jordan curve $S_0(x) = 0$ inside Δ_e such that $S_0 \times \mathbf{R}$ is the boundary of either a black hole or a white hole.*

The proof of Theorem 6.1 is based on the Poincaré–Bendixson theorem [11]. Suppose that (a) holds. Then the solution of (6.4) cannot reach S_1. Indeed, suppose that $\widehat{x}^{-}(\sigma_1) = y_1 \in S_1$ for some $\sigma_1 > 0$. Then $\widehat{x}^{-}(\sigma)$ leaves Ω_e if $\sigma > \sigma_1$. On the other hand, when σ increases, x_0 decreases. Therefore, $x = \widehat{x}^{-}(\sigma^{-}(x_0))$ leaves Ω_e when x_0 decreases, which contradicts (a_1). Since $x = \widehat{x}^{-}(\sigma)$ never reaches S_1, the limit set of the trajectory $x = \widehat{x}^{-}(\sigma)$ is contained inside Ω_e. Then, by the Poincaré–Bendixson theorem, there exists a limit cycle $S_0(x) = 0$, i.e., a Jordan curve that is a periodic solution of

$$\frac{d\widehat{x}}{d\sigma} = f^{-}(\widehat{x}(\sigma)).$$

Therefore, $S_0 \times \mathbf{R}$ is a characteristic surface and is the boundary of a black hole or a white hole. If (b) holds, then the solution of (6.3) never reach S_1. Therefore, again by the Poincaré–Bendixson theorem, there exists a black hole or a white hole.

Applying Theorem 6.1 to the Gordon equation, we get the following assertion.

Theorem 6.2. *Let S_Δ be an ergosphere, i.e., $|w|^2 = \frac{c^2}{n^2(x)}$. Suppose that $w(x)$ is not collinear with the normal to S_Δ for any $x \in S_\Delta$. Suppose that either*

$$(n^2(x) - 1)^{\frac{1}{2}}(v(x) \cdot N(x)) > 1 \quad \text{on} \quad S_1 \tag{6.8}$$

or

$$(n^2(x) - 1)^{\frac{1}{2}}(v(x) \cdot N(x)) < -1 \quad \text{on} \quad S_1, \tag{6.9}$$

where

$$v(x) = \left(1 - \frac{|w|^2}{c^2}\right)^{-\frac{1}{2}} \frac{w(x)}{c}$$

and $N(x)$ is the outward unit normal to S_1.

Then there exists a limit cycle $S_0(x) = 0$ and $S_0 \times \mathbf{R}$ is the boundary of a black hole or a white hole.

Remark 6.2. Note that black or white holes obtained by Theorems 6.1 and 6.2 are stable since the assumptions remain valid under slight deformations of the metric.

7 Rotating Black Holes. Examples

Example 7.1 (acoustic black hole [17]). Consider a fluid flow with velocity

$$v = (v^1, v^2) = \frac{A}{r}\widehat{r} + \frac{B}{r}\widehat{\theta}, \tag{7.1}$$

where

$$r = |x|, \quad \widehat{r} = \left(\frac{x_1}{|x|}, \frac{x_2}{|x|}\right), \quad \widehat{\theta} = \left(-\frac{x_2}{|x|}, \frac{x_1}{|x|}\right),$$

A and B are constants. The inverse of the metric tensor has the form

$$\begin{aligned} g^{00} &= \frac{1}{\rho c}, \quad g^{oj} = g^{j0} = \frac{1}{\rho c}v^j, \quad 1 \leqslant j \leqslant 2, \\ g^{jk} &= \frac{1}{\rho c}(-c^2\delta_{ij} + v^j v^k), \quad 1 \leqslant j, \ k \leqslant 2, \end{aligned} \tag{7.2}$$

where c is the sound speed and ρ is the density.

Consider the case $A > 0$, $B > 0$. Assume that $\rho = c = 1$. Then $r = \sqrt{A^2 + b^2}$ is an ergosphere. Consider the domain $\Omega_e = \{r_1 \leqslant r \leqslant \sqrt{A^2 + B^2}\}$, where $r_1 < A$. In the polar coordinates (r, θ), the differential equations (6.3) and (6.4) have the form

$$\frac{dr}{ds} = A^2 - r^2, \quad \frac{d\theta}{ds} = \frac{AB}{r} + \sqrt{A^2 + B^2 - r^2} \tag{7.3}$$

and

$$\frac{dr}{ds} = -1, \quad \frac{d\theta}{ds} = \frac{1 - \frac{B^2}{r^2}}{\frac{AB}{r} + \sqrt{A^2 + B^2 - r^2}}. \tag{7.4}$$

Then $r = A$ is a limit cycle and $\{r = A\} \times \mathbf{R}$ is the boundary of a white hole (cf. [7]).

Example 7.2. Consider a fluid flow with velocity

$$v = A(r)\hat{r} + B(r)\hat{\theta},$$

where $r_1 \leqslant r \leqslant r_0$, $A(r)$, $B(r)$ are smooth, $B(r) > 0$, $A^2(r_0) + B^2(r_0) = 1$, $A^2(r)+B^2(r) > 1$ on $[r_1, r_0)$, $A(r)+1$ has simple zeros $\alpha_1, \dots, \alpha_{m_1}$ on (r_1, r_0), $A(r) - 1$ has simple zeros $\beta_1, \dots, \beta_{m_2}$ on (r_1, r_0), $\alpha_j \neq \beta_k$ for all j, k, and $|A(r_1)| > 1$. Here, $r = r_0$ is an ergosphere. In the polar coordinates (r, θ), the differential equations (6.3) and (6.4) have the form

$$\frac{dr}{ds} = A(r) - 1, \quad \frac{d\theta}{ds} = \frac{A(r)B(r) + \sqrt{A^2(r) + B^2(r) - 1}}{A(r) + 1} \tag{7.5}$$

and

$$\frac{dr}{ds} = A(r) + 1, \quad \frac{d\theta}{ds} = \frac{A(r)B(r) - \sqrt{A^2(r) + B^2(r) - 1}}{A(r) - 1}. \tag{7.6}$$

Here, $r = \alpha_j$, $1 \leqslant j \leqslant m_1$, and $r = \beta_n$, $1 \leqslant k \leqslant m_2$, are limit cycles and there are $m_1 + m_2$ black and white holes.

Axially symmetric metrics. Consider Equation (2.1) in $\Omega \times \mathbf{R}$, where Ω is a three-dimensional domain. Let (r, θ, φ) be the spherical coordinates in $\mathbf{R}^3$. Suppose that g^{jk} are independent of φ.

Consider a characteristic surface S independent of φ and x_0, i.e., S depends on r and θ only. Then S satisfies the equation

$$a^{11}(r,\theta)\left(\frac{\partial S}{\partial r}\right)^2 + 2a^{12}(r,\theta)\frac{\partial S}{\partial r}\frac{\partial S}{\partial \theta} + a^{22}(r,\theta)\left(\frac{\partial S}{\partial \theta}\right)^2 = 0. \tag{7.7}$$

We assume that $a^{ij}(r,\theta)$ are also independent of φ, $1 \leqslant j, k \leqslant 2$.

Consider Equation (7.7) in a two-dimensional domain ω, where $\delta_1 \leqslant r \leqslant \delta_2$, $0 < \delta_3 < \theta < \pi - \delta_4$ if $(r, \theta) \in \omega$. Here, $\delta_j > 0$, $1 \leqslant j \leqslant 4$.

Assuming that ω and $a^{jk}(r, \theta)$, $1 \leqslant j, k \leqslant 2$, satisfy the assumption of Theorem 6.1, we can prove the existence of black or white holes with boundary $S_0 \times S^1 \times \mathbf{R}$, where $\varphi \in S^1$, $x_0 \in \mathbf{R}$, and S_0 is a Jordan curve in ω.

Such black (white) holes are called *rotating black (white) holes.*

8 Black Holes and Inverse Problems I

In this section, we consider the black or white hole bounded by $S_\Delta \times \mathbf{R}$, where S_Δ is an ergosphere. Suppose that Ω_{int} is a black hole, i.e., (5.3) holds. Let L be the operator (2.1). Consider $u(x_0, x)$ in $(\Omega \bigcap \Omega_{\text{ext}}) \times (0, T)$ such that

$$Lu = f, \quad (x_0, x) \in (\Omega \bigcap \Omega_{\text{ext}}) \times (0, T), \tag{8.1}$$

$$u(0, x) = 0, \quad \frac{\partial u(0, x)}{\partial x_0} = 0, \quad x \in (\Omega \bigcap \Omega_{\text{ext}}), \tag{8.2}$$

$$u\big|_{\partial\Omega_0\times(0,T)} = g. \tag{8.3}$$

We impose no boundary conditions on $S_\Delta \times (0,T)$ and, for the sake of simplicity, assume that there is no obstacles between $\partial\Omega_0$ and S_Δ. We estimate $u(x_0,x)$ in terms of g and f.

Denote $Hu = \sum_{j=1}^n g^{0j}(x)u_{x_j}$. Consider the equality

$$(Lu, g^{00}u_{x_0} + Hu) = (f, g^{00}u_{x_0} + Hu),$$

where (u,v) is the inner product in $L_2((\Omega\bigcap\Omega_{\text{int}})\times(0,T))$. We denote by $Q_p(u,v))$, $p \geqslant 1$, the expressions

$$(Q_p u, v) = \int_0^T \int_{\Omega\bigcap\,\Omega_{\text{int}}} \sum_{j,k=0}^n q_{jkp}(x)u_{x_j}v_{x_k}dxdx_0. \tag{8.4}$$

Denote

$$\begin{aligned} I_1 &= \Big(\sum_{j=1}^n \frac{1}{\sqrt{|g|}}\frac{\partial}{\partial x_j}\Big(\sqrt{|g|}g^{0j}\frac{\partial u}{\partial x_0}\Big) \\ &\quad + \sum_{j=0}^n \frac{1}{\sqrt{|g|}}\frac{\partial}{\partial x_0}\Big(\sqrt{|g|}g^{0j}\frac{\partial u}{\partial x_j}\Big), g^{00}u_{x_0} + Hu\Big) \\ &\overset{def}{=} I_{11} + I_{12} + I_{13}. \end{aligned} \tag{8.5}$$

We have

$$\begin{aligned} I_{11} &= \Big(\sum_{j=0}^n \frac{1}{\sqrt{|g|}}\frac{\partial}{\partial x_0}\Big(\sqrt{|g|}g^{0j}\frac{\partial u}{\partial x_j}\Big), g^{00}u_{x_0} + Hu\Big) \\ &= \frac{1}{2}\int_{\Omega\bigcap\,\Omega_{\text{int}}}\Big(\sum_{j=0}^n g^{0j}u_{x_j}\Big)^2 dx + \Big|_0^T Q_1(u,u), \end{aligned} \tag{8.6}$$

where $a\big|_0^T$ means $a(T) - a(0)$. Note that

$$\begin{aligned} I_{12} &= \Big(\sum_{j=1}^n \frac{1}{\sqrt{|g|}}\frac{\partial}{\partial x_j}\Big(\sqrt{|g|}g^{0j}\frac{\partial u}{\partial x_0}\Big), Hu\Big) \\ &= \frac{1}{2}\int_{\Omega\cap\Omega_{\text{ext}}} (Hu)^2 dx\,\big|_0^T + Q_2(u,u). \end{aligned} \tag{8.7}$$

Furthermore,

$$I_{13} = \Big(\frac{1}{\sqrt{|g|}}\sum_{j=1}^n \frac{\partial}{\partial x_j}\Big(\sqrt{|g|}g^{0j}\frac{\partial u}{\partial x_0}\Big), g^{00}u_{x_0}\Big)$$

$$= \frac{1}{2}\int_{\Omega\cap\Omega_{\text{ext}}} \sum_{j=1}^{n} \frac{\partial}{\partial x_j}(g^{0j}g^{00}u_{x_0}^2)dxdx_0 + Q_3(u,u). \tag{8.8}$$

By the divergence theorem,

$$I_{13} = -\int_0^T \int_{S_\Delta} \frac{1}{2}\Big(\sum_{j=1}^{n} g^{0j}(x)\nu_j(x)\Big) g^{00}u_{x_0}^2 dsdx_0$$
$$+ \int_0^T \int_{\partial\Omega_0} \frac{1}{2}\Big(\sum_{j1}^{n} g^{0j}(x)N_j(x)\Big) g^{00}u_{x_0}^2 dsdx_0 + Q_3(u,u), \tag{8.9}$$

where ds is the area element on S_Δ and $\partial\Omega_0$ respectively, $N(x) = (N_1, \dots, N_n)$ is the outward unit normal to $\partial\Omega_0$, and $\nu = (\nu_1, \dots, \nu_n)$ is the outward normal to S_Δ. Note that ν is the inward normal with respect to $\Omega \bigcap \Omega_{\text{ext}}$. Therefore,

$$I_1 = I_{11} + I_{12} + I_{13}$$
$$= \frac{1}{2}\int_{\Omega\bigcap\,\Omega_{\text{ext}}} \Big[\Big(\sum_{j=0}^{n} g^j u_{x_j}\Big)^2 + (Hu)^2\Big] dx\Big|_0^T$$
$$- \int_0^T \int_{S_\Delta} \frac{1}{2}\Big(\sum_{j=1}^{n} g^{0j}(x)\nu_j(x)\Big) g^{00}u_{x_0}^2 dsdx_0$$
$$+ \int_0^T \int_{\partial\Omega_0} \frac{1}{2}\Big(\sum_{j=1}^{n} g^{0j}(x)N_j(x)\Big) g^{00}u_{x_0}^2 dsdx_0 + Q_4(u,u). \tag{8.10}$$

Now, consider

$$I_2 = \Big(\sum_{j,k=1}^{n} \frac{1}{\sqrt{|g|}}\frac{\partial}{\partial x_j}\Big(\sqrt{|g|}g^{jk}\frac{\partial u}{\partial x_k}\Big), g^{00}u_{x_0}\Big).$$

Integrating by parts in x_j and taking into account that

$$-\sum g^{jk}u_{x_k}u_{x_0x_j} = -\frac{1}{2}\frac{\partial}{\partial x_0}\Big(\sum_{j,k=1}^{n} g^{jk}u_{x_j}u_{x_k}\Big),$$

we get

$$I_2 = -\frac{1}{2}\int_{\Omega\bigcap\,\Omega_{\text{ext}}} g^{00}\sum_{j,k=1}^{n} g^{jk}(x)u_{x_j}u_{x_k}dx\Big|_0^T$$
$$- \int_0^T \int_{S_\Delta} \Big(\sum_{j,k=1}^{n} g^{jk}u_{x_j}(x)\nu_k(x)g^{00}u_{x_0}\Big) dsdx_0$$

$$+\int_0^T \int_{\partial\Omega_0} \Big(\sum_{j,k=1}^n g^{jk}(x) u_{x_j} N_k(x) g^{00} u_{x_0} \Big) ds dx_0 + Q_5(u,u). \tag{8.11}$$

Since S_Δ is an ergosphere and a characteristic surface, we have (cf. (6.5))

$$\sum_{k=1}^n g^{jk}(x)\nu_k(x) = 0 \quad \text{on} \quad S_\Delta, \; j = 1, \ldots, n. \tag{8.12}$$

Let

$$I_3 = \Big(\sum_{j,k=1}^n \frac{1}{\sqrt{|g|}} \frac{\partial}{\partial x_j} \Big(\sqrt{|g|} g^{jk} \frac{\partial u}{\partial x_k} \Big), Hu \Big). \tag{8.13}$$

Integrating by parts in x_j, $1 \leqslant j \leqslant n$, we find

$$I_3 = -\Big(\sum_{j,k=1}^n g^{jk} u_{x_k}, H u_{x_j} \Big) - \int_0^T \int_{S_\Delta} \sum_{j,k=1}^n g^{jk} \nu_j u_{x_k} Hu \, ds dx_0$$
$$+ \int_0^T \int_{\partial\Omega_0} \sum_{j,k=1}^n g^{jk} N_j u_{x_k} Hu \, ds dx_0 + Q_6(u,u). \tag{8.14}$$

We have

$$\int_0^T \int_{\Omega \bigcap \Omega_{\text{ext}}} \frac{\partial}{\partial x_p} g^{0p} \sum_{j,k=1}^n g^{jk} u_{x_j} u_{x_k} dx dx_0$$
$$= -\int_0^T \int_{S_\Delta} g^{0p} \nu_p \sum_{j,k=1}^n g^{jk} u_{x_j} u_{x_k} ds dx_0$$
$$+ \int_0^T \int_{\partial\Omega_0} g^{0p} N_p \sum_{j,k=1}^n g^{jk} u_{x_j} u_{x_k} ds dx_0. \tag{8.15}$$

Using (8.15), we find

$$-\Big(\sum_{j,k=1}^n g^{jk} u_{x_k}, H u_{x_j} \Big) = \int_0^T \int_{S_\Delta} \Big(\sum_{p=1}^n g^{0p} \nu_p \Big) \sum_{j,k=1}^n g^{jk} u_{x_j} u_{x_k} ds dx_0$$
$$- \int_0^T \int_{\partial\Omega_0} \Big(\sum_{p=1}^n g^{0p} N_p \Big) \sum_{j,k=1}^n g^{jk} u_{x_j} u_{x_k} ds dx_0 + Q_7(u,u). \tag{8.16}$$

Note that the first integral in (8.16) is nonnegative since

$$\sum_{p=1}^n g^{0p} \nu_p < 0$$

on S_Δ and

$$\sum_{j,k=1}^{n} g^{jk} u_{x_j} u_{x_k} \leqslant 0$$

in $\Omega \bigcap \overline{\Omega}_{\text{int}}$ since the matrix $[g^{jk}]_{j,k=1}^{n}$ has one zero eigenvalue and $n-1$ negative eigenvalues on S_Δ.

Now, we estimate the integrals over $\partial\Omega_0 \times (0,T)$. Let $\alpha(x) \in C_0^\infty(\mathbf{R}^n)$ be such that $\alpha(x) = 1$ near $\partial\Omega_0$ and $\alpha(x) = 0$ near S_Δ. By (8.1), (8.2), and (8.3), $v = \alpha u$ satisfies

$$Lv = \alpha f + L_1 u \quad \text{in} \quad \Omega_0 \times (0,T), \tag{8.17}$$

$$v|_{x_0=0} = 0, \quad v_{x_0}|_{x_0=0} = 0, \quad x \in \Omega_0, \tag{8.18}$$

$$v|_{\partial\Omega_0 \times (0,T)} = g. \tag{8.19}$$

Here, g is the same as in (8.3) and ord $L_1 \leqslant 1$.

Since L is strictly hyperbolic and $\partial\Omega_0$ is not characteristic, the following estimate for the solution of the problem (8.17), (8.18), (8.19) holds (cf. for example, [12]):

$$\begin{aligned} &\|v_{x_0}(T,\cdot)\|_0^2 + \|v(T,\cdot)\|_1^2 + \Big[\frac{\partial v}{\partial N}\Big]_0^2 \\ &\qquad \leqslant C_T\Big([g]_1^2 + \int_0^T \|f(x_0,\cdot)\|_0^2 dx_0 \\ &\qquad + \int_0^T (\|u(x_0,\cdot)\|_1^2 + \|u_{x_0}(x_0,\cdot)\|_0^2) dx_0\Big), \end{aligned} \tag{8.20}$$

where $[w]_m$ is the norm in $H^m(\partial\Omega_0 \times (0,T))$ and $\|w(x_0,\cdot)\|_m$ is the norm in $H^m(\Omega_0)$, x_0 is fixed. All the integrals over $\partial\Omega_0 \times (0,T)$ in I_2 and I_3 have the form

$$I_4 = \int_0^T \int_{\partial\Omega_0} \sum_{j,k=0}^{n} p_{jk} u_{x_j} u_{x_k} ds dx_0.$$

Therefore,

$$\begin{aligned} |I_4| &\leqslant C\Big([g]_1^2 + \Big[\frac{\partial v}{\partial N}\Big]_0^2\Big) \\ &\leqslant C\Big([g]_1^2 + \int_0^T \|f(x_0,\cdot)\|_0^2 dx_0 \\ &+ \int_0^T (\|u(x_0,\cdot)\|_1^2 + \|u_{x_0}(x_0,\cdot)\|_0^2) dx_0\Big), \end{aligned} \tag{8.21}$$

where we used (8.20) to estimate $\Big[\dfrac{\partial v}{\partial N}\Big]_0^2$. Note that the norm

$$\int_{\Omega\bigcap\Omega_{\text{ext}}}\Big[(u_{x_0}+Hu)^2+(Hu)^2-\sum_{j,k=1}^n g^{jk}u_{x_j}u_{x_k}\Big]dx \tag{8.22}$$

is equivalent to $\|u(x_0,\cdot)\|_1^2+\|u_{x_0}(x_0,\cdot)\|_0^2$. Furthermore,

$$\begin{aligned}&|(f,g^{00}u_{x_0}+Hu)|\\&\quad\leqslant\frac{1}{2}\int_0^T\|f(y_0,\cdot)\|_0^2dy_0+C\int_0^T(\|u_{x_0}\|_0^2+\|u(y_0,\cdot)\|_1^2)dy_0.\end{aligned} \tag{8.23}$$

Combining (8.10), (8.11), (8.14), (8.16), and (8.21), taking into account (8.12), and applying all the estimates to the interval $(0,t_0)$ instead of $(0,T)$, $t_0\leqslant T$, we get

$$\begin{aligned}&C(\|u(t_0,\cdot)\|_1^2+\|u_{x_0}(t_0,\cdot)\|_0^2)\\&-\int_0^{t_0}\int_{S_\Delta}\Big(\sum_{j=1}^n g^{0j}\nu_j(x)\Big)\Big(g^{00}u_{x_0}^2-\sum_{j,k=1}^n g^{jk}u_{x_j}u_{x_k}\Big)dsdx_0\\&\leqslant C\int_0^T\|f(y_0,\cdot)\|_0^2dy_0+C\int_0^{t_0}(\|u(y_0,\cdot)\|_1^2+\|u_{x_0}(y_0,\cdot)\|_0^2)dy_0+C[g]_1^2.\end{aligned} \tag{8.24}$$

Note that the inequality

$$b(t)\leqslant C\int_0^t b(\tau)d\tau+d$$

implies $b(t)\leqslant e^{ct}d$. Therefore,

$$\begin{aligned}&C(\|u(t_0,\cdot)\|_1^2+\|u_{x_0}(t_0,\cdot)\|_0^2)\\&-\int_0^{t_0}\int_{S_\Delta}\Big(\sum_{j=1}^n g^{0j}\nu_j(x)\Big)\Big(g^{00}u_{x_0}^2-\sum_{j,k=1}^n g^{jk}u_{x_j}u_{x_k}\Big)dsdx_0\\&\leqslant C\int_0^T\|f(y_0,\cdot)\|_0^2dy_0+C[g]_1^2,\end{aligned} \tag{8.25}$$

where $0\leqslant t_0\leqslant T$.

We have proved the following assertion.

Theorem 8.1. *Let $u(x_0,x)$ be a solution of the problem* (8.1), (8.2), (8.3) *in $\Omega\bigcap\Omega_{\text{ext}}$. Let the ergosphere S_Δ be a characteristic surface, and let* (5.3) *hold. Then $u(x_0,x)$ satisfies* (8.25).

Note that

$$g^{00}u_{x_0}^2-\sum_{j,k=1}^n u_{x_j}u_{x_k}\geqslant 0$$

on $S_\Delta \times [0,T]$ since S_Δ is an ergosphere.

Theorem 8.1 implies that the domain of dependence of $(\Omega \bigcap \Omega_{\text{ext}}) \times \mathbf{R}$ is contained in $(\overline{\Omega} \bigcap \overline{\Omega}_{\text{ext}}) \times \mathbf{R}$. Suppose that u is a solution of (2.1) and $\operatorname{supp} u \subset \overline{\Omega}_{\text{int}}$ for $x_0 \leqslant t_0$, i.e., $u = 0$ for $x \in (\Omega_{\text{ext}} \bigcap \Omega) \times (-\infty, t_0)$. Then Theorem 8.1 implies that $u = 0$ for $(\overline{\Omega} \bigcap \overline{\Omega}_{\text{ext}}) \times [t_0, +\infty)$, i.e., $\operatorname{supp} u \subset \overline{\Omega}_{\text{int}} \times \mathbf{R}$. Therefore, the domain of influence of $\Omega_{\text{int}} \times \mathbf{R}$ is contained in $\overline{\Omega}_{\text{int}}) \times \mathbf{R}$, i.e., is a black hole.

Now, we discuss nonuniqueness for the inverse problem in the presence of a black hole. Consider two initial-boundary value problems (2.1), (2.5), (2.6) for operators L_1 and L_2 that differ only in Ω_{int}. Since $L_1 = L_2$ in Ω_{ext} and we assume that f is the same for L_1 and L_2, from Theorem 8.1 it follows that $u_1 = u_2$ in $(\Omega \bigcap \Omega_{\text{ext}}) \bigcap \mathbf{R}$, where u_1 and u_2 are solutions of the corresponding initial-boundary value problems. Therefore, $\Lambda_1 = \Lambda_2$ on $\partial\Omega_0 \times \mathbf{R}$. Since $L_1 \neq L_2$ in Ω_{int}, we obtain nonuniquenes in the inverse problem.

Consider the case where S_Δ is a characteristic surface and (5.2) holds. Suppose that $\Omega_{\text{int}} \bigcap \Omega$ contains an obstacle Ω_1 (it is possible that there are not obstacles or there are more than one obstacle, but for the sake of definiteness we discuss the case of one obstacle). Integrating by parts as in the proof of Theorem 8.1 and using (5.2) instead of (5.3), we get the following assertion.

Theorem 8.2. *Consider the initial-boundary value problem*

$$Lu = f \quad \text{in} \quad (\Omega_{\text{int}} \bigcap \Omega) \times (0,T), \tag{8.26}$$

$$u|_{\partial\Omega_1 \times (0,T)} = g, \quad u(0,x) = 0, \quad u_{x_0}(0,x) = 0 \quad \text{in} \quad \Omega \bigcap \Omega_{\text{int}}. \tag{8.27}$$

Suppose that the ergosphere S_Δ is a characteristic surface and (5.2) *holds. Then an estimate of the form* (8.25) *holds in $(\Omega_{\text{int}} \bigcap \Omega) \times (0,T)$ with the following modifications: the integral over $S_\Delta \times (0,T)$ is taken with the sign "plus," $\|u\|_s$ are the norms in $H^s(\Omega_{\text{int}} \bigcap \Omega)$, and $[g]_1$ is the norm in $H^1(\partial\Omega_1 \times (0,T))$.*

From Theorem 8.2 it follows that the domain of dependence of $(\Omega_{\text{int}} \bigcap \Omega) \times \mathbf{R}$ is contained in $(\overline{\Omega}_{\text{int}} \bigcap \overline{\Omega}) \times \mathbf{R}$. Therefore, if $u(x_0, x)$ is a solution of (2.1) and $\operatorname{supp} u \subset \overline{\Omega}_{\text{ext}} \bigcap \overline{\Omega}$ for $x_0 \leqslant t_0$, then $\operatorname{supp} u \subset \overline{\Omega}_{\text{ext}} \bigcap \overline{\Omega}$ for all $x_0 > t_0$, i.e., $\Omega_{\text{int}} \times \mathbf{R}$ is a white hole. If $u(x_0, x)$ is a solution of the problem (2.1), (2.5), (2.6), then $u(0,x) = 0$ in $\Omega_{\text{int}} \bigcap \Omega$, $u|_{\partial\Omega_1 \times \mathbf{R}} = 0$. Then, by Theorem 8.2, $u = 0$ in $(\Omega_{\text{int}} \bigcap \Omega) \times \mathbf{R}$. Therefore, we can change the coefficients of L in $\Omega_{\text{int}} \bigcap \Omega$ without changing the solution of (2.1), (2.5), (2.6), i.e., in the case of a white hole, we again have the nonuniqueness of a solution of the inverse problem.

Let $u(x_0, x)$ be a solution of the problem

$$Lu = f \quad \text{in} \quad \Omega_{\text{ext}} \times (0,T), \tag{8.28}$$

$$u(0,x)=\varphi_0(x),\quad u_{x_0}(0,x)=\varphi_1(x),\quad x\in\Omega_{\text{ext}},\tag{8.29}$$

i.e., we consider (8.28), (8.29) in unbounded domain $\Omega_{\text{ext}}=\mathbf{R}^n\setminus\overline{\Omega}_{\text{int}}$. We assume that $[g^{jk}(x)]_{j,k=1}^n$ are smooth and have bounded derivatives of any order, $g^{00}(x)\geqslant C_0>0$,

$$\sum_{j,k=1}^n g^{jk}(x)\xi_j\xi_k\leqslant -C_0\sum_{j=1}^n\xi_j^2,\quad x\in\overline{\Omega}_{\text{ext}},$$

is large, and (5.3) holds. Repeating the proof of Theorem 8.1 (with simplifications since we do not have the boundary condition (8.3)), for any $T>0$ we get

$$\begin{aligned}&\max_{0\leqslant x_0\leqslant T}\left(\|u(x_0,\cdot)\|_{1,\Omega_{\text{ext}}}^2+\|u_{x_0}(x_0,\cdot)\|_{0,\Omega_{\text{ext}}}^2\right)\\&-\int_0^T\int_{S_\Delta}\Big(u_{x_0}^2-\sum_{j,k=1}^n g^{jk}u_{x_j}u_{x_k}\Big)\Big(\sum_{j=1}^n g^{j0}\nu_j\Big)dsdy_0\\&\leqslant C([\varphi_0]_{1,\Omega_{\text{ext}}}^2+[\varphi_1]_{0,\Omega_{\text{ext}}}^2)+C\int_0^T\|f(x_0,\cdot)\|_{0,\Omega_{\text{ext}}}^2dx_0.\end{aligned}\tag{8.30}$$

Therefore, the following assertion holds.

Theorem 8.3. *Let* $u(x_0,x)$ *satisfy* (8.28) *and* (8.29). *Suppose that the ergosphere* S_Δ *is a characteristic surface and* (5.3) *holds. Then the estimate* (8.30) *holds for any* $T>0$.

From Theorem 8.3 it follows that $D_+(\Omega_{\text{int}}\times\mathbf{R})\subset\overline{\Omega}_{\text{int}}\times\mathbf{R}$, i.e., $\Omega_{\text{int}}\times\mathbf{R}$ is a black hole.

An important problem is to determine black or white holes by the boundary measurements on $\partial\Omega_0\times(0,T)$ or $\Gamma\times(0,T_0)$, where Γ is an open part of $\partial\Omega_0$. Let S_Δ be an ergosphere inside Ω_0, but not necessary a characteristic surface. The following assertion is a direct application of the proof of Theorem 2.1.

Theorem 8.4 (cf. [8]). *The* DN *operator* Λ *given on* $\Gamma\times(0,+\infty)$ *determines* S_Δ *up to a diffeomorphism* (2.8).

To determine S_Δ, it is required to take measurements on $\Gamma\times(0,+\infty)$. It is not enough to know the Cauchy data on $\Gamma\times(0,T)$ for any finite T. This phenomenon can be explained as follows. The proof of Theorem 2.1 allows us to recover metric tensor $[g^{jk}]$ (up to a diffeomorphism) gradually starting from the boundary $\partial\Omega_0$. The recovery of the metric at some point $x^{(1)}$ inside Ω_0 requires some observation time T_1. When $x^{(1)}$ is far from Γ, the observation time increases. When the point $x^{(1)}$ approaches S_Δ, i.e., $g_{00}(x^{(1)})\to 0$, the observation time tends to infinity. One can see this from the fact that either forward time-like ray or backward time-like ray tends to infinity in x_0 as $x^{(1)} \to S_\Delta$ (cf. [9]).

9 Black Holes and Inverse Problems II

In this section, we consider black or white holes inside an ergosphere.

Suppose that $S_0 \times \mathbf{R}$ is a characteristic surface, $n \geqslant 2$, and $S_0 \subset S_\Delta$, where S_Δ is an ergosphere. Suppose that the condition (5.3) on S_0 holds. Consider $v(x_0, x)$ in $\Omega_{\text{ext}} \times (0,T)$ such that

$$Lv = f, \quad x \in \Omega_{\text{ext}} \times (0,T), \tag{9.1}$$

$$v(0,x) = \varphi_0(x), \quad v_{x_0}(0,x) = \varphi_1(x), \quad x \in \Omega_{\text{ext}}. \tag{9.2}$$

We want to estimate $v(x_0,x)$ in $\Omega_{\text{ext}} \times (0,T)$ in terms of φ_0, φ_1, and f. Let $\widehat{\varphi}_0$, $\widehat{\varphi}_1$, and $\widehat{f}$ be smooth extensions of φ_0, φ_1 to $\mathbf{R}^n$ and f to $\mathbf{R}^n \times (0,T)$ such that

$$\begin{aligned} &\widehat{E}(\widehat{\varphi}_0, \widehat{\varphi}_1) \leqslant 2E(\varphi_0,\varphi_1), \\ &\|\widehat{f}\|_{0,\mathbf{R}^n\times(0,T)} \leqslant 2\|f\|_{0,\Omega_{\text{ext}}\times(0,T)}, \end{aligned} \tag{9.3}$$

where

$$E(\varphi_0,\varphi_1) = \int_{\Omega_{\text{ext}}} \Big(\sum_{j=1}^n \varphi_{0x_j}^2 + \varphi_1^2 \Big) dx$$

and $\widetilde{E}(\widetilde{\varphi}_0, \widetilde{\varphi}_1)$ is a similar integral over $\mathbf{R}^n$. Since L is strictly hyperbolic, there exists $\widehat{u}$ in $\mathbf{R}^n \times (0,T)$ such that

$$\begin{aligned} &L\widehat{u} = \widehat{f} \quad \text{in} \quad R^n \times (0,T), \\ &\widehat{u}(0,x) = \widehat{\varphi}_0(x), \quad \widehat{u}_{x_0}(0,x) = \widehat{\varphi}_1(x), \quad x \in \mathbf{R}^n, \end{aligned} \tag{9.4}$$

and

$$\begin{aligned} \max_{0\leqslant x_0 \leqslant T} &(\|\widehat{u}(x_0,\cdot)\|^2_{1,\mathbf{R}^n} + \|\widehat{u}_{x_0}(x_0,\cdot)\|^2_{0,\mathbf{R}^n}) \\ &\leqslant C\widehat{E}(\widehat{\varphi}_0,\widehat{\varphi}_1) + C\int_0^T \int_{R^n} |\widehat{f}|^2 dx dx_0. \end{aligned} \tag{9.5}$$

Replacing $v = u + \widehat{u}$, we find that $u(x_0,x)$ satisfies

$$\begin{aligned} &Lu = 0 \quad \text{in} \quad \Omega_{\text{ext}} \times (0,T), \\ &u(0,x) = u_{x_0}(0,x) = 0, \quad x \in \Omega_{\text{ext}}. \end{aligned} \tag{9.6}$$

Therefore, it remains to show that if $u(x_0,x)$ satisfies (9.6), then $u(x_0,x) = 0$ in $\Omega_{\text{ext}} \times (0,T)$. Note that we could use the same approach in Section 8 too. Instead, we use a technique similar to [2, 5]. Let T be small. Denote by Γ_1 the characteristic surface different from $S_0 \times \mathbf{R}$ and passing through $S_0 \times \{x_0 = T\}$. Let D_T be the domain bounded by Γ_1, $\Gamma_2 = S_0 \times [-\varepsilon, T]$, and $\Gamma_3 = \{x_0 = -\varepsilon\}$. For an arbitrary point $x^{(0)} \in S_0$ denote by D_{0T} the intersection of D_T with

$$\Sigma(x^{(0)}) = \{(x_0, x) : |x - x^{(0)} - (x - x^{(0)}) \cdot \nu(x^{(0)})\nu(x^{(0)})| < \varepsilon\},$$

where $\nu(x^{(0)})$ is the outward unit normal to S_0 at $x^{(0)}$. Suppose that $\alpha_j(x_0, x) \in C^\infty(D_T)$, $\sum_{j=1}^N \alpha_j \equiv 1$ in D_T, and supp $\alpha_j \subset D_{jT}$, where D_{jT} corresponds to $x^{(j)} \in S_0$ instead of $x^{(0)}$, $\alpha_j = 0$ in a neighborhood of the boundary of $\Sigma(x^{(0)})$).

Let α_0 be any of α_j, $1 \leqslant j \leqslant N$. Denote $u_0 = \alpha_0 u$. Then

$$Lu_0 = f_0, \quad u_0(0, x) = u_{x_0}(0, x) = 0, \tag{9.7}$$

where $f_0 = L'u$, ord $L' \leqslant 1$, and supp $f_0 \subset D_{0T}$. We introduce local coordinates in a neighborhood

$$B_{\varepsilon,T} = \{(x_0, x) : x_0 \in [-\varepsilon, T], |x - x^{(0)}| < 2\varepsilon\}.$$

Let $s = \varphi(x)$ be a solution of the eiconal equation

$$\sum_{j,k=1}^n g^{jk}(x)\varphi_{x_j}\varphi_{x_k} = 0 \quad \text{in} \quad B_{\varepsilon,T}. \tag{9.8}$$

Since S_0 is inside the ergosphere, $\varphi(x)$ exists if ε and T are small. We choose $s = \varphi(x)$ such that $\varphi(x) = 0$ is the equation of S_0 near $x^{(0)}$.

Let $\tau = \psi(x_0, x)$ be a solution of the eiconal equation

$$\sum_{j,k=0}^n g^{jk}(x)\psi_{x_j}\psi_{x_k} = 0 \tag{9.9}$$

with the initial data

$$\psi(x_0, x)|_{\Gamma_2} = T - x_0. \tag{9.10}$$

Finally, denote by $y_j = \varphi_j(x_0, x)$, $1 \leqslant j \leqslant n - 1$, a solution of the equation

$$\sum_{j,k=0}^n g^{jk}(x)\psi_{x_j}\varphi_{px_k} = 0 \quad \text{near} \quad x^{(0)}, \quad 1 \leqslant p \leqslant n - 1, \tag{9.11}$$

with the initial condition

$$\varphi_p(x_0, x)|_{\Gamma_2} = s_p(x), \quad 1 \leqslant p \leqslant n - 1, \tag{9.12}$$

where $s_1(x), \ldots, s_{n-1}(x)$ are coordinates on S_0 near $x^{(0)}$,

$$\frac{Dx}{D(s, s_1, \ldots, s_{n-1})} \neq 0 \quad \text{in } B_\varepsilon.$$

Note that φ_p does not depend on x_0 and $\psi(x_0, x) = T - x_0 + \psi_1(x)$, where $\psi_1(x)$ does not depend on x_0 either.

We make the change of coordinates in D_{0T}

$$s = \varphi(x), \quad \tau = \psi(x_0, x), \quad y_j = \varphi_j(x), \quad 1 \leqslant j \leqslant n-1. \tag{9.13}$$

Note that

$$\frac{D(x_0, x)}{D(s, \tau, y')} \neq 0$$

in B_ε, where $y' = (y_1, \ldots, y_{n-1})$. We write Lu_0 in the (s, τ, y') coordinates (cf. [3, 6]):

$$\begin{aligned}
\widehat{L}\widehat{u}_0 = & \frac{1}{\sqrt{|\widehat{g}|}} \frac{\partial}{\partial s} \sqrt{|\widehat{g}|}\widehat{g}^{s\tau}(s, \tau, y') \frac{\partial \widehat{u}_0}{\partial \tau} \\
& + \frac{1}{\sqrt{|\widehat{g}|}} \frac{\partial}{\partial \tau} \sqrt{|\widehat{g}|}\widehat{g}^{s\tau}(s, \tau, y') \frac{\partial \widehat{u}_0}{\partial s} \\
& + \sum_{j=1}^{n-1} \frac{1}{\sqrt{|\widehat{g}|}} \frac{\partial}{\partial s} \sqrt{|\widehat{g}|}\widehat{g}^{sj}(s, \tau, y') \frac{\partial \widehat{u}_0}{\partial y_j} \\
& + \sum_{j=1}^{n-1} \frac{1}{\sqrt{|\widehat{g}|}} \frac{\partial}{\partial y_j} \sqrt{|\widehat{g}|}\widehat{g}^{sj}(s, \tau, y') \frac{\partial \widehat{u}_0}{\partial s} \\
& + \sum_{j,k=1}^{n-1} \frac{1}{\sqrt{|\widehat{g}|}} \frac{\partial}{\partial y_j} \sqrt{|\widehat{g}|}\widehat{g}^{jk}(s, \tau, y') \frac{\partial \widehat{u}_0}{\partial y_k} \stackrel{\text{def}}{=} \widehat{L}_1\widehat{u}_0 + \widehat{L}_2\widehat{u}_0,
\end{aligned} \tag{9.14}$$

where L_2 is the last sum in (9.14) and L_1 is the remaining sum. Note that $\widehat{g}^{ss} = \widehat{g}^{\tau\tau} = \widehat{g}^{\tau j} = 0$, $1 \leqslant j \leqslant n-1$, because of (9.8), (9.9), and (9.11). In (9.14), $\widehat{u}_0(s, \tau, y') = u_0(x_0, x)$, where (x_0, x) and (s, τ, y') are related by (9.13). Since T and ε are small, we can introduce the (s, τ, y') coordinates in D_{0T}. Denote by $\widehat{D}_{0T}$ the image of D_{0T} in the (s, τ, y') coordinates.

Let $\partial\widehat{\Sigma}_0$ be the image of $\partial\Sigma(x^{(0)})$ in the (s, τ, y')coordinates. Since $u_0 = 0$ near $\partial\Sigma(x^{(0)})$, we have $\widehat{u}_0 = \widehat{\alpha}_0\widehat{u} = 0$ near $\partial\widehat{\Sigma}_0$. Note also that $u_0 = u_{0x_0} = 0$ for $x_0 = 0$ and we extend u_0 by zero for $x_0 < 0$. Since $\varphi_x(x^{(0)})$ is the outward normal to Ω_{int}, we have $s = \varphi(x) \geqslant 0$ in $\overline{\Omega}_{\text{ext}}$ near S_0. Since $\tau = \psi(T, x) = 0$ on Γ_2 and $\psi_{x_0}|_{\Gamma_2} = -1$, we conclude that $\tau \leqslant 0$ in $\widehat{D}_{0T}$.

Denote by $(\widehat{u}, \widehat{v})$ the L^2 inner product in $\widehat{D}_{0T}$. Consider

$$\Big(\widehat{L}\widehat{u}_0 - \widehat{f}_0, \widehat{g}^{s\tau}\widehat{u}_{0\tau} + \sum_{j=1}^{n-1} \widehat{g}^{sj}\widehat{u}_{0y_j} - \widehat{g}^{s\tau}\widehat{u}_{0s}\Big) = 0. \tag{9.15}$$

Let

$$I_1 = \Big(\widehat{L}_1\widehat{u}_0, \widehat{g}^{s\tau}\widehat{u}_{0\tau} + \sum_{j=1}^{n-1}\widehat{g}^{sj}\widehat{u}_{0y_j}\Big)$$
$$= \int_{\widehat{D}_{0T}} \frac{\partial}{\partial s}\Big(\widehat{g}^{s\tau}\widehat{u}_{0\tau} + \sum_{j=1}^{n-1}\widehat{g}^{sj}\widehat{u}_{0y_j}\Big)^2 ds d\tau dy' + Q_1(\widehat{u}_0, \widehat{u}_0), \tag{9.16}$$

where Q_1 admits the estimate

$$|Q_1(\widehat{u}_0, \widehat{u}_0)| \leqslant C\int_{\widehat{D}_{0T}} \Big(\widehat{u}_{0s}^2 + \widehat{u}_{0\tau}^2 + \sum_{j=1}^{n-1}\widehat{u}_{0y_j}^2\Big) ds d\tau dy'. \tag{9.17}$$

Therefore,

$$I_1 = -\int_{\Gamma_2}\Big(\widehat{g}^{s\tau}\widehat{u}_{0\tau} + \sum_{j=1}^{n-1}\widehat{g}^{s0}\widehat{u}_{y_j}\Big)^2 d\tau dy' + Q_1(\widehat{u}_0, \widehat{u}_0). \tag{9.18}$$

Denote

$$I_2 = \Big(\frac{1}{\sqrt{|\widehat{g}|}}\frac{\partial}{\partial s}\sqrt{|\widehat{g}|}\widehat{g}^{s\tau}(s,\tau,y')\frac{\partial\widehat{u}_0}{\partial\tau} + \frac{1}{\sqrt{|\widehat{g}|}}\frac{\partial}{\partial\tau}\sqrt{|\widehat{g}|}\widehat{g}^{s\tau}(s,\tau,y')\frac{\partial\widehat{u}_0}{\partial s}, -\widehat{g}^{s\tau}\widehat{u}_{0s}\Big). \tag{9.19}$$

We have

$$I_2 = -\int_{\widehat{D}_{0T}}\frac{\partial}{\partial\tau}\left((\widehat{g}^{s\tau})^2\widehat{u}_{0s}^2\right) ds d\tau dy' + Q_2(\widehat{u}_0, \widehat{u}_0). \tag{9.20}$$

Therefore,

$$I_2 = -\int_{\Gamma_1}(\widehat{g}^{s\tau})^2\widehat{u}_{0s}^2 ds dy' + Q_2(\widehat{u}_0, \widehat{u}_0). \tag{9.21}$$

Denote

$$I_3 = (\widehat{L}_2\widehat{u}_0, \widehat{g}^{s\tau}(\widehat{u}_{0\tau} - \widehat{u}_{0s})). \tag{9.22}$$

Integrating by parts in y_j (note that $\widehat{u}_0 = 0$ near $\partial\widehat{\Sigma}_0$ and $u_0 = 0$ for $x_0 < 0$), we get

$$I_3 = -\int_{\widehat{D}_T}\sum_{j,k=1}^{n-1}\widehat{g}^{s\tau}\widehat{g}^{jk}(\widehat{u}_{0sy_j} - \widehat{u}_{0\tau y_j})\widehat{u}_{0y_k} ds d\tau dy' + Q_3$$
$$= \frac{1}{2}\int_{\Gamma_1}\Big(\sum_{j,k=1}^{n-1}\widehat{g}^{jk}\widehat{u}_{0y_j}\widehat{u}_{0y_k}\Big)\widehat{g}^{s\tau} ds dy'$$
$$+ \frac{1}{2}\int_{\Gamma_2}\Big(\sum\widehat{g}^{jk}\widehat{u}_{0y_j}\widehat{u}_{0y_k}\Big)\widehat{g}^{s\tau} d\tau dy' + Q_4. \tag{9.23}$$

Note that

$$I_4 = \Big(\widehat{L}_2\widehat{u}_0, \sum_{j=1}^{n-1} \widehat{g}^{sj}\widehat{u}_{0y_j}\Big) = Q_5(\widehat{u}_0, \widehat{u}_0) \tag{9.24}$$

since it can be represented as the divergence of a quadratic form in $\widehat{u}_{0y_j}$ (cf. (8.15)). Moreover,

$$\Big|\Big(\widehat{f}_0, \widehat{g}^{s\tau}(\widehat{u}_{0\tau} - \widehat{u}_{0s}) + \sum_{j=1}^{n-1} \widehat{g}^{js}\widehat{u}_{0y_j}\Big)\Big| \leqslant C\int_{\widehat{D}_{0T}} |\widehat{f}_0|^2 ds d\tau dy' + Q_6(\widehat{u}_0, \widehat{u}_0).$$

Therefore,

$$\begin{aligned}
&\frac{1}{2}\int_{\Gamma_2}\Big(-\sum_{j,k=1}^{n-1} \widehat{g}^{jk}\widehat{u}_{0y_j}\widehat{u}_{0y_k}\widehat{g}^{s\tau}\Big) d\tau dy' \\
&+\frac{1}{2}\int_{\Gamma_1}\Big(-\sum_{j,k=1}^{n-1} \widehat{g}^{jk}\widehat{u}_{0y_j}\widehat{u}_{0y_k}\widehat{g}^{s\tau}\Big) ds dy' \\
&+\int_{\Gamma_2}\Big(\widehat{g}^{s\tau}\widehat{u}_{0\tau} + \sum_{j=1}^{n-1}\widehat{g}^{sj}\widehat{u}_{0y_j}\Big)^2 d\tau dy' + \int_{\Gamma_1}(\widehat{g}^{s\tau})^2\widehat{u}_{0s}^2 ds dy' \\
&\leqslant Q_7(\widehat{u}_0, \widehat{u}_0) + C\int_{\widehat{D}_{0T}} |\widehat{f}_0|^2 ds d\tau dy'
\end{aligned} \tag{9.25}$$

Note that $\widehat{g}^{s\tau} > 0$ and

$$-\sum_{j,k=1}^{n-1} \widehat{g}^{jk}\widehat{u}_{0y_j}\widehat{u}_{0y_k} \geqslant C\sum_{j=1}^{n-1}\widehat{u}_{0y_j}^2 \quad \text{in } \widehat{D}_{0T}. \tag{9.26}$$

Therefore,

$$\int_{\Gamma_1}\Big[(\widehat{g}^{s\tau})^2\widehat{u}_{0s}^2 - \frac{1}{2}\sum_{j,k=1}^{n-1} \widehat{g}^{jk}\widehat{u}_{0y_j}\widehat{u}_{0y_k}\widehat{g}^{s\tau}\Big] ds dy' \tag{9.27}$$

is equivalent to

$$\int_{\Gamma_1}\Big(\widehat{u}_{os}^2 + \sum_{j=1}^{n-1}\widehat{u}_{0y'}^2\Big) ds dy',$$

i.e., it is equivalent to the norm $\|\widehat{u}_0\|_{1,\Gamma_1}^2$ in $H^1(\Gamma_1)$. Similarly,

$$\int_{\Gamma_2}\Big[\Big(\widehat{g}^{s\tau}\widehat{u}_{0\tau} + \sum_{j=1}^{n-1}\widehat{g}^{sj}\widehat{u}_{0y_j}\Big)^2 - \frac{1}{2}\sum_{j,k=1}^{n-1} \widehat{g}^{jk}\widehat{u}_{0y_j}\widehat{u}_{0y_k}\widehat{g}^{s\tau}\Big] d\tau dy' \tag{9.28}$$

is equivalent to the norm

$$\|u_0\|_{1,\Gamma_2}^2 = \int_{\Gamma_2} \Big(\widehat{u}_{0\tau}^2 + \sum_{j=1}^{n-1} \widehat{u}_{0y_j}^2\Big) d\tau dy'$$

in $H^1(\Gamma_2)$. Therefore, (9.25) is equivalent to

$$\|\widehat{u}_0\|_{1,\Gamma_1}^2 + \|\widehat{u}_0\|_{1,\Gamma_2}^2 \leqslant Q_8(\widehat{u}_0, \widehat{u}_0) + C \int_{\widehat{D}_{0T}} |\widehat{f}_0|^2 ds d\tau dy'. \tag{9.29}$$

Denote by $D_{0T,t}$ the intersection of D_{0T} with the half-space $x_0 \geqslant t$. Integrating by parts in the integral

$$0 = \int_{D_{0T,t}} (Lu_0 - f_0)\Big(g^{00} u_{0x_0} + \sum_{j=1}^{n} g^{0j}(x) u_{0x_j}\Big) dx dx_0, \tag{9.30}$$

we obtain (cf. [3, 5])

$$\begin{aligned}
&\int_{D_{0T} \bigcap \{x_0 = t\}} \Big[\Big(\sum_{j=0}^{n} g^{j0} u_{0x_j}(t,x)\Big)^2 \\
&\quad + \Big(\sum_{j=1}^{n} g^{j0} u_{0x_j}(t,x)\Big)^2 - \sum_{j,k=1}^{n} g^{jk} u_{0x_j}(t,x) u_{0x_k}(t,x)\Big] dx \\
&\leqslant C(\|u_0\|_{1,\Gamma_{1t}}^2 + \|u_0\|_{1,\Gamma_{2t}}^2) + C \int_{D_{0T,t}} \Big(\sum_{j=0}^{n} u_{0x_j}^2\Big) dx dx_0 \\
&\quad + C \int_{D_{0T,t}} |f_0|^2 dx dx_0,
\end{aligned} \tag{9.31}$$

where Γ_{jt} is the intersection of Γ_j with $x_0 \geqslant t$, $j = 1, 2$. Note that the integral on the left-hand side of (9.31) is equivalent to

$$\int_{D_{0T} \bigcap \{x_0 = t\}} \Big(\sum_{j=0}^{n} u_{0x_j}^2(t,x)\Big) dx. \tag{9.32}$$

Writing (9.29) in the (x_0, x) coordinates and combining with (9.31), we get

$$\begin{aligned}
&\max_{0 \leqslant t \leqslant T} \int_{D_{0T} \bigcap \{x_0 = t\}} \Big(\sum_{j=0}^{n} u_{0x_j}^2\Big) dx \\
&\qquad \leqslant C \int_{D_{0T}} \Big(\sum_{j=0}^{n} u_{0x_j}^2\Big) dx_0 dx + C \int_{D_{0T}} |f_0|^2 dx_0 dx.
\end{aligned} \tag{9.33}$$

Let $\{\alpha_j(x)\}_{j=1,\dots,N}$ be as above. Denote $u_j = \alpha_j u$. Applying (9.33) with $u_j = \alpha_j u$ instead of $u_0 = \alpha_0 u$ and using that $\sum_{j=1}^{N} \alpha_j = 1$ in D_T, we get

$$\max_{0\leqslant t\leqslant T}\int_{D_T\bigcap\{x_0=t\}}\Big(\sum_{j=0}^{n}u_{x_j}^2(t,x)\Big)dx$$
$$\leqslant C\int_{D_T}\Big(\sum_{j=0}^{n}u_{x_j}^2(x_0,x)\Big)dx_0dx+C\sum_{j=1}^{N}\int_{D_{jT}}|f_j|^2dx_0dx, \tag{9.34}$$

where $f_j=(L\alpha_j-\alpha_jL)u$,

$$|f_j|\leqslant C\sum_{j=0}^{n}|u_{x_j}|. \tag{9.35}$$

Therefore,

$$\max_{0\leqslant t\leqslant T}\int_{D_T\bigcap\{x_0=t\}}\Big(\sum_{j=0}^{n}u_{x_j}^2(t,x)\Big)dx$$
$$\leqslant C\int_{D_T}\Big(\sum_{j=0}^{n}u_{x_j}^2(x_0,x)\Big)dx_0dx$$
$$\leqslant CT\max_{0\leqslant t\leqslant T}\int_{D_T\bigcap\{x_0=t\}}\Big(\sum_{j=0}^{n}u_{x_j}^2(t,x)\Big)dx. \tag{9.36}$$

Since T is small, $u=0$ in D_T. Take any $T_1<T$. Then there exists $\delta_1>0$ such that $S_{0\delta_1}\times[0,T_1]\subset D_T$, where $S_{0\delta_1}$ is a δ_1-neighborhood of S_0. Therefore, $u(x_0,x)=0$ in $S_{0\delta_1}\times[0,T_1]$ and we can extend $u(x_0,x)$ by zero in $\Omega_{\text{int}}\times[0,T_1]$. Then $Lu=0$ in $\mathbf{R}^n\times(0,T_1)$ and $u(0,x)=u_{x_0}(0,x)=0$ in $\mathbf{R}^n$. By the uniqueness of a solution to the hyperbolic Cauchy problem (cf. (9.5)), we have $u=0$ in $\mathbf{R}^n\times(0,T_1)$. Repeating the same arguments on $(T_1,2T_1)$ etc., we find that $u=0$ in $\Omega_{\text{ext}}\times(0,T)$ for any $T>0$. Therefore, $v=\widehat{u}$ in $\Omega_{\text{ext}}\times(0,T)$, where $v(x_0,x)$ satisfies (9.1), (9.2) and $\widehat{u}$ satisfies (9.4) in $\mathbf{R}^n\times(0,T)$. Then (9.5) implies that

$$\max_{0\leqslant x_0\leqslant T}\big(\|v(x_0,\cdot)\|_{1,\Omega_{\text{ext}}}^2+\|v_{x_0}(x_0,\cdot)\|_{0,\Omega_{\text{ext}}}^2\big)$$
$$\leqslant C(\|\varphi_0\|_{1,\Omega_{\text{ext}}}^2+\|\varphi_1\|_{0,\Omega_{\text{ext}}}^2+\int_0^T\|f(x_0,\cdot)\|_{0,\Omega_{\text{ext}}}^2dx_0). \tag{9.37}$$

Thus, we have proved an analogue of Theorem 8.3.

Theorem 9.1. *Let S_0 be a characteristic surface inside the ergosphere S_Δ, and let $v(x_0,x)$ satisfy* (9.1) *and* (9.2). *Suppose that* (5.3) *holds. Then $v(x_0,x)$ satisfies* (9.37).

Note that (9.37) implies that $D_+(\Omega_{\text{int}}\times\mathbf{R})\subset\overline{\Omega}_{\text{int}}\times\mathbf{R}$, i.e., $\Omega_{\text{int}}\times\mathbf{R}$ is a black hole.

As in the case of Theorem 8.1, we can take into account the boundary condition on $\partial\Omega_0$ and prove the estimate

$$\max_{0\leqslant x_0\leqslant T}\left(\|v(x_0,\cdot)\|^2_{1,\Omega_{\text{ext}}\bigcap\Omega}+\|v_{x_0}(x_0,\cdot)\|^2_{0,\Omega_{\text{ext}}\bigcap\Omega}\right)$$
$$\leqslant C\int_0^T\|f(x_0,\cdot)\|^2_{0,\Omega_{\text{ext}}\bigcap\Omega}dx_0+[g]^2_{1,\partial\Omega_0\times(0,T)}. \tag{9.38}$$

If (5.2) holds, the domain D_T is contained in $\Omega_{\text{int}}\times(0,T)$. In this case, arguments similar to the proof of Theorem 9.1 lead to an estimate of the form (9.38) in $(\Omega_{\text{int}}\bigcap\Omega)\times(0,T)$, i.e., in this case, $\Omega_{\text{int}}\times\mathbf{R}$ is a white hole.

References

1. Belishev, M.: Boundary control in reconstruction of manifolds and metrics (the BC method). Inverse Probl. **13**, R1-R45 (1997)
2. Courant, R., Hilbert, D.: Methods of Mathematical Physics, II, Interscience, New York et al. (1962)
3. Eskin, G.: A new approach to the hyperbolic inverse problems. Inverse Probl. **22**, no. 3, 815–831 (2006)
4. Eskin, G.: A new approach to the hyperbolic inverse problems II: global step. Inverse Probl. **23**, no. 6, 2343–2356 (2007)
5. Eskin, G.: Inverse hyperbolic problems with time-dependent coefficients. Commun. Partial Differ. Equ. **32**, 1737–1758 (2007)
6. Eskin, G.: Optical Aharonov-Bohm effect: inverse hyperbolic problem approach. Commun. Math. Phys. **284**, 317–343 (2008)
7. Eskin, G.: Inverse hyperbolic problems and optical black holes. ArXiv:0809.3987 (2008)
8. Eskin, G.: Perturbations of the Kerr black hole in the class of axisymmetric artificial black holes. ArXiv:0905.4129 (2009)
9. Eskin, G., Ralston, J.: On the determination boundaries for hyperbolic equations. ArXiv:0902.4497 (2009)
10. Gordon, W.: Ann. Phys. (Leipzig) **72**, 421 (1923)
11. Hartman, F.: Ordinary Differential Equations. John Wiley and Sons, New York (1964)
12. Hörmander, L.: The Analysis of Linear Partial Differential Operators III, Springer, Berlin (1985)
13. Katchalov, A., Kurylev, Y., Lassas, M.: Inverse Boundary Spectral Problems. Chapman and Hall, Boca Baton (2001)
14. Kozlov, V., Maz'ya, V., Fomin, A.: The inverse problem of coupled thermoelasticity. Inverse Probl. **10**, 153–160 (1994)
15. Novello, M., Visser, M., Volovik, G. (Eds.): Artificial Black Holes: World Scientific, Singapore (2002)
16. Tataru, D.: Unique continuation for solutions to PDE. Commun. Partial Differ. Equ. **20**, 855–884 (1995)
17. Visser, M.: Acoustic black holes, horizons, ergospheres and Hawking radiation. Classical Quantum Gravity **15**, no. 6, 1767–1791 (1998)

Global Green's Function Estimates

Michael W. Frazier and Igor E. Verbitsky

Abstract Under certain conditions, we obtain global pointwise estimates for Neumann series associated with an integral operator with general quasimetric kernel. The estimates involve the first and second iterations of the original kernel. As a consequence we deduce sharp bilateral bounds of Green's function for the fractional Schrödinger operator $(-\Delta)^{\alpha/2} - q$ with general attractive potential $q \geqslant 0$ on the entire Euclidean space $\mathbb{R}^n$ for $0 < \alpha < n$, or a bounded nontangentially accessible domain $\Omega \subseteq \mathbb{R}^n$ for $0 < \alpha \leqslant 2$, under a certain "smallness condition" on q. Most of our results are new even in the classical case $\alpha - 2$.

1 Introduction

We obtain global exponential bounds for Green's functions for a broad class of differential and integral equations with possibly singular coefficients, data, and boundaries of the domains involved. Of greatest interest is the Schrödinger operator with potential q, defined by $\mathcal{L} = -\triangle - q$, on a domain $\Omega \subseteq \mathbb{R}^n$ for $n \geqslant 3$. More generally, we consider the nonlocal operator $\mathcal{L}_\alpha = (-\triangle)^{\alpha/2} - q$ and the associated Green's function (defined on a domain Ω as in, for example, [12]), and the Dirichlet problem

$$\mathcal{L}_\alpha u = (-\triangle)^{\frac{\alpha}{2}} u - q\, u = \varphi \quad \text{in } \Omega, \quad u = 0 \quad \text{on } \Omega^c. \tag{1.1}$$

Michael W. Frazier
Mathematics Department, University of Tennessee, Knoxville, Tennessee 37922, USA
e-mail: frazier@math.utk.edu

Igor E. Verbitsky
Department of Mathematics, University of Missouri, Columbia, Missouri 65211, USA
e-mail: verbitskyi@missouri.edu

A. Laptev (ed.), *Around the Research of Vladimir Maz'ya III: Analysis and Applications*, International Mathematical Series 13, DOI 10.1007/978-1-4419-1345-6_5,

Here q and φ are locally integrable functions or locally finite measures. We allow $0 < \alpha < n$ if Ω is the entire space $\mathbb{R}^n$, and $0 < \alpha \leqslant 2$ for a bounded domain Ω, with $n \neq 2$ if $\alpha = 2$.

We consider domains Ω with a Green's operator $G^{(\alpha)}$ for $(-\triangle)^{\alpha/2}$ with a nonnegative Green's function $G^{(\alpha)}(x,y)$, where $G^{(\alpha)}(x,y)$ is the kernel of the integral operator $G^{(\alpha)}$. Our theory is applicable to any bounded domain Ω with the uniform Harnack boundary principle, established originally by Jerison and Kenig [23] for nontangentially accessible (NTA) domains. This principle is now known to hold for a large class of domains in $\mathbb{R}^n$, $n \geqslant 2$, including uniform domains in the classical case $\alpha = 2$, and even more general domains with the interior corkscrew condition if $0 < \alpha < 2$ (cf. [4, 19, 37]).

For $\Omega = \mathbb{R}^n$, by $G^{(\alpha)}$ we mean the Riesz potential I_α with kernel $c_{n,\alpha}|x - y|^{\alpha-n}, 0 < \alpha < n$. We often write G and $G(x,y)$ instead of $G^{(\alpha)}$ and $G^{(\alpha)}(x,y)$ when the dependence on α is understood.

The literature on Green's functions of elliptic differential operators, and their relation to spectral theory, stochastic processes, and nonlinear problems is immense. We refer to [6, 2, 7, 14, 20, 25, 30] for a variety of results and estimates, and further references. We were specifically motivated by work in [10, 24] on the nonlinear equation

$$(-\triangle)^{\frac{\alpha}{2}} u = q\, u^s + \varphi, \quad u \geqslant 0, \tag{1.2}$$

for $s > 1$ and arbitrary nonnegative q, φ. In [24], a necessary and matching sufficient condition were found for the existence of positive weak solutions, together with sharp pointwise estimates, for a wide class of differential and integral equations, including (1.2). An alternative proof for the Laplacian operator, which gives a better constant in the necessity part, was obtained in [10]. Curiously, neither of their methods apply directly to (1.1) in the end-point case $s = 1$.

Applying Green's operator to both sides of (1.1), we obtain the equivalent problem

$$u = G(q\, u) + f \quad \text{on } \Omega, \tag{1.3}$$

where $f = G(\varphi)$. In this paper, our estimates depend only on $|q|$, so we define

$$d\omega(y) = |q(y)|\, dy \tag{1.4}$$

and the operator T by

$$T(u)(x) = G(|q| \cdot u)(x) = \int_\Omega G(x,y) u(y)\, d\omega(y). \tag{1.5}$$

In fact, we can replace q in (1.1) by a nonnegative measure ω on Ω and obtain (1.5) in the same way. For $q \geqslant 0$, (1.3) takes the form

$$u = Tu + f. \tag{1.6}$$

A formal solution of (1.6) is $u = (I - T)^{-1}(f) = \sum_{j=0}^{\infty} T^j f$. Let $G_1(x, y) = G(x, y)$. For $j > 1$ define G_j inductively by

$$G_j(x, y) = \int_{\Omega} G(x, z) G_{j-1}(z, y)\, d\omega(z). \tag{1.7}$$

Then $G_j(x, y)$ is the kernel of T^j with respect to the measure $d\omega$, i.e.,

$$T^j u(x) = \int_{\Omega} G_j(x, y) u(y)\, d\omega(y).$$

We define the Neumann series

$$V(x, y) = \sum_{j=1}^{\infty} G_j(x, y), \quad x, y \in \Omega. \tag{1.8}$$

When $q \geqslant 0, V$ is called the *minimal Green's function* for $\mathcal{L}_\alpha = (-\triangle)^{\frac{\alpha}{2}} - q$. Our main objective is to obtain estimates for V. Note that for $q \geqslant 0$,

$$u(x) = \sum_{j=0}^{\infty} T^j f(x) = f(x) + \int_{\Omega} V(x, y) f(y)\, d\omega(y).$$

For $f = G(\varphi)$ we obtain

$$u(x) = \int_{\Omega} V(x, y) \varphi(y)\, dy, \tag{1.9}$$

for, say, $\varphi \geqslant 0$. Hence V plays the same role in the solution of (1.1) that the Green's kernel G plays when $q = 0$. To see (1.9), note that

$$\begin{aligned}
u(x) &= G\varphi(x) + \int_{\Omega} V(x, y) G\varphi(y)\, d\omega(y) \\
&= \int_{\Omega} G(x, z)\varphi(z)\, dz + \int_{\Omega} V(x, y) \int_{\Omega} G(y, z)\varphi(z)\, dz\, d\omega(y) \\
&= \int_{\Omega} \left(G(x, z) + \int_{\Omega} \sum_{j=1}^{\infty} G_j(x, y) G(y, z)\, d\omega(y) \right) \varphi(z)\, dz \\
&= \int_{\Omega} \left(G(x, z) + \sum_{j=2}^{\infty} G_j(x, z) \right) \varphi(z)\, dz = \int_{\Omega} V(x, z)\varphi(z)\, dz.
\end{aligned}$$

Our estimates do not involve cancellations resulting from q. Hence the lower bounds require the assumption $q \geqslant 0$. Otherwise, the assumptions needed are minimal.

Theorem 1.1. *Suppose that $\Omega \subseteq \mathbb{R}^n$ is either the entire space or a bounded domain satisfying the uniform boundary Harnack principle. Let $0 < \alpha < n$ if Ω is $\mathbb{R}^n$; otherwise, let $0 < \alpha \leqslant 2$, with $\alpha \neq 2$ if $n = 2$. Suppose that q is either a nonnegative locally integrable function or a nonnegative locally finite measure on Ω. Then there exists $c_1 > 0$, depending only on α and Ω, such that*

$$V(x,y) = \sum_{j=1}^{\infty} G_j^{(\alpha)}(x,y) \geqslant G^{(\alpha)}(x,y) e^{c_1 G_2^{(\alpha)}(x,y)/G^{(\alpha)}(x,y)}. \tag{1.10}$$

The estimate (1.10) in the classical case $\alpha = 2$ is implicit in the estimates of the conditional gauge in [14, Theorems 7.6 and 7.19–7.21] (cf. also similar lower estimates for negative q, under the additional assumption that q is in the local Kato class, in [18]).

The main goal of this paper is to show that there is an upper estimate for V of the same nature as the lower estimate in Theorem 1.1, under a certain smallness condition on ω. To motivate our condition, suppose that $q \geqslant 0$ and there exists $f \geqslant 0$ such that there is a nonnegative solution u, which is not identically 0, to the equation $u = Tu + f$. Then $Tu(x) \leqslant u(x)$ for almost every $x \in \Omega$. Hence, by the Schur lemma, the operator norm $\|T\| = \|T\|_{L^2(\omega)\to L^2(\omega)}$ of T on $L^2(\omega)$ satisfies $\|T\| \leqslant 1$. On the other hand, if $f \in L^2(\omega)$ and $\|T\| < 1$, then $\sum_{j=0}^{\infty} T^j f$ converges in $L^2(\omega)$ to a solution of $u = Tu + f$. The goal in this paper is to obtain pointwise estimates. For this purpose, we consider certain "weak boundedness" conditions. We use this terminology in a general sense; our condition is not the same as the weak boundedness condition considered in the famous $T1$ theorem.

Our weak boundedness conditions involve a function m defined on Ω, as follows. If $\Omega = \mathbb{R}^n$, let $m(x) = 1$ for all x. If Ω is a bounded $C^{1,1}$ domain, let $m(x) = \operatorname{dist}(x, \partial\Omega)^{\alpha/2}$. If Ω is any bounded domain that obeys the uniform boundary Harnack principle, let x_0 be a fixed point in Ω and let $m(x) = \min(1, G(x, x_0))$. We will note at the end of Section 2 that, in all the cases we consider, $d(x,y) = m(x)m(y)/G(x,y)$ satisfies the quasitriangle inequality $d(x,y) \leqslant \varkappa(d(x,z) + d(z,y))$ for all $x, y, z \in \Omega$, for some $\varkappa > 0$. We define the quasimetric balls

$$B_t(x) = \{z \in \Omega : d(x,z) < t\}$$

for $x \in \Omega$ and $t > 0$.

Define $\|\omega\|_*$ to be the infimum of all positive constants C such that

$$\int_{B\times B} G(x,y) m(x) m(y)\, d\omega(x)\, d\omega(y) \leqslant C \int_B m^2(z)\, d\omega(z) \tag{1.11}$$

for all quasimetric balls $B \subseteq \Omega$, and let $\|\omega\|_{wb}$ be the infimum of all positive constants C such that

$$\int_{E\times E} G(x,y)m(x)m(y)\,d\omega(x)\,d\omega(y) \leqslant C\int_E m^2(z)\,d\omega(z) \tag{1.12}$$

for all measurable sets $E \subseteq \Omega$. Conditions of this type for $\Omega = \mathbb{R}^n$ and $G(x,y) = |x-y|^{\alpha-n}$, $m(x) = 1$ were introduced by Adams [1, Theorem 7.2.1] and earlier, in an equivalent capacitary form, by Maz'ya [28, Sections 8.3 and 8.9].

It is obvious that $\|\omega\|_* \leqslant \|\omega\|_{wb}$. Since

$$\langle T(m\chi_E), m\chi_E\rangle_{L^2(\omega)} \leqslant \|T\|\ \|m\chi_E\|^2_{L^2(\omega)},$$

one obtains

$$\|\omega\|_{wb} \leqslant \|T\|. \tag{1.13}$$

The weak boundedness condition $\|\omega\|_{wb} < \infty$ is related to conditions studied in [1, Theorem 7.2.1] and [28, Theorem 8.3].

We say that a measure $d\nu$ on Ω is *d-doubling* if

$$\sup_{\{x\in\Omega, t>0:\nu(B_t(x))>0\}} \frac{\nu(B_{2\varkappa t}(x))}{\nu(B_t(x))} < \infty.$$

For the upper estimate for V, we do not need to assume that $q \geqslant 0$, but the bounds are expressed in terms of $d\omega(x) = |q(x)|dx$.

Theorem 1.2. *Suppose that $\Omega \subseteq \mathbb{R}^n$ is either the entire space or a bounded domain satisfying the uniform boundary Harnack principle. Let $0 < \alpha < n$ if Ω is $\mathbb{R}^n$; otherwise, let $0 < \alpha \leqslant 2$, with $\alpha \neq 2$ if $n = 2$. Suppose that q is either a locally integrable function or a locally finite nonnegative measure on Ω. Define m as above, depending on Ω. Then there exists $\epsilon > 0$ such that if*

$$m^2\,d\omega \text{ is a d-doubling measure and } \|\omega\|_* < \epsilon \tag{1.14}$$

or

$$\|\omega\|_{wb} < \epsilon, \tag{1.15}$$

then there exist nonnegative constants c_2 and C_2, depending only on α and Ω, such that

$$V(x,y) = \sum_{j=1}^{\infty} G_j^{(\alpha)}(x,y) \leqslant C_2 G^{(\alpha)}(x,y)e^{c_2 G_2^{(\alpha)}(x,y)/G^{(\alpha)}(x,y)}. \tag{1.16}$$

In the case (1.14), ϵ depends on α, Ω and the doubling constant for $m^2\,\omega$, whereas, in the case (1.15), ϵ depends only on α and Ω.

A natural question is to determine which potentials q are sufficiently mild that Green's kernel V for the operator $(-\triangle)^{\alpha/2} - q$ is pointwise equivalent to the Green's kernel $G^{(\alpha)}$ for the case $q = 0$. Under the smallness conditions in Theorem 1.2, the bilateral estimates from Theorems 1.1 and 1.2 show that

$V = \sum_{j=1}^{\infty} G_j \sim G$ if and only if the apparently much simpler condition $G_2 \leqslant CG$ holds for some constant C. Equivalences of this type have been observed under different assumptions, for example, in [14, 19, 30, 34]. In Section 5, we show how some Kato type conditions imply that $G_2 \leqslant CG$.

As an easy consequence of our estimates, we find that the seemingly weaker conditions $\|\omega\|_* < \infty$ (when ω is d-doubling) and $\|\omega\|_{wb} < \infty$ are each equivalent to the boundedness of T on $L^2(\omega)$.

Theorem 1.3. *Suppose that $\Omega \subseteq \mathbb{R}^n$ is either the entire space or a bounded domain satisfying the uniform boundary Harnack principle. Let $0 < \alpha < n$ if Ω is $\mathbb{R}^n$; otherwise, let $0 < \alpha \leqslant 2$, with $\alpha \neq 2$ if $n = 2$. Suppose that q is either a locally integrable function or a locally finite nonnegative measure on Ω. Define m as above depending on Ω. Then*

(i) $\|\omega\|_{wb} \approx \|T\|$

and

(ii) *if $m^2\, d\omega$ is a d-doubling measure, then* $\|\omega\|_* \approx \|T\|$.

The statement of Theorem 1.3 is analogous to the results of [26, 36] showing that, for certain operators, norm boundedness is equivalent to a testing condition of Sawyer type. In many cases, the assumption that $m^2 d\omega$ is a d-doubling measure in the second statement can be removed. Moreover, the boundary behavior of $m(x)$ in the condition (1.12) is not important since it suffices instead to check the testing condition

$$\int_{B\times B} |x-y|^{\alpha-n} d\omega(x)\, d\omega(y) \leqslant C\, \omega(B),$$

on all Euclidean balls B when $\Omega = \mathbb{R}^n$, or on balls B lying inside each Whitney cube Q associated with Ω, where C does not depend on $B \subseteq Q$. This can be shown using Theorem 1.3 and a localization argument in the case $\alpha = 2$ for any bounded domain satisfying the Hardy inequality (cf. [29]), in particular for Lipschitz and NTA domains. In the case $0 < \alpha < 2$, a similar localization argument is applicable at least to smooth domains (cf. [32]). For $\Omega = \mathbb{R}^n$ this is easy to deduce from [26].

Our results also yield estimates in cases where V and G are not pointwise equivalent. We consider one such an example in which reasonably explicit estimates can be obtained from Theorems 1.1 and 1.2.

Example 1.1. Suppose that $0 < \alpha < n$. Define the operator $\mathcal{L}$ in $\mathbb{R}^n$ by

$$\mathcal{L} = (-\triangle)^{\alpha/2} - A/|x|^{\alpha},$$

where A is a constant. Then there exists $\epsilon > 0$ such that if $0 < A < \epsilon$, there exist positive constants b_1, b_3, depending only on n, α, and A, and b_2, b_4, depending only on n and α, such that the minimal Green's function $V(x,y)$ of $\mathcal{L}$ satisfies

$$b_1 \frac{\left(\max\left\{\frac{|x|}{|y|}, \frac{|y|}{|x|}\right\}\right)^{b_2 A}}{|x-y|^{n-\alpha}} \leqslant V(x,y) \leqslant b_3 \frac{\left(\max\left\{\frac{|x|}{|y|}, \frac{|y|}{|x|}\right\}\right)^{b_4 A}}{|x-y|^{n-\alpha}} \tag{1.17}$$

for all $x, y \in \mathbb{R}^n$.

Such estimates, along with asymptotics of Green's functions, are well known in the classical case $\alpha = 2$ (cf. [38, Lemma 5], [17], and [30]).

Theorems 1.1, 1.2, and 1.3 are obtained from general results below (Theorems 2.3, 4.4, and 5.1, respectively) for general quasimetrically modifiable kernels on measure spaces (cf. Definitions 2.1 and 4.1).

Analogous results hold for many other differential and integral operators, as long as their Green's function is globally comparable to Green's function for $(-\triangle)^{\alpha/2}$ on Ω. We observe that perturbation theorems that guarantee such comparability are known for wide classes of elliptic operators (cf., for example, [7, 34, 39]).

In [16], we followed the general approach (as in [11] and [21]) of first studying dyadic models of these equations, where the proofs and basic estimates are more transparent. This is reminiscent of a similar approach that recently has been used in studies of nonlinear problems, including quasilinear, Hessian, and subelliptic equations (cf. [24, 32, 33]).

In [15], a different approach is presented in which the same types of estimates (still in the context of general quasimetrically modifiable kernels on measure spaces) are proved but with the assumption that $\|T\| < 1$ replacing the smallness assumptions on $\|\omega\|_*$ or $\|\omega\|_{wb}$.

2 Lower Bounds of Green's Functions

Definition 2.1. Let (Ω, ω) be a measure space. Suppose that K is a map from $\Omega \times \Omega$ into $(0, +\infty]$ such that $K(\cdot, y)$ is ω-measurable for all $y \in \Omega$. Define

$$d(x,y) = 1/K(x,y). \tag{2.1}$$

We say that K is a *quasimetric kernel* on Ω with quasimetric constant $\varkappa > 1/2$ if

(i) K is symmetric: $K(x,y) = K(y,x)$ for all $x, y \in \Omega$,

(ii) $K(x,y) < +\infty$ if $x \neq y$,

(iii) d satisfies the quasitriangle inequality with constant $\varkappa$:

$$d(x,y) \leqslant \varkappa\,(d(x,z) + d(z,y)) \tag{2.2}$$

for all $x, y, z \in \Omega$.

For $x \in \Omega$ and all $r > 0$ we define the *quasimetric balls*

$$B_r(x) = \{z \in \Omega : d(z,x) < r\}. \tag{2.3}$$

We say that ω is *locally finite* if $\omega(B_r(x)) < \infty$ for all $x \in \Omega$ and $r \in (0,\infty)$.

We remark that we do not require that $K(x,x) = +\infty$, and so it might happen that $d(x,x) > 0$. Hence some of the quasimetric balls $B_r(x)$ might be empty.

Inductively we define K_j for $j \in \mathbb{N}$ by letting $K_1 = K$ and for $j \geqslant 2$

$$K_j(x,y) = \int_\Omega K(x,z)K_{j-1}(z,y)\,d\omega(z). \tag{2.4}$$

Note that all of the kernels K_j are symmetric:

$$K_j(x,y) = K_j(y,x). \tag{2.5}$$

For $t \geqslant 0$ and $x \in \Omega$ we define

$$G_t(x) = \int_0^t \frac{\omega(B_r(x))}{r^2}\,dr. \tag{2.6}$$

Next we collect some simple facts which will be useful.

Lemma 2.1. *Suppose that ω is σ-finite. For any $t > 0$ and $x \in \Omega$*

$$\int_{B_t(x)} \frac{d\omega(z)}{d(x,z)} = G_t(x) + \frac{\omega(B_t(x))}{t}. \tag{2.7}$$

For $m = 0,1,2,\dots, y \in \Omega$, and $a > 0$

$$\int_a^\infty G_t^m(y)\omega(B_t(y))\frac{dt}{t^3} \leqslant \frac{1}{m+1}\int_a^\infty G_t^{m+1}(y)\frac{dt}{t^2} \tag{2.8}$$

and

$$\int_a^\infty G_t^{m+1}(y)\frac{dt}{t^2} \leqslant (m+1)\int_a^\infty G_t^m(y)\omega(B_t(y))\frac{dt}{t^3} + \frac{G_a^{m+1}(y)}{a}. \tag{2.9}$$

Proof. By the Fubini theorem,

$$\begin{aligned} G_t(x) &= \int_0^t \frac{\omega(B_s(x))}{s^2}\,ds = \int_0^t \int_{B_s(x)} \frac{1}{s^2}\,d\omega(z)\,ds \\ &= \int_{B_t(x)} \int_{d(x,z)}^t \frac{1}{s^2}\,ds\,d\omega(z) = \int_{B_t(x)} \frac{d\omega(z)}{d(x,z)} - \frac{\omega(B_t(x))}{t}, \end{aligned}$$

which proves (2.7). For $b > a$ integration by parts yields

$$\int_a^b G_t^m(y)\frac{\omega(B_t(y))}{t^2}\frac{dt}{t} = \left.\frac{G_t^{m+1}(y)}{(m+1)t}\right|_a^b + \int_a^b \frac{G_t^{m+1}(y)}{m+1}\frac{dt}{t^2}.$$

Then (2.9) follows by letting $b \to \infty$. To prove (2.8), we can assume that the right-hand side is finite. Note that

$$\left.\frac{G_t^{m+1}(y)}{t}\right|_a^b \leqslant \frac{G_b^{m+1}(y)}{b} = \frac{1}{\log 2}\int_b^{2b} \frac{G_b^{m+1}(y)}{b}\frac{dt}{t} \leqslant \frac{2}{\log 2}\int_b^{2b}\frac{G_t^{m+1}(y)}{t}\frac{dt}{t},$$

which goes to 0 as $b \to \infty$ if the right-hand side of (2.8) is finite. Hence (2.8) holds. □

Theorem 2.1. *Let K be a quasimetric kernel on a σ-finite measure space (Ω, ω) with quasimetric constant $\varkappa$. Define K_j for $j \in \mathbb{N}$ by (2.4) and G_t by (2.6). Then*

$$\sum_{j=1}^{\infty} K_j(x,y) \geqslant \int_{d(x,y)}^{\infty} e^{\frac{1}{4\varkappa}(G_t(x)+G_t(y))}\frac{dt}{t^2}. \tag{2.10}$$

Proof. We first claim that for $j \in \mathbb{N}$,

$$\int_{d(x,y)}^{\infty} G_t^j(y)\frac{dt}{t^2} \leqslant 2\varkappa j\int_\Omega\int_{d(z,y)}^{\infty} K(x,z)G_t^{j-1}(y)\frac{dt}{t^2}\,d\omega(z). \tag{2.11}$$

By (2.9),

$$\int_{d(x,y)}^{\infty} G_t^j(y)\frac{dt}{t^2} \leqslant j\int_{d(x,y)}^{\infty} G_t^{j-1}(y)\omega(B_t(y))\frac{dt}{t^3} + \frac{G_{d(x,y)}^j(y)}{d(x,y)}.$$

Note that for $d(x,y) < t$ and $z \in B_t(y)$, we have

$$d(x,z) \leqslant \varkappa(d(x,y)+d(y,z)) < 2\varkappa t \quad \text{or} \quad \frac{1}{t} \leqslant \frac{2\varkappa}{d(x,z)} = 2\varkappa K(x,z).$$

Hence

$$\begin{aligned} j\int_{d(x,y)}^{\infty} G_t^{j-1}(y)\omega(B_t(y))\frac{dt}{t^3} &= j\int_{d(x,y)}^{\infty} G_t^{j-1}(y)\int_{B_t(y)}\frac{d\omega(z)}{t}\frac{dt}{t^2} \\ &\leqslant 2\varkappa j\int_{d(x,y)}^{\infty} G_t^{j-1}(y)\int_{B_t(y)} K(x,z)\,d\omega(z)\frac{dt}{t^2}. \end{aligned}$$

For $t < d(x,y)$ and $z \in B_t(y)$ we have

$$d(x,z) \leqslant \varkappa(d(x,y)+d(y,z)) \leqslant \varkappa(d(x,y)+t) < 2\varkappa\, d(x,y)$$

or

$$\frac{1}{d(x,y)} < \frac{2\varkappa}{d(x,z)} = 2\varkappa K(x,z).$$

Therefore, by the fundamental theorem of calculus and the fact that $G_0 = 0$,

$$\begin{aligned}
\frac{G^j_{d(x,y)}(y)}{d(x,y)} &= \frac{1}{d(x,y)} \int_0^{d(x,y)} jG_t^{j-1}(y)\omega\left(B_t(y)\right) \frac{dt}{t^2} \\
&= j \int_0^{d(x,y)} G_t^{j-1}(y) \int_{B_t(y)} \frac{d\omega(z)}{d(x,y)} \frac{dt}{t^2} \\
&\leqslant 2\varkappa j \int_0^{d(x,y)} G_t^{j-1}(y) \int_{B_t(y)} K(x,z)\, d\omega(z) \frac{dt}{t^2}.
\end{aligned}$$

Putting these estimates together,

$$\int_{d(x,y)}^{\infty} G_t^j(y) \frac{dt}{t^2} \leqslant 2\varkappa j \int_0^{\infty} G_t^{j-1}(y) \int_{B_t(y)} K(x,z)\, d\omega(z) \frac{dt}{t^2},$$

and (2.11) follows by the Fubini theorem.

We claim that

$$K_j(x,y) \geqslant \frac{1}{(j-1)!(2\varkappa)^{j-1}} \int_{d(x,y)}^{\infty} G_t(y)^{j-1} \frac{dt}{t^2} \tag{2.12}$$

for all $j \in \mathbb{N}$. For $j = 1$

$$K(x,y) = \frac{1}{d(x,y)} = \int_{d(x,y)}^{\infty} \frac{dt}{t^2}. \tag{2.13}$$

Now, suppose that (2.12) holds for j. Then

$$\begin{aligned}
K_{j+1}(x,y) &= \int_{\Omega} K(x,z) K_j(z,y)\, d\omega(z) \\
&\geqslant \frac{1}{(j-1)!(2\varkappa)^{j-1}} \int_{\Omega} K(x,z) \int_{d(z,y)}^{\infty} G_t(y)^{j-1} \frac{dt}{t^2}\, d\omega(z) \\
&\geqslant \frac{1}{j!(2\varkappa)^j} \int_{d(x,y)}^{\infty} G_t(y)^j \frac{dt}{t^2}
\end{aligned}$$

by (2.11), which proves (2.12) by induction.

Summing (2.12) over j yields

$$\sum_{j=1}^{\infty} K_j(x,y) \geqslant \int_{d(x,y)}^{\infty} \frac{e^{\frac{1}{2\varkappa} G_t(y)}}{t^2}\, dt.$$

By symmetry (i.e., (2.5)), we have the same estimate with $G_t(y)$ replaced by $G_t(x)$ on the right-hand side. Averaging, we obtain

$$\sum_{j=1}^{\infty} K_j(x,y) \geqslant \int_{d(x,y)}^{\infty} \frac{\frac{1}{2}\left(e^{\frac{1}{2\varkappa}G_t(x)} + e^{\frac{1}{2\varkappa}G_t(y)}\right)}{t^2}\, dt.$$

Using $\frac{1}{2}(a+b) \geqslant \sqrt{ab}$, we obtain (2.10). □

The next lemma is of interest in its own right.

Lemma 2.2. *Let K be a quasimetric kernel on a σ-finite measure space (Ω, ω) with quasimetric constant $\varkappa$. Define K_j for $j \in \mathbb{N}$ by (2.4) and G_t by (2.6). Then for all $x, y \in \Omega$*

$$\frac{1}{4\varkappa}\int_{d(x,y)}^{\infty} G_t(x)+G_t(y)\,\frac{dt}{t^2} \leqslant K_2(x,y) \leqslant 4\varkappa \int_{d(x,y)}^{\infty} G_t(x)+G_t(y)\,\frac{dt}{t^2}. \quad (2.14)$$

Proof. The first inequality follows from the symmetry of K_2 and the case $j=2$ of (2.12). Let $s_0 = d(x,y)/(2\varkappa)$. Then

$$\begin{aligned} K_2(x,y) &= \int_{\Omega} K(x,z)K(z,y)\,d\omega(z) = \int_{\Omega} \frac{K(x,z)}{d(z,y)}\,d\omega(z) \\ &= \int_{\Omega} K(x,z)\int_{d(z,y)}^{\infty}\frac{dt}{t^2}\,d\omega(z) = \int_0^{\infty}\int_{B_t(y)} K(x,z)\,d\omega(z)\,\frac{dt}{t^2} \\ &= I + II \end{aligned}$$

by the Fubini theorem, where

$$\begin{aligned} I &= \int_0^{s_0}\int_{B_t(y)} K(x,z)\,d\omega(z)\,\frac{dt}{t^2}, \\ II &= \int_{s_0}^{\infty}\int_{B_t(y)} K(x,z)\,d\omega(z)\,\frac{dt}{t^2}. \end{aligned}$$

For I, note that for $s < s_0$ and $z \in B_s(y)$ we have

$$2\varkappa s_0 = d(x,y) \leqslant \varkappa\,(d(x,z)+d(z,y)) \leqslant \varkappa\,(d(x,z)+s) \leqslant \varkappa d(x,z) + \varkappa s_0.$$

Hence $s_0 < d(x,z)$ or $K(x,z) < 1/s_0$. Therefore,

$$\begin{aligned} I \leqslant \frac{1}{s_0}\int_0^{s_0}\frac{\omega\,(B_t(y))}{s^2}\,dt = \frac{G_{s_0}(y)}{s_0} &= 2\varkappa\int_{2\varkappa s_0}^{\infty}\frac{G_{s_0}(y)}{s^2}\,ds \\ &\leqslant 2\varkappa\int_{d(x,y)}^{\infty}\frac{G_t(y)}{t^2}\,dt \end{aligned}$$

since G_t is an increasing function of t.

Write

$$II \leqslant III + IV,$$

where

$$III = \int_{s_0}^{\infty} \int_{B_t(y)\setminus B_t(x)} K(x,z)\, d\omega(z)\, \frac{dt}{t^2},$$

$$IV = \int_{s_0}^{\infty} \int_{B_t(x)} K(x,z)\, d\omega(z)\, \frac{dt}{t^2}.$$

For $z \notin B_t(x)$ we have $K(x,z) = 1/d(x,z) < 1/t$, so

$$III \leqslant \int_{s_0}^{\infty} \int_{B_t(y)} d\omega(z)\, \frac{dt}{t^3} = \int_{s_0}^{\infty} \omega\left(B_t(y)\right) \frac{dt}{t^3} \leqslant \int_{s_0}^{\infty} G_t(y)\, \frac{dt}{t^2}$$

by the case $m = 0$ of (2.8). Since G_t is increasing in t, and by a change of variable,

$$III \leqslant \int_{s_0}^{\infty} G_{2\varkappa t}(y)\, \frac{dt}{t^2} = 2\varkappa \int_{2\varkappa s_0}^{\infty} G_t(y)\, \frac{dt}{t^2} = 2\varkappa \int_{d(x,y)}^{\infty} G_t(y)\, \frac{dt}{t^2}.$$

In view of (2.7) and (2.8), we have

$$IV \leqslant \int_{s_0}^{\infty} G_t(x)\, \frac{dt}{t^2} + \int_{s_0}^{\infty} \omega(B_t(x))\, \frac{dt}{t^3} \leqslant 2 \int_{s_0}^{\infty} G_t(x)\, \frac{dt}{t^2} \leqslant 4\varkappa \int_{d(x,y)}^{\infty} G_t(x)\, \frac{dt}{t^2}$$

by the same arguments as for III. Adding these estimates gives the right inequality of (2.14). □

Theorem 2.2. *Let K be a quasimetric kernel on a σ-finite measure space (Ω, ω) with quasimetric constant $\varkappa$. Define K_j for $j \in \mathbb{N}$ by (2.4) and G_t by (2.6). Then*

$$\sum_{j=1}^{\infty} K_j(x,y) \geqslant K(x,y) e^{\left(\frac{1}{16\varkappa^2}\right) K_2(x,y)/K(x,y)} \tag{2.15}$$

for all $x, y \in \Omega$.

Proof. Let $t_0 = d(x,y) = 1/K(x,y)$ for all $x, y \in \Omega$. By (2.14),

$$\frac{1}{16\varkappa^2} \frac{K_2(x,y)}{K(x,y)} \leqslant t_0 \int_{t_0}^{\infty} \frac{G_t(x) + G_t(y)}{4\varkappa}\, \frac{dt}{t^2}.$$

The measure $t_0 dt/t^2$ is a probability measure on the space (t_0, ∞), so exponentiating both sides of the last inequality and applying the Jensen inequality implies

$$K(x,y)e^{\left(\frac{1}{16\varkappa^2}\right)K_2(x,y)/K(x,y)} \leqslant \int_{d(x,y)}^{\infty} e^{\left(\frac{1}{4\varkappa}\right)(G_t(x)+G_t(y))} \frac{dt}{t^2} \leqslant \sum_{j=1}^{\infty} K_j(x,y)$$

by Theorem 2.1. □

Theorem 2.2 admits a simple extension which substantially increases its applicability.

Definition 2.2. Let (Ω, ω) be a measure space. We say that a map $K : \Omega \times \Omega \to (0, +\infty]$ is *quasimetrically modifiable* with constant $\varkappa$ if there exists an ω-measurable function $m : \Omega \to (0, \infty)$ such that $K(x,y)/(m(x)m(y))$ is a quasimetric kernel on Ω with quasimetric constant $\varkappa$. We call m the *modifier* of K.

Theorem 2.3. *Let K be a quasimetrically modifiable kernel with modifier m and constant $\varkappa$ on σ-finite measure space (Ω, ω). Define K_j for $j \in \mathbb{N}$ by (2.4) for $j \geqslant 1$. Then*

$$\sum_{j=1}^{\infty} K_j(x,y) \geqslant K(x,y)e^{\left(\frac{1}{16\varkappa^2}\right)K_2(x,y)/K(x,y)} \tag{2.16}$$

for all $x, y \in \Omega$.

Proof. Define a measure ν on Ω by $d\nu = m^2 d\omega$. Define

$$H(x,y) = \frac{K(x,y)}{m(x)\,m(y)}$$

and define the iterates H_j of H with respect to ν inductively for $j > 1$ by

$$H_j(x,y) = \int_{\Omega} H(x,z)H_{j-1}(z,y)\,d\nu(z). \tag{2.17}$$

Note that

$$H_j(x,y) = \frac{K_j(x,y)}{m(x)m(y)} \tag{2.18}$$

for all $j \in \mathbb{N}$: (2.18) holds by definition for $j = 1$, and, inductively, assuming (2.18) with j replaced by $j-1$, we have

$$\int_{\Omega} H(x,z)H_{j-1}(z,y)\,d\nu(z) = \int_{\Omega} \frac{K(x,z)}{m(x)m(z)} \frac{K_{j-1}(z,y)}{m(z)m(y)} m^2(z)d\omega(z)$$

$$= \frac{K_j(x,y)}{m(x)m(y)}.$$

Hence, applying Theorem 2.2 to the kernel H and the measure ν, we obtain

$$\sum_{j=1}^{\infty} K_j(x,y) = m(x)m(y)\sum_{j=1}^{\infty} H_j(x,y)$$
$$\geqslant m(x)m(y)H(x,y)e^{\left(\frac{1}{16\varkappa^2}\right)H_2(x,y)/H(x,y)}$$
$$= K(x,y)e^{\left(\frac{1}{16\varkappa^2}\right)K_2(x,y)/K(x,y)}.$$

The proof is complete. □

Theorem 1.1 follows from Theorem 2.3 and known facts concerning the Green's kernel $G^{(\alpha)}$ for $(-\triangle)^{\alpha/2}$. If $0 < \alpha < n$ and $\Omega = \mathbb{R}^n$, $G^{(\alpha)}(x,y) = c_{n,\alpha}|x-y|^{\alpha-n}$ is obviously a quasimetric kernel. The σ-finiteness of ω follows from the assumption that $q \in L^1_{loc}$ or that ω is locally finite. Hence Theorem 2.15 applies to yield Theorem 1.1 in these cases.

As was pointed out in [24], even in the classical case, the Green's kernel is not necessarily a quasimetric kernel when $n \geqslant 2$. However, for bounded $C^{1,1}$ domains it is well known that for $0 < \alpha \leqslant 2$, with $\alpha \neq 2$ when $n = 2$,

$$G^{(\alpha)}(x,y) \approx \frac{\delta(x)^{\alpha/2}\delta(y)^{\alpha/2}}{|x-y|^{n-\alpha}(|x-y| + \delta(x) + \delta(y))^{\alpha}}, \tag{2.19}$$

where $\delta(x) = \mathrm{dist}(x, \partial\Omega)$ for $x \in \Omega$ and "$\approx$" means that the ratio of the two sides is bounded from above and below by positive constants depending only on Ω and α (for $\alpha = 2$ see [39] for the upper estimate and [40] for the lower estimate; for $0 < \alpha \leqslant 2$ see [12]). It follows easily that $G^{(\alpha)}$ is a quasimetric kernel with modifier $m(x) = \delta(x)^{\alpha/2}$. The kernel $G^{(\alpha)}(x,y)/(\delta(x)\delta(y))^{\alpha/2}$ is the so-called Naïm kernel (cf. [31] and also [5]).

For Lipschitz, or more general NTA (or uniform) domains, $G^{(\alpha)}$ is a quasimetric kernel on Ω with modifier $m(x) = \min(1, G(x, x_0))$, where x_0 is a fixed pole in Ω. This was first conjectured in [24], for $\alpha = 2$, and proved by Ancona in response to a question of the second author discussed at the Mittag–Leffler Institute in November, 1999. It was rediscovered later, as a result of the work of Aikawa [3], Bogdan [9], and Riahi [35] for the classical Green's function. A slightly stronger version of the quasimetric property, for $0 < \alpha \leqslant 2$, was ultimately shown by Hansen [19] to be essentially equivalent to the uniform boundary Harnack principle (cf. [4, 8, 19, 37]). These results guarantee that Theorem 1.1 follows from Theorem 2.3.

We remark that the idea of modifying a kernel to obtain the quasimetric property goes back to [24] and [34].

3 Weak Boundedness and Local Estimates

In this section, we continue to assume that K is a quasimetric kernel with constant $\varkappa$ on a σ-finite measure space (Ω, ω), G_t is defined by (2.6), and

$d = 1/K$. Define

$$\|\omega\|_* = \sup \frac{1}{\omega(B)} \iint_{B\times B} K(x,y)\, d\omega(x)\, d\omega(y), \tag{3.1}$$

where the supremum is taken over all balls $B = B_t(z)$ such that $\omega(B) > 0$, and

$$\|\omega\|_{wb} = \sup \frac{1}{\omega(E)} \iint_{E\times E} K(x,y)\, d\omega(x)\, d\omega(y), \tag{3.2}$$

where the supremum is taken over all $t > 0$ and ω-measurable $E \subseteq \Omega$ such that $\omega(E) > 0$. Define T by

$$Tf(x) = \int_\Omega K(x,y) f(y)\, d\omega(y).$$

Let $\|T\|$ denote the operator norm of T on $L^2(\omega)$.

Lemma 3.1.

$$\|\omega\|_* \leqslant \|\omega\|_{wb} \leqslant \|T\|. \tag{3.3}$$

If ω is locally finite, then

$$\sup_{t>0, x\in\Omega} \frac{\omega(B_t(x))}{t} \leqslant 2\varkappa \|\omega\|_*. \tag{3.4}$$

Proof. The first inequality in (3.3) is trivial. For the second,

$$\iint_{E\times E} K(x,y)\, d\omega(x)\, d\omega(y) = \langle T\chi_E, \chi_E\rangle_{L^2(\omega)} \leqslant \|T\| \|\chi_E\|^2_{L^2(\omega)} = \|T\|\omega(E).$$

For (3.4), let $t > 0$ and $z \in \Omega$. For $x, y \in B_t(z)$ we have $d(x,y) \leqslant \varkappa(d(x,z) + d(z,y)) < 2\varkappa t$. Hence

$$\begin{aligned}
\frac{\omega(B_t(z))^2}{2\varkappa t} &= \iint_{B_t(z)\times B_t(z)} \frac{1}{2\varkappa t}\, d\omega(x)\, d\omega(y) \\
&< \iint_{B_t(z)\times B_t(z)} \frac{1}{d(x,y)}\, d\omega(x)\, d\omega(y) \leqslant \omega(B_t(z)) \|\omega\|_*,
\end{aligned}$$

which yields (3.4) by the local finiteness assumption. □

Note that if $\|\omega\|_* < \infty$, then ω is σ-finite by (3.4).

We say that a measure $d\nu$ on Ω is a *doubling measure* if $D_\nu < \infty$, where

$$D_\nu = \sup \frac{\nu(B_{2\varkappa t}(x))}{\nu(B_t(x))}, \tag{3.5}$$

where the supremum is taken over all $x \in \Omega$ and all $t > 0$ such that $\nu(B_t(x)) > 0$.

Lemma 3.2. *Suppose that ω is a doubling measure. Then*

$$\int_{B_t(x)} G_t^m \, d\omega \leqslant m! \, (2\varkappa D_\omega \|\omega\|_*)^m \, \omega(B_t(x)), \tag{3.6}$$

for all $x \in \Omega, t > 0$, and $m \in \mathbb{N} \cup \{0\}$. Also, if $0 \leqslant \epsilon < 1/(2\varkappa D_\omega \|\omega\|_*)$, then

$$\int_{B_t(x)} e^{\epsilon G_t} \, d\omega \leqslant \frac{\omega\left(B_t(x)\right)}{1 - 2\varkappa D_\omega \|\omega\|_* \epsilon}, \tag{3.7}$$

for all $x \in \Omega$ and all $t > 0$.

Proof. The inequality (3.6) is trivial for $m = 0$. We prove, by induction, that for $m \geqslant 1$

$$\int_{B_t(x)} G_t^m \, d\omega \leqslant m! D_\omega^{-1} \, (2\varkappa D_\omega \|\omega\|_*)^m \, \omega(B_{2\varkappa t}(x)) \tag{3.8}$$

for all $x \in \Omega$ and all $t > 0$. Assuming the inequality (3.8), we obtain (3.6) by the definition of D_ω. To prove (3.8) for $m = 1$, we use (2.7) to obtain

$$\int_{B_t(x)} G_t(y) \, d\omega(y) \leqslant \int_{B_t(x)} \int_{B_t(y)} K(y, z) \, d\omega(z) \, d\omega(y).$$

For $y \in B_t(x)$ we have $B_t(y) \subseteq B_{2\varkappa t}(x)$ by the quasitriangle inequality (2.2). Hence the last expression is bounded by

$$\int_{B_{2\varkappa t}(x)} \int_{B_t(x)} K(y, z) \, d\omega(y) \, d\omega(z).$$

Since $\varkappa \geqslant 1/2$, we can enlarge $B_t(x)$ to $B_{2\varkappa t}(x)$ to obtain

$$\int_{B_t(x)} G_t(y) \, d\omega(y) \leqslant \|\omega\|_* \, \omega(B_{2\varkappa t}(x)) \tag{3.9}$$

for all $x \in \Omega$ and all $t > 0$. Since $2\varkappa \geqslant 1$, we obtain (3.8) for $m = 1$. We observe for future reference that the proof of (3.9) does not require ω to be a doubling measure.

Inductively, suppose that (3.8) holds for $m - 1$. By the fundamental theorem of calculus and the Fubini theorem,

$$\begin{aligned}\int_{B_t(x)} G_t^m(y) \, d\omega(y) &= m \int_{B_t(x)} \int_0^t G_r^{m-1}(y) \frac{\omega(B_r(y))}{r^2} \, dr \, d\omega(y) \\ &= m \int_0^t \int_{B_t(x)} G_r^{m-1}(y) \int_{B_r(y)} d\omega(z) \, d\omega(y) \, \frac{dr}{r^2}.\end{aligned}$$

For $y \in B_t(x)$ and $0 \leqslant r \leqslant t$ we have $B_r(y) \subseteq B_{2\varkappa t}(x)$ by the quasitriangle inequality (2.2). Since $z \in B_r(y)$ if and only if $y \in B_r(z)$, we obtain

$$\int_{B_t(x)} G_r^{m-1}(y) \int_{B_r(y)} d\omega(z)\, d\omega(y) \leqslant \int_{B_{2\varkappa t}(x)} \int_{B_r(z)} G_r^{m-1}(y)\, d\omega(y)\, d\omega(z)$$

$$\leqslant (m-1)! D_\omega^{-1} (2\varkappa D_\omega \|\omega\|_*)^{m-1} \int_{B_{2\varkappa t}(x)} \omega(B_{2\varkappa r}(z))\, d\omega(z),$$

by the induction hypothesis. Substituting this estimate above, we come to

$$\int_{B_t(x)} G_t^m(y)\, d\omega(y) \leqslant m! D_\omega^{-1} (2\varkappa D_\omega \|\omega\|_*)^{m-1} \int_0^t \int_{B_{2\varkappa t}(x)} \omega(B_{2\varkappa r}(z))\, d\omega(z) \frac{dr}{r^2}$$

$$= m! D_\omega^{-1} (2\varkappa D_\omega \|\omega\|_*)^{m-1} \int_{B_{2\varkappa t}(x)} \int_0^t \frac{\omega(B_{2\varkappa r}(z))}{r^2}\, dr\, d\omega(z).$$

By a change of variables,

$$\int_0^t \omega(B_{2\varkappa r}(z))/r^2\, dr = 2\varkappa G_{2\varkappa t}(z).$$

By (3.9), we see that

$$\int_{B_{2\varkappa t}(x)} \int_0^t \frac{\omega(B_{2\varkappa r}(z))}{r^2}\, dr\, d\omega(z) = 2\varkappa \int_{B_{2\varkappa t}(x)} G_{2\varkappa t}(z)\, d\omega(z)$$

$$\leqslant 2\varkappa \|\omega\|_* \omega(B_{4\varkappa^2 t}(x)) \leqslant 2\varkappa D_\omega \|\omega\|_* \omega(B_{2\varkappa t}(x)).$$

Substituting this estimate above completes the induction and hence the proof of (3.6). The estimate (3.7) follows by writing out the Taylor series for $e^{\epsilon G_t}$, applying (3.6), and summing the resulting geometric series. □

We note the following consequences of our estimates.

Corollary 3.1. *Let K be a quasimetric kernel on a locally finite measure space (Ω, ω) with quasimetric constant $\varkappa$. Define G_t by (2.6) and K_2 by (2.4). If $\|\omega\|_* < \infty$, then for a.e. $y \in \Omega$*

(i) *$G_t(y) < \infty$ for all $t \in (0, \infty)$,*

(ii) *$K(x, y) < \infty$ for a.e. $x \in \Omega$,*

(iii) *$K_2(x, y) < \infty$ for a.e. $x \in \Omega$.*

Proof. Note that for any $x \in \Omega$ and $0 < t_0 < t_1 < \infty$,

$$G_{t_1}(x) = G_{t_0}(x) + \int_{t_0}^{t_1} \frac{\omega(B_t(x))}{t^2}\, dt \leqslant G_{t_0}(x) + 2\varkappa \|\omega\|_* \log(t_1/t_0) \quad (3.10)$$

by (3.4). Hence $G_t(x)$ is either finite for all $t \in (0,\infty)$ or infinite for all $t > 0$. Hence to prove (i), it suffices to prove that $\omega(\{x \in \Omega : G_1(x) = \infty\}) = 0$.

Select an arbitrary $z \in \Omega$. For $\ell \in \mathbb{N}$ and $x \in B_\ell(z)$ we have, by (3.9),

$$\int_{B_\ell(z)} G_1(x)\, d\omega(x) \leqslant \int_{B_\ell(z)} G_\ell(x)\, d\omega(x) \leqslant \|\omega\|_* \omega(B_{2\varkappa\ell}(z)) \leqslant 4\varkappa^2 \ell \|\omega\|_*^2 < \infty$$

by (3.4). Hence $G_1(x) < \infty$ for a.e. $x \in B_\ell(z)$. Then (i) follows by taking the union over $\ell > 1$.

For (ii), if $x \neq y$, then $K(x,y) < \infty$ by condition (ii) in Definition 2.1. If $K(y,y) = \infty$, then $d(y,y) = 0$. Therefore, $y \in B_r(y)$ for all $r > 0$. By (3.4), $\omega(\{y\}) \leqslant \omega(B_r(y)) \leqslant 2\varkappa\|\omega\|_* r$ for all $r > 0$, and hence $\omega(\{y\}) = 0$. Note that (ii) holds for every $y \in \Omega$, not just a.e. y.

By (i) and (ii), for a.e. $y \in \Omega$ we have $G_t(y) < \infty$ for all $t \in (0,\infty)$ and $K(x,y) < \infty$ for a.e $x \in \Omega$. For such y we claim that $K_2(x,y) < \infty$ for a.e. $x \in \Omega$, as follows. By Lemma 2.2 followed by (3.10),

$$\begin{aligned} K_2(x,y) &\leqslant 4\varkappa \int_{d(x,y)}^{\infty} G_t(x) + G_t(y)\, \frac{dt}{t^2} \\ &\leqslant 4\varkappa \int_{d(x,y)}^{\infty} G_{d(x,y)}(x) + G_{d(x,y)}(y) + 4\varkappa\|\omega\|_* \log(t/d(x,y))\, \frac{dt}{t^2} \\ &= 4\varkappa K(x,y) \left(G_{d(x,y)}(x) + G_{d(x,y)}(y) + 4\varkappa\|\omega\|_*\right) < \infty \quad \omega - a.e. \end{aligned}$$

by (i) and the choice of y. □

The upper estimate for K_2 at the end of the last proof is of interest in its own right, especially in view of the lower estimate:

$$K_2(x,y) \geqslant \frac{1}{4\varkappa} \int_{d(x,y)}^{\infty} G_t(x) + G_t(y)\, \frac{dt}{t^2} \geqslant \frac{K(x,y)}{4\varkappa} \left(G_{d(x,y)}(x) + G_{d(x,y)}(y)\right)$$

by (2.14) and the monotinicity of G_t. The lower estimate does not require any assumption that ω is doubling or satisfy $\|\omega\|_* < \infty$.

Even if ω is not doubling, estimates similar to those in Lemma 3.2 can be obtained if $\|\omega\|_{wb} < \infty$.

Lemma 3.3. *Suppose that ω is a measure satisfying $\|\omega\|_{wb} < \infty$. Then*

$$\int_{B_t(x)} G_t^m\, d\omega \leqslant m!\, (2\varkappa\|\omega\|_{wb})^m\, \omega(B_{2\varkappa t}(x)) \tag{3.11}$$

for all $x \in \Omega, t > 0$ and $m \in \mathbb{N} \cup \{0\}$. Also, if $0 \leqslant \epsilon < 1/(2\varkappa\|\omega\|_{wb})$, then

$$\int_{B_t(x)} e^{\epsilon G_t}\, d\omega \leqslant \frac{\omega\left(B_{2\varkappa t}(x)\right)}{1 - 2\varkappa\epsilon\|\omega\|_{wb}} \tag{3.12}$$

for all $x \in \Omega$ *and all* $t > 0$.

Proof. Since $\varkappa \geqslant 1/2$, the case $m = 0$ of (3.11) is trivial.

For any ω-measurable set E we define the measure ω_E on Ω by setting

$$d\omega_E = \chi_E \, d\omega.$$

Also define

$$G_t^E(x) = \int_0^t \frac{\omega_E(B_r(x))}{r^2} \, dr$$

for $t \geqslant 0$ and $x \in \Omega$. Observe that for any ω-measurable set F,

$$\iint_{F\times F} K(x,y) \, d\omega_E(x) \, d\omega_E(y) = \iint_{E\cap F\times E\cap F} K(x,y) \, d\omega(x) \, d\omega(y)$$
$$\leqslant \|\omega\|_{wb} \, \omega(E \cap F) = \|\omega\|_{wb} \, \omega_E(F).$$

Hence

$$\|\omega_E\|_{wb} \leqslant \|\omega\|_{wb}. \tag{3.13}$$

We claim that for any ω-measurable set E, any $x \in \Omega, t > 0$, and $m \in \mathbb{N}$

$$\int_{B_t(x)} \left(G_t^E\right)^m \, d\omega_E \leqslant m!(2\varkappa)^{m-1} \|\omega\|_{wb}^m \omega_E(B_{2\varkappa t}(x)). \tag{3.14}$$

If so, then (3.11) follows by taking $E = \Omega$ and noting that $2\varkappa \geqslant 1$.

By (3.9) with ω replaced by ω_E and G_t replaced by G_t^E,

$$\int_{B_t(x)} G_t^E \, d\omega_E \leqslant \|\omega_E\|_* \, \omega_E(B_{2\varkappa t}(x)) \leqslant \|\omega\|_{wb} \, \omega_E(B_{2\varkappa t}(x)) \tag{3.15}$$

since $\|\omega_E\|_* \leqslant \|\omega_E\|_{wb} \leqslant \|\omega\|_{wb}$ by (3.13). Hence (3.14) holds for $m = 1$.

Assume that (3.14) holds for $m - 1$. Fix $E, x \in \Omega$, and $t > 0$. Define

$$F = E \cap B_{2\varkappa t}(x).$$

Observe that if $y \in B_t(x)$ and $0 < r < t$, then $B_r(y) \subseteq B_{2\varkappa t}(x)$ by the $\varkappa$-triangle inequality, and hence $\omega_E(B_r(y)) = \omega_F(B_r(y))$. Therefore, $G_t^E(y) = G_t^F(y)$ for $y \in B_t(x)$. Since $\omega_E = \omega_F$ on $B_t(x)$, we obtain

$$\int_{B_t(x)} \left(G_t^E\right)^m (y) \, d\omega_E(y) = \int_{B_t(x)} \left(G_t^F\right)^m (y) \, d\omega_F(y).$$

Now the proof parallels the proof of Lemma 3.2. We have

$$\int_{B_t(x)} \left(G_t^F\right)^m (y) \, d\omega_F(y) = m \int_0^t \int_{B_t(x)} \left(G_r^F\right)^{m-1} (y) \int_{B_r(y)} d\omega_F(z) \, d\omega_F(y) \, \frac{dr}{r^2}$$

$$\begin{aligned}
&\leqslant m \int_0^t \int_{B_{2\varkappa t}(x)} \int_{B_r(z)} \left(G_r^F\right)^{m-1}(y)\, d\omega_F(y)\, d\omega_F(z)\, \frac{dr}{r^2} \\
&\leqslant m!(2\varkappa)^{m-2} \|\omega\|_{wb}^{m-1} \int_{B_{2\varkappa t}(x)} \int_0^t \frac{\omega_F(B_{2\varkappa r}(z))}{r^2}\, dr\, d\omega_F(z) \\
&= m!(2\varkappa)^{m-1} \|\omega\|_{wb}^{m-1} \int_{B_{2\varkappa t}(x)} G_{2\varkappa t}^F(z)\, d\omega_F(z) \\
&\leqslant m!(2\varkappa)^{m-1} \|\omega\|_{wb}^{m} \omega_F(B_{4\varkappa^2 t}(x)),
\end{aligned}$$

using (3.15) at the last step, and the induction hypothesis previously. However,

$$\omega_F(B_{4\varkappa^2 t}(x)) = \omega_E(B_{4\varkappa^2 t}(x) \cap B_{2\varkappa t}(x)) = \omega_E(B_{2\varkappa t}(x))$$

since $2\varkappa \geqslant 1$. Hence we obtain (3.14) for m. This completes the induction and hence the proof of (3.14).

Now, (3.12) follows from (3.11) by expanding the exponential and summing on m as in the proof of (3.7). □

We remark that the proof of Lemma 3.3 shows that we can replace $\|\omega\|_{wb}$ by a similar quantity defined as in (3.2), except that the supremum is taken over all sets E which can be written as a finite intersection of balls.

For $x \in \Omega$ and $0 \leqslant s \leqslant t \leqslant \infty$, let

$$G_{t,s}(x) = G_t(x) - G_s(x) = \int_s^t \frac{\omega(B_r(x))}{r^2}\, dr. \tag{3.16}$$

In the next lemma, we require a smoothness condition on d. Ultimately (Theorem 4.3), this condition will be removed.

Lemma 3.4. *Suppose that there exist constants $0 < \gamma \leqslant 1$ and $\varkappa_1 > 0$ such that $d(x,y) = 1/K(x,y)$ obeys*

$$|d(x,z) - d(y,z)| \leqslant \varkappa_1\, d(x,y)^\gamma \left[d(x,z) + d(y,z)\right]^{1-\gamma}, \quad x, y, z \in \Omega. \tag{3.17}$$

Then there exists $c_0 > 0$, depending only on $\varkappa$, $\varkappa_1$, and γ, such that for all $x \in \Omega$, $0 < s < t \leqslant \infty$, and $y \in B_s(x)$,

$$|G_{t,s}(y) - G_{t,s}(x)| \leqslant c_0 \sup_{x \in \Omega, t>0} \frac{\omega(B_t(x))}{t}. \tag{3.18}$$

Proof. Let $B = \sup_{x\in\Omega,t>0} \omega(B_t(x))/t$. We set $\omega_1 = \omega\, \chi_{\Omega \setminus B_{2\varkappa t}(x)}$, $\omega_2 = \omega\, \chi_{B_{2\varkappa t}(x) \setminus B_{2\varkappa s}(x)}$, and $\omega_3 = \omega\, \chi_{B_{2\varkappa s}(x)}$. Then

$$|G_{t,s}(y) - G_{t,s}(x)| = \left| \int_s^t \frac{\omega(B_r(y))}{r^2}\, dr - \int_s^t \frac{\omega(B_r(x))}{r^2}\, dr \right| \leqslant \sum_{i=1}^3 A_i,$$

where

$$A_i = \left| \int_s^t \frac{\omega_i(B_r(y))}{r^2}\, dr - \int_s^t \frac{\omega_i(B_r(x))}{r^2}\, dr \right|.$$

For $y \in B_s(x)$ and $s < t$ we have $B_t(y) \subseteq B_{2\varkappa t}(x)$, so $A_1 = 0$. Also, for any y

$$\int_s^t \frac{\omega_3(B_r(y))}{r^2}\, dr \leqslant \omega\left(B_{2\varkappa s}(x)\right) \int_s^\infty \frac{1}{r^2}\, dr \leqslant 2\varkappa B.$$

Hence $A_3 \leqslant 2\varkappa B$.

Similarly, for any y

$$\int_t^\infty \frac{\omega_2(B_r(y))}{r^2}\, dr \leqslant \omega\left(B_{2\varkappa t}(x)\right) \int_t^\infty \frac{1}{r^2}\, dr \leqslant 2\varkappa B.$$

Hence the estimate $A_2 \leqslant (c_1 + 4\varkappa)B$ will follow if we prove

$$\left| \int_s^\infty \frac{\omega_2(B_r(y))}{r^2}\, dr - \int_s^\infty \frac{\omega_2(B_r(x))}{r^2}\, dr \right| \leqslant c_1 B. \tag{3.19}$$

However, for $y \in B_s(x)$ and $r < s$ we have $B_r(y) \subseteq B_{2\varkappa s}(x)$, so $\omega_2(B_r(y)) = 0$. Hence

$$\int_s^\infty \frac{\omega_2(B_r(y))}{r^2}\, dr = \int_0^\infty \frac{\omega_2(B_r(y))}{r^2}\, dr = \int_0^\infty \int_{B_r(y)} \frac{1}{r^2}\, d\omega_2(z)\, dr$$

$$= \int_\Omega \int_{d(y,z)}^\infty \frac{1}{r^2}\, dr\, d\omega_2(z) = \int_\Omega \frac{d\omega_2(z)}{d(y,z)} = \int_{B_{2\varkappa t}(x) \setminus B_{2\varkappa s}(x)} \frac{d\omega(z)}{d(y,z)}.$$

Therefore,

$$\left| \int_s^\infty \frac{\omega_2(B_r(y))}{r^2}\, dr - \int_s^\infty \frac{\omega_2(B_r(x))}{r^2}\, dr \right|$$

$$= \left| \int_{B_{2\varkappa t}(x) \setminus B_{2\varkappa s}(x)} \frac{1}{d(y,z)} - \frac{1}{d(x,z)}\, d\omega(z) \right|.$$

Since $y \in B_s(x)$ and $s < t$, for $z \notin B_{2\varkappa s}(x)$

$$2\varkappa s \leqslant d(x,z) \leqslant \varkappa(d(x,y) + d(y,z)) \leqslant \varkappa s + \varkappa d(y,z).$$

It follows that $d(y,z) \geqslant s > d(x,y)$, and hence $d(x,z) \leqslant 2\varkappa d(y,z)$. Similarly, $d(x,z) \geqslant 2\varkappa s \geqslant d(x,y)$ and $d(y,z) \leqslant \varkappa(d(x,y) + d(x,z)) \leqslant 2\varkappa d(x,z)$. We obtain

$$d(x,y) \leqslant \min[d(x,z),\, d(y,z)], \quad \frac{1}{2\varkappa} d(y,z) \leqslant d(x,z) \leqslant 2\varkappa d(y,z).$$

It is clear that

$$\left|\frac{1}{d(y,z)} - \frac{1}{d(x,z)}\right| \leqslant 2\varkappa \frac{|d(y,z) - d(x,z)|}{d(x,z)^2}.$$

By (3.17),

$$\begin{aligned}|d(y,z) - d(x,z)| &\leqslant \varkappa_1 d(x,y)^\gamma [d(x,z) + d(y,z)]^{1-\gamma} \\ &\leqslant \varkappa_1 (2\varkappa + 1)^{1-\gamma} s^\gamma d(x,z)^{1-\gamma}.\end{aligned}$$

Combining these estimates and recalling that $d(x,y) < s$, we get

$$\left|\frac{1}{d(y,z)} - \frac{1}{d(x,z)}\right| \leqslant c\, s^\gamma \, d(x,z)^{-1-\gamma},$$

where $c = \varkappa_1(2\varkappa)(2\varkappa+1)^{1-\gamma}$. Substituting this gives

$$\int_{B_{2\varkappa t}(x)\setminus B_{2\varkappa s}(x)} \left|\frac{1}{d(y,z)} - \frac{1}{d(x,z)}\right| d\omega(z) \leqslant c\, s^\gamma \int_{\Omega\setminus B_{2\varkappa s}(x)} \frac{d\omega(z)}{d(x,z)^{1+\gamma}}$$

$$= c\, s^\gamma \int_{\Omega\setminus B_{2\varkappa s}(x)} \int_{d(x,z)}^\infty \frac{1}{r^{2+\gamma}}\, dr\, d\omega(z) \leqslant c\, s^\gamma \int_{2\varkappa s}^\infty \omega(B_r(x)) \frac{1}{r^{2+\gamma}}\, dr$$

$$\leqslant c\, s^\gamma B \int_{2\varkappa s}^\infty \frac{1}{r^{1+\gamma}}\, dr = \frac{c}{\gamma (2\varkappa)^\gamma} B.$$

The proof is complete. □

The proof of Lemma 3.4 shows that it suffices to require that (3.17) hold for $d(x,y) \leqslant \min[d(x,z),\, d(y,z)]$.

4 Upper Bounds of Green's Functions

We continue to assume that K is a quasimetric kernel with constant $\varkappa$ on a σ-finite measure space (Ω, ω), G_t is defined by (2.6), $G_{t,s}$ is defined as in (3.16), $d = 1/K$, and other notation is as above.

Lemma 4.1. *Suppose that there exists $A_\omega > 0$ such that for all $t > 0$, $n \in \mathbb{N} \cup \{0\}$ and all $x \in \Omega$*

$$\int_{B_t(x)} G_t^n \, d\omega \leqslant n! A_\omega^n \omega(B_{2\varkappa t}(x)) \tag{4.1}$$

and

$$\frac{\omega\left(B_t(x)\right)}{t} \leqslant A_\omega. \tag{4.2}$$

Also, suppose that there exists $c_0 > 0$ such that for all $x \in \Omega, 0 < s < t \leqslant \infty$, and all $z \in B_s(x)$

$$|G_{t,s}(z) - G_{t,s}(x)| \leqslant c_0 A_\omega. \tag{4.3}$$

Let $c_\varkappa = 16\varkappa^2 (1 + \varkappa e^{c_0})$. Then for all $x, y \in \Omega$ and all $m, n \in \mathbb{N} \cup \{0\}$

$$\int_\Omega K(x,z) \int_{d(y,z)}^\infty G_t^m(y) G_t^n(z) \frac{dt}{t^2} d\omega(z)$$

$$\leqslant n! c_\varkappa \int_{d(x,y)}^\infty \left(A_\omega^n \frac{G_t^{m+1}(y)}{m+1} + G_t^m(y) \sum_{j=1}^{n+1} A_\omega^{n+1-j} \frac{G_t^j(x)}{j!} \right) \frac{dt}{t^2}. \tag{4.4}$$

Proof. Let $s_0 = d(x,y)/(2\varkappa)$. By the Fubini theorem, the left-hand side of (4.4) is

$$\int_0^\infty \int_{B_t(y)} K(x,z) G_t^m(y) G_t^n(z)\, d\omega(z) \frac{dt}{t^2} = I + II,$$

where

$$I = \int_0^{s_0} G_t^m(y) \int_{B_t(y)} G_t^n(z) K(x,z)\, d\omega(z) \frac{dt}{t^2}$$

and

$$II = \int_{s_0}^\infty G_t^m(y) \int_{B_t(y)} G_t^n(z) K(x,z)\, d\omega(z) \frac{dt}{t^2}.$$

For I, note that for $t < s_0$ and $z \in B_t(y)$ we have

$$2\varkappa s_0 = d(x,y) \leqslant \varkappa (d(x,z) + d(z,y)) \leqslant \varkappa (d(x,z) + t) \leqslant \varkappa d(x,z) + \varkappa s_0.$$

Hence $s_0 < d(x,z)$, or $K(x,z) < 1/s_0$. Therefore, using (4.1) and the facts that G_t is a nondecreasing function of t and $\varkappa \geqslant 1/2$,

$$I \leqslant \frac{1}{s_0} \int_0^{s_0} G_t^m(y) \int_{B_t(y)} G_t^n(z)\, d\omega(z) \frac{dt}{t^2} \leqslant \frac{n! A_\omega^n}{s_0} \int_0^{s_0} G_t^m(y) \frac{\omega(B_{2\varkappa t}(y))}{t^2}\, dt$$

$$\leqslant \frac{n! A_\omega^n}{s_0} \int_0^{s_0} G_{2\varkappa t}^m(y) \frac{\omega(B_{2\varkappa t}(y))}{t^2}\, dt = \frac{n! 2\varkappa A_\omega^n}{s_0} \int_0^{2\varkappa s_0} G_t^m(y) \frac{\omega(B_t(y))}{t^2}\, dt$$

$$= \frac{n! 2\varkappa A_\omega^n}{(m+1)s_0} G_{2\varkappa s_0}^{m+1}(y) = \frac{n! 4\varkappa^2 A_\omega^n}{m+1} \frac{G_{d(x,y)}^{m+1}(y)}{d(x,y)}$$

$$= \frac{n! 4\varkappa^2 A_\omega^n}{m+1} \int_{d(x,y)}^\infty G_{d(x,y)}^{m+1}(y) \frac{dt}{t^2} \leqslant \frac{n! 4\varkappa^2 A_\omega^n}{m+1} \int_{d(x,y)}^\infty G_t^{m+1}(y) \frac{dt}{t^2}.$$

Write $II \leqslant III + IV$, where

$$III = \int_{s_0}^\infty G_t^m(y) \int_{B_t(y) \setminus B_t(x)} G_t^n(z) K(x,z)\, d\omega(z) \frac{dt}{t^2}$$

and

$$IV = \int_{s_0}^{\infty} G_t^m(y) \int_{B_t(x)} G_t^n(z) K(x,z)\, d\omega(z)\, \frac{dt}{t^2}.$$

For $z \in B_t(y) \setminus B_t(x)$ we have $K(x,z) = 1/d(x,z) \leqslant 1/t$. Hence, using (4.1), we obtain

$$\begin{aligned} III &\leqslant \int_{s_0}^{\infty} G_t^m(y) \int_{B_t(y)} G_t^n(z)\, d\omega(z)\, \frac{dt}{t^3} \leqslant n! A_\omega^n \int_{s_0}^{\infty} G_t^m(y) \omega\left(B_{2\varkappa t}(y)\right) \frac{dt}{t^3} \\ &\leqslant n! A_\omega^n \int_{s_0}^{\infty} G_{2\varkappa t}^m(y) \omega\left(B_{2\varkappa t}(y)\right) \frac{dt}{t^3} = n! 4\varkappa^2 A_\omega^n \int_{d(x,y)}^{\infty} G_t^m(y) \omega\left(B_t(y)\right) \frac{dt}{t^3} \\ &\leqslant \frac{n! 4\varkappa^2 A_\omega^n}{m+1} \int_{d(x,y)}^{\infty} G_t^{m+1}(y)\, \frac{dt}{t^2} \end{aligned}$$

by (2.8).

Write $IV = V + VI$, where

$$V = \int_{s_0}^{\infty} G_t^m(y) \int_{B_t(x)} \frac{G_t^n(z)}{t}\, d\omega(z)\, \frac{dt}{t^2}$$

and

$$VI = \int_{s_0}^{\infty} G_t^m(y) \int_{B_t(x)} G_t^n(z) \left(\frac{1}{d(x,z)} - \frac{1}{t} \right) d\omega(z)\, \frac{dt}{t^2}.$$

By (4.1),

$$V \leqslant n! A_\omega^n \int_{s_0}^{\infty} G_t^m(y) \omega\left(B_{2\varkappa t}(x)\right) \frac{dt}{t^3}.$$

Note that

$$\begin{aligned} \frac{\omega\left(B_{2\varkappa t}(x)\right)}{t} &= 2 \int_t^{2t} \omega\left(B_{2\varkappa t}(x)\right) \frac{ds}{s^2} \leqslant 2 \int_t^{2t} \omega\left(B_{2\varkappa s}(x)\right) \frac{ds}{s^2} \\ &= 4\varkappa \int_{2\varkappa t}^{4\varkappa t} \omega\left(B_s(x)\right) \frac{ds}{s^2} \leqslant 4\varkappa G_{4\varkappa t}(x). \end{aligned}$$

Therefore,

$$\begin{aligned} V &\leqslant n! 4\varkappa A_\omega^n \int_{s_0}^{\infty} G_t^m(y) G_{4\varkappa t}(x)\, \frac{dt}{t^2} \leqslant n! 4\varkappa A_\omega^n \int_{s_0}^{\infty} G_{4\varkappa t}^m(y) G_{4\varkappa t}(x)\, \frac{dt}{t^2} \\ &= n! 16\varkappa^2 A_\omega^n \int_{4\varkappa s_0}^{\infty} G_t^m(y) G_t(x)\, \frac{dt}{t^2} \leqslant n! 16\varkappa^2 A_\omega^n \int_{d(x,y)}^{\infty} G_t^m(y) G_t(x)\, \frac{dt}{t^2} \end{aligned}$$

since $4\varkappa s_0 = 2d(x,y)$.

To estimate VI, we first consider the interior integral

$$VII = \int_{B_t(x)} G_t^n(z) \left(\frac{1}{d(x,z)} - \frac{1}{t} \right) d\omega(z)$$

$$= \int_{B_t(x)} G_t^n(z) \int_{d(x,z)}^t \frac{ds}{s^2} \, d\omega(z) = \int_0^t \int_{B_s(x)} G_t^n(z) \, d\omega(z) \, \frac{ds}{s^2}$$

by the Fubini theorem. For $z \in B_s(x)$ and $0 < s < t$

$$G_t(z) = G_s(z) + G_{t,s}(z) - G_{t,s}(x) + G_{t,s}(x) \leqslant G_s(z) + c_0 A_\omega + G_{t,s}(x)$$

by (4.3). Hence

$$\int_{B_s(x)} G_t^n(z) \, d\omega(z) \leqslant \int_{B_s(x)} (G_s(z) + c_0 A_\omega + G_{t,s}(x))^n \, d\omega(z)$$

$$= \sum_{j=0}^n \frac{n!}{j!(n-j)!} (c_0 A_\omega + G_{t,s}(x))^j \int_{B_s(x)} G_s^{n-j}(z) \, d\omega(z)$$

$$\leqslant \sum_{j=0}^n \frac{n!}{j!(n-j)!} (c_0 A_\omega + G_{t,s}(x))^j \, (n-j)! A_\omega^{n-j} \omega\left(B_{2\varkappa s}(x)\right)$$

$$= \sum_{j=0}^n \frac{n!}{j!} (c_0 A_\omega + G_{t,s}(x))^j \, A_\omega^{n-j} \omega\left(B_{2\varkappa s}(x)\right)$$

by (4.1). Hence

$$VII \leqslant \sum_{j=0}^n \frac{n!}{j!} A_\omega^{n-j} \int_0^t (c_0 A_\omega + G_{t,s}(x))^j \, \omega\left(B_{2\varkappa s}(x)\right) \frac{ds}{s^2} = VIII + IX,$$

where

$$VIII = \sum_{j=0}^n \frac{n!}{j!} A_\omega^{n-j} \int_{t/(2\varkappa)}^t (c_0 A_\omega + G_{t,s}(x))^j \, \omega\left(B_{2\varkappa s}(x)\right) \frac{ds}{s^2}$$

and

$$IX = \sum_{j=0}^n \frac{n!}{j!} A_\omega^{n-j} \int_0^{t/(2\varkappa)} (c_0 A_\omega + G_{t,s}(x))^j \, \omega\left(B_{2\varkappa s}(x)\right) \frac{ds}{s^2}.$$

For $t/(2\varkappa) < s < t$ (4.2) implies

$$G_{t,s}(x) \leqslant G_{t,t/(2\varkappa)} = \int_{t/(2\varkappa)}^t \omega\left(B_r(x)\right) \frac{dr}{r^2} \leqslant A_\omega \int_{t/(2\varkappa)}^t \frac{dr}{r} = A_\omega \log(2\varkappa).$$

Hence, with $c_1 = c_0 + \ln(2\varkappa)$,

$$\int_{t/(2\varkappa)}^{t} (c_0 A_\omega + G_{t,s}(x))^j \,\omega\left(B_{2\varkappa s}(x)\right) \frac{ds}{s^2}$$
$$\leqslant (c_1 A_\omega)^j \int_{t/(2\varkappa)}^{t} \omega\left(B_{2\varkappa s}(x)\right) \frac{ds}{s^2} \leqslant (c_1 A_\omega)^j \int_0^t \omega\left(B_{2\varkappa s}(x)\right) \frac{ds}{s^2}$$
$$= 2\varkappa\, (c_1 A_\omega)^j \int_0^{2\varkappa t} \omega\left(B_s(x)\right) \frac{ds}{s^2} = 2\varkappa\, (c_1 A_\omega)^j \, G_{2\varkappa t}(x).$$

Thus,

$$VIII \leqslant 2\varkappa \sum_{j=0}^{n} \frac{n!}{j!} A_\omega^{n-j} c_1^j A_\omega^j G_{2\varkappa t}(x) = 2\varkappa n! A_\omega^n G_{2\varkappa t}(x) \sum_{j=0}^{n} \frac{c_1^j}{j!}$$
$$\leqslant 2e^{c_1} \varkappa n! A_\omega^n G_{2\varkappa t}(x).$$

Therefore,

$$\int_{s_0}^{\infty} G_t^m(y) VIII \,\frac{dt}{t^2} \leqslant 2e^{c_1} \varkappa n! A_\omega^n \int_{s_0}^{\infty} G_t^m(y) G_{2\varkappa t}(x) \,\frac{dt}{t^2}$$
$$\leqslant 2e^{c_1} \varkappa n! A_\omega^n \int_{s_0}^{\infty} G_{2\varkappa t}^m(y) G_{2\varkappa t}(x) \,\frac{dt}{t^2} = 4e^{c_1} \varkappa^2 n! A_\omega^n \int_{d(x,y)}^{\infty} G_t^m(y) G_t(x) \,\frac{dt}{t^2}.$$

For $0 < s < t/(2\varkappa)$ we have

$$G_{t,s}(x) - G_{t,2\varkappa s}(x) = \int_s^{2\varkappa s} \frac{\omega\left(B_r(x)\right)}{r^2}\, dr \leqslant A_\omega \int_s^{2\varkappa s} \frac{1}{r}\, dr = \log(2\varkappa) A_\omega,$$

and hence

$$c_0 A_\omega + G_{t,s}(x) \leqslant c_0 A_\omega + \log(2\varkappa) A_\omega + G_{t,2\varkappa s}(x) = c_1 A_\omega + G_{t,2\varkappa s}(x).$$

Therefore,

$$\int_0^{t/(2\varkappa)} (c_0 A_\omega + G_{t,s}(x))^j \,\frac{\omega\left(B_{2\varkappa s}(x)\right)}{s^2}\, ds$$
$$\leqslant \int_0^{t/(2\varkappa)} (c_1 A_\omega + G_{t,2\varkappa s}(x))^j \,\frac{\omega\left(B_{2\varkappa s}(x)\right)}{s^2}\, ds$$
$$= 2\varkappa \int_0^{t} (c_1 A_\omega + G_{t,s}(x))^j \,\frac{\omega\left(B_s(x)\right)}{s^2}\, ds$$
$$= 2\varkappa \sum_{\ell=0}^{j} \frac{j!}{(j-\ell)!\ell!} (c_1 A_\omega)^{j-\ell} \int_0^t G_{t,s}^\ell(x) \frac{\omega\left(B_s(x)\right)}{s^2}\, ds$$

$$= 2\varkappa \sum_{\ell=0}^{j} \frac{j!}{(j-\ell)!\ell!} (c_1 A_\omega)^{j-\ell} \Big(- \frac{G_{t,s}^{\ell+1}(x)}{\ell+1} \Big) \Big|_0^t$$

$$= 2\varkappa \sum_{\ell=0}^{j} \frac{j!}{(j-\ell)!\ell!} (c_1 A_\omega)^{j-\ell} \frac{G_t^{\ell+1}(x)}{\ell+1}$$

$$= 2\varkappa \sum_{\ell=0}^{j} \frac{j!}{(j-\ell)!(\ell+1)!} (c_1 A_\omega)^{j-\ell} G_t^{\ell+1}(x)$$

because

$$\frac{d}{ds} G_{t,s}(x) = -\omega\left(B_s(x)\right)/s^2,$$

$G_{t,t} = 0$, and $G_{t,0} = G_t$. Therefore,

$$IX \leqslant 2\varkappa \sum_{j=0}^{n} \frac{n!}{j!} A_\omega^{n-j} \sum_{\ell=0}^{j} \frac{j!}{(j-\ell)!(\ell+1)!} (c_1 A_\omega)^{j-\ell} G_t^{\ell+1}(x)$$

$$= 2\varkappa n! \sum_{\ell=0}^{n} \frac{G_t^{\ell+1}(x)}{(\ell+1)!} \sum_{j=\ell}^{n} \frac{c_1^{j-\ell}}{(j-\ell)!} \leqslant 2\varkappa e^{c_1} n! \sum_{\ell=0}^{n} A_\omega^{n-\ell} \frac{G_t^{\ell+1}(x)}{(\ell+1)!}$$

$$= 2\varkappa e^{c_1} n! \sum_{\ell=1}^{n+1} A_\omega^{n+1-\ell} \frac{G_t^{\ell}(x)}{\ell!}.$$

Hence

$$\int_{s_0}^{\infty} G_t^m(y) IX \frac{dt}{t^2} \leqslant 2\varkappa e^{c_1} n! \int_{s_0}^{\infty} G_t^m(y) \sum_{\ell=1}^{n+1} A_\omega^{n+1-\ell} \frac{G_t^\ell(x)}{\ell!} \frac{dt}{t^2}$$

$$\leqslant 4\varkappa^2 c^{c_1} n! \int_{2\varkappa s_0}^{\infty} G_{t/(2\varkappa)}^m(y) \sum_{\ell=1}^{n+1} A_\omega^{n+1-\ell} \frac{G_{t/(2\varkappa)}^\ell(x)}{\ell!} \frac{dt}{t^2}$$

$$\leqslant 4\varkappa^2 e^{c_1} n! \int_{d(x,y)}^{\infty} G_t^m(y) \sum_{\ell=1}^{n+1} A_\omega^{n+1-\ell} \frac{G_t^\ell(x)}{\ell!} \frac{dt}{t^2}$$

by the monotinicity of G_t and the fact that $d(x,y) = 2\varkappa s_0$.

Since

$$VI = \int_{s_0}^{\infty} G_t^m(y) VII \frac{dt}{t^2} \leqslant \int_{s_0}^{\infty} G_t^m(y) \left(VIII + IX\right) \frac{dt}{t^2},$$

these estimates yield the result. □

For $\ell = 0, 1, 2, \ldots$ we define

$$P_{t,\ell}(x,y)=\frac{(G_t(x)+G_t(y))^\ell}{\ell!}. \tag{4.5}$$

Lemma 4.2. *Under the assumptions of Lemma* 4.1,

$$\int_\Omega K(x,z)\int_{d(z,y)}^\infty P_{t,\ell}(z,y)\,\frac{dt}{t^2}\,d\omega(z)\leqslant c_\varkappa\int_{d(x,y)}^\infty\sum_{j=0}^{\ell}P_{t,\ell-j+1}(x,y)A_\omega^j\,\frac{dt}{t^2}. \tag{4.6}$$

Proof. By the binomial theorem and (4.4),

$$\begin{aligned}
&\int_\Omega K(x,z)\int_{d(y,z)}^\infty P_{t,\ell}(z,y)\,\frac{dt}{t^2}\,d\omega(z)\\
&=\frac{1}{\ell!}\sum_{n=0}^{\ell}\frac{\ell!}{(\ell-n)!n!}\int_\Omega K(x,z)\int_{d(y,z)}^\infty G_t^{\ell-n}(y)G_t^n(z)\,\frac{dt}{t^2}\,d\omega(z)\\
&\leqslant c_\varkappa\int_{d(x,y)}^\infty\sum_{n=0}^{\ell}\frac{1}{(\ell-n)!}A_\omega^n\,\frac{G_t^{\ell-n+1}(y)}{\ell-n+1}\\
&\quad+\sum_{n=0}^{\ell}\frac{1}{(\ell-n)!}G_t^{\ell-n}(y)\sum_{j=1}^{n+1}A_\omega^{n+1-j}\,\frac{G_t^j(x)}{j!}\,\frac{dt}{t^2}\\
&=c_\varkappa\int_{d(x,y)}^\infty\sum_{n=0}^{\ell}A_\omega^n\,\frac{G_t^{\ell-n+1}(y)}{(\ell-n+1)!}+\sum_{n=0}^{\ell}\frac{1}{(\ell-n)!}G_t^{\ell-n}(y)\sum_{j=0}^{n}A_\omega^j\,\frac{G_t^{n+1-j}(x)}{(n+1-j)!}\,\frac{dt}{t^2}\\
&=c_\varkappa\int_{d(x,y)}^\infty\sum_{j=0}^{\ell}A_\omega^j\,\frac{G_t^{\ell-j+1}(y)}{(\ell-j+1)!}+\sum_{j=0}^{\ell}A_\omega^j\sum_{n=j}^{\ell}\frac{G_t^{\ell-n}(y)G_t^{n+1-j}(x)}{(\ell-n)!(n+1-j)!}\,\frac{dt}{t^2}\\
&=c_\varkappa\int_{d(x,y)}^\infty\sum_{j=0}^{\ell}A_\omega^j\sum_{n=j-1}^{\ell}\frac{G_t^{\ell-n}(y)G_t^{n+1-j}(x)}{(\ell-n)!(n+1-j)!}\,\frac{dt}{t^2}.
\end{aligned}$$

By the change of index $m=n-j+1$,

$$\begin{aligned}
\sum_{n=j-1}^{\ell}\frac{G_t^{\ell-n}(y)G_t^{n+1-j}(x)}{(\ell-n)!(n+1-j)!}&=\sum_{m=0}^{\ell-j+1}\frac{G_t^{\ell-j+1-m}(y)G_t^m(x)}{(\ell-j+1-m)!m!}\\
&=\frac{(G_t(x)+G_t(y))^{\ell-j+1}}{(\ell-j+1)!}=P_{t,\ell-j+1}(x,y).
\end{aligned}$$

The proof is complete. □

We can eliminate the constant $c_\varkappa$ in (4.6) by rescaling. Let

$$Q_{t,j}(x,y) = c_\varkappa^j P_{t,j}(x,y) = \frac{(c_\varkappa(G_t(x)+G_t(y)))^j}{j!}, \tag{4.7}$$

and let $B_\omega = c_\varkappa A_\omega$. In terms of $Q_{t,j}$ and B_ω, (4.6) becomes

$$\int_\Omega K(x,z) \int_{d(z,y)}^\infty Q_{t,\ell}(z,y)\,\frac{dt}{t^2}\,d\omega(z) \leqslant \int_{d(x,y)}^\infty \sum_{j=0}^{\ell} Q_{t,\ell-j+1}(x,y)B_\omega^j\,\frac{dt}{t^2}. \tag{4.8}$$

By definition,

$$K_1(x,y) = K(x,y) = \frac{1}{d(x,y)} = \int_{d(x,y)}^\infty \frac{dt}{t^2} = \int_{d(x,y)}^\infty Q_{t,0}(x,y)\,\frac{dt}{t^2}.$$

By (4.8) in the case $\ell = 0$ (or by Lemma 2.2, with a different constant),

$$\begin{aligned} K_2(x,y) &= \int_\Omega K(x,z)K(z,y)\,d\omega(z) \\ &\leqslant \int_\Omega K(x,z) \int_{d(z,y)}^\infty Q_{t,0}(z,y)\,\frac{dt}{t^2}\,d\omega(z) \leqslant \int_{d(x,y)}^\infty Q_{t,1}(x,y)\,\frac{dt}{t^2}. \end{aligned}$$

Similarly,

$$K_3(x,y) \leqslant \int_{d(x,y)}^\infty Q_{t,2}(x,y) + B_\omega Q_{t,1}(x,y)\,\frac{dt}{t^2}.$$

By induction,

$$K_j(x,y) \leqslant \int_{d(x,y)}^\infty Q_{t,j-1}(x,y) + O(B_\omega)\,\frac{dt}{t^2}$$

for all $j \in \mathbb{N}$. Thus, we can estimate the terms in $\sum_{j=1}^\infty K_j(x,y)$ that do not involve B_ω by

$$\sum_{j=1}^\infty Q_{t,j-1}(x,y) = \sum_{j=0}^\infty \frac{c_\varkappa^j\,(G_t(x)+G_t(y))^j}{j!} = e^{c_\varkappa(G_t(x)+G_t(y))},$$

i.e., if we formally set $B_\omega = 0$, or equivalently $A_\omega = 0$, we obtain a certain exponential estimate for $\sum_{j=1}^\infty K_j$. Our goal is to show that we obtain an estimate of the same type, although with a different constant, for small enough A_ω.

For $j = 1,2,3,\ldots$ and $k = 0,1,\ldots,j$ we define constants $c_{j,k} \geqslant 0$ inductively in j as follows. For $j = 1$, let

$$c_{1,0} = 1, \quad \text{and } c_{1,1} = 0.$$

Suppose that $c_{j-1,k}$ is defined for $k = 0,1,\ldots,j-1$. For $k = 0,1,\ldots,j-1$ we set

$$c_{j,k} = \sum_{\ell=0}^{k} c_{j-1,\ell}$$

and define $c_{j,j} = 0$.

Lemma 4.3. *Suppose that K is a quasimetric kernel on a locally finite measure space (Ω, ω) with quasimetric constant $\varkappa$, satisfying* (3.17) *for some $\varkappa_1$ and $\gamma \in (0,1]$. Define $\|\omega\|_*$ by* (3.1) *and $\|\omega\|_{wb}$ by* (3.2). *Suppose that either*

(i) *ω is a doubling measure and $\|\omega\|_* < \infty$*

or

(ii) $\|\omega\|_{wb} < \infty$.

In case (i), *define D_ω by* (3.5) *and $A_\omega = 2\varkappa D_\omega \|\omega\|_*$. In case* (ii), *let $A_\omega = 2\varkappa \|\omega\|_{wb}$. Let $c_\varkappa = 16\varkappa^2(1+\varkappa e^{c_0})$, where c_0 is the constant in* (3.17). *Let $B_\omega = c_\varkappa A_\omega$. Define $Q_{t,j}$ by* (4.7). *Then for $j \in \mathbb{N}$ with $j \geqslant 2$*

$$K_j(x,y) \leqslant \int_{d(x,y)}^{\infty} \sum_{k=1}^{j-1} c_{j-1,j-k-1} B_\omega^{j-k-1} Q_{t,k}(x,y) \frac{dt}{t^2}. \tag{4.9}$$

Proof. First suppose that (i) holds. By Lemma 3.1,

$$\frac{\omega(B_t(x))}{t} \leqslant 2\varkappa \|\omega\|_* \leqslant 2\varkappa D_\omega \|\omega\|_* = A_\omega,$$

so (4.2) holds. By Lemma 3.2, (4.1) holds. By Lemma 3.4 and (4.2), we have (4.3). Now, suppose that (ii) holds. By (3.4) and (3.3), (4.2) holds. By Lemma 3.3, (4.1) holds. By Lemma 3.4 and (4.2), we have (4.3). In each case, therefore, the conditions of Lemmas (4.1) and (4.2) hold, and therefore we have (4.8).

For $j = 2$ the conclusion is the estimate for K_2 above. Suppose that (4.9) holds for j. Note that we can start the sum on k in (4.9) at $k = 0$ instead of $k = 1$ because of the convention that $c_{j-1,j-1} = 0$. Then, by (4.8),

$$\begin{aligned}
K_{j+1}(x,y) &= \int_\Omega K(x,z) K_j(z,y)\, d\omega(z) \\
&\leqslant \sum_{k=0}^{j-1} c_{j-1,j-k-1} B_\omega^{j-k-1} \int_\Omega K(x,z) \int_{d(x,y)}^{\infty} Q_{t,k}(z,y) \frac{dt}{t^2}\, d\omega(z) \\
&\leqslant \sum_{k=0}^{j-1} c_{j-1,j-k-1} B_\omega^{j-k-1} \int_{d(x,y)}^{\infty} \sum_{n=0}^{k} Q_{t,k-n+1}(x,y) B_\omega^n \frac{dt}{t^2} \\
&= \int_{d(x,y)}^{\infty} \sum_{k=0}^{j-1} c_{j-1,j-k-1} B_\omega^{j-k-1} \sum_{m=1}^{k+1} B_\omega^{k-m+1} Q_{t,m}(x,y) \frac{dt}{t^2}
\end{aligned}$$

$$= \int_{d(x,y)}^{\infty} \sum_{m=1}^{j} \left(\sum_{k=m-1}^{j-1} c_{j-1,j-k-1} \right) B_\omega^{j-m} Q_{t,m}(x,y) \, \frac{dt}{t^2}.$$

Noting that

$$\sum_{k=m-1}^{j-1} c_{j-1,j-k-1} = \sum_{\ell=0}^{j-m} c_{j-1,\ell} = c_{j,j-m},$$

we obtain (4.9) for $j+1$. By induction, the proof is complete. □

Lemma 4.4 (Shattuck). *For all* $j = 1, 2, 3, \ldots$ *and* $k = 0, 1, \ldots, j$

$$c_{j,k} \leqslant \frac{3^k j^k}{2^k k!}. \tag{4.10}$$

Proof. We first prove $c_{j,k} \leqslant \binom{j+k}{k}$ by induction on $m = j+k$. The result for $m = 1$ holds because $c_{1,0} = 1$. Suppose that the result holds for $j+k \leqslant m-1$. Now, suppose that $j + k = m$. Then

$$c_{j,k} = \sum_{\ell=0}^{k} c_{j-1,\ell} = \sum_{\ell=0}^{k-1} c_{j-1,\ell} + c_{j-1,k} = c_{j,k-1} + c_{j-1,k}$$

$$\leqslant \binom{j+k-1}{k-1} + \binom{j+k-1}{k} = \binom{j+k}{k}.$$

This completes the induction. When $k = j$, (4.10) is trivial because $c_{j,j} = 0$. Now suppose that $k \leqslant j-1$. By the inequality between the arithmetic and geometric mean,

$$\frac{(j+k)!}{j!} = (j+1)(j+2)\cdots(j+k) \leqslant \left[\frac{1}{k}\sum_{\ell=1}^{k}(j+\ell)\right]^k = \left[j + \frac{k+1}{2}\right]^k \leqslant \left[\frac{3j}{2}\right]^k$$

since $k \leqslant j-1$. Dividing by $k!$ and using $c_{j,k} \leqslant \binom{j+k}{k}$ yields (4.10). □

For $|t| < 1/(3ec_\varkappa)$ we define

$$\gamma(t) = \sum_{n=2}^{\infty} \frac{(3c_\varkappa n)^n}{n!} t^n.$$

A standard calculus exercise shows that the series converges for t as stated. For $t = 1/(3ec_\varkappa)$, however, the series diverges, by the Stirling formula.

Theorem 4.1. *Suppose that K is a quasimetric kernel on a σ-finite measure space (Ω, ω) with quasimetric constant $\varkappa$, satisfying* (3.17) *for some $\varkappa_1$ and $\gamma \in (0,1]$. Define $\|\omega\|_*$ by* (3.1) *and $\|\omega\|_{wb}$ by* (3.2). *Define D_ω by (3.5) and* *let* $c_\varkappa = 16\varkappa^2(1+\varkappa e^{c_0})$, *where* c_0 *is the constant in* (3.17). *Suppose that either*

(i) ω *is a doubling measure and* $\|\omega\|_* < 1/(6\varkappa e c_\varkappa D_\omega)$

or

(ii) $\|\omega\|_{wb} < 1/(6\varkappa e c_\varkappa)$.

In case (i), *let* $A_\omega = 2\varkappa D_\omega \|\omega\|_*$. *In case* (ii), *let* $A_\omega = 2\varkappa\|\omega\|_{wb}$. *Then*

$$\sum_{j=1}^{\infty} K_j(x,y) \leqslant (1+\gamma(A_\omega)) \int_{d(x,y)}^{\infty} e^{e^{1/e} c_\varkappa (G_t(x)+G_t(y))} \frac{dt}{t^2}. \tag{4.11}$$

In fact, the slightly more precise estimate

$$\sum_{j=1}^{\infty} K_j(x,y) \leqslant \int_{d(x,y)}^{\infty} \gamma(A_\omega)\left(e^{c_\varkappa(G_t(x)+G_t(y))}-1\right) + e^{e^{3c_\varkappa A_\omega} c_\varkappa (G_t(x)+G_t(y))} \frac{dt}{t^2} \tag{4.12}$$

holds.

Proof. We first consider the possibility that ω is not locally finite. Then there exists $x_0 \in \Omega$ and $t_0 \in (0,\infty)$ such that $\omega(B_{t_0}(x_0)) = \infty$. Let $x \in \Omega$ be arbitrary. Then for $t > 2\varkappa(t_0 + d(x,x_0))$ we have $B_{t_0}(x_0) \subseteq B_t(x)$, and hence $\omega(B_t(x)) = \infty$ for all $t > 2\varkappa(t_0 + d(x,x_0))$. Therefore, $G_t(x) = \infty$ for all $t > 2\varkappa(t_0 + d(x,x_0))$. Hence the right-hand sides of (4.11) and (4.12) are infinite, so the estimates hold trivially.

Now, we assume that ω is locally finite. By Lemma 4.3,

$$\sum_{j=2}^{\infty} K_j(x,y) \leqslant \int_{d(x,y)}^{\infty} \sum_{j=2}^{\infty}\sum_{k=1}^{j-1} c_{j-1,j-k-1} B_\omega^{j-k-1} Q_{t,k}(x,y)\frac{dt}{t^2}$$

$$\leqslant \int_{d(x,y)}^{\infty} \sum_{k=1}^{\infty}\sum_{j=k+1}^{\infty} c_{j-1,j-k-1} B_\omega^{j-k-1} Q_{t,k}(x,y)\frac{dt}{t^2}.$$

For $k \geqslant 1$, by Lemma (4.4), we have

$$\sum_{j=k+1}^{\infty} c_{j-1,j-k-1} B_\omega^{j-k-1} = \sum_{n=0}^{\infty} c_{k+n,n} B_\omega^n \leqslant \sum_{n=0}^{\infty} \frac{3^n(k+n)^n}{2^n n!} B_\omega^n$$

$$= \sum_{n=0}^{k} \frac{3^n(k+n)^n}{2^n n!} B_\omega^n + \sum_{n=k+1}^{\infty} \frac{3^n(k+n)^n}{2^n n!} B_\omega^n \leqslant \sum_{n=0}^{k} \frac{3^n k^n}{n!} B_\omega^n$$

$$+ \sum_{n=k+1}^{\infty} \frac{3^n n^n}{n!} B_\omega^n \leqslant e^{3kB_\omega} + \sum_{n=2}^{\infty} \frac{3^n n^n}{n!} B_\omega^n = e^{3kc_\varkappa A_\omega} + \gamma(A_\omega)$$

because $B_\omega = c_\varkappa A_\omega$. Since

$$K_1(x,y) = \int_{d(x,y)}^{\infty} dt/t^2,$$

we obtain

$$\sum_{j=1}^{\infty} K_j(x,y) \leqslant \int_{d(x,y)}^{\infty} 1 + \sum_{k=1}^{\infty} \left(e^{3kc_\varkappa A_\omega} + \gamma\left(A_\omega\right)\right) \frac{\left(c_\varkappa \left(G_t(x) + G_t(y)\right)\right)^k}{k!} \frac{dt}{t^2}$$

$$= \int_{d(x,y)}^{\infty} \gamma(A_\omega) \left(e^{c_\varkappa (G_t(x)+G_t(y))} - 1\right) + \sum_{k=0}^{\infty} \frac{\left(e^{3c_\varkappa A_\omega} c_\varkappa \left(G_t(x) + G_t(y)\right)\right)^k}{k!} \frac{dt}{t^2},$$

which yields (4.12). Using assumption (i) or (ii), we obtain $3c_\varkappa A_\omega \leqslant 1/e$, and (4.11) follows. □

We remark that a result like Theorem 4.1, but with different constants, can be proved in a manner analogous to the proof of Theorem 3.1 in [16]. However, the approach above yields the estimate (4.9) for K_j under the assumption that either ω is doubling and $\|\omega\|_* < \infty$ or that $\|\omega\|_{wb} < \infty$. The smallness of the norms is only required to obtain the convergence of the sum on j. The other proof required the small norm assumption to obtain estimates for K_j. Perhaps, the estimate (4.9) for the iterates of K is of interest in its own right.

Theorem 4.2. *Suppose that K is a quasimetric kernel on a σ-finite measure space (Ω, ω) with quasimetric constant $\varkappa$, satisfying* (3.17) *for some $\varkappa_1$ and $\gamma \in (0,1]$. Define $\|\omega\|_*$ by* (3.1) *and $\|\omega\|_{wb}$ by* (3.2). *Define D_ω by* (3.5). *Let $c_\varkappa = 16\varkappa^2(1 + \varkappa c^{c_0})$, where c_0 is the constant in* (3.17). *Suppose that either*

(i) *ω is a doubling measure and $\|\omega\|_* < 1/\left(6\varkappa e c_\varkappa D_\omega\right)$*

or

(ii) *$\|\omega\|_{wb} < 1/\left(6\varkappa e c_\varkappa\right)$.*

In case (i), *let $A_\omega = 2\varkappa D_\omega \|\omega\|_*$. In case* (ii), *let $A_\omega = 2\varkappa \|\omega\|_{wb}$. Then*

$$\sum_{j=1}^{\infty} K_j(x,y) \leqslant 3\left(1 + \gamma(A_\omega)\right) K(x,y) e^{4\varkappa e^{1/e} c_\varkappa K_2(x,y)/K(x,y)}. \tag{4.13}$$

Proof. As in the proof of Lemma 4.1, if ω is not locally finite, then for each $x \in \Omega$ there exists t_x such that $G_t(x) = \infty$ for all $t > t_x$. Hence the lower bound for K_2 in Lemma 2.2 implies that $K_2(x,y)$ is infinite for all $x, y \in \Omega$. Thus, (4.13) is trivial in this case.

Now, suppose that ω is locally finite. Let $t_0 = d(x,y) = 1/K(x,y)$. By $D_\omega \geqslant 1$ and $\|\omega\|_* \leqslant \|\omega\|_{wb}$, Theorem 4.1 and (3.10) imply that

$$\sum_{j=1}^{\infty} K_j(x,y) \leqslant (1+\gamma(A_\omega)) \int_{t_0}^{\infty} e^{e^{1/e} c_\varkappa (G_{t_0}(x)+G_{t_0}(y)+2A_\omega \ln(t/t_0))} \frac{dt}{t^2}$$

$$= (1+\gamma(A_\omega))\, e^{e^{1/e} c_\varkappa (G_{t_0}(x)+G_{t_0}(y))} \int_{t_0}^{\infty} \left(\frac{t}{t_0}\right)^{2e^{1/e} c_\varkappa A_\omega} \frac{dt}{t^2}$$

$$= \frac{(1+\gamma(A_\omega))\, e^{e^{1/e} c_\varkappa (G_{t_0}(x)+G_{t_0}(y))}}{\left(1-2e^{1/e} c_\varkappa A_\omega\right) t_0}$$

$$\leqslant 3\,(1+\gamma(A_\omega))\, K(x,y) e^{e^{1/e} c_\varkappa (G_{t_0}(x)+G_{t_0}(y))}$$

since $A_\omega \leqslant 1/(3ec_\varkappa)$. This last estimate for $\sum K_j$ is a variant of (4.11) that may be of interest in itself. By the monotinicity of G_t,

$$G_{t_0}(x) + G_{t_0}(y) \leqslant t_0 \int_{t_0}^{\infty} G_t(x) + G_t(y) \frac{dt}{t^2} \leqslant 4\varkappa K_2(x,y)/K(x,y)$$

because of (2.14) and the definition of t_0. Substituting this estimate in the estimate for $\sum_{j=1}^{\infty} K_j$ above yields (4.13). □

The hypothesis (3.17) is practically redundant since one can replace a quasimetric d with an equivalent quasimetric $\widetilde{d}$ which satisfies (3.17). The following lemma is well known (cf. [27] and [22, Proposition 14.5]). It is worth observing that it is valid even without the usual assumption that (Ω, d) is a homogeneous space and the assumption that $d(x,y) = 0$ if $x = y$, by a slight variation of the proof in [22].

Lemma 4.5. *Suppose that d defined by (2.1) obeys the assumptions of Definition 2.1. Then there exists a metric $\rho(x,y)$ which satisfies the triangle inequality and obeys*

$$c_4\, \rho(x,y)^{\frac{1}{\gamma}} \leqslant d(x,y) \leqslant c_3\, \rho(x,y)^{\frac{1}{\gamma}}, \quad x,y \in \Omega, \tag{4.14}$$

where $0 < \gamma \leqslant 1$, and c_3, c_4 are positive constants depending only on $\varkappa$. Then $\widetilde{d} = \rho^{\frac{1}{\gamma}}$ is a quasimetric equivalent to d which satisfies (3.17) with constants γ and $\varkappa_1$ depending only on $\varkappa$.

To see the last statement, apply the elementary inequality $|a^p - b^p| \leqslant p \max(a,b)^{p-1}|a-b|$ with $p = \gamma^{-1} \geqslant 1, a = \rho(x,z)$ and $b = \rho(y,z)$.

Lemma 4.5 allows us to remove the assumption (3.17) in Theorem 4.2, with some cost with regard to the constants.

Theorem 4.3. *Suppose that K is a quasimetric kernel on a σ-finite measure space (Ω, ω) with quasimetric constant $\varkappa$. Define $\|\omega\|_*$ by (3.1) and $\|\omega\|_{wb}$ by (3.2). Then there exists $\epsilon > 0$ such that if either*

(i) *ω is a doubling measure and $\|\omega\|_* < \epsilon$*

or

(ii) $\|\omega\|_{wb} < \epsilon$,

then there exist $c_2, C_2 > 0$ *such that*

$$\sum_{j=1}^{\infty} K_j(x,y) \leqslant C_2 K(x,y) e^{c_2 K_2(x,y)/K(x,y)}. \tag{4.15}$$

In case (i), ϵ *depends only on* $\varkappa$ *and the doubling constant* D_ω *defined by* (3.5), c_2 *depends only on* $\varkappa$, *and* C_2 *depends only on* $\varkappa, D_\omega$, *and* $\|\omega\|_*$. *In case* (ii), ϵ *and* c_2 *depend only on* $\varkappa$ *and* C_2 *depends only on* $\varkappa$ *and* $\|\omega\|_{wb}$.

Proof. Let $c_\varkappa = 16\varkappa^2(1 + \varkappa e^{c_0})$, where c_0 is the constant in (3.17). Let $\widetilde{\epsilon} = 1/(6\varkappa e c_\varkappa D_\omega)$ in case (i), and let $\widetilde{\epsilon} = 1/(6\varkappa e c_\varkappa)$ in case (ii). Let $\widetilde{c}_2 = 4\varkappa e^{1/e} c_\varkappa$, and let $\widetilde{C}_2 = 3(1 + \gamma(A_\omega))$ for A_ω as in Theorem 4.2.

Let $d = 1/K$. By Lemma 4.5, there exists $\widetilde{d}$ satisfying the condition (3.17) and constants c_4, c_3 depending only on $\varkappa$ such that

$$c_4 \widetilde{d}(x,y) \leqslant d(x,y) \leqslant c_3 \widetilde{d}(x,y)$$

for all $x, y \in \Omega$. Let $\widetilde{\varkappa}$ be the smallest constant such that $\widetilde{d}(x,y) \leqslant \widetilde{\varkappa}(\widetilde{d}(x,z) + \widetilde{d}(z,y))$ for all $x, y, z \in \Omega$. Then

$$\widetilde{\varkappa} \leqslant c_3 \varkappa / c_4. \tag{4.16}$$

Let $\widetilde{K} = 1/\widetilde{d}$. Then

$$c_3^{-1} \widetilde{K}(x,y) \leqslant K(x,y) \leqslant c_4^{-1} \widetilde{K}(x,y) \tag{4.17}$$

for all $x, y \in \Omega$. Define $\widetilde{\omega}$ by

$$d\widetilde{\omega} = c_4^{-1} d\omega. \tag{4.18}$$

Our goal is to apply Theorem 4.2 to $\widetilde{\omega}$ and $\widetilde{K}$.

In case (i), suppose that ω is a doubling measure. Let ℓ be the smallest nonnegative integer such that

$$c_3^2 \leqslant (2\varkappa)^\ell c_4^2.$$

Let

$$\epsilon = c_4 c_3^{-1} \widetilde{\epsilon} / D_\omega^{2\ell+2}. \tag{4.19}$$

Associated with $\widetilde{d}$ are the balls

$$\widetilde{B}_t(x) = \{z \in \Omega : \widetilde{d}(x,z) < t\}$$

for $x \in \Omega$ and $t > 0$. Note that the function $\widetilde{d}(x,\cdot)$ is ω-measurable since, by construction (cf. [22, Proposition 14.5]),

$$\widetilde{d}(x,y) = \inf\Big\{\sum_{i=0}^{n} d(x_i, x_{i+1})^\gamma\Big\},$$

where the infimum is taken over all finite chains of points x_i such that $x_0 = x$ and $x_{n+1} = y$. Hence $\widetilde{B}_t(x)$ is ω-measurable. Note that

$$\widetilde{B}_{t/c_3}(x) \subseteq B_t(x) \leqslant \widetilde{B}_{t/c_4}(x) \tag{4.20}$$

for all x, t. Then for all $t > 0$ and $x \in \Omega$

$$\widetilde{B}_{2\widetilde{\varkappa}t}(x) \subseteq B_{2c_3\widetilde{\varkappa}t}(x) \subseteq B_{2c_3^2\varkappa t/c_4}(x) \subseteq B_{(2\varkappa)^{\ell+1}c_4t}(x). \tag{4.21}$$

Hence

$$\omega(\widetilde{B}_{2\widetilde{\varkappa}t}(x)) \leqslant \omega\left(B_{(2\varkappa)^{\ell+1}c_4t}(x)\right) \leqslant D_\omega^{\ell+1}\omega(B_{c_4t}(x)) \leqslant D_\omega^{\ell+1}\omega(\widetilde{B}_t(x)).$$

Therefore,

$$\widetilde{D}_{\widetilde{\omega}} \equiv \sup_{x,t} \frac{\widetilde{\omega}(\widetilde{B}_{2\widetilde{\varkappa}t}(x))}{\widetilde{\omega}(\widetilde{B}_t(x))} = \sup_{x,t} \frac{\omega(\widetilde{B}_{2\widetilde{\varkappa}t}(x))}{\omega(\widetilde{B}_t(x))} \leqslant D_\omega^{\ell+1}.$$

Also, for all $t > 0$ and $z \in \Omega$

$$\begin{aligned}\iint_{\widetilde{B}_t(z)\times\widetilde{B}_t(z)} \widetilde{K}(x,y)\, d\widetilde{\omega}(x)\, d\widetilde{\omega}(y) &\leqslant \frac{c_3}{c_4^2}\iint_{B_{c_3t}(z)\times B_{c_3t}(z)} K(x,y)\, d\omega(x)\, d\omega(y)\\ &\leqslant \frac{c_3}{c_4^2}\|\omega\|_*\,\omega\left(B_{c_3t}(z)\right).\end{aligned}$$

By (4.21) and the fact that $\widetilde{\varkappa} \geqslant 1/2$, we obtain

$$B_{c_3t}(z) \subseteq B_{2c_3\widetilde{\varkappa}t}(z) \subseteq B_{(2\varkappa)^{\ell+1}c_4t}(z),$$

and hence

$$\begin{aligned}\iint_{\widetilde{B}_t(z)\times\widetilde{B}_t(z)} \widetilde{K}(x,y)\, d\widetilde{\omega}(x)\, d\widetilde{\omega}(y) &\leqslant \frac{c_3}{c_4^2} D_\omega^{\ell+1}\|\omega\|_*\,\omega\left(B_{c_4t}(z)\right)\\ \leqslant c_3c_4^{-2}D_\omega^{\ell+1}\|\omega\|_*\,\omega\left(\widetilde{B}_t(z)\right) &= c_3c_4^{-1}D_\omega^{\ell+1}\|\omega\|_*\,\widetilde{\omega}\left(\widetilde{B}_t(z)\right).\end{aligned}$$

Therefore,

$$\|\widetilde{\omega}\|_{\widetilde{*}} \equiv \sup_{z,t}\widetilde{\omega}\left(\widetilde{B}_t(z)\right)^{-1}\iint_{\widetilde{B}_t(z)\times\widetilde{B}_t(z)} \widetilde{K}(x,y)\, d\widetilde{\omega}(x)\, d\widetilde{\omega}(y) \leqslant c_3c_4^{-1}D_\omega^{\ell+1}\|\omega\|_*.$$

By (4.19), if $\|\omega\|_* < \epsilon$, then

$$\widetilde{D}_{\widetilde{\omega}}\|\widetilde{\omega}\|_{\widetilde{*}} \leqslant c_3c_4^{-1}D_\omega^{2\ell+2}\|\omega\|_* < \widetilde{\epsilon},$$

so condition (i) of Theorem (4.2) is met.

In case (ii), let

$$\epsilon = c_4 c_3^{-1} \widetilde{\epsilon}. \tag{4.22}$$

Suppose that $\|\omega\|_{wb} < \epsilon$. An argument like the one above, but simpler, shows that

$$\|\widetilde{\omega}\|_{wb} \equiv \sup_E (\widetilde{\omega}(E))^{-1} \iint_{E \times E} \widetilde{K}(x, y)\, d\widetilde{\omega}(x)\, d\widetilde{\omega}(y) \leqslant c_3 c_4^{-1} \|\omega\|_{wb} < \widetilde{\epsilon}.$$

Hence condition (ii) of Theorem (4.2) holds.

We define the kernels $\widetilde{K}_j$ for $j > 1$ inductively by

$$\widetilde{K}_j(x, y) = \int_\Omega \widetilde{K}(x, z) \widetilde{K}_{j-1}(z, y)\, d\widetilde{\omega}(z).$$

We can apply Theorem 4.2 to obtain

$$\sum_{j=1}^{\infty} \widetilde{K}_j(x, y) \leqslant \widetilde{C}_2 \widetilde{K}(x, y) e^{\widetilde{c}_2 \widetilde{K}_2(x,y)/\widetilde{K}(x,y)}. \tag{4.23}$$

Observe that $K_j \leqslant c_4^{-1} \widetilde{K}_j$ for all j: this fact holds for $j = 1$ by (4.17), and the general case follows by an easy induction, using (4.18). Similarly, $\widetilde{K}_2 \leqslant c_3^2 c_4^{-1} K_2$ and hence $\widetilde{K}_2/\widetilde{K} \leqslant c_3^2 c_4^{-2} K_2/K$. Therefore, (4.23) implies (4.15) with $C_2 = c_3 c_4^{-1} \widetilde{C}_2$ and $c_2 = c_3^2 c_4^{-2} \widetilde{c}_2$. □

For an extension of Theorem 4.3 to the quasimetrically modifiable case (cf. Definition 2.2), the balls and doubling condition need to be defined with respect to the quasimetric resulting from the modification, and the definitions of $\|\omega\|_*$ and $\|\omega\|_{wb}$ need to be modified accordingly.

Definition 4.1. Let K be a quasimetrically modifiable kernel on a σ-finite measure space (Ω, ω) with modifier m and constant $\varkappa$. Let

$$d(x, y) = \frac{m(x)m(y)}{K(x, y)}.$$

For $x \in \Omega$ and $t > 0$, let $B_t(x) = \{z \in \Omega : d(x, z) < t\}$. We say that a measure $d\sigma$ on Ω is *doubling* if $D_\sigma \equiv \sup_{\{x,t:\sigma(B_t(x))>0\}} \sigma(B_{2\varkappa t}(x))/\sigma(B_t(x)) < \infty$. Let $\|\omega\|_*$ be the infimum of all positive constants C such that

$$\iint_{B \times B} K(x, y) m(x) m(y)\, d\omega(x)\, d\omega(y) \leqslant C \int_B m^2(z)\, d\omega(z)$$

for all balls $B = B_t(x)$. Let $\|\omega\|_{wb}$ be the infimum of positive constants C such that

$$\iint_{E\times E} K(x,y)m(x)m(y)\,d\omega(x)\,d\omega(y) \leqslant C\int_E m^2(z)\,d\omega(z),$$

for all ω-measurable sets E.

These definitions show that if $\nu = m^2\,d\omega$ and $H(x,y) = \frac{K(x,y)}{m(x)\,m(y)}$, and we define $\|\nu\|_*$ and $\|\nu\|_{wb}$ as in (3.1) and (3.2), except with the quasimetric kernel H in place of K, then

$$\|\omega\|_* = \|\nu\|_* \text{ and } \|\omega\|_{wb} = \|\nu\|_{wb}. \tag{4.24}$$

Theorem 4.4. *Suppose that K is a quasimetrically modifiable kernel on (Ω,ω) with modifier m. If either*

(i) *$m^2\,d\omega$ is doubling and $\|\omega\|_* < \epsilon$*

or

(ii) $\|\omega\|_{wb} < \epsilon$,

where ϵ is the same as in Theorem 4.3 *in the corresponding cases, then* (4.15) *holds for the same values c_2 and $C_2 > 0$.*

Proof. Let ϵ be as determined in Theorem 4.3 for each case. Let $d\nu = m^2\,d\omega$, and let $H(x,y) = K(x,y)/(m(x)m(y))$. Our assumptions (i) and (ii) give the corresponding conditions (i) and (ii) in Theorem 4.3 for the quasimetric kernel H and measure ν. The conclusion follows from Theorem 4.3 and (2.18) just as in the proof of Theorem 2.3. □

Theorem 1.2 follows from Theorem 4.4 and the facts discussed at the end of Section 2 in the same way that Theorem 1.1 followed from Theorem 2.3. The independence of C_2 from $\|\omega\|_*$ or $\|\omega\|_{wb}$ is obtained by reducing ϵ to yield an interval for which $\gamma(A_\omega)$ is bounded.

We have observed a dichotomy relating to the local finiteness of ω in the quasimetric case. If ω is locally finite, then, under the assumption $\|\omega\|_* < \infty$, Corollary 3.1 shows that for a.e. x we have $K_2(x,y) < \infty$ for a.e. y. Hence, when the estimate (4.13) holds, we know that for a.e. x, $\sum_{j=1}^{\infty} K_j(x,y) < \infty$ for a.e. y. On the other hand, if ω is not locally finite, the first paragraph of the proof of Theorem 4.2 shows that $K_2(x,y) = \infty$ for all x,y. The same remarks apply to the quasimetrically modifiable case with $m^2\,d\omega$ in place of ω. It is worth noting that, in the cases covered in Theorems 1.2 and 1.3, we are always in the locally finite case where K_2 and V are finite a.e. This is not obvious because local finiteness is defined with respect to quasimetric balls, which can go out to the boundary of a domain, whereas the assumptions in Theorems 1.1, 1.2, and 1.3 are that the ω measure of a closed Euclidean ball inside Ω is finite.

Lemma 4.6. *Suppose that $\Omega \subseteq \mathbb{R}^n$ is either the entire space or a bounded domain satisfying the uniform boundary Harnack principle. Let $0 < \alpha < n$*

if Ω is $\mathbb{R}^n$; otherwise, let $0 < \alpha \leqslant 2$ with $\alpha \neq 2$ if $n = 2$. Suppose that q is either a nonnegative locally integrable function or a nonnegative locally finite measure on Ω. Define m as in the introduction, depending on Ω, and define the quasimetric balls

$$B_t(x) = \{z \in \Omega : d(x,z) = m(x)m(z)/G^{(\alpha)}(x,z) < t\}$$

for $x \in \Omega$ and $t \in (0,\infty)$. Let $d\nu = m^2\, d\omega$. Suppose that either

(i) *$m^2\, d\omega$ is d-doubling,*

or

(ii) $\|\omega\|_{wb} < \infty$.

Then $\nu(B_t(x)) < \infty$ for all $x \in \Omega$ and all $t \in (0,\infty)$.

Proof. For $\Omega = \mathbb{R}^n$ we have $m = 1$ and $G^{(\alpha)}(x,y) = c_{n\alpha}|x-y|^{\alpha-n}$. Hence the quasimetric balls are Euclidean in this case, and the result holds by the assumption that ω is locally finite in the Euclidean sense.

Now, suppose that Ω is a bounded domain. Let $D = \sup\{|x-y| : x, y \in \Omega\} < \infty$ be the Euclidean diameter of Ω. Observe that Ω also has a finite diameter with respect to d, as follows. If Ω is a $C^{1,1}$ domain, then by definition, $m(x) = \delta(x)^{\alpha/2}$ where $\delta(x) = \text{dist}(x, \partial\Omega)$. Hence, by (2.19), for any $x, y \in \Omega$

$$d(x,y) \leqslant C|x-y|^{n-\alpha}\left(|x-y| + \delta(x) + \delta(y)\right)^{\alpha} \leqslant CD^{n-\alpha}(3D)^{\alpha} = C3^{\alpha}D^n.$$

For $R > C3^{\alpha}D^n$ we have $\Omega = B_R(x)$ for any $x \in \Omega$. If Ω satisfies the uniform boundary Harnack principle, then $m(x) = \min(1, G(x,x_0))$ for some arbitrarily chosen $x_0 \in \Omega$. Since G is singular on the diagonal, $m(x_0) = 1$, so

$$d(x,x_0) = m(x)m(x_0)/G^{(\alpha)}(x,x_0) = \min(1, G^{(\alpha)}(x,x_0))/G^{(\alpha)}(x,x_0) \leqslant 1$$

for all $x \in \Omega$. Hence, in this case, $\Omega = B_R(x_0)$ for any $R > 1$ and $d(x,y) \leqslant 2\varkappa$ for all $x, y \in \Omega$. Therefore, what we need to show is that

$$\nu(\Omega) = \int_{\Omega} m^2\, d\omega < \infty$$

or that ν is a finite measure. Select $r > 0$ such that the Euclidean ball $E_r(x_0) = \{z \in \mathbb{R}^n : |z - x_0| < r\}$ has closure contained in Ω (here, x_0 is the point chosen in the definition of m in the general case of a domain with the uniform boundary Harnack principle; for a $C^{1,1}$ domain, $x_0 \in \Omega$ can be chosen arbitrarily). Hence $\omega(E_r(x_0)) < \infty$. Since m is bounded, we also have $\nu(E_r(x_0)) < \infty$. We claim that for ϵ small enough,

$$B_\epsilon(x_0) \subseteq E_r(x_0). \tag{4.25}$$

If so, then $\nu(B_\epsilon(x_0)) < \infty$. In case (i), then, applying the doubling property enough times yields $\nu(\Omega) = \nu(B_R(x_0)) < \infty$.

We now prove (4.25). If Ω is a $C^{1,1}$ domain, then, by (2.19), we have

$$d(x_0, y) \geqslant C|x_0 - y|^{n-\alpha} \left(|x_0 - y| + \delta(x_0) + \delta(y)\right)^\alpha \geqslant C\delta(x_0)^\alpha |x_0 - y|^{n-\alpha}.$$

Hence if $d(x_0, y) < \epsilon = r^{n-\alpha} C\delta(x_0)^\alpha$, then $|x - y| < r$, establishing (4.25). If Ω is, more generally, a domain satisfying the uniform boundary Harnack inequality, we use the general fact (cf. [39] for $\alpha = 2$ and [13, p.1329] for $0 < \alpha \leqslant 2$) that $G^\alpha(x, y) \leqslant C_{\alpha,n}|x - y|^{\alpha-n}$. Let $\epsilon = \min(1/2, r^{n-\alpha}/C_\alpha)$. As above, we have $d(x, x_0) = \min(1, G^{(\alpha)}(x, x_0))/G^{(\alpha)}(x, x_0)$. If $d(x, x_0) < \epsilon$, then since $\epsilon < 1$ we must have

$$d(x, x_0) = 1/G^{(\alpha)}(x, x_0) \geqslant |x_0 - y|^{n-\alpha}/C_\alpha.$$

Hence $|x_0 - y| < r$, and we have (4.25). This completes the proof under the doubling condition (i).

Now, suppose that $\|\omega\|_{wb} < \infty$. Let $E_k = \{x \in \Omega : \delta(x) \geqslant 1/k\}$ for $k \in \mathbb{N}$, where $\delta(x) = \text{dist}\,(x, \partial\Omega)$. Each E_k is compact. Hence, by the Euclidean local finiteness of ω, we have $\omega(E_k) < \infty$ for each k. Since m is bounded on Ω, we have $\nu(E_k) < \infty$. From above we have $d(x, y) \leqslant C < \infty$ for any $x, y \in \Omega$, in either the case of a $C^{1,1}$ domain with $m(x) = \delta(x)^{\alpha/2}$ or the case where $m(x) = \min(1, G(x, x_0))$. Hence

$$\frac{(\nu(E_n))^2}{C} = \frac{1}{C}\left(\int_{E_n} m^2\, d\omega\right)^2 \leqslant \iint_{E_n\times E_n} \frac{1}{d(x,y)} m^2(x) m^2(y)\, d\omega(x)\, d\omega(y)$$

$$= \iint_{E_n\times E_n} K(x,y)m(x)m(y)\, d\omega(x)\, d\omega(y) \leqslant \|\omega\|_* \int_{E_n} m^2\, d\omega = \|\omega\|_* \nu(E_n).$$

Since $\nu(E_n) < \infty$, we obtain $\nu(E_n) \leqslant C\|\omega\|_*$ for all $n \in \mathbb{N}$. Applying. for example, the monotone convergence theorem, we obtain $\nu(\Omega) < \infty$. □

5 Applications and Examples

We first give an example of some conditions which imply $G_2 \leqslant CG$, as promised in the Introduction. Let $0 < \alpha < n$. We say that q belongs to the *Kato class* on $\mathbb{R}^n$ if (cf., for example,, [6, 14, 13])

$$\lim_{r\to 0^+} \sup_{x\in\mathbb{R}^n} \int_{B_r(x)} \frac{d\omega(z)}{|x - z|^{n-\alpha}} = 0, \tag{5.1}$$

where $d\omega = |q(x)|\, dx$. The local Kato class consists of q such that $\chi_D\, q$ is in the Kato class for any bounded measurable set $D \subset \mathbb{R}^n$.

We need an additional condition at ∞ introduced by Zhao [41] for $\alpha = 2$ (cf. [13] for $0 < \alpha < 2$):

$$\lim_{A\to\infty} \sup_{x\in\mathbb{R}^n} \int_{\mathbb{R}^n\setminus B_A(0)} \frac{d\omega(z)}{|x-z|^{n-\alpha}} = 0. \tag{5.2}$$

Let $0 < \alpha < n$ for $\Omega = \mathbb{R}^n$, and let $0 < \alpha \leqslant 2$, $\alpha \neq 2$ if $n = 2$, for a bounded domain Ω. For $\Omega = \mathbb{R}^n$ or a bounded domain Ω satisfying the boundary Harnack inequality we have the $3G$-inequality: there exists $c > 0$ such that for all $x, y, z \in \Omega$

$$\frac{G(x,y)G(z,y)}{G(x,z)} \leqslant c\left(\frac{1}{|x-z|^{n-\alpha}} + \frac{1}{|y-z|^{n-\alpha}}\right). \tag{5.3}$$

This is obvious for $\Omega = \mathbb{R}^n$ (cf. [14, Theorem 6.5] for bounded Lipschitz domains in the case $\alpha = 2$ and [19] for domains with the boundary Harnack principle and general $\alpha \in (0, 2]$). If we assume that q is in the local Kato class, for a bounded domain Ω, and additionally (5.2) for $\Omega = \mathbb{R}^n$, and apply the $3G$-inequality, we obtain

$$\frac{G_2(x,y)}{G(x,y)} = \int_\Omega \frac{G(x,z)G(z,y)}{G(x,y)}\, d\omega(z) \leqslant c\int_\Omega |x-z|^{\alpha-n} + |y-z|^{\alpha-n}\, d\omega(z) \leqslant C$$

since ω is potentially bounded (cf. [14, Proposition 3.1 and Corollary] and [41] for $\alpha = 2$; [13] for $0 < \alpha < 2$).

We now give a result in the general context of quasimetrically modifiable kernels which yields Theorem 1.3.

Theorem 5.1. *Let K be a quasimetrically modifiable kernel with modifier m and constant $\varkappa$ on (Ω, ω). Define $\|\omega\|_*$ and $\|\omega\|_{wb}$ as in Definition 4.1. Define*

$$Tf(x) = \int_\Omega K(x,y)f(y)\, d\omega(y).$$

Let $\|T\|$ denote the operator norm of T on $L^2(\omega)$. Let ϵ be the constant obtained in Theorem 4.3. Then

(i) *if $m^2 d\omega$ is a doubling measure, then $\|\omega\|_* \approx \|T\|$, more precisely,*

$$\|\omega\|_* \leqslant \|T\| \leqslant \epsilon^{-1}\|\omega\|_*, \tag{5.4}$$

(ii) $\|\omega\|_{wb} \approx \|T\|$, *more precisely,*

$$\|\omega\|_{wb} \leqslant \|T\| \leqslant \epsilon^{-1}\|\omega\|_{wb}. \tag{5.5}$$

Proof. Trivially, $\|\omega\|_* \leqslant \|\omega\|_{wb}$ and $\|\omega\|_{wb} \leqslant \|T\|$, as for (1.13). Hence the first inequalities in (5.4) and (5.5) hold. For the reverse, we will show for (i) that $\|\omega\|_* < \epsilon$ implies $\|T\| \leqslant 1$ and for (ii) that $\|\omega\|_{wb} < \epsilon$ implies $\|T\| \leqslant 1$. These facts will establish (5.4) and (5.5).

We first prove the result under the assumption that $m = 1$, i.e., K is a quasimetric. By Corollary 3.1, we can select $z_0 \in \Omega$ such that $K(x, z_0)$ and $K_2(x, z_0)$ are both finite for a.e. $x \in \Omega$. Let

$$u(x) = \sum_{j=1}^{\infty} K_j(x, z_0).$$

Our assumption (i) or (ii) allows us to apply Theorem 4.3 to obtain

$$u(x) \leqslant C_2 K(x, z_0) e^{c_2 K_2(x,z_0)/K(x,z_0)}.$$

Hence $u < \infty$ ω-a.e. Note that

$$Tu(x) = \int_\Omega K(x,y) \sum_{j=1}^{\infty} K_j(y, z_0)\, d\omega(y) = \sum_{j=2}^{\infty} K_j(x, z_0) \leqslant u(x).$$

Note that $u \geqslant K(x, z_0) > 0$ for $x \in \Omega$. Hence the Schur lemma gives that T is bounded on $L^2(\omega)$ with $\|T\| \leqslant 1$. This completes the proof in the case where K is a quasimetric.

Now, suppose that K is quasimetrically modifiable with modifier m. Define the quasimetric kernel $H(x, y) = K(x, y)/(m(x)m(y))$ and set $d\nu = m^2\, d\omega$. Define $\widetilde{T}$ by

$$\widetilde{T}f(x) = \int_\Omega H(x,y) f(y)\, d\nu(y).$$

A calculation shows that

$$\|T\|_{L^2(\omega)\to L^2(\omega)} = \|\widetilde{T}\|_{L^2(\nu)\to L^2(\nu)}.$$

By (4.24) and our assumptions, ν is doubling and $\|\nu\|_* = \|\omega\|_* < \epsilon$ in case (i) and $\|\nu\|_{wb} = \|\omega\|_{wb} < \epsilon$ in case (ii). Applying the result proved for the quasimetric case, we find that $\|\widetilde{T}\|_{L^2(\nu)\to L^2(\nu)} \leqslant 1$, and hence $\|T\|_{L^2(\omega)\to L^2(\omega)} \leqslant 1$. □

Theorem 1.3 follows from Theorem 5.1, using the facts noted at the end of Section 2.

We turn to the proof of Example 1.1. For the remainder of this paper, we change notation, and $B_t(x)$ will denote the Euclidean ball $\{y \in \mathbb{R}^n : |x-y| < t\}$. The following lemma is useful.

Lemma 5.1. *Suppose that* $0 < \alpha < n$. *For* $x \in \mathbb{R}^n$ *and* $t > 0$, *let*

$$h(x,t) = \int_{B_t(0)} \frac{1}{|x-z|^\alpha} \frac{1}{|z|^{n-\alpha}} dz.$$

Then there exist $c_1, c_2 > 0$, *depending only on* n *and* α, *such that*

$$c_1((1+|x|/t)^{-\alpha} + \log^+(t/|x|)) \leqslant h(x,t) \leqslant c_2((1+|x|/t)^{-\alpha} + \log^+(t/|x|)).$$

Proof. By a dilation argument, $g(x,t) = g(x/t,1)$. Hence we assume that $t = 1$. If $|x| \geqslant 2$, then for all $z \in B_1(0)$ we have $|x|/2 \leqslant |x-z| \leqslant 3|x|/2$. Hence $h(x,1)$ is bounded from above and below by constant multiples of

$$\frac{1}{|x|^\alpha}\int_{B_1(0)} \frac{1}{|z|^{n-\alpha}}\,dz = \frac{c}{|x|^\alpha},$$

as needed.

Now, we suppose that $|x| < 2$ and establish the lower bound. For $z \in B_1(0)$ we have $|x-z| \leqslant 3$. Hence

$$h(x,1) \geqslant 3^{-\alpha}\int_{B_1(0)} |z|^{n-\alpha}\,dz = c.$$

Now, suppose that $|x| < 1$. For $z \notin B_{|x|}(0)$ we have $|z-x| \leqslant |z| + |x| \leqslant 2|z|$. Hence

$$h(x,1) \geqslant \int_{B_1(0)\setminus B_{|x|}(0)} \frac{2^{-\alpha}}{|z|^n}\,dz = c\int_{|x|}^1 \frac{1}{r}\,dr = c\log(1/|x|).$$

It remains to show the upper bound for $|x| < 2$. For $|z| < |x|/2$ we have $|x-z| \geqslant |x| - |z| \geqslant |x|/2$. Hence

$$\int_{B_{|x|/2}(0)} \frac{1}{|x-z|^\alpha}\frac{1}{|z|^{n-\alpha}}dz \leqslant \frac{2^\alpha}{|x|^\alpha}\int_{B_{|x|/2}(0)} |z|^{\alpha-n}\,dy = c.$$

Similarly, for $|z-x| < |x|/2$ we have $|z| \geqslant |x| - |z-x| \geqslant |x|/2$. Therefore,

$$\int_{B_{|x|/2}(x)} \frac{1}{|x-z|^\alpha}\frac{1}{|z|^{n-\alpha}}dz \leqslant \frac{2^{n-\alpha}}{|x|^{n-\alpha}}\int_{B_{|x|/2}(x)} |x-z|^{-\alpha}\,dz = c.$$

Let $A = B_1(0) \setminus \left(B_{|x|/2}(0) \cup B_{|x|/2}(x)\right)$. Let $A_1 = \{z \in A : |z-x| \geqslant |z|\}$ and let $A_2 = \{z \in A : |z-x| < |z|\}$. Then

$$\int_{A_1} \frac{1}{|x-z|^\alpha}\frac{1}{|z|^{n-\alpha}}dz \leqslant \int_{B_1(0)\setminus B_{|x|/2}(0)} |z|^{-n}\,dz \leqslant c\left(1+\log^+(1/|x|)\right).$$

Since $|x| < 2$, we have $B_1(0) \subseteq B_3(x)$. Hence

$$\int_{A_2} \frac{1}{|x-z|^\alpha}\frac{1}{|z|^{n-\alpha}}dz \leqslant \int_{B_3(x)\setminus B_{|x|/2}(x)} |x-z|^{-n}\,dz \leqslant c\left(1+\log^+(1/|x|)\right).$$

This estimate completes the proof of the lemma. □

Proof of Example 1.1. We apply Theorems 1.1 and 1.2. Here, $m(x) = 1$, $K(x,y) = c_{n,\alpha}|x-y|^{\alpha-n}$ is the Riesz kernel, and $d\omega(x) = A|x|^{-\alpha}dx$. In Theorems 1.1 and 1.2, the balls determined by the quasimetric $d(x,y) = 1/K(x,y) = c_{n,\alpha}^{-1}|x-y|^{n-\alpha}$ are Euclidean balls. Therefore, to check that the condition (1.14) holds, it is enough to consider Euclidean balls in the definitions. The measure ω is well known to be doubling. We check that there exists $C = C_{n,\alpha} > 0$ such that

$$\int_{B(a,R)} \int_{B(a,R)} \frac{1}{|x|^\alpha} \frac{1}{|x-y|^{n-\alpha}} \frac{1}{|y|^\alpha}\, dx\, dy \leqslant C \int_{B(a,R)} \frac{1}{|z|^\alpha}\, dz \tag{5.6}$$

for all $a \in \mathbb{R}^n$ and all $R > 0$. Then putting in the factors of A in the definition of ω shows that $\|\omega\|_* \leqslant CA$. Hence taking A small enough yields (1.14).

We now prove (5.6). By a simple scaling argument, we may assume that $R = 1$. By a translation argument, it is enough to show that

$$\int_{B(0,1)} \int_{B(0,1)} \frac{1}{|x-a|^\alpha} \frac{1}{|x-y|^{n-\alpha}} \frac{1}{|y-a|^\alpha}\, dx\, dy \leqslant C \int_{B(0,1)} \frac{1}{|z-a|^\alpha}\, dz \tag{5.7}$$

for all $a \in \mathbb{R}^n$.

If $|a| \geqslant 2$, then for $t \in B(0,1)$, we have $|a|/2 \leqslant |t-a| \leqslant 3|a|/2$. Hence the right-hand side of (5.7) is bounded from below by $c|a|^{-\alpha}$. The left-hand side is bounded from above by

$$\begin{aligned} c|a|^{-2\alpha} &\int_{B(0,1)} \int_{B(0,1)} \frac{1}{|x-y|^{n-\alpha}}\, dx\, dy \\ &\leqslant c|a|^{-2\alpha} \int_{B(0,1)} \int_{B(y,2)} \frac{1}{|x-y|^{n-\alpha}}\, dx\, dy \leqslant c|a|^{-2\alpha} \leqslant c|a|^{-\alpha} \end{aligned}$$

since $|a| \geqslant 2$, as desired.

Now, suppose that $|a| < 2$. For $z \in B(0,1)$, then $|z-a| \leqslant 3$, so the right-hand side of (5.7) is bounded from below by a constant depending on n and α. To show that the left-hand side is bounded from above by a constant, we first replace $B(0,1)$ by the larger ball $B(a,3)$, then translate the integrals back to the origin, and then use a dilation argument to replace the radius of 3 with 1. Hence it suffices to prove that

$$\int_{B(0,1)} \int_{B(0,1)} \frac{1}{|x|^\alpha} \frac{1}{|x-y|^{n-\alpha}} \frac{1}{|y|^\alpha}\, dx\, dy < \infty.$$

This estimate follows easily from Lemma 5.1 (with α and $n-\alpha$ interchanged). Hence Theorem 1.2 applies, and Theorem 1.1 applies automatically. Hence

$$G^{(\alpha)}(x,y) e^{c_1 G_2^{(\alpha)}(x,y)/G^{(\alpha)}(x,y)} \leqslant V(x,y) \leqslant C_2 G^{(\alpha)}(x,y) e^{c_2 G_2^{(\alpha)}(x,y)/G^{(\alpha)}(x,y)}.$$

Now, we estimate $G_2^{(\alpha)}(x,y)/G^{(\alpha)}(x,y) = c_{n,\alpha} A\Phi(x,y)$, where

$$\Phi(x,y) \equiv |x-y|^{n-\alpha} \int_{\mathbb{R}^n} \frac{1}{|x-z|^{n-\alpha}} \frac{1}{|y-z|^{n-\alpha}} \frac{1}{|z|^{\alpha}}\, dz.$$

A dilation argument shows that $\Phi(Rx, Ry) = \Phi(x,y)$ for $R > 0$. Hence we assume for the moment that $|x-y| = 1$. We claim that there exist positive constants a_1 and a_2 such that for $|x-y|=1$

$$\Phi(x,y) \geqslant a_1 \big((1+|x|)^{-\alpha} + (1+|y|)^{-\alpha} + \log^+(1/|x|) + \log^+(1/|y|)\big) \quad (5.8)$$

and

$$\Phi(x,y) \leqslant a_2 (1 + (1+|x|)^{-\alpha} + (1+|y|)^{-\alpha} + \log^+(1/|x|) + \log^+(1/|y|)). \quad (5.9)$$

If $z \in B(x, 1/2)$, then $|y-z| \geqslant |y-x| - |x-z| \geqslant 1 - 1/2 = 1/2$ and $|y-z| \leqslant |y-x| + |x-z| \leqslant 3/2$. Therefore,

$$\int_{B(x,1/2)} \frac{1}{|x-z|^{n-\alpha}} \frac{1}{|y-z|^{n-\alpha}} \frac{1}{|z|^{\alpha}}\, dz \approx \int_{B(x,1/2)} \frac{1}{|x-z|^{n-\alpha}} \frac{1}{|z|^{\alpha}}\, dz$$

$$= \int_{B(0,1/2)} \frac{1}{|t|^{n-\alpha}} \frac{1}{|x-t|^{\alpha}}\, dt \approx (1+|x|)^{-\alpha} + \log^+(1/|x|)$$

by Lemma 5.1. By symmetry,

$$\int_{B(y,1/2)} \frac{1}{|x-z|^{n-\alpha}} \frac{1}{|y-z|^{n-\alpha}} \frac{1}{|z|^{\alpha}}\, dz \approx (1+|y|)^{-\alpha} + \log^+(1/|y|).$$

Since $|x-y| = 1$, the balls $B(x,1/2)$ and $B(y,1/2)$ are disjoint, so these estimates establish (5.8).

These estimates also yield an upper bound of the needed form for the integrals over $B(x,1/2)$ and $B(y,1/2)$ of the integrand in the definition of $\Phi(x,y)$. Also,

$$\int_{B(x,3)\setminus(B(x,1/2)\cup B(y,1/2))} \frac{1}{|x-z|^{n-\alpha}} \frac{1}{|y-z|^{n-\alpha}} \frac{1}{|z|^{\alpha}}\, dz$$

$$\leqslant 2^{2(n-\alpha)} \int_{B(x,3)} |z|^{-\alpha}\, dz \leqslant c.$$

For $z \notin B(x,3)$ we have $|z-y| \geqslant |z-x| - |x-y| = |z-x| - 1 \geqslant \frac{2}{3}|z-x|$. Hence

$$\int_{\mathbb{R}^n \setminus B(x,3)} \frac{1}{|x-z|^{n-\alpha}} \frac{1}{|y-z|^{n-\alpha}} \frac{1}{|z|^{\alpha}}\, dz$$

$$\leqslant \left(\frac{3}{2}\right)^{n-\alpha} \int_{\mathbb{R}^n \setminus B(x,3)} \frac{1}{|x-z|^{2n-2\alpha}} \frac{1}{|z|^{\alpha}}\, dz.$$

Let $A_1 = \{z \in \mathbb{R}^n : |z-x| \geqslant 3 \text{ and } |z-x| \leqslant 3|z|\}$ and $A_2 = \{z \in \mathbb{R}^n : |z-x| \geqslant 3 \text{ and } |z-x| \geqslant 3|z|\}$. Then

$$\int_{A_1} \frac{1}{|x-z|^{2n-2\alpha}} \frac{1}{|z|^\alpha}\, dz \leqslant 3^\alpha \int_{\mathbb{R}^n \setminus B(x,3)} \frac{1}{|z-x|^{2n-\alpha}}\, dz \leqslant c$$

since $\alpha < n$.

Note that if $A_2 \neq \varnothing$, then $|x| \geqslant 2$ since for any $z \in A_2$ we have $|x| \geqslant |x-z| - |z| \geqslant 2|x-z|/3 \geqslant 2$. Hence the estimate is complete for $|x| < 2$. Suppose that $|x| \geqslant 2$. Then for $z \in B(0, |x|/2)$ we have $|z-x| \geqslant |x| - |z| \geqslant |x|/2$. Hence

$$\int_{A_2 \cap B(0,|x|/2)} \frac{1}{|x-z|^{2n-2\alpha}} \frac{1}{|z|^\alpha}\, dz \leqslant \left(\frac{2}{|x|}\right)^{2(n-\alpha)} \int_{B(0,|x|/2)} |z|^{-\alpha}\, dz$$
$$\leqslant c\, |x|^{\alpha-n} \leqslant c,$$

since $|x| \geqslant 2$. Finally, since $|z-x| \geqslant 3|z|$ for $z \in A_2$,

$$\int_{A_2 \setminus B(0,|x|/2)} \frac{1}{|x-z|^{2n-2\alpha}} \frac{1}{|z|^\alpha}\, dz \leqslant 3^{2(\alpha-n)} \int_{\mathbb{R}^n \setminus B(0,|x|/2)} |z|^{-(2n-\alpha)}\, dz$$
$$\leqslant c|x|^{-(n-\alpha)} \leqslant c$$

because $|x| \geqslant 2$. This completes the proof of (5.9) under the assumption $|x-y| = 1$.

Now, let $x, y \in \mathbb{R}^n$ be general. By the dilation property $\Phi(Rx, Ry) = \Phi(x,y)$ and the result in the case where $|x-y| = 1$,

$$\begin{aligned}
\Phi(x,y) &= \Phi\left(\frac{x}{|x-y|}, \frac{y}{|x-y|}\right) \\
&\leqslant a_2 + a_2 \left(1 + \frac{|x|}{|x-y|}\right)^{-\alpha} + a_2 \left(1 + \frac{|y|}{|x-y|}\right)^{-\alpha} \\
&+ a_2 \log^+ \left(\frac{|x-y|}{|x|}\right) + a_2 \log^+ \left(\frac{|x-y|}{|y|}\right) \leqslant \widetilde{c} + \widetilde{c} \log\left(\max\left\{\frac{|x|}{|y|}, \frac{|y|}{|x|}\right\}\right)
\end{aligned}$$

for some constant $\widetilde{c}$.

For the lower bound, by (5.8), we have

$$\Phi(x,y) = \Phi\left(\frac{x}{|x-y|}, \frac{y}{|x-y|}\right) \geqslant a_1 \log^+ \left(\frac{|x-y|}{|x|}\right) + a_1 \log^+ \left(\frac{|x-y|}{|y|}\right).$$

Hence $e^{c_1 c_{n,\alpha} A \Phi(x,y)} \geqslant e^{b_4 A \log\left(\max\left\{\frac{|x|}{|y|}, \frac{|y|}{|x|}\right\}\right)} = \left(\max\left\{\frac{|x|}{|y|}, \frac{|y|}{|x|}\right\}\right)^{b_4 A}$ for some b_4 depending only on α and n. Substituting these estimate above yields (1.17). □

Acknowledgments. We thank Mark Shattuck for finding and proving the bound in Lemma 4.4. This bound improves a bound we obtained earlier, and hence improves the constants in Theorem 4.1. Dr. Shattuck also gave an interpretation of the sequence $c_{j,k}$ in terms of Catalan paths and showed that the number 3/2 in the inequality (4.10) cannot be replaced by any number smaller than $4/e \approx 1.47$.

The second author is supported in part by NSF grant DMS-0556309.

References

1. Adams, D.R., Hedberg, L.I.: Function Spaces and Potential Theory. Springer (1996)
2. Agmon, S.: On positivity and decay of solutions of second order elliptic equations on Riemannian manifolds. In: Methods of Functional Analysis and Theory of Elliptic Equations' (Naples, 1982), Liguori, Naples, pp. 19–52 (1983)
3. Aikawa, H.: On the minimal thinness in a Lipschitz domain. Analysis **5**, 347–382 (1985)
4. Aikawa, H.: Boundary Harnack principle and Martin boundary for a uniform domain. J. Math. Soc. Japan **53**, 119–145 (2001)
5. Aikawa, H., Essén, M.: Potential Theory-Selected Topics. Springer (1996)
6. Aizenman, M., Simon, B.: Brownian motion and Harnack inequality for Schrödinger operators. Commun. Pure Appl. Math. **35**, 209–273 (1982)
7. Ancona, A.: First eigenvalues and comparison of Green's functions for elliptic operators on manifolds or domains. J. Anal. Math. **72**, 45–92 (1997)
8. Ancona, A.: Some results and examples about the behaviour of harmonic functions and Green's functions with respect to second order elliptic operators. Nagoya Math. J. **165**, 123–158 (2002)
9. Bogdan, K.: Sharp estimates for the Green function in Lipschitz domains. J. Math. Anal. Appl. **243**, 326–337 (2000)
10. Brezis, H., Cabré, X.: Some simple nonlinear PDE's without solutions. Boll. Unione Mat. Ital. Sez. 1-B **8**, 223–262 (1998)
11. Chang, S.-Y.A., Wilson, J.M., Wolff, T.H.: Some weighted norm inequalities concerning the Schrödinger operators. Comment. Math. Helv. **60**, 217–246 (1985)
12. Chen, Z.-Q., Song, R.: Estimates on Green functions and Poisson kernels for symmetric stable processes. Math. Ann. **312**, 465–501 (1998)
13. Chen, Z.-Q., Song, R.: General gauge and conditional gauge theorems. Ann. Prob. **30**, 1313–1339 (2002)
14. Chung, K.L., Zhao, Z.: From Brownian Motion to Schrödinger's Equation. Springer, Berlin etc. (1995)
15. Frazier, M., Nazarov, F., Verbitsky, I E.: Global Estimates for Kernels of Neumann Series, Green's Functions, and the Conditional Gauge. Preprint, 2008.
16. Frazier, M., Verbitsky, I E.: Solvability conditions for a discrete model of Schrödinger's equation. In: A. Cialdea (ed.) et al. Analysis, Partial Differential Equations and Applications. The Vladimir Maz'ya Anniversary Volume. Birkhäuser (2009)
17. Grigor'yan, A.: Heat kernels on weighted manifolds and applications. Contemp. Math. **398**, 93–191 (2006)

18. Grigor'yan, A., Hansen, W.: Lower estimates for a perturbed Green function. J. Anal. Math. **104**, 25–58 (2008)
19. Hansen, W.: Uniform boundary Harnack principle and generalized triangle property. J. Funct. Anal. **226**, 452–484 (2005)
20. Hansson, K., Maz'ya, V.G., Verbitsky, I.E.: Criteria of solvability for multidimensional Riccati's equations. Ark. Mat. **37**, 87–120 (1999)
21. Hedberg, L.I., Wolff, T.H.: Thin sets in nonlinear potential theory. Ann. Inst. Fourier (Grenoble) **33**, 161–187 (1983)
22. Heinonen, J.: Lectures on Analysis on Metric Spaces. Springer, New York (2001)
23. Jerison, D., Kenig, C.: Boundary behavior of harmonic functions in nontangentially accessible domains. Adv. Math. **46**, 80–147 (1982)
24. Kalton, N.J., Verbitsky, I.E.: Nonlinear equations and weighted norm inequalities. Trans. Am. Math. Soc. **351**, 3441–3497 (1999)
25. Kenig, C.E.: Harmonic Analysis Techniques for Second Order Elliptic Boundary Value Problems. Am. Math. Soc., Providence, RI (1994)
26. Kerman, R., Sawyer, E.: The trace inequality and eigenvalue estimates for Schrödinger operators. Ann. Inst. Fourier (Grenoble) **36**, 207–228 (1987)
27. Macias, R.A., Segovia, C.: Lipschitz functions on spaces of homogeneous type. Trans. Am. Math. Soc. **33**, 257–270 (1979)
28. Maz'ya, V.G.: Sobolev Spaces. Springer, Berlin etc. (1985)
29. Maz'ya, V.G., Verbitsky, I.E.: The Schrödinger operator on the energy space: boundedness and compactness criteria. Acta Math. **188**, 263–302 (2002)
30. Murata, M.: Structure of positive solutions to $(-\triangle + V)u = 0$ in $\mathbb{R}^n$. Duke Math. J. **53**, 869–943 (1986)
31. Naïm, L.: Sur le rôle de la frontière de R. S. Martin dans la théorie du potentiel. Ann. Inst. Fourier (Grenoble) **7**, 183–281 (1957)
32. Phuc, N.C., Verbitsky, I.E.: Quasilinear and Hessian equations of Lane-Emden type. Ann. Math. **168**, 859–914 (2008)
33. Phuc, N.C., Verbitsky, I.E.: Singular quasilinear and Hessian equations and inequalities, J. Funct. Anal. **256**, 1875–1906 (2009)
34. Pinchover, Y.: Maximum and anti-maximum principles and eigenfunction estimates via perturbation theory of positive solutions of elliptic equations. Math. Ann. **314**, 555–590 (1999)
35. Riahi, L.: The 3G-inequality for general Schrödinger operators on Lipschitz domains. Manuscr. Math. **116**, 211–227 (2005)
36. Sawyer, E.T., Wheeden, R.L.: Weighted inequalities for fractional integrals on Euclidean and homogeneous spaces. Am. J. Math. **114**, 813–874 (1992)
37. Song, R., Wu, J.-M.: Boundary Harnack principle for symmetric stable processes. J. Funct. Anal. **168**, 403–427 (1999)
38. Verzhbinskii, G.M., Maz'ya, V.G.: Closure in L_p of the Dirichlet-problem operator in a region with conical points. Sov. Math. (Izv. VUZ) **18**, 5–14 (1974)
39. Widman, K.-O.: Inequalities for the Green function and boundary continuity of the gradients of solutions of elliptic differential equations. Math. Scand. **21**, 13–67 (1967)
40. Zhao, Z.: Green function for Schrödinger operator and conditioned Feynman-Kac gauge. J. Math. Anal. Appl. **116**, 309–334 (1986)
41. Zhao, Z.: Subcriticality and gaugeability of the Schrödinger operator. Trans. Am. Math. Soc. **334**, 75–96 (1992)

On Spectral Minimal Partitions: the Case of the Sphere

Bernard Helffer, Thomas Hoffmann-Ostenhof, and Susanna Terracini

Abstract In continuation of previous work, we analyze the properties of spectral minimal partitions and focus in this paper our analysis on the case of the sphere. We prove that a minimal 3-partition for the sphere $\mathbb{S}^2$ is up to a rotation the so-called **Y**-partition. This question is connected to a celebrated conjecture of Bishop in harmonic analysis.

1 Introduction

Motivated by questions related to some conjecture of Bishop [4], we continue the analysis of spectral minimal partitions developed for planar domains in [21, 20, 6, 5] and analyze the case of the two-dimensional sphere $\mathbb{S}^2$. As in [21], eigenvalue problems in domains with corners come up. Related problems were investigated by Maz'ya, Plamenevskij and coworkers in important works.

Throught the paper, the Laplacian is the Laplace Beltrami operator on $\mathbb{S}^2$. As usual, we describe $\mathbb{S}^2$ in $\mathbb{R}^3_{x,y,z}$ by the spherical coordinates

$$x = \cos\phi \sin\theta, y = \sin\phi \sin\theta, z = \cos\theta, \quad \phi \in [-\pi, \pi[\,, \theta \in]0, \pi[\,, \tag{1.1}$$

Bernard Helffer
Département de Mathématiques, Bat. 425, Université Paris-Sud and CNRS, 91 405 Orsay Cedex, France
e-mail: Bernard.Helffer@math.u-psud.fr

Thomas Hoffmann-Ostenhof
Institut für Theoretische Chemie, Universität Wien, Währinger Strasse 17, A-1090 Wien, Austria and International Erwin Schrödinger Institute for Mathematical Physics, Boltzmanngasse 9, A-1090 Wien, Austria
e-mail: thoffman@esi.ac.at

Susanna Terracini
Università di Milano Bicocca, Via Cozzi, 53 20125 Milano, Italy
e-mail: susanna.terracini@unimib.it

A. Laptev (ed.), *Around the Research of Vladimir Maz'ya III: Analysis and Applications*, International Mathematical Series 13, DOI 10.1007/978-1-4419-1345-6_6,

and add two poles "North" and "South" corresponding to the two points $(0,0,1)$ and $(0,0,-1)$. If Ω is a regular bounded open set with piecewise $C^{1,+}$ boundary[1], we consider the Dirichlet Laplacian $H = H(\Omega)$ and we would like to analyze the question of the existence and of the properties of minimal partitions for open sets Ω on $\mathbb{S}^2$. When Ω is strictly contained in $\mathbb{S}^2$, the question is not fundamentally different of the case of planar domains, hence we will focus on the case of the whole sphere $\mathbb{S}^2$ and on the search for possible candidates for minimal k-partitions of the sphere for k small.

To be more precise, let us recall a few definitions that the reader can, for example, find in [21]. For $1 \leqslant k \in \mathbb{N}$ and $\Omega \subset \mathbb{S}^2$ we call a *spectral k-partition*[2] of Ω a family $\mathcal{D} = \{D_i\}_{i=1}^k$ of pairwise disjoint open regular domains such that

$$\bigcup_{i=1}^{k} D_i \subset \Omega. \tag{1.2}$$

It is called *strong* if

$$\mathrm{Int}\,\left(\overline{\bigcup_{i=1}^{k} D_i}\right) \setminus \partial\Omega = \Omega\,. \tag{1.3}$$

We denote by $\mathfrak{O}_k$ the set of such partitions. For $\mathcal{D} \in \mathfrak{O}_k$ we introduce

$$\Lambda(\mathcal{D}) = \max_i \lambda(D_i)\,, \tag{1.4}$$

where $\lambda(D_i)$ is the ground state energy of $H(D_i)$, and

$$\mathfrak{L}_k(\Omega) = \inf_{\mathcal{D}\in\mathfrak{O}_k} \Lambda(\mathcal{D})\,. \tag{1.5}$$

By a *spectral minimal k-partition* we mean a k-partition $\mathcal{D} \in \mathfrak{O}_k$ such that

$$\mathfrak{L}_k(\Omega) = \Lambda(\mathcal{D})\,.$$

More generally, we can consider (cf. [21]) for $p \in [1, +\infty[$

$$\Lambda^p(\mathcal{D}) = \left(\frac{1}{k}\sum_i \lambda(D_i)^p\right)^{\frac{1}{p}} \tag{1.6}$$

and

$$\mathfrak{L}_{k,p}(\Omega) = \inf_{\mathcal{D}\in\mathfrak{O}_k} \Lambda^p(\mathcal{D})\,. \tag{1.7}$$

We write $\mathfrak{L}_{k,\infty}(\Omega) = \mathfrak{L}_k(\Omega)$ and recall the monotonicity property

$$\mathfrak{L}_{k,p}(\Omega) \leqslant \mathfrak{L}_{k,q}(\Omega) \text{ if } p \leqslant q\,. \tag{1.8}$$

---

[1] i.e., piecewise $C^{1,\alpha}$ boundary for some $\alpha > 0$.

[2] We say "k-partition" for brevity.

The notion of p-minimal k-partition can be extended accordingly, by minimizing $\Lambda^p(\mathcal{D})$.

In this paper, we prove the following assertion.

Theorem 1.1. *Any minimal 3-partition of $\mathbb{S}^2$ is obtained, up to a fixed rotation, by the so-called* **Y***-partition whose boundary is given by the intersection of $\mathbb{S}^2$ with the three half-planes defined respectively by $\phi = 0, \frac{2\pi}{3}, \frac{-2\pi}{3}$. Hence*

$$\mathfrak{L}_3(\mathbb{S}^2) = \frac{15}{4}. \tag{1.9}$$

This theorem is immediately related to (actually, is a consequence of) a conjecture of Bishop (Conjecture 6) proposed in [4] stating that

Conjecture 1.2 (Bishop, 1992). *The minimal 3-partition for*

$$\frac{1}{3}\Big(\sum_{i=1}^{3} \lambda(D_i)\Big)$$

corresponds to the **Y***-partition.*

We can indeed observe that if for some (k, p) there exists a p-minimal k-partition $\mathcal{D}_{k,p}$ such that $\lambda(D_i) = \lambda(D_j)$ for all i, j, then, by the monotonicity property, $\mathcal{D}_{k,p}$ is a q-minimal partition for any $q \geqslant p$.

Remark 1.3. At the origin, Bishop's conjecture was motivated by the analysis of the properties of harmonic functions in conic sets. The whole paper by Friedland–Hayman [18] (see also references therein) which inspires our Section 7 is written in this context. The link between our problem of minimal partitions and the problem in harmonic analysis can be summarized in this way. If we consider a homogeneous Lipschitzian function of the form $u(x) = r^\alpha g(\theta, \phi)$ in $\mathbb{R}^3$, which is harmonic outside its nodal set and such that the complementary of the nodal set divides the sphere in three parts, then

$$\alpha(\alpha + 1) \geqslant \mathfrak{L}_3(\mathbb{S}^2).$$

Hence Theorem 1.1 (and more specifically (1.9)) implies $\alpha \geqslant 3/2$. This kind of property can be useful to improve some statements in [11, 12] (cf. the proofs of Lemma 2 in [11] and Lemma 4.1 in [12]).

A similar question was analyzed (with partial success) when looking in [20] at candidates of minimal 3-partitions of the unit disk $D(0, 1)$ in $\mathbb{R}^2$. The most natural candidate was indeed the Mercedes Star, which is the 3-partition given by three disjoint sectors with opening angle $2\pi/3$, i.e.,

$$D_1 = \{x \in \Omega \mid \omega \in]0, 2\pi/3[\} \tag{1.10}$$

and D_2, D_3 are obtained by rotating D_1 by $2\pi/3$, respectively by $4\pi/3$. Hence the Mercedes Star in [20] is replaced here by the **Y**-partition in Theorem 1.1.

We observe that the **Y**-partition can also be described by the inverse image of the Mercedes Star partition by the map $\mathbb{S}^2 \ni (x,y,z) \mapsto (x,y) \in D(0,1)$.

Here, let us mention the two main statements giving the proof of Theorem 1.1.

Proposition 1.4. *If $\mathcal{D} = (D_1, D_2, D_3)$ is a 3-minimal partition, then its boundary contains two antipodal points.*

The proof of Proposition 1.4 will be achieved in Section 5 and involves the Euler formula and the thorem of Lyusternik and Shnirelman.

Proposition 1.5. *If there exists a minimal 3-partition $\mathcal{D} = (D_1, D_2, D_3)$ of $\mathbb{S}^2$ with two antipodal points in $\bigcup_i \partial D_i$, then it is (possibly, after a rotation) the* **Y***-partition.*

The proof of Proposition 1.5 will be done in Section 6 by lifting this 3-partition on the double covering $\mathbb{S}^2_{\mathcal{C}}$ of $\ddot{\mathbb{S}}^2$, where $\ddot{\mathbb{S}}^2$ is the sphere minus two antipodal points.

More precisely, following what has been done in the approach of the Mercedes Star conjecture in [20], the steps for the proof of Theorem 1.1 (or towards Conjecture 1.2 if we were able to show that the minimal partition for Λ^1 has all the $\lambda(D_j)$ equal) are the following.

1. One has to prove that minimal partitions on $\mathbb{S}^2$ exist and share the same properties as for planar domains: regularity and equal angle meeting property. This will be done in Section 2.
2. One can observe that the minimal 3-partition cannot be a nodal partition. This is a consequence of Theorem 2.8 in Section 2 and of the fact that the multiplicity of the second eigenvalue (i.e., the first nonzero one) is more than 2, actually 3.
3. The Euler formula implies that there exists only one possible type of minimal 3-partitions. Its boundary consists of two points x_1 and x_2 and three arcs joining these two points. This will be deduced in Subsection 4.1.
4. The next point is to show that a minimal partition has in its boundary two antipodal points.
5. The next point is that any minimal 3-partition which contains two antipodal points in its boundary can be lifted in a symmetric 6-partition on the double covering $\mathbb{S}^2_{\mathcal{C}}$. More precisely, if $\mathcal{D} = (D_1, D_2, D_3)$ and Π is the canonical projection of $\mathbb{S}^2_{\mathcal{C}}$ onto $\ddot{\mathbb{S}}^2$, we get the 6-partition $\mathcal{D}_{\mathcal{C}}$ by considering
$$\mathcal{D}_{\mathcal{C}} = (D_1^+, D_2^+, D_3^+, D_1^-, D_2^-, D_3^-)\,,$$
where for $j = 1,2,3$ D_j^+ and D_j^- denote the two components of $\Pi^{-1}(D_j)$. If $\mathcal{I}$ denotes the map on $\mathbb{S}^2_{\mathcal{C}}$ defined for $m \in \mathbb{S}^2_{\mathcal{C}}$ by
$$\Pi(\mathcal{I}(m)) = \Pi(m) \text{ with } \mathcal{I}(m) \neq m\,, \tag{1.11}$$

we observe that

$$\mathcal{I}(D_j^+) = D_j^- .$$

6. The last point is to show that on this double covering a minimal symmetric 6-partition is necessarily the double **Y**-partition, which is the inverse image in $\mathbb{S}^2_{\mathcal{C}}$ of the **Y**-partition.

All these points will be detailed in the following sections together with analogous questions in the case of minimal 4-partitions.

In the last section, we describe what can be said towards the proof of Bishop's conjecture and about the large k behavior of $\mathfrak{L}_k$ using mainly the tricky estimates of Friedland–Hayman [18].

2 Definitions, Notation, and Extension of Previous Results to the Sphere

We first recall more notation, definitions and results essentially extracted from [21], but which have to be extended from the case of planar domains to the case of domains in $\mathbb{S}^2$.

Definition 2.1. An open domain $D \subset \mathbb{S}^2$ is said to be *regular* if it satisfies an interior cone condition and ∂D is the union of a finite number of simple regular arcs $\gamma_i(\overline{I_i})$, with $\gamma_i \in C^{1,+}(\overline{I}_i)$ with no mutual nor self intersections, except possibly at the endpoints.

For a given set $\Omega \subset \mathbb{S}^2$, we are interested in the eigenvalue problem for $H(\Omega)$, the Dirichlet realization of the Laplace–Beltrami operator in Ω. We denote for any open domain Ω by $\lambda(\Omega)$ the lowest eigenvalue of $H(\Omega)$. We define for any eigenfunction u of $H(\Omega)$

$$N(u) = \overline{\{x \in \Omega \mid u(x) = 0\}} \tag{2.1}$$

and call the components of $\Omega \setminus N(u)$ the *nodal domains* of u. The number of nodal domains of such a function is denoted by $\mu(u)$.

If $\mathcal{D}$ is a strong partition, we say that $D_i, D_j \in \mathcal{D}$ are *neighbors* if

$$\mathrm{Int}\,(\overline{D_i \cup D_j}) \setminus \partial\Omega \text{ is connected} \tag{2.2}$$

and write in this case $D_i \sim D_j$. We then define a *graph* $G(\mathcal{D})$ by associating to each $D_i \in \mathcal{D}$ a vertex v_i and to each pair $D_i \sim D_j$ we associate an edge $e_{i,j}$.

Attached to a regular partition $\mathcal{D}$ we can associate its boundary $N = N(\mathcal{D})$ which is the closed set in $\overline{\Omega}$ defined by

$$N(\mathcal{D}) = \overline{\bigcup_i (\partial D_i \cap \Omega)}. \tag{2.3}$$

This leads us to introduce the set $\mathcal{M}(\Omega)$ of the regular closed sets.

Definition 2.2. A closed set $N \subset \overline{\Omega}$ belongs to $\mathcal{M}(\Omega)$ if N meets the following requirements.

(i) There are finitely many distinct critical points $x_i \in \Omega \cap N$ and associated positive integers $\nu(x_i)$ with $\nu(x_i) \geqslant 3$ such that, in a sufficiently small neighborhood of each of the x_i, N is the union of $\nu(x_i)$ disjoint (away from x_i non self-crossing) smooth arcs with one end at x_i (and each pair defining at x_i a positive angle in $]0, 2\pi[$) and such that in the complement of these points in Ω, N is locally diffeomorphic to a smooth arc. We denote by $X(N)$ the set of these critical points.

(ii) $\partial\Omega \cap N$ consists of a (possibly empty) finite set of points z_i, such that at each z_i $\rho(z_i)$, with $\rho(z_i) \geqslant 1$ arcs hit the boundary. Moreover, for each $z_i \in \partial\Omega$, then N is near z_i the union of $\rho(z_i)$ distinct smooth arcs which hit z_i with strictly positive distinct angles. We denote by $Y(N)$ the set of these critical points.

Conversely, if N is a regular closed set, then the family $\mathcal{D}(N)$ of connected components of $\Omega \setminus N$ belongs (by definition) to $\mathcal{R}(\Omega)$, hence regular and strong.

Definition 2.3. We say that a closed set has the *equal angle meeting property* (eamp) if the arcs meet with equal angles at each critical point $x_i \in N \cap \Omega$ and also with equal angles at the $z_i \in N \cap \partial\Omega$. For the boundary points z_i we mean that the two arcs in the boundary are included.

We say that the partition is *eamp-regular* if it is regular and satisfies the equal angle meeting property.

The following assertion was proved in [13, 14, 15].

Theorem 2.4. *For any k there exists a minimal* **eamp**-*regular strong k-partition.*

The following assertion was proved in [21].

Theorem 2.5. *Any minimal spectral k-partition admits a representative which is eamp-regular and strong.*

A basic result concerns the regularity (up to the boundary if any) of the nodal partition associated to an eigenfunction.

We first observe that the results about minimal partitions for plane domains can be transferred to the sphere $\mathbb{S}^2$. It is indeed enough to use the stereographic projection on the plane which gives an elliptic operator on the plane with analytic coefficients. This map is a conformal map, hence respecting the angles. The regularity questions being local there are no particular problem for recovering the equal angle meeting property.

A natural question is whether a minimal partition is the nodal partition induced by an eigenfunction. The next theorem gives a simple criterion for a partition to be associated to a nodal set. For this we need some additional definitions.

We recall that the graph $G(\mathcal{D})$ is *bipartite* if its vertices can be colored by two colors (two neighbours having different colors). In this case, we say that the partition is *admissible*. We recall that a collection of nodal domains of an eigenfunction is always admissible.

We have now the following converse theorem [21].

Theorem 2.6. *An admissible minimal k-partition is nodal, i.e., associated to the nodal set of an eigenfunction of $H(\Omega)$ corresponding to an eigenvalue equal to $\mathfrak{L}_k(\Omega)$.*

This theorem was already obtained for planar domains in [19] by adding a strong a priori regularity and the assumption that Ω is simply connected. Any subpartition of cardinality 2 corresponds indeed to a second eigenvalue and the criterion of pair compatibility (cf. [19]) can be applied.

A natural question is now to determine how general is the situation described in Theorem 2.6. As for partitions in planar domains, this can only occur in very particular cases when $k > 2$. If $\lambda_k(\Omega)$ denotes the kth eigenvalue of the Dirichlet realization of the Laplacian in an open set Ω of $\mathbb{S}^2$, the Courant theorem says:

Theorem 2.7. *The number of nodal domains $\mu(u)$ of an eigenfunction u associated with $\lambda_k(\Omega)$ satisfies $\mu(u) \leqslant k$.*

Then we say, as in [21], that u is *Courant-sharp* if $\mu(u) = k$. For any integer $k \geqslant 1$, we denote by $L_k(\Omega)$ the smallest eigenvalue whose eigenspace contains an eigenfunction with k nodal domains. In general, we have

$$\lambda_k(\Omega) \leqslant \mathfrak{L}_k(\Omega) \leqslant L_k(\Omega)\,. \tag{2.4}$$

The next result of [21] gives the full picture of the equality cases.

Theorem 2.8. *Suppose that $\Omega \subset \mathbb{S}^2$ is regular. If $\mathfrak{L}_k(\Omega) = L_k(\Omega)$ or $\lambda_k(\Omega) = \mathfrak{L}_k(\Omega)$, then $\lambda_k(\Omega) = \mathfrak{L}_k(\Omega) = L_k(\Omega)$, and any minimal k-partition is nodal and admits a representative which is the family of nodal domains of some eigenfunction u associated to $\lambda_k(\Omega)$.*

This theorem will be quite useful for showing, for example, that for $k = 3$ and $k = 4$ a k-minimal partition of $\mathbb{S}^2$ cannot be nodal. This will be further discussed in Section 3 (cf. Theorem 3.7).

3 The Courant Nodal Theorem with Inversion Symmetry

We collect here some easy useful observations for the analysis of the sphere. These considerations already appear in [24], but the application to minimal partitions is new. We consider the Courant nodal theorem for $H(\Omega)$, where Ω is an open connected set in $\mathbb{S}^2$. Let

$$\mathbb{S}^2 \ni (x,y,z) \mapsto I(x,y,z) = (-x,-y,-z) \tag{3.1}$$

denotes the inversion map and assume that

$$I\Omega = \Omega\,. \tag{3.2}$$

Note that $\Omega = \mathbb{S}^2$ satisfies the condition. These assumptions imply that we can write $H(\Omega)$ as a direct sum

$$H(\Omega) = H_S(\Omega) \bigoplus H_A(\Omega)\,, \tag{3.3}$$

where $H_S(\Omega)$ and $H_A(\Omega)$ are respectively the restrictions of $H(\Omega)$ to the I-symmetric (respectively, antisymmetric) L^2-functions in Ω in $D(H(\Omega))$.

For simplicity, we just write H_S, H_A. For the spectrum of $H(\Omega)$, σ, we have

$$\sigma = \sigma_S \cup \sigma_A$$

so that $\sigma_S = \{\lambda_k^S\}_{k=1}^{\infty}$ and analogously $\sigma_A = \{\lambda_k^A\}_{k=1}^{\infty}$. It is of independent interest to investigate how σ_S and σ_A are related. It is obvious that

$$\lambda_1^S < \lambda_1^A \leqslant \lambda_2^A \text{ and } \lambda_1^S < \lambda_2^S\,,$$

by standard spectral theory.

In the present situation, we can ask the question of a theorem à la Courant separately for the eigenfunctions of H_S and H_A.

First we note the following easy properties:

1. Suppose that u is an eigenfunction and that u is either symmetric or antisymmetric. Then $I\,N(u) = N(u)$, i.e., the nodal set is symmetric with respect to inversion.
2. If u^A is an eigenfunction of H_A, then for each nodal domain D_i of u^A, ID_i is a distinct nodal domain of u^A. Hence the nodal domains come in pairs and $\mu(u^A)$ is even.
3. If u^S is a symmetric eigenfunction, then there are two classes of nodal domains:

   - the symmetric domains,

$$D_{i,S} = ID_{i,S}\,, \tag{3.4}$$

   - the symmetric pairs of domains $D_{i,S}^1, D_{i,S}^2$ so that

$$ID_{i,S}^1 = D_{i,S}^2\,. \tag{3.5}$$

Theorem 3.1. *Suppose that Ω satisfies the symmetry assumption* (3.2). *Then, if (λ_k^A, u^A) is a spectral pair for $H_A(\Omega)$, we have*

$$\mu(u^A) \leqslant 2k \,. \tag{3.6}$$

If (λ_k^S, u^S) is a spectral pair for $H_S(\Omega)$ and if we denote by $\ell(k)$ the number of pairs of nodal domains of u^S satisfying (3.5) *and $m(k)$ the number of domains satisfying* (3.4), *then we have*

$$\ell(k) + m(k) \leqslant k \tag{3.7}$$

and

$$\mu(u^S) \leqslant k + \ell(k) \,. \tag{3.8}$$

Remark 3.2. Of course, the original Courant theorem holds, but the above result gives additional informations.

Proof. We just have to mimick the proof of the original Courant theorem. Let us first show (3.6). We can of course add the condition that

$$\lambda_{k-1}^A < \lambda_k^A \,.$$

Assume for contradiction that for some u_k^A we have $\mu(u_k^A) > 2k$. To each pair $(D_i,\ ID_i)$ of nodal domains of u_k^A, we associate the corresponding ground states, so that

$$H(D_i)\phi_i = \lambda_k^A \phi_i,\ \phi_i \in W_0^{1,2}(D_i)\,,$$

and, with $I\phi_i = -\phi_i \circ I$,

$$H(ID_i, V) I\phi_i = \lambda_k^A I\phi_i \,.$$

We use the variational principle in the form domain of H_A. We have

$$\lambda_k^A = \inf_{\varphi^A \perp \mathcal{Q}_A^{k-1}} \frac{\displaystyle\int_\Omega (|\nabla \varphi^A|^2)\, d\mu_{\mathbb{S}^2}}{\displaystyle\int_\Omega |\varphi^A|^2\, d\mu_{\mathbb{S}^2}}, \tag{3.9}$$

where $I\varphi^A = -\varphi^A$ and $\varphi^A \in W_0^{1,2}(\Omega)$. Here, $\mathcal{Q}_A^{k-1}$ is just the space spanned by the first $(k-1)$ eigenfunctions of H_A. We proceed now as in the proof of the Courant nodal theorem. Hence, in other words, we have just replaced in this proof the single domains by pairs of domains.

The proof of (3.8) is similar. □

We can also find some immediate consequences concerning the relation between σ_A and σ_S. Take, for instance, a spectral pair (u^A, λ_j^A) and assume that $\mu(u^A) = 2k$. Then we can construct from the $2k$ ground states of each connected component k symmetric ones, each one being supported in a symmetric pair of components. By the variational principle, this time for H_S we obtain

$$\lambda_k^S = \inf_{\varphi^S \perp \mathcal{Q}_S^{k-1}} \frac{\displaystyle\int_\Omega |\nabla \varphi^S|^2 \, d\mu_{\mathbb{S}^2}}{\displaystyle\int_\Omega |\varphi^S|^2 d\mu_{\mathbb{S}^2}}. \tag{3.10}$$

This implies:

Proposition 3.3. *Any eigenvalue λ^A of H_A, whose corresponding eigenspace contains an eigenfunction with $2k$ nodal domains, satisfies $\lambda^A \geqslant \lambda_k^S$.*

A similar argument can be also made for the symmetric case if $\ell(k) > 0$. This gives us new versions of Courant-sharp properties.

If we call *pair symmetric partition* a partition which is invariant by the symmetry, but such that no element of the partition is invariant, we have the following Courant-sharp analog.

Theorem 3.4. *If, for some eigenvalue λ_k^A of $H_A(\Omega)$, there exists an eigenfunction u^A such that $\mu(u^A) = 2k$, then the corresponding family of nodal domains is a minimal pair symmetric partition.*

Note that, if the labelling of the eigenvalue (counted as eigenvalue of $H(\Omega)$) is $> 2k$), then it is not a minimal $(2k)$-partition of Ω.

Remark 3.5. Let us finally mention as connected result (cf., for example, [3]), that if Ω satisfies (3.2), then $\lambda_2(\Omega) = \lambda_1^A(\Omega)$.

Application. It is known that the eigenfunctions are the restriction to $\mathbb{S}^2$ of the homogeneous harmonic polynomials. Moreover, the eigenvalues are $\ell(\ell+1)$ $(\ell \geqslant 0)$ with multiplicity $(2\ell+1)$. Then the Courant nodal theorem says that for a spherical harmonic u_ℓ corresponding to $\ell(\ell+1)$ one should have

$$\mu(u_\ell) \leqslant \ell^2 + 1\,. \tag{3.11}$$

As observed in [24], one can, using the fact that

$$u_\ell(-x) = (-1)^\ell u_\ell(x)\,, \tag{3.12}$$

improve this result by using a variant of the Courant nodal theorem with symmetry (cf. Theorem 3.1) and this leads to the improvement

$$\mu(u_\ell) \leqslant \ell(\ell-1) + 2\,. \tag{3.13}$$

Let us briefly sketch the proof of (3.13). If ℓ is odd, any eigenfunction is odd with respect to inversion. Hence the number of nodal domains is even $\mu(u_\ell) = 2n_\ell$ and there are no nodal domains invariant by inversion. Using the Courant nodal theorem for H_A, we get with $\ell = 2p+1$ that

$$\begin{aligned} n_\ell &\leqslant \sum_{q=0}^{p-1}(2(2q+1)+1) \quad +1 = 2p(p-1)+3p+1 \\ &= p(2p+1)+1 = \frac{1}{2}\ell(\ell-1)+1\,. \end{aligned}$$

If ℓ is even, we can only write

$$\mu(u_\ell) = 2n_\ell + p_\ell\,,$$

where p_ℓ is the cardinality of the nodal domains which are invariant by inversion.

Using the Courant nodal theorem for H_S, we get with $\ell = 2p$, that

$$\begin{aligned} n_\ell + p_\ell &\leqslant \Big(\sum_{q=0}^{p-1}(4q+1)\Big) +1 = 2p(p-1)+p+1 \\ &= p(2p-1)+1 = \frac{1}{2}\ell(\ell-1)+1\,. \end{aligned}$$

Using this improved estimate, we immediately obtain the following assertion.

Proposition 3.6. *The only cases where u_ℓ can be Courant-sharp are for $\ell = 0$ and $\ell = 1$.*

This proposition has the following consequence.

Theorem 3.7. *A minimal k-partition of $\mathbb{S}^2$ is nodal if and only if $k \leqslant 2$.*

Note that in [24, 25] the more sophisticated conjecture (verified for $\ell \leqslant 6$) is proposed:

Conjecture 3.8.

$$\mu(u_\ell) \leqslant \begin{cases} \frac{1}{2}(\ell+1)^2, & \ell \textit{ is odd} \\ \frac{1}{2}\ell(\ell+2), & \ell \textit{ is even}. \end{cases}$$

Remark 3.9. As indicated by D. Jakobson to one of us, there is also a probabilistic version of this conjecture [27]. Karpushkin [23] has also the following bound for the number of components:

$$\mu(u_\ell) \leqslant \begin{cases} (\ell-1)^2+2, & \ell \text{ is odd,} \\ (\ell-1)^2+1, & \ell \text{ is even.} \end{cases}$$

This is for ℓ large slightly better than what we obtained with the refined Courant-sharp theorem. Let us also mention the recent paper [16] and references therein.

Remark 3.10. Considering the Laplacian on the double covering $\mathbb{S}^2_{\mathcal{C}}$ of $\ddot{\mathbb{S}}^2 :=$ $\mathbb{S}^2 \setminus \{\text{North}, \text{South}\}$, Theorems 3.1 and 3.4 hold, where I is replaced by $\mathcal{I}$

(introduced in (1.11)) corresponding to the map $\phi \mapsto \phi+2\pi$. The $\mathcal{I}$-symmetric eigenfunctions can be identified to the eigenfunctions of $H(\mathbb{S}^2)$ by $u_S(x) = u(\pi(x))$ and the restriction $H_{\mathcal{A}}(\mathbb{S}^2_{\mathcal{C}})$ of $H(\mathbb{S}^2_{\mathcal{C}})$ to the $\mathcal{I}$-antisymmetric space leads to a new spectrum, which will be analyzed in Section 6.

4 On Topological Properties of Minimal 3-Partitions of the Sphere

As in the case of planar domains, a classification of the possible types of minimal partitions could simplify the analysis. The case of the whole sphere $\mathbb{S}^2$ shows some difference with for example the case of the disk.

4.1 Around the Euler formula

As in the case of domains in the plane [20], we will use the following result.

Proposition 4.1. *Let Ω be an open set in $\mathbb{S}^2$ with piecewise $C^{1,+}$ boundary, and let $N \in \mathcal{M}(\Omega)$ be such that the associate $\mathcal{D}$ consists of μ domains $D_1, \dots, D_\mu$. Let b_0 be the number of components of $\partial\Omega$, and let b_1 be the number of components of $N \cup \partial\Omega$. Denote by $\nu(x_i)$ and $\rho(z_i)$ the numbers associated to the $x_i \in X(N)$, respectively $z_i \in Y(N)$. Then*

$$\mu = b_1 - b_0 + \sum_{x_i \in X(N)} \left(\frac{\nu(x_i)}{2} - 1\right) + \frac{1}{2} \sum_{z_i \in Y(N)} \rho(z_i) + 1\,. \tag{4.1}$$

Remark 4.2. In the case $\Omega = \mathbb{S}^2$, the statement simply reads

$$\mu = b_1 + \sum_{x_i \in X(N)} \left(\frac{\nu(x_i)}{2} - 1\right) + 1\,, \tag{4.2}$$

where b_1 is the number of components of N.

4.2 Application to 3- and 4-partitions.

The case of 3-partitions. Let us analyze in this spirit the topology of minimal 3-partitions of $\mathbb{S}^2$.

First we recall that a minimal 3-partition cannot be nodal. The multiplicity of the second eigenvalue of $-\Delta_{\mathbb{S}^2}$ is indeed 3. Hence our minimal 3-partition cannot be admissible.

Let us look now to the information given by the Euler formula. We argue like in [20] for the case of the disk. We recall that, at any critical point x_c of N,

$$\nu(x_c) \geqslant 3\,. \tag{4.3}$$

Hence (4.2) implies that $b_1 \leqslant 2$ and we have

$$1 \leqslant b_1 \leqslant 2\,. \tag{4.4}$$

When $b_1 = 1$, we get as a unique solution $\#X(N) = 2$. So $X(N)$ consists of two points x_1 and x_2 such that $\nu(x_i) = 3$ for $i = 1$ and 2. The other case where $\#X(N) = 1$ leads indeed to an even number of half-lines arriving to the unique critical point and to an admissible (hence excluded) partition.

When $b_1 = 2$, we find that $X(N)$ is empty and the partition should be admissible which is excluded. Hence we have proved the following assertion.

Proposition 4.3. *If $\mathcal{D}$ is a regular nonadmissible strong 3-partition of $\mathbb{S}^2$, then $X(N)$ consists of two points x_1 and x_2 such that $\nu(x_i) = 3$, and N consists of three noncrossing (except at their ends) arcs joining the two points x_1 and x_2.*

In particular, this can be applied to minimal 3-partitions of $\mathbb{S}^2$.

The case of 4-partitions. We can analyze in the same way the case of nonadmissible 4-partitions. The Euler formula leads to the following classification.

Proposition 4.4. *If $\mathcal{D}$ is a regular nonadmissible strong 4-partition of $\mathbb{S}^2$, then we are in one of the following cases:*

- *$X(N)$ consists of four points x_i $(i = 1, \dots, 4)$ such that $\nu(x_i) = 3$, and N consists of six noncrossing (except at their ends) segments, each one joining two points x_i and x_j $(i \neq j)$.*
- *$X(N)$ consists of three points x_i $(i = 1, 2, 3)$ such that $\nu(x_1) = \nu(x_2) = 3$, $\nu(x_3) = 4$ and N consists of five noncrossing (except at their ends) segments joining two critical points.*
- *$X(N)$ consists of two points x_i $(i = 1, 2)$ such that $\nu(x_i) = 3$, and N consists of three noncrossing (except at their ends) segments joining the two points x_1 and x_2 and of one closed line.*
- *$X(N)$ consists of two points x_i $(i = 1, 2)$ such that $\nu(x_1) = 3$, $\nu(x_2) = 5$ and N consists of four noncrossing (except at their ends) segments joining the two critical points and of one noncrossing (except at his ends) segment starting from one critical point and coming back to the same one.*

Note that the spherical tetrahedron corresponds to the first type and we recall from Theorem 3.7 that minimal 4-partitions are not admissible.

5 Lyusternik–Shnirelman Theorem and Proof of Proposition 1.4

As was shown in the previous section, $N(\mathcal{D})$ consists of two points $x_1 \neq x_2$ and 3 mutually noncrossing arcs γ_1, γ_2, γ_3 connecting x_1 and x_2. This means that each D_i has a boundary which is a closed curve which is away from x_1, x_2 smooth.

We first recall the well-known theorem of Lyusternik and Shnirelman from 1930, that can be found for instance in [26, p.23]. It states the following assertion.

Theorem 5.1. *Suppose that $S_1, S_2, \ldots, S_d$ are closed subsets of $\mathbb{S}^{d-1}$ such that $\bigcup_{i=1}^{d} S_i = \mathbb{S}^{d-1}$. Then there is at least one S_i that contains a pair of antipodal points.*

We use this theorem in the case $d = 3$ and apply it with S_1, S_2, S_3 defined by

$$S_i = \overline{D}_i. \tag{5.1}$$

Proof of Proposition 1.4. It suffices to show that

$$N(\mathcal{D}) \cap I\, N(\mathcal{D}) \neq \varnothing\,, \tag{5.2}$$

where we recall that I is the antipodal map.

By Theorem 5.1, we know that there is an $S_i = \overline{D}_i$ which contains a pair of antipodal points. After relabelling the D_i's, we can assume that

$$I\,\overline{D_1} \cap \overline{D_1} \neq \varnothing\,, \tag{5.3}$$

and the goal is to show that

$$I\,\partial D_1 \cap \partial D_1 \neq \varnothing\,. \tag{5.4}$$

Our D_i's have the properties of 3-minimal partitions established in the previous section. In particular, ∂D_1 has one component and is also the boundary of ∂D_{13}, where $D_{13} = \mathrm{Int}\,(\overline{D_2} \cup \overline{D_3})$.

The proof is by contradiction. Let us assume that

$$I\,\partial D_1 \cap \partial D_1 = \varnothing\,. \tag{5.5}$$

Then there are two possible cases

Case a. $I\,\partial D_1 \subset D_1$
Case b. $I\,\partial D_1 \cap \overline{D_1} = \varnothing$

Let us start with Case a. Again, there are two possibilities.

Case a1. $I\,D_1 \subset\subset D_1$ ($\subset\subset$ means compactly included).
Case a2. $I\,D_2 \subset\subset D_1$

But Case a1 contradicts the fact that I is an isometry and Case a2 is in contradiction with $\lambda(D_2) = \lambda(D_1)$. Of course, we could have taken D_3 instead of D_2.

We now look at Case b. Then $I\,\partial D_1$ is delimiting in $\mathbb{S}^2$ two components and D_1 is compactly supported in one component. One of the component is $I\,D_1$. But (5.3) implies that D_1 is in this last component:

$$D_1 \subset\subset I\,D_1\,.$$

This can not be true for two isometric domains. Hence we have a contradiction with (5.5) in all the cases. This achieves the proof of the proposition. □

6 The Laplacian on $\mathbb{S}^2_{\mathcal{C}}$

6.1 Spherical harmonics with half integers

These spherical harmonics appear from the beginning of Quantum mechanics in connection with the representation theory [28]. We refer to [17] (Problem 56 (NB2) in the first volume together with Problem 133 in the second volume). We are looking for eigenfunctions of the Friedrichs extension of

$$\mathbf{L}^2 = -\frac{1}{\sin^2\theta}\frac{\partial^2}{\partial\phi^2} - \frac{1}{\sin\theta}\frac{\partial}{\partial\theta}\sin\theta\,\frac{\partial}{\partial\theta} \tag{6.1}$$

in $L^2(\sin\theta d\theta\, d\phi)$, satisfying

$$\mathbf{L}^2 Y_{\ell m} = \ell(\ell+1)Y_{\ell m}\,. \tag{6.2}$$

The standard spherical harmonics corresponding to $\ell \geqslant 0$ are defined for an integer $m \in \{-\ell, \ldots, \ell\}$ by

$$Y_{\ell m}(\theta,\phi) = c_{\ell,m}\exp im\phi\frac{1}{\sin^m\theta}(-\frac{1}{\sin\theta}\frac{d}{d\theta})^{\ell-m}\sin^{2\ell}\theta\,, \tag{6.3}$$

where $c_{\ell,m}$ is an explicit normalization constant.

For future extensions, we prefer to take this as a definition for $m \geqslant 0$ and then to observe that

$$Y_{\ell,-m} = \widehat{c}_{\ell,m}\overline{Y_{\ell,m}}\,. \tag{6.4}$$

For $\ell = 0$, we get $m = 0$ and the constant. For $\ell = 1$, we obtain, for $m = 1$, the function $(\theta,\phi) \mapsto \sin\theta\exp i\phi$ and for $m = -1$, the function $\sin\theta\exp -i\phi$ and for $m = 0$ the function $\cos\theta$, which shows that the multiplicity is 3 for the eigenvalue 2.

Of course, concerning nodal sets, we look at the real-valued functions $(\theta,\phi) \mapsto \sin\theta\cos\phi$ and $(\theta,\phi) \mapsto \sin\theta\sin\phi$ for $|m| = 1$.

As observed a long time ago, these formulas still define eigenfunctions for pairs (ℓ, m) with ℓ a positive half-integer (and not integer), $m \in \{-\ell, \dots, \ell\}$ and $m - \ell$ integer.

For definiteness, we prefer (in the half-integer case) to only consider the pairs with $\ell > 0$ and $m > 0$ and to complete the set of eigenfunctions by introducing

$$\widehat{Y}_{\ell,m} = \overline{Y_{-\ell,m}}. \tag{6.5}$$

These functions are only defined on the double covering $\mathbb{S}^2_{\mathcal{C}}$ of $\ddot{\mathbb{S}}^2 := \mathbb{S}^2 \setminus \{\{\theta = 0\} \cup \{\theta = \pi\}\}$, which can be defined by extending ϕ to the interval $]-2\pi, 2\pi]$.

When restricted to $\phi \in]-\pi, \pi]$, they correspond to the antiperiodic problem with respect to period 2π in the ϕ variable.

To show the completeness, it is enough to show that, for given $m > 0$, the orthogonal family (indexed by $\ell \in \{m + \mathbb{N}\}$) of functions

$$\theta \mapsto \psi_{\ell,m}(\theta) := \frac{1}{\sin^m \theta} \Big(-\frac{1}{\sin\theta} \frac{d}{d\theta} \Big)^{\ell - m} \sin^{2\ell} \theta$$

span all $L^2(]0, \pi[, \sin\theta \, d\theta)$.

For this purpose, we consider $\chi \in C_0^\infty(]0, \pi[)$ and assume that

$$\int_0^\pi \chi(\theta) \psi_{\ell,m}(\theta) \sin\theta d\theta = 0 \quad \forall \ell \in \{m + \mathbb{N}\} .$$

We would like to deduce that this implies $\chi = 0$. After a change of variable $t = \cos\theta$ and an integration by parts, we see that this problem is equivalent to the problem to show that, if

$$\int_{-1}^1 \psi(t) \, [(1 - t^2)^\ell]^{(\ell - m)} \, dt = 0 \quad \forall \ell \in \{m + \mathbb{N}\} ,$$

then $\psi = 0$.

Observing that the space spanned by the functions $(1 - t^2)^{-m}((1 - t^2)^\ell)^{(\ell - m)}$ (which are actually polynomials of exact order ℓ) is the space of all polynomials we can conclude the completeness.

Hence we have obtained the following assertion.

Theorem 6.1. *The spectrum of the Laplace–Beltrami operator on $\mathbb{S}^2_{\mathcal{C}}$ can be described by the eigenvalues $\mu_\ell = \ell(\ell + 1)$ ($\ell \in \mathbb{N}/2$), each eigenvalue being of multiplicity $(2\ell + 1)$. Moreover, the $Y_{\ell,m}$, as introduced in* (6.3), (6.4), *and* (6.5), *define an orthonormal basis for the eigenspace E_{μ_ℓ}.*

In particular, for $\ell = \frac{1}{2}$ we get a basis of two orthogonal real eigenfunctions $\sin\frac{\phi}{2}(\sin\theta)^{\frac{1}{2}}$ and $\cos\frac{\phi}{2}(\sin\theta)^{\frac{1}{2}}$ for the eigenspace associated with $\frac{3}{4}$. For $\ell = \frac{3}{2}$ the multiplicity is 4 and the functions $\sin\frac{3\phi}{2}(\sin\theta)^{\frac{3}{2}}$, $\cos\frac{3\phi}{2}(\sin\theta)^{\frac{3}{2}}$, $\sin\frac{\phi}{2}(\sin\theta)^{\frac{1}{2}}\cos\theta$, and $\cos\frac{\phi}{2}(\sin\theta)^{\frac{1}{2}}\cos\theta$ form a basis for the eigenspace associated with the eigenvalue $\frac{15}{4}$.

6.2 Covering argument and minimal partition

Here, we give one part of the proof of Proposition 1.5.

Lemma 6.2. *Assume that there exist a 3-minimal partition $\mathcal{D} = (D_1, D_2, D_3)$ of $\mathbb{S}^2$ containing two antipodal points in its boundary. Consider the associated punctured $\ddot{\mathbb{S}}^2$ and the corresponding double covering $\mathbb{S}^2_{\mathcal{C}}$. Denote by Π the projection of $\mathbb{S}^2_{\mathcal{C}}$ on $\ddot{\mathbb{S}}^2$. Then $\Pi^{-1}(D_i)$ consists of two components and $\pi^{-1}(\mathcal{D})$ defines a 6-partition of $\mathbb{S}^2_{\mathcal{C}}$ which is pairwise symmetric.*

The only point to observe is that, according to the property of a minimal partition established in Proposition 4.3, the boundary of the partition necessarily contains a "broken" line joining the two antipodal points.

Using the minimax principle, one immediately gets that, *under the assumption of the lemma,*

$$\mathfrak{L}_3(\mathbb{S}^2) \geqslant \lambda^3_{AS}(\mathbb{S}^2_{\mathcal{C}}), \tag{6.6}$$

where λ^3_{AS} is the third eigenvalue of the Laplace–Beltrami operator on $\mathbb{S}^2_{\mathcal{C}}$ restricted to the antisymmetric spectrum. We will compute $\lambda^3_{AS}(\mathbb{S}^2_{\mathcal{C}})$ in the next subsection.

6.3 Covering argument and Courant-sharp eigenvalues

In the case of the double covering $\mathbb{S}^2_{\mathcal{C}}$ of $\ddot{\mathbb{S}}^2$, we have seen that we have to add the antisymmetric (or antiperiodic) spectrum (corresponding to the map Π, which writes in spherical coordinates as $\phi \mapsto \phi + 2\pi$). This adds the eigenvalue $\frac{3}{4} = \frac{1}{2}(1 + \frac{1}{2})$ with multiplicity 2 and the eigenvalue $\frac{15}{4} = \frac{3}{2}(1 + \frac{3}{2})$ with multiplicity 4. Hence $\frac{15}{4}$ is the 7th eigenvalue of the Laplacian on $\mathbb{S}^2_{\mathcal{C}}$ hence not Courant-sharp, but it is the third antisymmetric eigenvalue

$$\lambda^3_{AS} = \frac{15}{4}\,.$$

Hence, observing that the nodal set of an eigenfunction associated to λ^3_{AS} has six nodal domains which are pairwise symmetric and giving by projection the Y-partition, we immediately obtain, under the assumption of the lemma,

$$\mathfrak{L}_3(\mathbb{S}^2) = \frac{15}{4}\,.$$

But Proposition 1.5 says more. For getting this result, we have to prove the following proposition.

Proposition 6.3. *Let $\mathfrak{L}^{AS}_{2\ell}(\mathbb{S}^2_{\mathcal{C}})$ be the infimum obtained over the pairwise symmetric (by Π) (2ℓ)-partitions of $\mathbb{S}^2_{\mathcal{C}}$. If*

$$\mathfrak{L}_{2\ell}^{AS}(\mathbb{S}_C^2) = \lambda_{AS}^{\ell},$$

then λ_{AS}^{ℓ} *is Courant-sharp in the sense of the antisymmetric spectrum and any minimal pairwise symmetric* (2ℓ)*−partition is nodal.*

The proof is the same as that of Theorem 1.17 in [21] and Theorem 2.6 in [20], with the difference that we consider everywhere antisymmetric states.
Applying this proposition for $\ell = 3$, we have the proof of Proposition 1.5.

7 On Bishop's Approach for Minimal 2-Partitions and Extensions to Strong k-Partitions

7.1 The main result for $k = 2$

For 2-partitions it is immediate to show that the minimal 2-partitions realizing $\mathfrak{L}_2(\mathbb{S}^2)$ are given by the two hemispheres. One is indeed in the Courant-sharp situation. The case of $\mathfrak{L}_{2,p}(\mathbb{S}^2)$ for $p < \infty$ is more difficult. Bishop [4] described, how one can show that the minimal 2-partitions realizing $\mathfrak{L}_{2,1}(\mathbb{S}^2)$ are also given by the two hemispheres. It is then easy to see that it implies the property for any $p \in [1, +\infty[$. Hence the following assertion holds.

Theorem 7.1. *For any* $p \in [1, +\infty]$, $\mathfrak{L}_{2,p}(\mathbb{S}^2)$ *is realized by the partition of* $\mathbb{S}^2$ *by two hemispheres.*

The proof is based on two theorems due to Sperner [29] and Friedland–Hayman [18] respectively. We discuss their proof because it has some consequences for the analysis of minimal 3 and 4-partitions.

7.2 The lower bounds of Sperner and Friedland–Hayman

For a given domain D on the unit sphere S^{m-1} in $\mathbb{R}^m$, Sperner proved the following theorem, which plays on the sphere the same role as the Faber-Krahn inequality plays in $\mathbb{R}^m$:

Theorem 7.2. *Among all sets* $E \subset \mathbb{S}^{m-1}$ *with given* $(m-1)$*-dimensional surface area* $\sigma_m S$ *on (with* σ_m *being the area of* $\mathbb{S}^{m-1}$*), a spherical cap has the smallest characteristic constant.*

Here, the characteristic constant for a domain D is related to the ground state energy by

$$\lambda(D) = \alpha(D)(\alpha(D) + m - 2), \tag{7.1}$$

with $\alpha(D) \geqslant 0$.

We introduce for short

$$\alpha(S, m) = \alpha(\mathcal{SC}(\sigma_m S)), \tag{7.2}$$

where $\mathcal{SC}(\sigma_m S)$ is a spherical cap of surface area $\sigma_m S$.

This theorem is not sufficient in itself for the problem. The second ingredient[3] is a lower bound of $\alpha(S, m)$ by various convex decreasing functions. In [18], the following assertion was proved.[4]

Theorem 7.3. *We have the following lower bound:*

$$\alpha(S, 3) \geqslant \Phi_3(S), \tag{7.3}$$

where Φ_3 is the convex decreasing function defined by

$$\Phi_3 = \max(\widehat{\Phi}_3, \Phi_\infty), \tag{7.4}$$

$$\Phi_\infty(S) = \begin{cases} \frac{1}{2}\log\left(\dfrac{1}{4S}\right) + \dfrac{3}{2}, & 0 < S \leqslant 1/4, \\ 2(1-S), & 1/4 \leqslant S < 1. \end{cases} \tag{7.5}$$

$$\widehat{\Phi}_3(S) = \begin{cases} 2(1-S), & 1/2 \leqslant S < 1, \\ \dfrac{1}{2} j_0 \left(\dfrac{1}{S} - \dfrac{1}{2}\right)^{\frac{1}{2}} - \dfrac{1}{2}, & S < 1/2, \end{cases} \tag{7.6}$$

and j_0 being the first zero of the Bessel function of order 0*:*

$$j_0 \sim 2.4048. \tag{7.7}$$

7.3 Bishop's proof for 2-partitions

With these two ingredients, we observe (following a remark of C. Bishop) that for a 2-partition we have necessarily

$$\alpha(D_1) + \alpha(D_2) \geqslant 2, \tag{7.8}$$

the equality being obtained for two hemispheres.

The minimization for the sum corresponds to

$$\inf\left(\alpha(D_1)(\alpha(D_1)+1) + \alpha(D_2)(\alpha(D_2)+1)\right). \tag{7.9}$$

This infimum is surely larger or equal to

[3] See, for example, [1, p.441] and [4].

[4] We write the result only for $m = 3$, but (7.3) holds for any $m \geqslant 3$.

$$\inf_{\alpha_1+\alpha_2\geqslant 2\,,\,\alpha_1\geqslant 0,\alpha_2\geqslant 0} (\alpha_1(\alpha_1+1)+\alpha_2(\alpha_2+1))\,.$$

It is then easy to see that the infimum is obtained for $\alpha_1 = \alpha_2 = 1$,

$$\inf_{\alpha_1+\alpha_2\geqslant 2\,,\,\alpha_1\geqslant 0,\alpha_2\geqslant 0} (\alpha_1(\alpha_1+1)+\alpha_2(\alpha_2+1)) = 4\,. \tag{7.10}$$

This gives a lower bound for $\mathfrak{L}_{2,1}(\mathbb{S}^2)$ which is equal to the upper bound of $\mathfrak{L}_2(\mathbb{S}^2)$ and which is attained for the two hemispheres. This achieves the proof of Theorem 7.1.

Remark 7.4. A natural question is to determine under which condition the infimum of $\Lambda^1(\mathcal{D})$ for $\mathcal{D} \in \mathfrak{O}_2$ is realized for a pair (D_1, D_2) such that $\lambda(D_1) = \lambda(D_2)$. Let us illustrate the question by a simple example. If we consider two disks C_1 and C_2 such that $\lambda(C_1) < \lambda(C_2) \leqslant \lambda_2(C_1)$, it is not too difficult to see that if we take Ω as the union of these two disks and of a thin channel joining the two disks, then $\mathfrak{L}_2(\Omega) = \lambda_2(\Omega)$ will be very close to $\lambda(C_2)$ and the infimum of $\Lambda^1(\mathcal{D})$ will be less than $\frac{1}{2}(\lambda(C_1)+\lambda(C_2))$. Hence we will have strict inequality if the channel is small enough. We refer to [8, 9, 2, 22] for the spectral analysis of this type of situation. These authors are actually more interested in the symmetric situation where tunneling plays an important role. This example is also considered in [10].

7.4 Application to general k-partitions

One can also discuss what can be obtained in the same spirit for k-partitions ($k \geqslant 3$). This will not lead to the proof of Bishop's conjecture, but give rather accurate lower bounds corresponding in a slightly different context to the ones proposed by Friedland–Hayman [18] for harmonic functions in cones of $\mathbb{R}^m$.

Let us first mention the easy result extending (7.10).

Lemma 7.5. *Let $k \in \mathbb{N}^*$ and $\rho > 0$. If*

$$T^{k,\rho} := \Big\{\alpha \in \overline{\mathbb{R}}_+^k \mid \sum_{j=1}^k \alpha_j \geqslant \rho\Big\},$$

then

$$\frac{1}{k}\inf_{\alpha\in T^{k,\rho}} \sum_{j=1}^k \alpha_j(\alpha_j+1) \geqslant \frac{\rho}{k}\Big(\frac{\rho}{k}+1\Big)\,.$$

For a k-partition $\mathfrak{D} = (D_1,\ldots,D_k)$ the corresponding characteristic numbers satisfy the inequality

$$\sum_j \alpha(D_j) \geqslant \sum_j \Phi_3(S_j) \tag{7.11}$$

with

$$\sigma_3 \, S_j = \text{ Area } (D_j) \; (j = 1, \ldots, k) \,, \tag{7.12}$$

and

$$\sum_j S_j = 1 \,. \tag{7.13}$$

Using the convexity of Φ_3, we obtain

Proposition 7.6. *If $\mathfrak{D} = (D_j)_{j=1,\ldots,k}$ is a strong k-partition of $\mathbb{S}^2$, then*

$$\frac{1}{k} \sum_{j=1}^{k} \alpha(D_j) \geqslant \Phi_3(1/k) \,. \tag{7.14}$$

Applying Lemma 7.5 with $\rho = k\Phi_3(1/k)$, this leads together with (7.14) and (1.8) to the lower bound of $\mathfrak{L}_{k,1}(\mathbb{S}^2)$:

Proposition 7.7.

$$\mathfrak{L}_k(\mathbb{S}^2) \geqslant \mathfrak{L}_{k,1}(\mathbb{S}^2) \geqslant \Phi_3(1/k)\,(1 + \Phi_3(1/k)) \,. \tag{7.15}$$

Let us see what it gives coming back to the definition of Φ_3.

Corollary 7.8.

$$\mathfrak{L}_k(\mathbb{S}^2) \geqslant \mathfrak{L}_{k,1}(\mathbb{S}^2) \geqslant \gamma_k \,, \tag{7.16}$$

with

$$\gamma_k := \Phi_\infty(1/k)\,(1 + \Phi_\infty(1/k)) \,, \tag{7.17}$$

and

$$\Phi_\infty(1/k) := \begin{cases} \dfrac{2(k-1)}{k}, & k \leqslant 4 \,, \\ 2 \log\left(\dfrac{k}{4}\right) + \dfrac{3}{2}, & k > 4 \,. \end{cases} \tag{7.18}$$

In particular,

$$\gamma_2 = 2 \,, \; \gamma_3 = \frac{28}{9} \,, \; \gamma_4 = \frac{15}{4} \,. \tag{7.19}$$

We note that γ_2 is optimal and $\gamma_3 < \frac{15}{4}$. Hence for $k = 3$ the lower bound is not optimal and does not lead to a proof of Bishop's conjecture. Let us now consider the estimates associated with Φ_3.

Corollary 7.9.

$$\mathfrak{L}_k(\mathbb{S}^2) \geqslant \mathfrak{L}_{k,1}(\mathbb{S}^2) \geqslant \delta_k \,, \tag{7.20}$$

with

$$\delta_k := \widehat{\Phi}_3(1/k)\,(1 + \widehat{\Phi}_3(1/k)) \,, \tag{7.21}$$

In particular,

$$\delta_3 = \frac{5}{8} j_0^2 - \frac{1}{4}, \quad \delta_4 = \frac{7}{8} j_0^2 - \frac{1}{4}. \tag{7.22}$$

7.5 Discussion for the cases $k = 3$ and $k = 4$.

We observe that $\delta_3 > \gamma_3$ and $\delta_4 > \gamma_4$. Simple computations [5] show that

$$\widehat{\Phi}_3(1/3) \sim 1.401, \tag{7.23}$$

which is higher than $\Phi_\infty(1/3) = \frac{4}{3}$, and

$$\widehat{\Phi}_3(1/4) \sim 1.748, \tag{7.24}$$

which is higher than $\Phi_\infty(1/4) = \frac{3}{2}$. This leads to the lower bound:

Proposition 7.10.

$$\mathfrak{L}_4(\mathbb{S}^2) \geqslant \mathfrak{L}_{4,1}(\mathbb{S}^2) > 15/4 = \mathfrak{L}_3(\mathbb{S}^2). \tag{7.25}$$

In particular, the best lower bound of $\mathfrak{L}_{4,1}(\mathbb{S}^2)$ is approximately

$$\delta_4 \sim 4.8035. \tag{7.26}$$

Note that by a third method, one can find in [18] (Theorem 5 and Table 1, p. 155, computed by J.G. Wendel) another convex function $\widetilde{\Phi}$ such that $\alpha(D) \geqslant \widetilde{\Phi}(S)$ and

$$\widetilde{\Phi}(1/3) \sim 1.41167. \tag{7.27}$$

Note that $\widetilde{\Phi}(1/4) < \widehat{\Phi}_3(1/4)$, so this improvment occurs only for 3-partitions.

In the case of $\mathbb{S}^2$, unlike the case of the square or the disk, the minimal 4-partition is not nodal (as proved in Theorem 3.7). Note that this implies that the 4-minimal partition realizing $\mathfrak{L}_{4,p}(\Omega)$ for $p \in [1, +\infty]$ is neither nodal.

As already mentioned in [18], there is at least a natural candidate which is the spherical regular tetrahedron. Numerical computations[6] give, for the corresponding 4-partition $\mathcal{D}_4^{Tetra}$,

$$\Lambda(\mathcal{D}_4^{Tetra}) \sim 5.13. \tag{7.28}$$

Hence we obtain

$$\frac{15}{4} < \mathfrak{L}_4(\mathbb{S}^2) \leqslant \Lambda(\mathcal{D}_4^{Tetra}) < 6 = L_4(\mathbb{S}^2). \tag{7.29}$$

---

[5] already done in [18].

[6] transmitted to us by M. Costabel.

It is interesting to compare it with (7.26).

According to a personal communication of M. Dauge, one can also observe that the largest circle inside a face of the tetrahedron is actually a nodal line corresponding to an eigenfunction with eigenvalue 6 (up to a rotation, this is the (restriction to $\mathbb{S}^2$ of the) harmonic polynomial $\mathbb{S}^2 \ni (x,y,z) \mapsto x^2+y^2-2z^2$). This gives directly the comparison $\Lambda(\mathcal{D}_4^{Tetra}) < 6$.

7.6 Large k lower bounds

We can push the argument by looking at the asymptotic as $k \to +\infty$ of δ_k. This gives

$$\mathfrak{L}_{1,k}(\mathbb{S}^2) \geqslant \frac{1}{4} j_0^2 k - \frac{1}{8} j_0^2 - \frac{1}{4} . \tag{7.30}$$

We note that at least for k large this is much better than the trivial lower bound

$$\mathfrak{L}_{1,k}(\mathbb{S}^2) \geqslant \frac{1}{k} \mathfrak{L}_k(\mathbb{S}^2) . \tag{7.31}$$

We discuss in Remark 7.12 below an independent improvment.

Multiplying (7.30) by 4π, the area of $\mathbb{S}^2$, and dividing by k, we obtain

$$\text{Area}(\mathbb{S}^2) \liminf_{k\to+\infty} \frac{\mathfrak{L}_{1,k}(\mathbb{S}^2)}{k} \geqslant \pi j_0^2 . \tag{7.32}$$

But πj_0^2 is the groundstate energy $\lambda(D^1)$ of the Laplacian on the disk D^1 in $\mathbb{R}^2$ of area 1. Although not written explicitly[7] in [21, 6], the Faber-Krahn inequality gives for planar domains

$$|\Omega| \frac{\mathfrak{L}_{k,1}(\Omega)}{k} \geqslant \lambda(D^1) = \pi j_0^2 . \tag{7.33}$$

We have not verified the details, but we think that as in [6] for planar domains, we will have

$$\text{Area}(\mathbb{S}^2) \limsup_{k\to+\infty} \frac{\mathfrak{L}_k(\mathbb{S}^2)}{k} \leqslant \lambda(\text{Hexa}^1) , \tag{7.34}$$

where Hexa^1 denotes the regular hexagon of area 1.

As for the case of plane domains, it is natural to conjecture (cf., for example, [6, 11], but we first heard of this question from van den Berg five years ago) that:

Conjecture 7.11.

$$\lim_{k\to+\infty} \frac{\mathfrak{L}_k(\mathbb{S}^2)}{k} = \lim_{k\to+\infty} \frac{\mathfrak{L}_{k,1}(\mathbb{S}^2)}{k} = \lambda(Hexa^1) .$$

[7] The authors mention only the lower bound for $\mathfrak{L}_k(\Omega)$.

The first equality in the conjecture corresponds to the idea, which is well illustrated in the recent paper by Bourdin-Bucur-Oudet [7] that, asymptotically as $k \to +\infty$, a minimal k-partition for Λ^p will correspond to D_j's such that the $\lambda(D_j)$ are equal.

Remark 7.12. If Ω is a regular bounded open set in $\mathbb{R}^2$ or in $\mathbb{S}^2$, then

$$\frac{1}{k}\sum_{j=1}^{k}\lambda_j(\Omega) \leqslant \mathfrak{L}_{k,1}(\Omega)\,. \tag{7.35}$$

The proof is "fermionic." It is enough to apply the minimax characterization for the groundstate energy $\lambda^{\mathrm{Fermi},k}$ of the Dirichlet realization of the Laplacian on Ω^k (in $(\mathbb{R}^2)^k$ or in $(\mathbb{S}^2)^k$) restricted to the Fermionic space $\wedge^k L^2(\Omega)$ which is

$$\lambda^{Fermi,k} = \sum_{j=1}^{k}\lambda_j(\Omega)\,.$$

For any k-partition $\mathcal{D}$ of Ω, we can consider the Slater determinant of the normalized groundstates ϕ_j of each D_j and observe that the corresponding energy is $k\Lambda^1(\mathcal{D})$.

This suggests the following conjecture (which is proved for $p = 1$ and $p = +\infty$):

$$\Big(\frac{1}{k}\sum_{j=1}^{k}\lambda_j(\Omega)^p\Big)^{\frac{1}{p}} \leqslant \mathfrak{L}_{k,p}(\Omega) \quad \forall k \geqslant 1 \quad \forall p \in [1,+\infty]\,. \tag{7.36}$$

The case where Ω is the union of two disks considered in Remark 7.4 gives an example for $k = 2$, where (7.35) becomes equality. In this case, we have $\lambda_1(\Omega) = \lambda(C_1)$ and $\lambda_2(\Omega) = \lambda(C_2)$.

Acknowledgments. This work was motivated by questions of A. Lemenant about Bishop's conjecture. Discussions with (and numerical computations of) M. Costabel and M. Dauge were also quite helpful. Many thanks also to A. Ancona and D. Jakobson for indicating to us useful references.

References

1. Alt, H.W., Caffarelli, L.A., Friedman, A.: Variational problems with two phases and their free boundaries. Trans. Am. Math. Soc. **282**, no. 2, 431–461 (1984)
2. Anné, C.: A note on the generalized Dumbbell problem. Proc. Am. Math. Soc. **123**, no. 8, 2595–2599 (1995)

3. Besse, A.: Manifolds All of whose Geodesics are Closed. Springer, Berlin etc. (1978)
4. Bishop, C.J.: Some questions concerning harmonic measure. In: Dahlberg, B. (ed.) et al., Partial Differential Equations with Minimal Smoothness and Applications. Proceedings of an IMA Participating Institutions (PI) conference, held Chicago, IL, USA, March 21-25, 1990, pp. 89–07. Springer, New York etc. (1992)
5. Bonnaillie-Noël, V., Helffer, B., Hoffmann-Ostenhof, T.: Aharonov-Bohm Hamiltonians, Isospectrality and Minimal Partitions. Preprint (2008) [To appear in J. Phys. A]
6. Bonnaillie-Noël, V., Helffer, B., Vial, G.: Numerical Simulations for Nodal Domains and Spectral Minimal Partitions. Preprint (2007) [To appear in COCV]
7. Bourdin, B., Bucur, D., Oudet, E.: Optimal Partitions for Eigenvalues. Preprint (2009)
8. Brown, R.M., Hislop, P.D., Martinez, A.: Lower bounds on the interaction between cavities connected by a thin tube. Duke Math. J. **73**, No 1, 163–176 (1994)
9. Brown, R.M., Hislop, P.D., Martinez, A.: Lower bounds on eigenfunctions and the first eigenvalue gap. Inc. Math. Sci. Eng. **192**, 33–49 (1993)
10. Bucur, D., Buttazzo, G., Henrot, A.: Existence results for some optimal partition problems Adv. Math. Sci. Appl. **8**, 571–579 (1998)
11. Caffarelli, L.A., Lin, F.H.: An optimal partition problem for eigenvalues. J. Sci. Comput. **31**, no. 1/2, 5–18 (2006)
12. Caffarelli, L.A., Lin, F.H.: Singularly perturbed elliptic systems and multi-valued harmonic functions with free boundaries. J. Am. Math. Soc. **21**, no. 3, 847–862 (2008)
13. Conti, M., Terracini, S., Verzini, G.: An optimal partition problem related to nonlinear eigenvalues. J. Funct. Anal. **198**, no. 1, 160–196 (2003)
14. Conti, M., Terracini, S., Verzini, G.: A variational problem for the spatial segregation of reaction-diffusion systems. Indiana Univ. Math. J. **54**, no. 3, 779–815 (2005)
15. Conti, M., Terracini, S., Verzini, G.: On a class of optimal partition problems related to the Fucik spectrum and to the monotonicity formula. Calc. Var. Partial Differ. Equ. **22**, no. 1, 45–72 (2005)
16. Eremenko, A., Jakobson, D., Nadirashvili, N.: On nodal sets and nodal domains on $\mathbb{S}^2$. Ann. Inst. Fourier **57**, no. 7, 2345–2360 (2007)
17. Flügge, S.: Practical Quantum Mechanics I, II. Springer, Berlin etc. (1971)
18. Friedland, S., Hayman, W.K.: Eigenvalue inequalities for the Dirichlet problem on spheres and the growth of subharmonic functions. Comment. Math. Helv. **51**, 133–161 (1976)
19. Helffer, B., Hoffmann-Ostenhof, T.: Converse spectral problems for nodal domains. Moscow Math. J. **7**, 67–84 (2007)
20. Helffer, B., Hoffmann-Ostenhof, T.: On spectral minimal partitions: new properties and applications to the disk. CRM Lecture Notes (2009) [To appear]
21. Helffer, B., Hoffmann-Ostenhof, T., Terracini, S.: Nodal domains and spectral minimal partitions. Ann. Inst. H. Poincaré Anal. Non-Linéaire **26**, 101–138 (2009)
22. Jimbo, S., Morita, Y.: Remarks on the behavior of certain eigenvalues on a perturbed domain with several thin channels. Commun. Partial Differ. Equ. **17**, no. 3-4, 523–552 (1992)

23. Karpushkin, V.N.: On the number of components of the complement to some algebraic curves. Russ. Math. Surv. **57**, 1228–1229 (2002)
24. Leydold, J.: Nodal Properties of Spherical Harmonics. PHD Disser. 1993 (Vienna University).
25. Leydold, J.: On the number of nodal domains of spherical harmonics. Topology **35**, 301–321 (1996)
26. Matousek, J.: Using the Borsuk-Ulam Theorem. Springer, Berlin (2003)
27. Nazarov, F, Sodin, M.: On the Number of Nodal Domains of Random Spherical Harmonics. Preprint (2008)
28. Pauli, W.: Uber ein Kriterium fur Einoder Zweiwertigkeit der Eigenfunktionen in der Wellenmechanik (German). Helv. Phys. Acta **12**, 147–168 (1939)
29. Sperner, E.: Zur symmetrisierung von Functionen auf Sphären. (German) Math. Z. **134**, 317–327 (1973)

Weighted Sobolev Space Estimates for a Class of Singular Integral Operators

Dorina Mitrea, Marius Mitrea, and Sylvie Monniaux

In celebration of the distinguished career of our esteemed friend, V.G. Maz'ya

Abstract The aim of this paper is to prove the boundedness of a category of integral operators mapping functions from Besov spaces on the boundary of a Lipschitz domain $\Omega \subseteq \mathbb{R}^n$ into functions belonging to weighted Sobolev spaces in Ω. The model we have in mind is the Poisson integral operator

$$(\mathrm{PI}f)(x) := -\int_{\partial\Omega} \partial_{\nu(y)} G(x,y) f(y)\, d\sigma(y), \quad x \in \Omega,$$

where $G(\cdot,\cdot)$ is the Green function for the Dirichlet Laplacian in Ω, ∂_ν is the normal derivative, and σ is the surface area on $\partial\Omega$, in the case where $\Omega \subseteq \mathbb{R}^n$ is a bounded Lipschitz domain satisfying a uniform exterior ball condition.

1 Introduction

The main result of this paper is the following theorem.

Theorem 1.1. *Let Ω be a bounded Lipschitz domain in $\mathbb{R}^n$, $n \geqslant 2$. Denote by σ the surface measure on $\partial\Omega$, and set*

Dorina Mitrea
University of Missouri at Columbia, Columbia, MO 65211, USA
e-mail: `mitread@missouri.edu`

Marius Mitrea
University of Missouri at Columbia, Columbia, MO 65211, USA
e-mail: `mitream@missouri.edu`

Sylvie Monniaux
LATP - UMR 6632 Faculté des Sciences, de Saint-Jérôme - Case Cour A Université Aix-Marseille 3, F-13397 Marseille Cédex 20, France
e-mail: `sylvie.monniaux@univ-cezanne.fr`

A. Laptev (ed.), *Around the Research of Vladimir Maz'ya III: Analysis and Applications*, International Mathematical Series 13, DOI 10.1007/978-1-4419-1345-6_7,

$$\delta(x) := \operatorname{dist}(x, \partial\Omega), \quad x \in \mathbb{R}^n. \tag{1.1}$$

Consider the integral operator

$$\mathcal{Q}f(x) := \int_{\partial\Omega} q(x,y) f(y)\, d\sigma(y), \quad x \in \Omega, \tag{1.2}$$

satisfying the following conditions:

(1) $\mathcal{Q}1$ *=constant in* Ω,

(2) *there exists* $N \in \mathbb{N}_0 := \mathbb{N} \cup \{0\}$ *and* $\varepsilon \in [0,1)$ *such that for each* $k \in \{0,1,\dots,N\}$

$$|\nabla_x^{k+1} q(x,y)| \leqslant c_o\, \delta(x)^{-k-\varepsilon} |x-y|^{-n+\varepsilon} \tag{1.3}$$

for all $x \in \Omega$ *and almost every* $y \in \partial\Omega$, *for some constant* $c_o = c_o(\Omega, k) > 0$.
Assume that

$$\frac{n-1}{n-\varepsilon} < p \leqslant \infty, \quad (n-1)\Big(\frac{1}{p} - 1\Big)_+ < s < 1 - \varepsilon. \tag{1.4}$$

Then for each $k \in \{0,1,2,\dots,N\}$ *there exists* $C = C(\Omega, p, s, k) > 0$ *such that*

$$\|\delta^{k+1-\frac{1}{p}-s} |\nabla^{k+1} \mathcal{Q}f|\|_{L^p(\Omega)} + \sum_{j=0}^{k} \|\nabla^j \mathcal{Q}f\|_{L^p(\Omega)} \leqslant C \|f\|_{B_s^{p,p}(\partial\Omega)} \tag{1.5}$$

for every $f \in B_s^{p,p}(\partial\Omega)$.

Above, $|\nabla^k u| := \sum_{|\beta| \leqslant k} |\partial^\beta u|$ and $(a)_+ := \max\{a, 0\}$. Also, $B_s^{p,p}(\partial\Omega)$ denotes the (diagonal) Besov scale on $\partial\Omega$ (cf. Section 2 for more details).

The primary motivation for considering this type of result comes from the study of the Dirichlet problem

$$\Delta u = 0 \quad \text{in } \Omega, \quad u \in B^{p,q}_{s+1/p}(\Omega), \quad \operatorname{Tr} u = f \in B_s^{p,q}(\partial\Omega), \tag{1.6}$$

where $B_\alpha^{p,q}(\Omega)$ denotes the scale of Besov spaces in Ω and Tr is the boundary trace operator, via the potential theoretic representation

$$u(x) = -\int_{\partial\Omega} \partial_{\nu(y)} G(x,y) f(y)\, d\sigma(y), \quad x \in \Omega, \tag{1.7}$$

where ν is the outward unit normal to $\partial\Omega$ and $G(x,y)$ is the Green function for the Dirichlet Laplacian in Ω. In this scenario, the *a priori* estimate

$$\|u\|_{B^{p,q}_{s+1/p}(\Omega)} \leqslant C \|f\|_{B_s^{p,q}(\partial\Omega)} \tag{1.8}$$

is equivalent to the boundedness of the Poisson integral operator

$$\mathrm{PI} : B^{p,q}_s(\partial\Omega) \longrightarrow B^{p,q}_{s+\frac{1}{p}}(\Omega) \tag{1.9}$$

defined as

$$(\mathrm{PI}f)(x) := -\int_{\partial\Omega} \partial_{\nu(y)} G(x,y) f(y)\, d\sigma(y), \quad x \in \Omega. \tag{1.10}$$

In the case where $\Omega \subseteq \mathbb{R}^n$ is a smooth bounded domain, it is well known that

$$|\nabla_y G(x,y)| \leqslant C\,|x-y|^{1-n}, \quad x, y \in \Omega, \tag{1.11}$$

from which it can be deduced that, for every $k \in \mathbb{N}_0$,

$$|\nabla_x^{k+1} \nabla_y G(x,y)| \leqslant C \frac{|x-y|^{-n}}{\min\{|x-y|,\, \delta(x)\}^k} \quad \forall\, x \in \Omega,\ \forall\, y \in \overline{\Omega}. \tag{1.12}$$

As a consequence, in the case $\partial\Omega \in C^\infty$, the integral operator (1.10) has kernel $q(x,y) := -\partial_{\nu(y)} G(x,y)$ which satisfies conditions (1)–(2) of Theorem 1.1 and hence

$$\begin{aligned} \|\delta^{k+1-\frac{1}{p}-s} |\nabla^{k+1} \mathrm{PI}\, f|\|_{L^p(\Omega)} &+ \sum_{j=0}^{k} \|\nabla^j \mathrm{PI}\, f\|_{L^p(\Omega)} \\ &\leqslant C \|f\|_{B^{p,p}_s(\partial\Omega)} \quad \forall\, f \in B^{p,p}_s(\partial\Omega). \end{aligned} \tag{1.13}$$

In order to pass from the weighted Sobolev space estimate (1.13) to the Besov estimate implicit in (1.9), we need an auxiliary regularity result which we now describe. Let L be a homogeneous, elliptic differential operator of even order with (possibly matrix-valued) constant coefficients. Fix a Lipschitz domain $\Omega \subset \mathbb{R}^n$. Denote by $\operatorname{Ker} L$ the space of functions u satisfying $Lu = 0$ in Ω. Then for $0 < p \leqslant \infty$ and $s \in \mathbb{R}$ denote by $\mathbb{H}^p_s(\Omega; L)$ the space of functions $u \in \operatorname{Ker} L$ subject to the size/smoothness condition

$$\|u\|_{\mathbb{H}^p_s(\Omega;L)} := \|\delta^{\langle s\rangle - s} |\nabla^{\langle s\rangle} u|\|_{L^p(\Omega)} + \sum_{j=0}^{\langle s\rangle - 1} \|\nabla^j u\|_{L^p(\Omega)} < \infty. \tag{1.14}$$

Hereinafter, for a given $s \in \mathbb{R}$ we set

$$\langle s\rangle := \begin{cases} s, & s \in \mathbb{N}_0, \\ [s]+1, & s > 0,\ s \notin \mathbb{N}, \\ 0, & s < 0, \end{cases} \tag{1.15}$$

where $[\cdot]$ is the integer-part function, i.e., $\langle s\rangle$ is the smallest nonnegative integer greater than or equal to s. Let $F^{p,q}_\alpha(\Omega)$ denote the scale of Triebel–Lizorkin spaces in Ω (again, cf. Section 2 for definitions).

Theorem 1.2. *Let L be as above, and let Ω be a bounded Lipschitz domain in $\mathbb{R}^n$. Then for any $s \in \mathbb{R}$ and $p, q \in (0, \infty)$,*

$$\mathbb{H}^p_s(\Omega; L) = F^{p,q}_s(\Omega) \cap \operatorname{Ker} L. \tag{1.16}$$

Consequently,

$$F^{p,q}_s(\Omega) \cap \operatorname{Ker} L = B^{p,p}_s(\Omega) \cap \operatorname{Ker} L \tag{1.17}$$

for $s \in \mathbb{R}$ and $p, q \in (0, \infty)$. Finally, for $p = \infty$

$$\mathbb{H}^\infty_{k+s}(\Omega; L) = B^{\infty,\infty}_{k+s}(\Omega) \cap \operatorname{Ker} L \tag{1.18}$$

for any $k \in \mathbb{N}_0$ and $s \in (0, 1)$.

For $1 < p, q < \infty$, $s > 0$, this theorem was proved in [9] for $L = \Delta$ and in [1] for $L = \Delta^2$. The present formulation was stated in [14, 10]. The boundedness of the operator (1.9) directly follows from Theorem 1.2 and (1.13).

Consider the case of an irregular $\partial\Omega$. In this case, the estimate (1.11) is not necessarily satisfied. Indeed, if (1.11) holds, then the Green operator

$$\mathbb{G}v(x) := \int_\Omega G(x, y)v(y)\, dy, \quad x \in \Omega, \tag{1.19}$$

behaves itself like a fractional integral operator of order one. Thus, in particular,

$$\mathbb{G} : L^p(\Omega) \longrightarrow L^{p^*}(\Omega) \tag{1.20}$$

would be bounded whenever

$$1 < p < n \quad \text{and} \quad \frac{1}{p^*} = \frac{1}{p} - \frac{1}{n}, \tag{1.21}$$

by the Hardy–Littlewood–Sobolev fractional integration theorem (cf., for example, [22]). However, Dahlberg [4] showed that for a Lipschitz domain $\Omega \subseteq \mathbb{R}^n$ the operator (1.20) is bounded only if $1 < p < p_n + \varepsilon$ (with $\varepsilon = \varepsilon(\Omega) > 0$), where

$$p_n := \frac{3n}{n+3} \quad \text{for} \quad n \geqslant 3 \quad \text{and} \quad p_2 := \frac{4}{3}. \tag{1.22}$$

By means of counterexamples, Dahlberg also showed that this result is sharp. Thus, as a consequence, the estimate (1.11) cannot hold in a general Lipschitz domain. Hence extra regularity properties need to be imposed.

Recall that $\Omega \subseteq \mathbb{R}^n$ satisfies the *uniform exterior ball condition* (henceforth abbreviated as UEBC) if, outside Ω, one call "roll" a ball of a fixed size along the boundary. It is easy to show that any convex domain satisfies

UEBC. Parenthetically, we also note that a bounded open set Ω has $C^{1,1}$ boundary if and only if Ω and $\mathbb{R}^n \setminus \overline{\Omega}$ satisfy UEBC. However, UEBC alone does allow the boundary to develop irregularities which are "outwardly directed." Grüter and Widman [7] showed that if $\Omega \subset \mathbb{R}^n$ is a bounded open domain satisfying UEBC, then there exists $C = C(\Omega) > 0$ such that the Green function for the Dirichlet Laplacian satisfies the following estimates for all $x, y \in \Omega$:

(i) $G(x,y) \leqslant C \operatorname{dist}(x, \partial\Omega)|x-y|^{1-n}$;

(ii) $G(x,y) \leqslant C \operatorname{dist}(x, \partial\Omega) \operatorname{dist}(y, \partial\Omega)|x-y|^{-n}$;

(iii) $|\nabla_x G(x,y)| \leqslant C|x-y|^{1-n}$;

(iv) $|\nabla_x G(x,y)| \leqslant C \operatorname{dist}(y, \partial\Omega)|x-y|^{-n}$;

(v) $|\nabla_x \nabla_y G(x,y)| \leqslant C|x-y|^{-n}$.

Thus, it is possible to run the above program (based on Theorems 1.1 and 1.2) in order to conclude that for a bounded Lipschitz domain $\Omega \subseteq \mathbb{R}^n$ satisfying UEBC the Poisson integral operator (1.9) is bounded whenever

$$0 < q \leqslant \infty, \quad \frac{n-1}{n} < p \leqslant \infty \text{ and } (n-1)\Big(\frac{1}{p}-1\Big)_+ < s < 1. \tag{1.23}$$

In addition, a similar result is valid for

$$\mathrm{PI} : B^{p,p}_s(\partial\Omega) \longrightarrow F^{p,q}_{s+\frac{1}{p}}(\Omega) \tag{1.24}$$

provided that $p, q < \infty$ (cf. Theorem 3.4).

The layout of the paper is as follows. Section 2 contains a background material pertaining to Lipschitz domains; smoothness spaces defined first in the whole Euclidean space $\mathbb{R}^n$, then in open subsets of $\mathbb{R}^n$ and, finally, on Lipschitz surfaces of codimension one in $\mathbb{R}^n$, as well as basic interpolation results and Green function estimates. In Section 3, we deduce a number of estimates depending on the geometric properties of a domain, which are then used to prove the main result, Theorem 1.1. We mention that a result similar to Theorem 1.1 holds for matrix-valued kernels $q(\cdot,\cdot)$ and vector-valued functions f (in this case, condition (1) should read: $\mathcal{Q}$ maps constant vectors defined on $\partial\Omega$ into constant vectors in Ω). A result similar to Theorem 1.1, but for a more restrictive class of operators was proved in [19, 14].

2 Preliminaries

Recall that an open, bounded set Ω in $\mathbb{R}^n$ is called a bounded Lipschitz domain if there exists a finite open covering $\{\mathcal{O}_j\}_{1 \leqslant j \leqslant N}$ of $\partial\Omega$ with the property that, for every $j \in \{1, ..., N\}$, $\mathcal{O}_j \cap \Omega$ coincides with the portion of $\mathcal{O}_j$ lying above the graph of a Lipschitz function $\varphi_j : \mathbb{R}^{n-1} \to \mathbb{R}$ (where $\mathbb{R}^{n-1} \times \mathbb{R}$ is a new system of coordinates obtained from the original one via a rigid motion). As is known, for a Lipschitz domain Ω (bounded or unbounded), the surface measure $d\sigma$ is well defined on $\partial\Omega$ and there exists an outward pointing normal vector $\nu = (\nu_1, \cdots, \nu_n)$ at almost every point on $\partial\Omega$. In particular, this allows one to define the Lebesgue scale in the usual fashion, i.e., for $0 < p \leqslant \infty$

$$L^p(\partial\Omega) := \Big\{ f : \partial\Omega \to \mathbb{R} : f \text{ measurable, and} \\ \|f\|_{L^p(\partial\Omega)} := \Big(\int_{\partial\Omega} |f|^p \, d\sigma \Big)^{1/p} < \infty \Big\}.$$

The Besov and Triebel–Lizorkin scales for a Lipschitz domain Ω are defined by restrictions of the corresponding Besov and Triebel–Lizorkin spaces on $\mathbb{R}^n$, so we start by briefly reviewing the latter. One convenient point of view is offered by the classical Littlewood–Paley theory (cf., for example, [20, 23, 24]). More specifically, let Ξ be the collection of all systems $\{\zeta_j\}_{j=0}^{\infty}$ of Schwartz functions with the following properties:

(i) there exist positive constants A, B, C such that

$$\begin{aligned} &\operatorname{supp}(\zeta_0) \subset \{x : |x| \leqslant A\}; \\ &\operatorname{supp}(\zeta_j) \subset \{x : B2^{j-1} \leqslant |x| \leqslant C2^{j+1}\} \quad \text{if } j \in \mathbb{N}; \end{aligned} \tag{2.1}$$

(ii) for every multiindex α there is a positive finite constant C_α such that

$$\sup_{x\in\mathbb{R}^n} \sup_{j\in\mathbb{N}} 2^{j|\alpha|} |\partial^\alpha \zeta_j(x)| \leqslant C_\alpha; \tag{2.2}$$

(iii)

$$\sum_{j=0}^{\infty} \zeta_j(x) = 1 \text{ for every } x \in \mathbb{R}^n. \tag{2.3}$$

Fix a family $\{\zeta_j\}_{j=0}^{\infty} \in \Xi$. Also, let $\mathcal{F}$ and $S'(\mathbb{R}^n)$ denote the Fourier transform and the class of tempered distributions in $\mathbb{R}^n$ respectively.. Then the Triebel–Lizorkin space $F_s^{p,q}(\mathbb{R}^n)$ is defined for $s \in \mathbb{R}$, $0 < p < \infty$ and $0 < q \leqslant \infty$ as

$$F_s^{p,q}(\mathbb{R}^n) := \Big\{ f \in S'(\mathbb{R}^n) : \\ \|f\|_{F_s^{p,q}(\mathbb{R}^n)} := \Big\| \Big(\sum_{j=0}^{\infty} |2^{sj} \mathcal{F}^{-1}(\zeta_j \mathcal{F} f)|^q \Big)^{1/q} \Big\|_{L^p(\mathbb{R}^n)} < \infty \Big\} \tag{2.4}$$

(with a natural interpretation when $q = \infty$). The case $p = \infty$ is somewhat special, in that a suitable version of (2.4) needs to be used (cf., for example, [20, p. 9]).

If $s \in \mathbb{R}$ and $0 < p, q \leqslant \infty$, then the Besov space $B^{p,q}_s(\mathbb{R}^n)$ can be defined as

$$B^{p,q}_s(\mathbb{R}^n) := \Big\{ f \in S'(\mathbb{R}^n) : \\ \|f\|_{B^{p,q}_s(\mathbb{R}^n)} := \Big(\sum_{j=0}^{\infty} \|2^{sj} \mathcal{F}^{-1}(\zeta_j \mathcal{F} f)\|^q_{L^p(\mathbb{R}^n)} \Big)^{1/q} < \infty \Big\}. \tag{2.5}$$

Different choices of the system $\{\zeta_j\}_{j=0}^{\infty} \in \Xi$ yield the same spaces (2.4)–(2.5) equipped with equivalent norms. Furthermore, the class of Schwartz functions in $\mathbb{R}^n$ is dense in both $B^{p,q}_s(\mathbb{R}^n)$ and $F^{p,q}_s(\mathbb{R}^n)$ provided $s \in \mathbb{R}$ and $0 < p, q < \infty$.

Next, we discuss the adaptation of certain smoothness classes to the situation where the Euclidean space is replaced with the boundary of a Lipschitz domain Ω. Consider three parameters p, q, s such that

$$0 < p, q \leqslant \infty, \quad (n-1)\Big(\frac{1}{p} - 1\Big)_+ < s < 1 \tag{2.6}$$

and assume that $\Omega \subset \mathbb{R}^n$ is the upper-graph of a Lipschitz function $\varphi : \mathbb{R}^{n-1} \to \mathbb{R}$. We then define $B^{p,q}_s(\partial\Omega)$ as the space of locally integrable functions f on $\partial\Omega$ for which the assignment $\mathbb{R}^{n-1} \ni x \mapsto f(x, \varphi(x))$ belongs to $B^{p,q}_s(\mathbb{R}^{n-1})$, the classical Besov space in $\mathbb{R}^{n-1}$. We equip this space with the (quasi-) norm

$$\|f\|_{B^{p,q}_s(\partial\Omega)} := \|f(\cdot, \varphi(\cdot))\|_{B^{p,q}_s(\mathbb{R}^{n-1})}. \tag{2.7}$$

As far as Besov spaces with a negative amount of smoothness are concerned, in the same context as above, we set

$$f \in B^{p,q}_{s-1}(\partial\Omega) \Longleftrightarrow f(\cdot, \varphi(\cdot))\sqrt{1 + |\nabla\varphi(\cdot)|^2} \in B^{p,q}_{s-1}(\mathbb{R}^{n-1}), \tag{2.8}$$

$$\|f\|_{B^{p,q}_{s-1}(\partial\Omega)} := \|f(\cdot, \varphi(\cdot))\sqrt{1 + |\nabla\varphi(\cdot)|^2}\|_{B^{p,q}_{s-1}(\mathbb{R}^{n-1})}. \tag{2.9}$$

As is known, the case $p = q = \infty$ corresponds to the usual (inhomogeneous) Hölder spaces $C^s(\partial\Omega)$ defined by the requirement

$$\|f\|_{C^s(\partial\Omega)} := \|f\|_{L^\infty(\partial\Omega)} + \sup_{\substack{x \neq y \\ x, y \in \partial\Omega}} \frac{|f(x) - f(y)|}{|x - y|^s} < +\infty, \tag{2.10}$$

i.e.,

$$B^{\infty,\infty}_s(\partial\Omega) = C^s(\partial\Omega), \quad s \in (0, 1). \tag{2.11}$$

All the definitions then readily extend to the case of (bounded) Lipschitz domains in $\mathbb{R}^n$ via a standard partition of unity argument. These Besov spaces have been defined in such a way that a number of basic properties from the Euclidean setting carry over to spaces defined on $\partial\Omega$ in a rather direct fashion. We continue by recording an interpolation result which is going to be useful for us here (for a proof see [14, 10]). To state it, recall that $(\cdot,\cdot)_{\theta,q}$ and $[\cdot,\cdot]_\theta$ stand for the real and complex interpolation brackets.

Proposition 2.1. *Suppose that Ω is a bounded Lipschitz domain in $\mathbb{R}^n$. Assume that $0 < p, q, q_0, q_1 \leqslant \infty$ and*

$$\begin{aligned} &\textit{either}\ ,(n-1)\Big(\frac{1}{p}-1\Big)_+ < s_0 \neq s_1 < 1, \\ &\textit{or}\ \ -1+(n-1)\Big(\frac{1}{p}-1\Big)_+ < s_0 \neq s_1 < 0. \end{aligned} \tag{2.12}$$

Then with $0 < \theta < 1$, $s = (1-\theta)s_0 + \theta s_1$

$$(B^{p,q_0}_{s_0}(\partial\Omega), B^{p,q_1}_{s_1}(\partial\Omega))_{\theta,q} = B^{p,q}_s(\partial\Omega). \tag{2.13}$$

Furthermore, if $s_0 \neq s_1$ and $0 < p_i, q_i \leqslant \infty$, $i = 0, 1$, satisfy $\min\{q_0, q_1\} < \infty$ *as well as either of the following two conditions:*

$$\begin{aligned} &\textit{either}\ (n-1)\Big(\tfrac{1}{p_i}-1\Big)_+ < s_i < 1,\ \ i = 0,1, \\ &\textit{or}\ \ -1+(n-1)\Big(\tfrac{1}{p_i}-1\Big)_+ < s_i < 0,\ \ i = 0,1, \end{aligned} \tag{2.14}$$

then

$$[B^{p_0,q_0}_{s_0}(\partial\Omega), B^{p_1,q_1}_{s_1}(\partial\Omega)]_\theta = B^{p,q}_s(\partial\Omega), \tag{2.15}$$

where $0 < \theta < 1$, $s := (1-\theta)s_0 + \theta s_1$, $\frac{1}{p} := \frac{1-\theta}{p_0} + \frac{\theta}{p_1}$ and $\frac{1}{q} := \frac{1-\theta}{q_0} + \frac{\theta}{q_1}$.

We next discuss atomic decompositions of the diagonal Besov scale on $\partial\Omega$. We call $S = S_r = S_r(x)$ a *surface ball* provided that $x \in \partial\Omega$, $0 < r \leqslant \operatorname{diam}(\Omega)$, and $S_r = B(x,r) \cap \partial\Omega$. Also, for $\varkappa > 0$ and $S_r(x)$ surface ball we write $\varkappa S := B(x, \varkappa r) \cap \partial\Omega$. Recall that the tangential gradient is defined by $\nabla_{\tan} u := \nabla u - (\partial_\nu u)\nu$. A function $a_S \in \operatorname{Lip}(\partial\Omega)$ is called an *atom* for $B^{p,p}_s(\partial\Omega)$, $(n-1)/n < p \leqslant 1$, $(n-1)(\frac{1}{p}-1) < s < 1$, if

$$(1)\ \exists\, S = S_r,\ \ \text{surface ball, such that } \operatorname{supp}(a_S) \subseteq S, \tag{2.16}$$

$$(2)\ \|\nabla_{\tan} a_S\|_{L^\infty(\partial\Omega)} \leqslant r^{s-\frac{n-1}{p}-1}. \tag{2.17}$$

It is useful to observe that, by Fundamental Theorem of Calculus, (1) & (2) above also entail

$$\|a_S\|_{L^\infty(\partial\Omega)} \leqslant C r^{s-\frac{n-1}{p}}, \tag{2.18}$$

where C depends exclusively on the Lipschitz character of Ω. The following proposition, extending well-known results from the Euclidean setting, appeared in [14].

Proposition 2.2. *Let $\Omega \subset \mathbb{R}^n$ be a bounded Lipschitz domain. Fix $(n-1)/n < p \leqslant 1$ and $(n-1)(\frac{1}{p}-1) < s < 1$. Then*

$$\|f\|_{B_s^{p,p}(\partial\Omega)} \approx \inf\Big\{\Big(\sum_S |\lambda_S|^p\Big)^{1/p} : f = \sum_S \lambda_S a_S,$$
$$a_S \text{ are } B_s^{p,p}(\partial\Omega) \text{ atoms}, \{\lambda_S\}_S \in \ell^p\Big\}, \tag{2.19}$$

uniformly for $f \in B_s^{p,p}(\partial\Omega)$.

In (2.19), the infimum is taken over all possible representations of f as $\sum_S \lambda_S a_S$, for countable families of surface balls, and the series is assumed to converge absolutely in $L^1_{\text{loc}}(\partial\Omega)$.

Given an arbitrary open subset Ω of $\mathbb{R}^n$, we denote by $f|_\Omega$ the restriction of a distribution f in $\mathbb{R}^n$ to Ω. For $0 < p, q \leqslant \infty$ and $s \in \mathbb{R}$, both $B_s^{p,q}(\mathbb{R}^n)$ and $F_s^{p,q}(\mathbb{R}^n)$ are spaces of (tempered) distributions, hence it is meaningful to define

$$\begin{aligned} &A_s^{p,q}(\Omega) := \{f \text{ distribution in } \Omega : \exists\, g \in A_s^{p,q}(\mathbb{R}^n) \text{ such that } g|_\Omega = f\},\\ &\|f\|_{A_s^{p,q}(\Omega)} := \inf\{\|g\|_{A_s^{p,q}(\mathbb{R}^n)} : g \in A_s^{p,q}(\mathbb{R}^n),\ g|_\Omega = f\}, \quad f \in A_s^{p,q}(\Omega), \end{aligned} \tag{2.20}$$

where $A = B$ or $A = F$.

The existence of a universal extension operator for Besov and Triebel–Lizorkin spaces in an arbitrary Lipschitz domain $\Omega \subset \mathbb{R}^n$ was established by Rychkov [21]. This allows us transferring a number of properties of the Besov–Triebel–Lizorkin spaces in the Euclidean space $\mathbb{R}^n$ to the setting of a bounded Lipschitz domain $\Omega \subset \mathbb{R}^n$. If k is a nonnegative integer and $1 < p < \infty$, then

$$F_k^{p,2}(\Omega) = W^{k,p}(\Omega) := \{f \in L^p(\Omega) : \partial^\alpha f \in L^p(\Omega),\ |\alpha| \leqslant k\}, \tag{2.21}$$

the classical Sobolev spaces in Ω.

A proof of the following proposition can be found in [10].

Proposition 2.3. *Suppose Ω is a bounded Lipschitz domain in $\mathbb{R}^n$. Let $\alpha_0, \alpha_1 \in \mathbb{R}$, $\alpha_0 \neq \alpha_1$, $0 < q_0, q_1, q \leqslant \infty$, $0 < \theta < 1$, $\alpha = (1-\theta)\alpha_0 + \theta\alpha_1$. Then*

$$(F_{\alpha_0}^{p,q_0}(\Omega), F_{\alpha_1}^{p,q_1}(\Omega))_{\theta,q} = B_\alpha^{p,q}(\Omega), \quad 0 < p < \infty, \tag{2.22}$$
$$(B_{\alpha_0}^{p,q_0}(\Omega), B_{\alpha_1}^{p,q_1}(\Omega))_{\theta,q} = B_\alpha^{p,q}(\Omega), \quad 0 < p \leqslant \infty. \tag{2.23}$$

Furthermore, if $\alpha_0, \alpha_1 \in \mathbb{R}$, $0 < p_0, p_1 \leqslant \infty$, and $0 < q_0, q_1 \leqslant \infty$ are such that

$$\text{either} \quad \max\{p_0, q_0\} < \infty, \quad \text{or} \quad \max\{p_1, q_1\} < \infty, \tag{2.24}$$

then

$$[F_{\alpha_0}^{p_0,q_0}(\Omega), F_{\alpha_1}^{p_1,q_1}(\Omega)]_\theta = F_\alpha^{p,q}(\Omega), \tag{2.25}$$

where $0 < \theta < 1$, $\alpha = (1-\theta)\alpha_0 + \theta\alpha_1$, $\frac{1}{p} = \frac{1-\theta}{p_0} + \frac{\theta}{p_1}$ and $\frac{1}{q} = \frac{1-\theta}{q_0} + \frac{\theta}{q_1}$.

On the other hand, if $\alpha_0, \alpha_1 \in \mathbb{R}$, $0 < p_0, p_1, q_0, q_1 \leqslant \infty$ *are such that*

$$\min\{q_0, q_1\} < \infty, \tag{2.26}$$

then

$$[B^{p_0,q_0}_{\alpha_0}(\Omega), B^{p_1,q_1}_{\alpha_1}(\Omega)]_\theta = B^{p,q}_\alpha(\Omega), \tag{2.27}$$

where θ, α, p, q *are as above.*

Let Ω be a bounded Lipschitz domain in $\mathbb{R}^n$. The Green function for the Laplacian in Ω is a unique function $G : \Omega \times \Omega \to [0, +\infty]$ satisfying

$$G(\cdot, y) \in W^{1,2}(\Omega \setminus B_r(y)) \cap \mathring{W}^{1,1}(\Omega) \quad \forall\, y \in \Omega, \ \forall\, r > 0, \tag{2.28}$$

($\mathring{W}^{1,1}(\Omega)$ denotes the closure in $W^{1,1}(\Omega)$ of smooth compactly supported functions in Ω), and

$$\int_\Omega \langle \nabla_x G(x, y), \nabla \varphi(x)\rangle\, dx = \varphi(y) \quad \forall\, \varphi \in C^\infty_c(\Omega). \tag{2.29}$$

Thus,

$$\begin{aligned} G(x, y)\Big|_{x \in \partial\Omega} &= 0 \quad \text{for every } y \in \Omega, \\ -\Delta G(\cdot, y) &= \delta_y \quad \text{for each fixed } y \in \Omega, \end{aligned} \tag{2.30}$$

where the restriction to the boundary is taken in the sense of Sobolev trace theory and δ_y is the Dirac distribution in Ω with mass at y (cf., for example, [7, 11]). As is well known, the Green function is symmetric, i.e.,

$$G(x, y) = G(y, x) \quad \forall\, x, y \in \Omega, \tag{2.31}$$

so that, by the second line in (2.30),

$$-\Delta G(x, \cdot) = \delta_x \quad \text{for each fixed } x \in \Omega. \tag{2.32}$$

Definition 2.4. An open set $\Omega \subset \mathbb{R}^n$ satisfies a *uniform exterior ball condition* (UEBC) if there exists $r > 0$ with the following property: For every $x \in \partial\Omega$ there exists a point $y = y(x) \in \mathbb{R}^n$ such that

$$\overline{B_r(y)} \setminus \{x\} \subseteq \mathbb{R}^n \setminus \Omega \ \text{ and } \ x \in \partial B_r(y). \tag{2.33}$$

The largest radius r satisfying the above property will be referred to as the UEBC *constant* of Ω.

The relevance of the above concept is apparent from the following result of Grüter and Widman (which is contained in [7, Theorem 3.3]).

Theorem 2.5. *Let* $\Omega \subset \mathbb{R}^n$ *be open and satisfy a* UEBC. *Then there exists* $C = C(\Omega) > 0$ *such that the Green function for the Laplacian satisfies*

$$|\nabla_x \nabla_y G(x,y)| \leqslant C|x-y|^{-n} \quad \textit{for all } x,y \in \Omega. \tag{2.34}$$

We record here a version of the interpolation theorem of E. Stein for analytic families of operators which will be needed in the sequel. This particular variant appeared in [2].

Theorem 2.6. *Let (A_0, A_1) be an interpolation pair of complex Banach spaces. We set $\mathcal{X} = A_0 \bigcap A_1$ and $\mathcal{X}_\theta = [A_0, A_1]_\theta$ for $0 \leqslant \theta \leqslant 1$. Analogously, let (B_0, B_1) be another interpolation pair of complex Banach spaces. We set $\mathcal{Y} = B_0 \bigcap B_1$ and $\mathcal{Y}_\theta = [B_0, B_1]_\theta$ for $0 \leqslant \theta \leqslant 1$.*

Let L_z be a family of linear operators defined in $\mathcal{X}$, with values in $\mathcal{Y}$, indexed by a complex parameter z, with $0 \leqslant \Re z \leqslant 1$. Assume that $l(L_z f)$ is continuous and bounded in $0 \leqslant \Re z \leqslant 1$, and analytic in $0 < \Re z < 1$ for every $f \in \mathcal{X}$ and every continuous linear functional l on $\mathcal{Y}$. Assume further that for $\Re z = 0$ and $f \in \mathcal{X}$

$$\|L_z f\|_{\mathcal{Y}_0} \leqslant c_0 \|f\|_{\mathcal{X}_0} \tag{2.35}$$

and for $\Re z = 1$ and $f \in \mathcal{X}$

$$\|L_z f\|_{\mathcal{Y}_1} \leqslant c_1 \|f\|_{\mathcal{X}_1}. \tag{2.36}$$

Then for $0 < \Re z = \theta < 1$ there exists $c = c(s, q_0, q_1, c_0, c_1)$ such that

$$\|L_z f\|_{\mathcal{Y}_\theta} \leqslant c \|f\|_{\mathcal{X}_\theta} \tag{2.37}$$

uniformly for $f \in \mathcal{X}$.

3 Geometric Estimates and the Proof of the Main Result

In the proof of Theorem 1.1, we use a couple of geometric lemmas which we discuss below (recall the notation in (1.1)).

Lemma 3.1. *Let $\Omega \subseteq \mathbb{R}^n$ be a Lipschitz domain. Then for each point $y \in \partial\Omega$ and parameters $\alpha < 1$, $N < n - \alpha$ there exists a finite constant $C = C(\Omega, N, \alpha) > 0$ such that*

$$\int_{\Omega \cap B(y,r)} \frac{\delta(x)^{-\alpha}}{|x-y|^N}\, dx \leqslant C r^{n-\alpha-N} \quad \forall r > 0. \tag{3.1}$$

Furthermore, if $N > n-1$ and $1 > \alpha > n - N$, then

$$\int_{\Omega \setminus B(y,r)} \frac{\delta(x)^{-\alpha}}{|x-y|^N}\, dx \leqslant C r^{n-\alpha-N} \quad \forall r > 0, \tag{3.2}$$

for some $C = C(\Omega, N, \alpha) > 0$.

This is the basic geometric result on which our entire subsequent analysis is based. The proof is straightforward if $\Omega = \mathbb{R}^n_+$ and is reduced to this special case in the case of a general Lipschitz domain by localizing and flattening the boundary via a bi-Lipschitz map (which does not distort distances by more than a fixed factor). We omit the details, but parenthetically mention that Lemma 3.1 continues to hold for a more general class of domains (more specifically, for domains satisfying an interior corkscrew condition and such that $\partial\Omega$ is Ahlfors regular of dimension $n-1$; for definitions, background, and pertinent references the interested reader is referred to [8, 11]).

Lemma 3.2. *Let $\Omega \subseteq \mathbb{R}^n$ be a Lipschitz domain. Then for any points $y, z \in \partial\Omega$ and parameters $c > 1$, $\beta < n$, $M > n - \beta$ there exists a finite constant $C = C(\Omega, c, N, \alpha) > 0$ such that*

$$\int_{\Gamma(z)} \frac{\delta(x)^{-\beta}}{|x-y|^M}\, dx \leqslant C|y-z|^{n-\beta-M}, \tag{3.3}$$

where

$$\Gamma(z) := \{x \in \Omega : |x-z| < c\,\delta(x)\}. \tag{3.4}$$

Proof. Once again, it is possible to show that the conclusion in Lemma 3.2 remains valid if $\Omega \subseteq \mathbb{R}^n$ is a domain satisfying an interior corkscrew condition and such that $\partial\Omega$ is Ahlfors regular of dimension $n-1$. We shall, however, not pursue this avenue here.

To start the proof in earnest, fix $c > 1$, $z, y \in \partial\Omega$ and set $r := |z-y|$. For $j \in \mathbb{N}$ introduce

$$\Gamma_j(z) := \{x \in \Gamma(z) : 2^{j-1} r < |x-z| < 2^j r\} \tag{3.5}$$

and define

$$I_j := \int_{\Gamma_j(z)} \frac{\delta(x)^{-\beta}}{|x-y|^M}\, dx. \tag{3.6}$$

Note that for $x \in \Gamma_j(z)$ we have $|x-y| \leqslant |x-z| + |z-y| \leqslant (2^j+1)r \leqslant 2^{j+1}r$ and $\delta(x) \approx |x-z| \approx 2^j r$, where the notation $a \approx b$ means that there exist c_1, $c_2 > 0$ such that $c_1 a \leqslant b \leqslant c_2 a$. Keeping these in mind, we can write for each $\alpha \in \mathbb{R}$

$$I_j \leqslant C(2^j r)^{\alpha-\beta} \int_{B(y, 2^{j+1}r)\cap\Omega} \frac{\delta(x)^{-\alpha}}{|x-y|^M}\, dx \quad \forall j \in \mathbb{N}. \tag{3.7}$$

Now, we choose $\alpha < \min\{1, n-M\}$ and apply Lemma 3.1 to the integral in (3.7) to further obtain

$$I_j \leqslant C(2^j r)^{\alpha-\beta}(2^j r)^{n-\alpha-M} = 2^{j(n-M-\beta)} r^{n-M-\beta} \quad \forall j \in \mathbb{N}. \tag{3.8}$$

Next, we note that from our hypothesis $n-M-\beta<0$, so $\sum_{j=1}^{\infty} 2^{j(n-M-\beta)} < \infty$ which, in the combination with (3.8), gives that there exists $C>0$ such that

$$\int_{\Gamma(z)\setminus B(z,r/2)} \frac{\delta(x)^{-\beta}}{|x-y|^M}\,dx \leqslant C|y-z|^{n-\beta-M}. \tag{3.9}$$

It remains to estimate

$$\int_{\Gamma(z)\cap B(z,r/2)} \frac{\delta(x)^{-\beta}}{|x-y|^M}\,dx.$$

Observe that if $x \in \Gamma(z) \cap B(z,r/2)$, then $|x-y| \approx |z-y| = r$. Thus, it suffices to prove that

$$\int_{\Gamma(z)\cap B(z,r/2)} \delta(x)^{-\beta}\,dx \leqslant Cr^{n-\beta}. \tag{3.10}$$

For each $j=0,1,2,\dots$ we consider

$$\Gamma^j(z) := \{x \in \Gamma(z) : 2^{-j-1}r \leqslant |x-z| \leqslant 2^{-j}r\}. \tag{3.11}$$

If $x \in \Gamma^j(z)$, we have $\delta(x) \approx |x-z| \approx 2^{-j}r$. Thus,

$$\int_{\Gamma^j(z)} \delta(x)^{-\beta}\,dx \leqslant C(2^{-j}r)^{-\beta}(2^{-j}r)^n \quad \forall j=0,1,2,\dots. \tag{3.12}$$

Furthermore,

$$\begin{aligned}\int_{\Gamma(z)\cap B(z,r/2)} \delta(x)^{-\beta}\,dx &= \sum_{j=0}^{\infty}\int_{\Gamma^j(z)} \delta(x)^{-\beta}\,dx \\ &\leqslant Cr^{n-\beta}\sum_{j=0}^{\infty} 2^{j(\beta-n)} \leqslant Cr^{n-\beta},\end{aligned} \tag{3.13}$$

as desired, where for the last inequality in (3.13) we used the fact that $\beta-n<0$. This proves (3.10). The proof of Lemma 3.2 is complete. □

After these preliminaries, we are ready to prove Theorem 1.1.

Proof of Theorem 1.1. Consider first the case $p=1$, in which scenario we prove that for each $k \in \{0,1,\dots,N\}$ there exists $C=C(\Omega,k)>0$ such that

$$\|\delta^{k-s}|\nabla^{k+1}\mathcal{Q}f|\,\|_{L^1(\Omega)} \leqslant C\|f\|_{B_s^{1,1}(\partial\Omega)} \quad \forall f \in B_s^{1,1}(\partial\Omega). \tag{3.14}$$

Fix $f \in B_s^{1,1}(\partial\Omega)$. By property (1), the operator $\nabla^{k+1}\mathcal{Q}$ annihilates constants. Hence

$$(\nabla^{k+1}\mathcal{Q}f)(x)=\int_{\partial\Omega}\nabla^{k+1}q(x,y)(f(y)-f(z))\,d\sigma(y)\ \ \forall x\in\Omega,\ z\in\partial\Omega. \quad (3.15)$$

Combining (3.15) with *(2)*, we obtain

$$|(\nabla^{k+1}\mathcal{Q}f)(x)|\leqslant\int_{\partial\Omega}\delta(x)^{-k-\varepsilon}|x-y|^{-n+\varepsilon}|f(y)-f(z)|\,d\sigma(y) \quad (3.16)$$

for all $x\in\Omega$ and $z\in\partial\Omega$. Next, fix $c>1$ and for each $x\in\Omega$ define the set

$$E_x:=\{z\in\partial\Omega:\ |x-z|<c\delta(x)\}. \quad (3.17)$$

Now, consider $x^*\in\partial\Omega$ such that $|x-x^*|=\delta(x)$. Then for every $z\in E_x$ we have $|z-x^*|\leqslant|x-z|+|x-x^*|<(c+1)\delta(x)$. Moreover, if $0<\theta<c-1$, then for every $z\in\partial\Omega$ such that $|z-x^*|<\theta\delta(x)$ we have $|x-z|\leqslant|x-x^*|+|x^*-z|<c\delta(x)$. Thus,

$$B(x^*,\theta\delta(x))\cap\partial\Omega\subseteq E_x\subseteq B(x^*,(c+1)\delta(x))\cap\partial\Omega. \quad (3.18)$$

From (3.18) it follows that

$$\sigma(E_x)\approx\delta(x)^{n-1} \quad (3.19)$$

(note that for (3.19) to hold we only need $\partial\Omega$ to be Ahlfors regular). Next, we take the integral average over E_x of (3.16) and use (3.19) to conclude that, for all $x\in\Omega$,

$$|(\nabla^{k+1}\mathcal{Q}f)(x)|\leqslant C\delta(x)^{1-n-k-\varepsilon}\int_{E_x}\int_{\partial\Omega}\frac{|f(y)-f(z)|}{|x-y|^{n-\varepsilon}}\,d\sigma(y)\,d\sigma(z). \quad (3.20)$$

Multiplying the left- and right-hand sides of (3.20) by $\delta(x)^{k-s}$ and then integrating over Ω with respect to x, we obtain

$$\begin{aligned}&\int_\Omega\delta(x)^{k-s}|\nabla_x^{k+1}\mathcal{Q}f(x)|\,dx\\&\leqslant C\int_\Omega\delta(x)^{1-n-s-\varepsilon}\int_{E_x}\int_{\partial\Omega}|x-y|^{-n+\varepsilon}|f(y)-f(z)|\,d\sigma(y)\,d\sigma(z)\,dx\\&=C\int_{\partial\Omega}\int_{\partial\Omega}|f(y)-f(z)|\Big(\int_{\Gamma(z)}\frac{\delta(x)^{1-n-s-\varepsilon}}{|x-y|^{n-\varepsilon}}\,dx\Big)\,d\sigma(y)\,d\sigma(z), \quad (3.21)\end{aligned}$$

where

$$\Gamma(z)=\{x\in\Omega:\ |x-z|<c\delta(x)\}. \quad (3.22)$$

At this point, we use Lemma 3.2 with $\beta = -1+n+s+\varepsilon$ and $M = n-\varepsilon$ (note that since $p=1$, we have $0<s<1-\varepsilon$, so $\beta<n$ and $n-M-\beta<0$ as needed). By Lemma 3.2, the integral over $\Gamma(z)$ in (3.21) is bounded by $C|x-y|^{-(n-1+s)}$. The latter used back in (3.21) yields (3.14) since

$$\|f\|_{B^{1,1}_s(\partial\Omega)} \approx \|f\|_{L^1(\partial\Omega)} + \int_{\partial\Omega}\int_{\partial\Omega} \frac{|f(x)-f(y)|}{|x-y|^{n-1+s}}\, d\sigma(x)\, d\sigma(y). \qquad (3.23)$$

Consider the case $p=\infty$. The goal is to show that

$$\|\delta^{k+1-s}|\nabla^{k+1}\mathcal{Q}f|\|_{L^\infty(\Omega)} \leqslant C\|f\|_{B^{\infty,\infty}_s(\partial\Omega)}, \quad 0\leqslant k\leqslant N. \qquad (3.24)$$

For this purpose, we assume that $x\in\Omega$ is arbitrary and again denote by $x^*\in\partial\Omega$ a point such that $|x-x^*|=\delta(x)$. Then

$$\delta(x)^{k+1-s}|\nabla^{k+1}\mathcal{Q}f(x)| = \delta(x)^{k+1-s}\left|\int_{\partial\Omega}\nabla_x^{k+1}q(x,y)(f(y)-f(x^*))\, d\sigma(y)\right|. \qquad (3.25)$$

Since $f\in B^{\infty,\infty}_s(\partial\Omega) = C^s(\partial\Omega)$, we have

$$|f(y)-f(x^*)| \leqslant \|f\|_{B^{\infty,\infty}_s(\partial\Omega)}|y-x^*|^s \quad \forall\, y\in\partial\Omega. \qquad (3.26)$$

To proceed, we split the integral in (3.25) into two parts, I_1 and I_2, corresponding to $y\in B(x^*,cr)\cap\partial\Omega$ and $y\in\partial\Omega\setminus B(x^*,cr)$ respectively, where $r:=|x-x^*|$ and $c=c(\partial\Omega)>0$ is a suitable constant. Using (3.26) and (1.3), we obtain

$$\begin{aligned}
|I_1| &\leqslant C\|f\|_{B^{\infty,\infty}_s(\partial\Omega)} r^{k+1-s}\int_{B(x^*,cr)\cap\partial\Omega}\frac{|y-x^*|^s}{r^{k+\varepsilon}|x-y|^{n-\varepsilon}}\, d\sigma(y)\\
&\leqslant C\|f\|_{B^{\infty,\infty}_s(\partial\Omega)} r^{1-s-\varepsilon}\int_{B(x^*,cr)\cap\partial\Omega}|x-y|^{s-n+\varepsilon}\, d\sigma(y)\\
&\leqslant C\|f\|_{B^{\infty,\infty}_s(\partial\Omega)} r^{1-s-\varepsilon}\int_{B(x^*,cr)\cap\partial\Omega} r^{s-n+\varepsilon}\, d\sigma(y)\\
&= C\|f\|_{B^{\infty,\infty}_s(\partial\Omega)}.
\end{aligned} \qquad (3.27)$$

If $y\in B(x^*,cr)\cap\partial\Omega$, then $|y-x^*|\leqslant cr\leqslant c|x-y|$ and $|x-y|\geqslant r$, which are used to obtain the second and third inequalities in (3.27) (for the third one we also recall that $s-n+\varepsilon<0$). Turning our attention to I_2, we observe that if $y\in\partial\Omega\setminus B(x^*,cr)$, then $|y-x^*|\leqslant|x-x^*|+|x-y|\leqslant 2|x-y|$, which yields

$$\begin{aligned}
|I_2| &\leqslant C\|f\|_{B^{\infty,\infty}_s(\partial\Omega)} r^{k+1-s}\int_{\partial\Omega\setminus B(x^*,cr)}\frac{|y-x^*|^s}{r^{k+\varepsilon}|x-y|^{n-\varepsilon}}\, d\sigma(y)\\
&\leqslant C\|f\|_{B^{\infty,\infty}_s(\partial\Omega)} r^{1-s-\varepsilon}\int_{\partial\Omega\setminus B(x^*,cr)}|y-x^*|^{s-n+\varepsilon}\, d\sigma(y)
\end{aligned}$$

$$\leqslant C\|f\|_{B^{\infty,\infty}_s(\partial\Omega)} r^{1-s-\varepsilon}\int_{cr}^{\infty}\rho^{s-2+\varepsilon}\,d\rho \leqslant C\|f\|_{B^{\infty,\infty}_s(\partial\Omega)}. \tag{3.28}$$

Now, (3.24) follows by combining (3.27) and (3.28). Thus, the case $p=\infty$ is complete.

To treat the case $1<p<\infty$, we use what we have proved so far and Theorem 2.6. More precisely, for $s_0, s_1 \in (0,1)$ we consider the family of operators

$$L_z f := \delta^{k+z-[(1-z)s_0+zs_1]}|\nabla^{k+1}\mathcal{Q}f| \tag{3.29}$$

such that

$$\begin{aligned}\Re e z = 0 &\Rightarrow |L_0 f| = \delta^{k-s_0}|\nabla^{k+1}\mathcal{Q}f|,\\ \Re e z = 1 &\Rightarrow |L_1 f| = \delta^{k+1-s_1}|\nabla^{k+1}\mathcal{Q}f|.\end{aligned}$$

Our results for $p=1$ and $p=\infty$ lead to the conclusion that the operators

$$\begin{aligned}L_0 &: B^{1,1}_{s_0}(\partial\Omega)\to L^1(\Omega),\\ L_1 &: B^{\infty,\infty}_{s_1}(\partial\Omega)\to L^\infty(\Omega)\end{aligned}$$

are well defined and are bounded for any $s_0, s_1\in(0,1)$. Pick $0<s_0<s_1<1-\varepsilon$, otherwise arbitrary, so that (2.15) applies. In this scenario, Theorem 2.6 can be used, and we can conclude that for each $0\leqslant k\leqslant N$ the operator

$$\delta^{k+1-\frac{1}{p}-s}|\nabla^{k+1}\mathcal{Q}f| : B^{p,p}_s(\partial\Omega)\longrightarrow L^p(\Omega) \tag{3.30}$$

is well define, linear and bounded for every $s\in(0,1)$ and $p\in[1,\infty]$. This takes care of the estimate for the "higher order term" on the left-hand side of (1.5). The "lower order terms" $\nabla^j\mathcal{Q}$ in (1.5) can be handle in a simpler, more straightforward fashion, so we omit the argument.

It remains to analyze the case where $\frac{n-1}{n-\varepsilon}<p<1$, in which scenario (by once again focusing only on the higher order term in the left-hand side of (1.5)), it suffices to prove that for each $k\in\{0,1,\dots,N\}$ there exists some finite constant $C>0$ such that

$$\|\delta^{k+1-\frac{1}{p}-s}|\nabla^{k+1}\mathcal{Q}a\|_{L^p(\Omega)}\leqslant C \quad \text{for every } B^{p,p}_s(\partial\Omega)\text{-atom } a. \tag{3.31}$$

For this purpose, we assume that

$$\begin{aligned}&\operatorname{supp} a\subseteq S_r(x_0) \quad \text{for some} \quad x_0\in\partial\Omega \text{ and } r>0,\\ &\|\nabla_{\tan}a\|_{L^\infty(\partial\Omega)}\leqslant r^{s-1-(n-1)\frac{1}{p}}.\end{aligned} \tag{3.32}$$

Next, we proceed with the rescaling $\widetilde{a}(x):=r^\tau a(x)$ for $x\in\partial\Omega$, with $\tau\in\mathbb{R}$ to be specified soon. Since $s<1$, we have $1-p(2-s)<1-p$, so we can pick $\theta\in(1-p(2-s),1-p)$. Fix such a θ and select $\tau:=\frac{n-\theta}{p}-n$. Then $-1+s+\frac{1-\theta}{p}\in(s,1)$ and $\widetilde{a}$ is a $B^{1,1}_{-1+s+\frac{1-\theta}{p}}(\partial\Omega)$-atom. In particular, there exists $C = C(\Omega, p, s, \theta) > 0$ such that

$$\|\widetilde{a}\|_{B^{1,1}_{-1+s+\frac{1-\theta}{p}}(\partial\Omega)} \leqslant C, \tag{3.33}$$

and, based on what we proved for $p = 1$,

$$\|\delta^{k+1-s-\frac{1-\theta}{p}}|\nabla^{k+1}\mathscr{Q}\widetilde{a}\|_{L^1(\Omega)} \leqslant C\|\widetilde{a}\|_{B^{1,1}_{-1+s+\frac{1-\theta}{p}}(\partial\Omega)} \leqslant C. \tag{3.34}$$

Applying the Hölder inequality, we can write

$$\begin{aligned}
&\int_{B(x_0,2r)\cap\Omega} \Big(\delta(x)^{k+1-\frac{1}{p}-s}|\nabla^{k+1}\mathscr{Q}a(x)|\Big)^p\, dx \\
&= r^{-\tau p}\int_{B(x_0,2r)\cap\Omega} \delta(x)^{kp+p-1-sp}|\nabla^{k+1}\mathscr{Q}\widetilde{a}(x)|^p\, dx \\
&\leqslant r^{-\tau p}\Bigg(\int_{B(x_0,2r)\cap\Omega} \delta(x)^{k+1-s-\frac{1}{p}+\frac{\theta}{p}}|\nabla^{k+1}\mathscr{Q}\widetilde{a}(x)|\, dx\Bigg)^p \\
&\quad\times\Bigg(\int_{B(x_0,2r)\cap\Omega} \delta(x)^{-\frac{\theta}{1-p}}\, dx\Bigg)^{1-p}.
\end{aligned} \tag{3.35}$$

Since $\theta < 1-p$, it follows that $-\frac{\theta}{1-p} > -1$ so that

$$\int_{B(x_0,2r)\cap\Omega} \delta(x)^{-\frac{\theta}{1-p}}\, dx \leqslant Cr^{n-1}\int_0^{cr} t^{-\frac{\theta}{1-p}}\, dt = Cr^{n-\frac{\theta}{1-p}}. \tag{3.36}$$

Thus, combining (3.35), (3.36), and (3.34), we obtain

$$\begin{aligned}
&\int_{B(x_0,2r)\cap\Omega} \Big(\delta(x)^{k+1-\frac{1}{p}-s}|\nabla^{k+1}\mathscr{Q}a(x)|\Big)^p\, dx \\
&\leqslant C\Bigg(\int_{B(x_0,2r)\cap\Omega} \delta(x)^{k+1-s-\frac{1}{p}+\frac{\theta}{p}}|\nabla^{k+1}\mathscr{Q}\widetilde{a}(x)|\, dx\Bigg)^p \leqslant C.
\end{aligned} \tag{3.37}$$

Next, we turn our attention to the contribution away from the support of the atom. For notational simplicity, we assume that $x_0 = 0$ (which can always be arranged via a translation). Taking into account $\|a\|_{L^\infty(\partial\Omega)} \leqslant Cr^{s-(n-1)\frac{1}{p}}$ (cf. (2.18)) and recalling *(2)* again, we have

$$\begin{aligned}
|\nabla^{k+1}\mathscr{Q}a(x)| &\leqslant C\int_{B(0,r)\cap\partial\Omega} \frac{|a(y)|}{\delta(x)^{k+\varepsilon}|x-y|^{n-\varepsilon}}\, d\sigma(y) \\
&\leqslant C\,\frac{r^{s+(n-1)(1-\frac{1}{p})}}{\delta(x)^{k+\varepsilon}|x|^{n-\varepsilon}} \quad \text{if } x \in \Omega\setminus B(0,2r).
\end{aligned} \tag{3.38}$$

At this point, we use (3.38) and (3.2) from Lemma 3.1 with $\alpha = 1-p+sp+\varepsilon p$ and $N = np - \varepsilon p$ to conclude that

$$\int_{\Omega\setminus B(0,2r)} \delta(x)^{(k+1-\frac{1}{p}-s)p}|\nabla^{k+1}\mathcal{Q}a(x)|^p\,dx \leqslant C. \tag{3.39}$$

Now, the estimate (3.31) follows from (3.37) and (3.39), completing the proof of the case $\frac{n-1}{n-\varepsilon} < p < 1$. This completes the proof of the estimate (1.5) for the full range of indices s and p. □

We now discuss a setting where the kernel of the Poisson integral operator satisfies the estimate (1.3).

Theorem 3.3. *Let $\Omega \subseteq \mathbb{R}^n$ be a Lipschitz domain satisfying a* UEBC. *Then for every $k \in \mathbb{N}_0$ there exists a finite constant $C = C(\Omega, k) > 0$ such that $G(\cdot,\cdot)$, the Green function for the Laplacian in Ω, satisfies*

$$|\nabla_x^{k+1}\nabla_y G(x,y)| \leqslant C\frac{|x-y|^{-n}}{\min\{|x-y|,\,\delta(x)\}^k} \quad \forall\, x\in\Omega,\ \forall\, y\in\overline{\Omega}\setminus E, \tag{3.40}$$

for some set $E \subseteq \partial\Omega$ with $\sigma(E) = 0$, where $\delta(x) = \operatorname{dist}(x, \partial\Omega)$ for all $x \in \Omega$.

Proof. We first claim that it suffices to prove (3.40) for every $x, y \in \Omega$. Indeed, assume that the latter has been proved. Then, keeping $x \in \Omega$, by the Fatou theorem proved in [3] for bounded harmonic functions in Lipschitz domains, it follows that there exists $E_x \subseteq \partial\Omega$ with $\sigma(E_x) = 0$ such that (3.40) holds for every $y \in \overline{\Omega}\setminus E_x$. Fixing a countable dense subset D of Ω and setting $E := \bigcup_{x\in D} E_x$, we see that $\sigma(E) = 0$ and (3.40) holds for every $x \in D$ and $y \in \overline{\Omega}\setminus E$. Keeping now $y \in \overline{\Omega}\setminus E$ fixed, by density then (3.40) holds for every $x \in \Omega$.

The case $k = 0$ is contained in Theorem 2.5. Thus, it remains to prove the estimate in (3.40) for every $x, y \in \Omega$ and $k \geqslant 1$.

Step I. Proof of the statement for $k = 1$. Here, we distinguish two cases: $x, y \in \Omega$ with $\delta(x) \leqslant |x-y|$ and $x, y \in \Omega$ with $\delta(x) > |x-y|$.

Case (a). $x, y \in \Omega$ with $\delta(x) \leqslant |x-y|$. In this scenario, $\min\{|x-y|, \delta(x)\} = \delta(x)$. So, we need to show that

$$|\nabla_x^2\nabla_y G(x,y)| \leqslant C\delta(x)^{-1}|x-y|^{-n}. \tag{3.41}$$

Consider $D := B(x, \delta(x)/2) \subseteq \Omega$ so that $y \notin D$ and, if we set $d := \operatorname{dist}(y, \partial D)$, then $d = |x-y| - \frac{1}{2}\delta(x) \geqslant \frac{1}{2}|x-y|$. Thus,

$$\frac{1}{2}|x-y| \leqslant d \leqslant |x-y|. \tag{3.42}$$

Note that $\nabla_x\nabla_y G(\cdot, y)$ is harmonic in D and, by Theorem 2.5, we have

$$|\nabla_x \nabla_y G(x,y)| \leqslant C|x-y|^{-n} \quad \text{for all} \quad x, y \in \Omega. \tag{3.43}$$

Hence $|\nabla_x \nabla_y G(x,y)| \leqslant Cd^{-n}$ if x, y are as in case (a). The latter, combined with interior estimates for the harmonic function $\nabla_x \nabla_y G(\cdot, y)$ in D, implies

$$\begin{aligned} |\nabla_x^2 \nabla_y G(x,y)| &\leqslant C\delta(x)^{-1} \sup_{x \in D} |\nabla_x \nabla_y G(x,y)| \\ &\leqslant C\delta(x)^{-1} d^{-n} \leqslant C\delta(x)^{-1}|x-y|^{-n}, \end{aligned} \tag{3.44}$$

where for the last inequality in (3.44) we used (3.42). This completes the proof of (3.41).

Case (b). $x, y \in \Omega$ with $\delta(x) > |x-y|$.
Now, we have $\min\{|x-y|, \delta(x)\} = |x-y|$, and we seek to prove that

$$|\nabla_x^2 \nabla_y G(x,y)| \leqslant C|x-y|^{-n-1}. \tag{3.45}$$

Consider the harmonic function $\nabla_x \nabla_y G(\cdot, y)$ in $B(x, \frac{1}{2}|x-y|)$ which, by Theorem 2.5, is bounded by $C|x-y|^{-n}$ in this ball. This and interior estimates further imply that, under the current assumptions on x and y,

$$|\nabla_x^2 \nabla_y G(x,y)| \leqslant C|x-y|^{-1}|x-y|^{-n} = C|x-y|^{-n-1}. \tag{3.46}$$

This completes the proof of (3.45) and, consequently, the proof of Step I.

Step II. Proof of the fact that if (3.40) holds for some $k \in \mathbb{N}_0$ when $x, y \in \Omega$, then (3.40) also holds for $k+1$ when $x, y \in \Omega$.

Case (a). $x, y \in \Omega$ with $\delta(x) \leqslant |x-y|$. Under the current assumptions on x and y, it follows that for every $z \in B(x, \frac{1}{2}\delta(x))$ we have $|z-y| \geqslant \frac{1}{2}\delta(x)$ and $\delta(z) \geqslant \frac{1}{2}\delta(x)$, so that $\min\{\delta(z), |z-y|\} \geqslant \frac{1}{2}\delta(x)$. Also, $|z-y| \geqslant |x-y| - |z-x| \geqslant |x-y| - \frac{1}{2}\delta(x) \geqslant \frac{1}{2}|x-y|$. Hence, by invoking the induction hypothesis, we obtain

$$\begin{aligned} |\nabla_x^k \nabla_y G(z,y)| &\leqslant C \frac{|z-y|^{-n}}{\min\{\delta(z), |z-y|\}^k} \\ &\leqslant C\delta(x)^{-k}|x-y|^{-n} \quad \forall\, z \in B\big(x, \frac{1}{2}\delta(x)\big). \end{aligned} \tag{3.47}$$

Thus, if we now use interior estimates for the harmonic function $\nabla_x^k \nabla_y G(\cdot, y)$ in $B(x, \frac{1}{2}\delta(x))$ combined with (3.47), we arrive at

$$\begin{aligned} |\nabla_x^{k+1} \nabla_y G(x,y)| &\leqslant C\delta(x)^{-1}\delta(x)^{-k}|x-y|^{-n} \\ &= \frac{C|x-y|^{-n}}{\min\{\delta(x), |x-y|\}^{k+1}}. \end{aligned} \tag{3.48}$$

This concludes the proof of case (a) in step II.

Case (b). $x, y \in \Omega$ with $\delta(x) > |x-y|$. Under the current assumptions on x and y, it follows that for every $z \in B(x, \frac{1}{2}|x-y|)$ we have $\delta(z) \geqslant \frac{1}{2}|x-y|$ and $|z-y| \geqslant |x-y| - |z-x| \geqslant |x-y| - \frac{1}{2}|x-y| \geqslant \frac{1}{2}|x-y|$, so that $\min\{\delta(z), |z-y|\} \geqslant \frac{1}{2}|x-y|$. The latter, together with the induction hypothesis, implies that

$$\begin{aligned} |\nabla_x^k \nabla_y G(z,y)| &\leqslant C \frac{|z-y|^{-n}}{\min\{\delta(z), |z-y|\}^k} \\ &\leqslant C|x-y|^{-n-k} \quad \forall\, z \in B(x, \frac{1}{2}|x-y|). \end{aligned} \tag{3.49}$$

Employing interior estimates for the harmonic function $\nabla_x^k \nabla_y G(\cdot, y)$ this time in $B(x, \frac{1}{2}|x-y|)$, and then recalling (3.49), we write

$$|\nabla_x^{k+1} \nabla_y G(x,y)| \leqslant C|x-y|^{-n-k-1} = C \frac{|x-y|^{-n}}{\min\{\delta(x), |z-y|\}^{k+1}}. \tag{3.50}$$

This completes the proof of case (b) in step II.

Combining steps I and II, we obtian the desired result. □

It should be remarked that, with a little more effort, the conclusion in the above theorem can be seen to hold in any nontangentially accessible domain $\Omega \subseteq \mathbb{R}^n$ (in the sense of D. Jerison and C. Kenig) with an Ahlfors regular boundary, and which satisfies a UEBC.

We conclude this section with an application to the mapping properties of the Poisson integral operator on Besov–Triebel–Lizorkin spaces, which has obvious implications for the solvability of the Dirichlet problem for the Laplacian in this context (the interested reader is referred to [5, 6, 9, 12, 15, 16, 17, 18, 13, 14] and the references therein).

Theorem 3.4. *Let Ω be a bounded Lipschitz domain in $\mathbb{R}^n$ with outward unit normal ν and surface measure σ on $\partial\Omega$. Let $G(\cdot,\cdot)$ denote the Green function for the Laplacian in Ω. Define*

$$(\mathrm{PI}f)(y) := -\int_{\partial\Omega} \partial_{\nu(x)} G(x,y) f(x)\, d\sigma(x), \quad y \in \Omega. \tag{3.51}$$

Then, if Ω satisfies a UEBC, it follows that the operators

(i) $\mathrm{PI} : B_s^{p,q}(\partial\Omega) \to B_{s+\frac{1}{p}}^{p,q}(\Omega)$

(ii) $\mathrm{PI} : B_s^{p,p}(\partial\Omega) \to F_{s+\frac{1}{p}}^{p,q}(\Omega)$

are bounded whenever $0 < p, q \leqslant \infty$ and $(n-1)(\frac{1}{p}-1)_+ < s < 1$, with the additional condition that $p, q \neq \infty$ in the case of Triebel–Lizorkin spaces.

Proof. By Theorem 3.3 (with the roles of x and y reversed), we know that for every $k \in \mathbb{N}_0$ there exists $C > 0$ and $E \subseteq \partial\Omega$ with $\sigma(E) = 0$ such that

$$\left|\nabla_y^{k+1}\left[\partial_{\nu(x)}G(x,y)\right]\right| \leqslant \frac{C}{\delta(y)^k|x-y|^n} \quad \forall y \in \Omega,\ \forall x \in \partial\Omega \setminus E. \tag{3.52}$$

Furthermore, it is not difficult to check that

$$\mathrm{PI}(1) = 1 \quad \text{in} \quad \Omega \quad \text{and} \quad \Delta \circ \mathrm{PI} = 0 \quad \text{in} \quad \Omega. \tag{3.53}$$

As a consequence, we can apply Theorem 1.1 in order to conclude that if $\frac{n-1}{n} < p \leqslant \infty$ and $(n-1)(\frac{1}{p}-1)_+ < s < 1$, then for every $k \in \mathbb{N}$ there exists $C > 0$ such that

$$\|\delta^{k+1-\frac{1}{p}-s}|\nabla^{k+1}\mathrm{PI}f\|_{L^p(\Omega)} + \sum_{j=0}^{k}\|\nabla^j \mathrm{PI}f\|_{L^p(\Omega)} \leqslant C\|f\|_{B_s^{p,p}(\partial\Omega)} \tag{3.54}$$

for all $f \in B_s^{p,p}(\partial\Omega)$. Next, we recall from Theorem 1.2 that if $0 < p, q \leqslant \infty$ and $\alpha \in \mathbb{R}$, then (recall (1.15))

$$\Delta u = 0 \text{ in } \Omega, \quad \delta^{\langle\alpha\rangle-\alpha}|\nabla^{\langle\alpha\rangle}u| \in L^p(\Omega) \Longrightarrow u \in F_\alpha^{p,q}(\Omega) \cap B_\alpha^{p,p}(\Omega) \tag{3.55}$$

with the extra assumption that $p, q < \infty$ in the case of the Triebel–Lizorkin spaces. Now, if we chose $\alpha := s + \frac{1}{p}$ and $k := \langle\alpha\rangle - 1$, then (3.53), (3.54), and (3.55) imply that the operator in (ii) is bounded. The boundedness of the operator in (i) now follows from this and real interpolation (cf. Propositions 2.1 and 2.3). □

Acknowledgment. The first author was supported in part by NSF FRG grant 0456306. The second author was supported in part by NSF grants DMS-0653180 and DMS-FRG 0456306

References

1. Adolfsson, V., Pipher, J.: The inhomogeneous Dirichlet problem for Δ^2 in Lipschitz domains. J. Funct. Anal. **159**, no. 1, 137–190 (1998)
2. Calderón, A., Torchinsky, A.: Parabolic maximal functions associated with a distribution II. Adv. Math. **24**, 101–171 (1977)
3. Dahlberg, B.: Estimates of harmonic measure. Arch. Rat. Mech. Anal. **65**, 275–288 (1977)
4. Dahlberg, B.E.J.: L^q-estimates for Green potentials in Lipschitz domains. Math. Scand. **44**, no. 1, 149–170 (1979)
5. Fabes, E., Mendez, O., Mitrea, M.: Boundary layers on Sobolev–Besov spaces and Poisson's equation for the Laplacian in Lipschitz domains. J. Funct. Anal. **159**, no. 2, 323–368 (1998)
6. Grisvard, P.: Elliptic Problems in Nonsmooth Domains. Pitman, Boston, MA (1985)
7. Grüter, M., Widman, K.-O.: The Green function for uniformly elliptic equations. Manuscripta Math. **37**, no. 3, 303–342 (1982)

8. Jerison, D., Kenig, C.E.: Boundary behavior of harmonic functions in nontangentially accessible domains. Adv. Math. **46**, no. 1, 80–147 (1982)
9. Jerison, D., Kenig, C.: The inhomogeneous Dirichlet problem in Lipschitz domains. J. Funct. Anal. **130**, no. 1, 161–219 (1995)
10. Kalton, N., Mayboroda, S., Mitrea, M.: Interpolation of Hardy–Sobolev–Besov–Triebel–Lizorkin spaces and applications to problems in partial differential equations. Contemp. Math. **445**, 121–177 (2007)
11. Kenig, C.E.: Harmonic Analysis Techniques for Second Order Elliptic Boundary Value Problems. Am. Math. Soc., Providence, RI (1994)
12. Kozlov, V.A., Maz'ya, V.G., Rossmann, J.: Elliptic Boundary Value Problems in Domains with Point Singularities. Am. Math. Soc., Providence, RI (1997)
13. Mayboroda, S., Mitrea, M.: Sharp estimates for Green potentials on nonsmooth domains. Math. Res. Lett. **11**, 481–492 (2004)
14. Mayboroda, S., Mitrea, M.: The solution of the Chang–Krein–Stein conjecture. In: Proc. Conf. Harmonic Analysis and its Applications (March 24-26, 2007), pp. 61–154. Tokyo Woman's Cristian University, Tokyo (2007)
15. Maz'ya, V.G.: Solvability in $\dot{W}_2^2$ of the Dirichlet problem in a region with a smooth irregular boundary (Russian). Vestn. Leningr. Univ. **22**, no. 7, 87–95 (1967)
16. Maz'ya, V.G.: The coercivity of the Dirichlet problem in a domain with irregular boundary (Russian). Izv. VUZ, Ser. Mat. no. 4, 64–76 (1973)
17. Maz'ya, V.G., Shaposhnikova, T.O.: Theory of Multipliers in Spaces of Differentiable Functions. Pitman, Boston etc. (1985) Russian edition: Leningrad. Univ. Press, Leningrad (1986)
18. Maz'ya, V., Mitrea, M., Shaposhnikova, T.: The Dirichlet Problem in Lipschitz Domains with Boundary Data in Besov Spaces for Higher Order Elliptic Systems with Rough Coefficients. Preprint (2008)
19. Mitrea, M., Taylor, M.: Potential theory on Lipschitz domains in Riemannian manifolds: Sobolev-Besov space results and the Poisson problem. J. Funct. Anal. **176**, no. 1, 1–79 (2000)
20. Runst, T., Sickel, W.: Sobolev Spaces of Fractional Order, Nemytskij Operators, and Nonlinear Partial Differential Operators. de Gruyter, Berlin–New York (1996)
21. Rychkov, V.: On restrictions and extensions of the Besov and Triebel–Lizorkin spaces with respect to Lipschitz domains. J. London Math. Soc. (2) **60**, no. 1, 237–257 (1999)
22. Stein, E.: Singular Integrals and Differentiability Properties of Functions. Princeton Univ. Press, Princeton, N.J. (1970)
23. Triebel, H.: Theory of Function Spaces. Birkhäuser, Berlin (1983)
24. Triebel, H.: Theory of Function Spaces II. Birkhäuser, Basel (1992)

On General Cwikel–Lieb–Rozenblum and Lieb–Thirring Inequalities

Stanislav Molchanov and Boris Vainberg

To our dear friend Vladimir Maz'ya

Abstract These classical inequalities allow one to estimate the number of negative eigenvalues and the sums $S_\gamma = \sum |\lambda_i|^\gamma$ for a wide class of Schrödinger operators. We provide a detailed proof of these inequalities for operators on functions in metric spaces using the classical Lieb approach based on the Kac–Feynman formula. The main goal of the paper is a new set of examples which include perturbations of the Anderson operator, operators on free, nilpotent, and solvable groups, operators on quantum graphs, Markov processes with independent increments. The study of the examples requires an exact estimate of the kernel of the corresponding parabolic semigroup on the diagonal. In some cases the kernel decays exponentially as $t \to \infty$. This allows us to consider very slow decaying potentials and obtain some results that are precise in the logarithmical scale.

1 Introduction

Let us recall the classical estimates concerning the negative eigenvalues of the operator $H = -\Delta + V(x)$ on $L^2(R^d)$, $d \geqslant 3$. Let $N_E(V)$ be the number of eigenvalues E_i of the operator H that are less than or equal to $E \leqslant 0$. In

Stanislav Molchanov
Department of Mathematics and Statistics, University of North Carolina at Charlotte, Charlotte, NC 28223, USA
e-mail: smolchan@uncc.edu

Boris Vainberg
Department of Mathematics and Statistics, University of North Carolina at Charlotte, Charlotte, NC 28223, USA
e-mail: brvainbe@uncc.edu

A. Laptev (ed.), *Around the Research of Vladimir Maz'ya III: Analysis and Applications*, International Mathematical Series 13, DOI 10.1007/978-1-4419-1345-6_8,

particular, $N_0(V)$ is the number of nonpositive eigenvalues. Let

$$N(V) = \#\{E_i < 0\}$$

be the number of strictly negative eigenvalues of the operator H. Then the Cwikel–Lieb–Rozenblum and Lieb–Thirring inequalities have the following form respectively (cf. [4, 15, 16, 17, 18, 23, 22]):

$$N(V) \leqslant C_d \int_{R^d} W^{\frac{d}{2}}(x)dx, \tag{1.1}$$

$$\sum_{i:E_i<0} |E_i|^\gamma \leqslant C_{d,\gamma} \int_{R^d} W^{\frac{d}{2}+\gamma}(x)dx. \tag{1.2}$$

Here, $W = |V_-|$, $V_-(x) = \min(V(x), 0)$, $d \geqslant 3$, $\gamma \geqslant 0$. The inequality (1.1) can be considered as a particular case of (1.2) with $\gamma = 0$. Conversely, the inequality (1.2) can be easily derived from (1.1) (cf. [22]). So, we mostly discuss the Cwikel–Lieb–Rozenblum inequality and its extensions, although some new results concerning the Lieb–Thirring inequality will also be stated.

A review of different approaches to the proof of (1.1) can be found in [25]. We remind only several results. Lieb [15, 16] and Daubechies [5] offered the following general form of (1.1) and (1.2). Let $H = H_0 + V(x)$, and let $V(x) = V_+(x) - V_-(x)$, $V_\pm \geqslant 0$. Then

$$N(V) \leqslant \frac{1}{g(1)} \int_0^\infty \frac{\pi(t)}{t} dt \int_X G(tW(x))\mu(dx). \tag{1.3}$$

$$\sum_{i:E_i<0} |E_i|^\gamma \leqslant \frac{1}{g(1)} \int_0^\infty \frac{\pi(t)}{t} dt \int_X G(tW(x))W^\gamma(x)\mu(dx). \tag{1.4}$$

Here, $W = V_- = \max(0, -V(x))$, G is a continuous, convex, nonnegative function which grows at infinity not faster than a polynomial, and is such that $z^{-1}G(z)$ is integrable at zero (hence $G(0) = 0$), and the integral (1.3) is finite. The function $g(\lambda)$, $\lambda \geqslant 0$, is defined by

$$g(\lambda) = \int_0^\infty z^{-1}G(z)e^{-z\lambda}dz, \quad \text{i.e.,}\ g(1) = \int_0^\infty z^{-1}G(z)e^{-z}dz. \tag{1.5}$$

Note that $\pi(t) = (2\pi t)^{-\frac{d}{2}}$ in the classical case of $H_0 = -\Delta$ on $L^2(R^d)$, and (1.1) follows from (1.3) in this case by substitution $t \to \tau = tW(x)$ if G is such that

$$\int_0^\infty z^{-1-\frac{d}{2}}G(z)dz < \infty.$$

The inequalities above are meaningful only for those W for which integrals converge. They become particularly transparent (cf. [16]) if $G(z) = 0$ for $z \leqslant \sigma$, $G(z) = z - \sigma$ for $z > \sigma$, $\sigma \geqslant 0$. Then (1.3), (1.4) take the form

$$N(V) \leqslant \frac{1}{c(\sigma)} \int_X W(x) \int_{\frac{\sigma}{W(x)}}^{\infty} \pi(t) dt \mu(dx), \tag{1.6}$$

$$\sum_{i:E_i<0} |E_i|^\gamma \leqslant \frac{1}{c(\sigma)} \int_X W^{\gamma+1}(x) \int_{\frac{\sigma}{W(x)}}^{\infty} \pi(t) dt \mu(dx), \tag{1.7}$$

where

$$c(\sigma) = e^{-\sigma} \int_0^\infty \frac{z e^{-z} dz}{z+\sigma}.$$

Daubechies [5] used the Lieb method to justify the estimates above for some pseudodifferential operators in R^d. She also mentioned there that the Lieb method works in a wider setting. A slightly different approach based on the Trotter formula was used by Rozenblum and Solomyak [24, 25]. They proved (1.3) for a wide class of operators in $L^2(X, \mu)$, where X is a measure space with a σ-finite measure $\mu = \mu(dx)$. They also suggested the following form of (1.3). Assume that the function $\pi(t)$ has different power asymptotics as $t \to 0$ and $t \to \infty$. Let

$$p_0(t,x,x) \leqslant c/t^{\alpha/2}, \quad t \leqslant h, \quad p_0(t,x,x) \leqslant c/t^{\beta/2}, \quad t > h,, \tag{1.8}$$

where $h > 0$ is arbitrary. The parameters α and β characterize the "local dimension" and the "global dimension" of X respectively. For example, $\alpha = \beta = d$ in the classical case of the Laplacian $H_0 = -\Delta$ in the Euclidean space $X = R^d$. If $H_0 = -\Delta$ is the difference Laplacian on the lattice $X = Z^d$, then $\alpha = 0$, $\beta = d$. If $X = S^n \times R^d$ is the product of n-dimensional sphere and R^d, then $\alpha = n + d$, $\beta = d$.

If $\alpha, \beta > 2$, then the inequality (1.3) implies (cf. [25]) that

$$N(V) \leqslant C(h)\Big[\int_{\{W(x)\leqslant h^{-1}\}} W^{\frac{\beta}{2}}(x)\mu(dx) + \int_{\{W(x)>h^{-1}\}} W^{\frac{\alpha}{2}}(x)\mu(dx)\Big]. \tag{1.9}$$

Note that the restriction $\beta > 2$ is essential here in the same way as the condition $d > 2$ in (1.1). We show that the assumption on α can be omitted, but the form of the estimate in (1.9) changes in this case.

The paper consists of two parts. In a shorter first part we give a detailed proof of the general form of the Cwikel–Lieb–Rozenblum (1.3) and Lieb–Thirring (1.4) inequalities for the Schrödinger operator in $L^2(X, \mu)$, where X is a metric space with a σ-finite measure μ. We use the Lieb method which is based on trace inequalities and the Kac–Feynman representation of the Schrödinger parabolic semigroup. This approach could be particularly preferable for readers with a background in probability theory. We do not go there beyond the results obtained in [24, 25]. This part has mostly a methodological character. We also show that the inequality (1.3) is valid for $N_0(V)$, not only for $N(V)$.

The main goal of the paper is a new set of examples. We consider operators which may have different power asymptotics of $\pi(t)$ as $t \to 0$ or $t \to \infty$ or exponential asymptotics as $t \to \infty$. The latter case allows us to consider the potentials which decay very slowly at infinity. This is particularly important in some applications, such as the Anderson model, where the borderline between operators with a finite and infinite number of eigenvalues is defined by the decay of the perturbation in the logarithmic scale.

The paper is organized as follows. The general statement will be proved in Theorem 2.1 in the next section. Theorems 2.2 and 2.3 at the end of that section are consequences of Theorem 2.1. They provide more transparent results under additional assumptions on the asymptotic (power or exponential) behavior of $\pi(t)$. Note that we consider all $\alpha \geqslant 0$ in (1.8). Sections 3–6 are devoted to examples. Some cases of a low local dimension α are studied in Section 3. Operators on lattices (cf. also ([25])) and graphs are considered there. Section 4 deals with perturbations of the Anderson operator. The Lobachevsky plane (cf. also ([25])) and pseudodifferential operators related to processes with independent increments are considered in Section 5. Section 6 is devoted to operators on free groups, continuous and discrete Heisenberg group (cf. also [9, 11]), continuous and discrete groups of affine transformations of the line. The Appendix contains a justification of the asymptotics of $\pi(t)$ for the quantum graph operator.

Note that in order to apply any of the estimates (1.3), (1.4) or (1.6)–(1.9), one needs an exact bound for $\pi(t)$ which can be a challenging problem in some cases.

2 General Cwikel–Lieb–Rozenblum and Lieb–Thirring Inequalities

We assume that X is a complete σ-compact metric space with Borel σ-algebra $\mathcal{B}(X)$ and a σ-finite measure $\mu(dx)$. Let H_0 be a selfadjoint nonnegative operator on $L^2(X, \mathcal{B}, \mu)$ with the following two properties:

(a) Operator $-H_0$ is the generator of a semigroup P_t acting on $C(X)$. The kernel $p_0(t,x,y)$ of P_t is continuous with respect to all the variables when $t > 0$ and satisfies the relations

$$\frac{\partial p_0}{\partial t} = -H_0 p_0, \quad t > 0, \quad p_0(0,x,y) = \delta_y(x), \quad \int_X p_0(t,x,y)\mu(dy) = 1, \quad (2.1)$$

i.e., p_0 is a fundamental solution of the corresponding parabolic problem. We assume that $p_0(t,x,y)$ is symmetric, nonnegative, and it defines a Markov process x_s, $s \geqslant 0$, on X with the transition density $p_0(t,x,y)$ with respect to the measure μ.

Note that this assumption implies that $p_0(t,x,x)$ is strictly positive for all $x \in X$, $t > 0$ since

$$p_0(t,x,x) = \int_X p_0^2(\frac{t}{2},x,y)\mu(dy) > 0. \tag{2.2}$$

(b) There exists a function $\pi(t)$ such that $p_0(t,x,x) \leqslant \pi(t)$ for $t \geqslant 0$ and all $x \in X$. We also assume that $\pi(t)$ has at most power singularity at $t \to 0$ and is integrable at infinity, i.e., there exists m such that

$$\int_0^\infty \frac{t^m}{1+t^m}\pi(t)dt < \infty. \tag{2.3}$$

Note that condition (b) implies that

$$p_0(t,x,y) \leqslant \pi(t), \quad x,y \in X. \tag{2.4}$$

In fact,

$$\begin{aligned} p_0(t,x,y) &= \int_X p_0\Big(\frac{t}{2},x,z\Big)p_0\Big(\frac{t}{2},z,y\Big)\mu(dz) \\ &\leqslant \Big(\int_X p_0^2\Big(\frac{t}{2},x,z\Big)\mu(dz)\Big)^{\frac{1}{2}}\Big(\int_X p_0^2\Big(\frac{t}{2},z,y\Big)\mu(dz)\Big)^{\frac{1}{2}}, \end{aligned}$$

which implies (2.4) due to (2.2). Let us note that (2.3), (2.4) imply that the process x_s is transient.

We decided to put an extra requirement on X to be a metric space in order to be able to assume that p_0 is continuous and use a standard version of the Kac–Feynman formula. This makes all the arguments more transparent. In fact, X is a metric space in all examples below. However, all the arguments can be modified to be applicable to the case where X is a measure space by using the L^2-theory of Markov processes based on Dirichlet forms.

Many examples of operators which satisfy conditions (a) and (b) will be given later. At this point, we would like to mention only a couple of examples. First, note that selfadjoint uniformly elliptic operators of second order satisfy conditions (a) and (b). Condition (b) holds with $\pi(t) = Ct^{-d/2}$ due to the Aronson inequality.

Another wide class of operators with conditions (a) and (b) consists of operators which satisfy condition (a) and are invariant with respect to transformations from a rich enough subgroup Γ of the group of isometries of X. The subgroup Γ has to be transitive, i.e., for some reference point $x_0 \in X$ and each $x \in X$ there exists an element $g_x \in \Gamma$ for which $g_x(x_0) = x$. Then $p_0(t,x,x) = p_0(t,x_0,x_0) = \pi(t)$. The simplest example of such an operator is given by $H_0 = -\Delta$ on $L^2(R^d, \mathcal{B}(R^d), dx)$. The group Γ in this case is the group of translations or the group of all Euclidean transformations (translations and rotations). Another example is given by $X = Z^d$ being a lattice and $-H_0$ a difference Laplacian. Other examples will be given later.

(c) Our next assumption mostly concerns the potential. We need to know that the perturbed operator $H = H_0 + V(x)$ is well defined and has pure discrete spectrum on the negative semiaxis. For this purpose it is enough to assume that the operator $V(x)(H_0 - E)^{-1}$ is compact for some $E > 0$. This assumption can be weakened. If the domain of H_0 contains a dense in $L^2(X, \mathcal{B}, \mu)$ set of bounded compactly supported functions, then it is enough to assume that $V_-(x)(H_0 - E)^{-1}$ is compact for some $E > 0$ and the positive part of the potential is locally integrable (cf. [1]). Other criteria of discreteness of the spectra can be found in [19, 20].

Typically (in particular, in all the examples below) H_0 is an elliptic operator, the kernel of the resolvent $(H_0 - E)^{-1}$ has singularity only at $x = y$, this singularity is weak, and the assumptions (c) holds if the potential has an appropriate behavior at infinity. Therefore, we do not need to discuss the validity of this assumption in the examples below.

Theorem 2.1. *Let $(X, \mathcal{B}, \mu)$ be a complete σ-compact metric space with the Borel σ-algebra $\mathcal{B}$ and a σ-finite measure μ on $\mathcal{B}$.*

Let $H = H_0 + V(x)$, where H_0 is a selfadjoint, nonnegative operator on $L^2(X, \mathcal{B}, \mu)$, the potential $V = V(x) = V_+ - V_-$, $V_\pm \geqslant 0$, is real valued, and conditions (a)–(c) *hold.*

Then

$$N_0(V) \leqslant \frac{1}{g(1)} \int_0^\infty \frac{\pi(t)}{t} \int_X G(tW(x))\mu(dx)dt \tag{2.5}$$

and

$$\sum_{i:E_i<0} |E_i|^\gamma \leqslant \frac{1}{g(1)} \int_0^\infty \frac{\pi(t)}{t} \int_X G(tW(x))W(x)^\gamma \mu(dx)dt, \tag{2.6}$$

where $W(x) = V_-(x)$ and the functions G and g are introduced in (1.3) *and* (1.5) *respectively.*

Remark 2.1. Note that (2.5) differs from (1.3) only by inclusion of the dimension of the null space of the operator H into the left-hand side of (2.5). This difference is not very essential, and the first goal of this part of the paper is to give an alternative proof of (1.3) suitable for readers with a background in probability theory.

Remark 2.2. If $G(z) = 0$ for $z \leqslant \sigma$, $G(z) = z - \sigma$ for $z > \sigma$, $\sigma \geqslant 0$, then (2.5), (2.6) take the form

$$N_0(V) \leqslant \frac{1}{c(\sigma)} \int_X W(x) \int_{\frac{\sigma}{W(x)}}^\infty \pi(t)dt\mu(dx), \tag{2.7}$$

$$\sum_{i:E_i<0} |E_i|^\gamma \leqslant \frac{1}{c(\sigma)} \int_X W^{\gamma+1}(x) \int_{\frac{\sigma}{W(x)}}^\infty \pi(t)dt\mu(dx), \tag{2.8}$$

where

$$c(\sigma) = e^{-\sigma} \int_0^\infty \frac{ze^{-z}dz}{z+\sigma}.$$

Some applications of these inequalities will be given below.

Remark 2.3. The inequalities (2.5), (2.6) are valid with $\pi(t)$ moved under sign of the interior integrals and replaced by $p_0(t,x,x)$. For example, (2.5) holds in the following form

$$N_0(V) \leqslant \frac{1}{g(1)} \int_0^\infty \frac{1}{t} \int_X p_0(t,x,x) G(tW(x)) \mu(dx) dt.$$

The same change can be made in (2.7), (2.8). A very minor change in the proof of the theorem is needed in order to justify this remark. Namely, one needs only to omit the last line in (2.23).

Proof of Theorem 2.1. Step 1. Since the eigenvalues E_i depend monotonically on the potential $V(x)$, without loss of generality one can assume that $V(x) = -W(x) \leqslant 0$.

First (Steps 1–6), we prove the inequality (2.5) for $N(V)$ instead of $N_0(V)$. Here, we can assume that $V(x) \in C_{\text{com}}(X)$. Indeed, when $N(V)$ is considered, the inequality (2.5) with $V(x) \in C_{\text{com}}(X)$ implies the same inequality with any V such that the integral in (2.5) converges (cf. [22]). Then (step 7), we'll show that the inequality (2.5) for $N(V)$ leads to the same inequality for $N_0(V)$. Finally (step 8), we remind the reader of standard arguments which allow us to derive (2.6) from (2.5).

Step 2. We denote by B and B_n the operators

$$B = W^{1/2}(H_0+\varkappa^2)^{-1}W^{1/2}, \;\; B_n = W^{1/2}(H_0+\varkappa^2+nW)^{-1}W^{1/2}, \;\; W = W(x).$$

If $N_{-\varkappa^2}(V) = \#\{E_i \leqslant -\varkappa^2 < 0\}$, λ_k are eigenvalues of the operator B and $n(\lambda, B) = \#\{k : \lambda_k \geqslant \lambda\}$, then the Birman–Schwinger principle implies

$$N_{-\varkappa^2}(V) = n(1, B). \tag{2.9}$$

Thus, if $F = F(\lambda)$, $\lambda \geqslant 0$, is a nonnegative strictly monotonically growing function, and $\{\mu_k\}$ is the set of eigenvalues of the operator $F(B)$, then

$$N_{-\varkappa^2}(V) \leqslant \sum_{k:\mu_k \geqslant F(1)} 1 \leqslant \frac{1}{F(1)} \sum_{k:\mu_k \geqslant F(1)} \mu_k \leqslant \frac{1}{F(1)} \text{TrF(B)}. \tag{2.10}$$

This inequality will be used with a function F of the form

$$F(\lambda) = \int_0^\infty P(e^{-z}) e^{\frac{-z}{\lambda}} dz, \quad P(t) = \sum_0^N c_n t^n, \tag{2.11}$$

The exponential polynomial $P(e^{-z})$, $z > 0$ will be chosen later, but it will be a nonnegative function with zero of order m at $z = 0$, i.e.,

$$P(e^{-z}) \leqslant C \frac{z^m}{1+z^m}, \quad z \geqslant 0, \tag{2.12}$$

where m is defined in the condition (b). Since $P(e^{-z}) \geqslant 0$, (2.11) implies that F is nonnegative and monotonic, and therefore (2.10) holds.

From (2.11) it follows that

$$F(\lambda) = \sum_{n=0}^{N} c_n \frac{\lambda}{1 + n\lambda},$$

and the obvious relation $B_n = B(1 + nB)^{-1}$ implies that

$$F(B) = \sum_{n=0}^{N} c_n B_n = W^{\frac{1}{2}} \sum_{n=0}^{N} c_n (H_0 + \varkappa^2 + nW)^{-1} W^{\frac{1}{2}}.$$

For an arbitrary operator K, we denote its kernel by $K(x, y)$. The kernel of the operator $F(B)$ can be expressed trough the fundamental solutions $p = p_n(t, x, y)$ of the parabolic problem

$$p_t = (H_0 + nW(x))p, \ t > 0, \ \ p(0, x, y) = \delta_y(x).$$

Namely,

$$F(B)(x, y) = W^{\frac{1}{2}}(x) \int_0^\infty e^{-\varkappa^2 t} \sum_{n=0}^{N} c_n p_n(t, x, y) dt W^{\frac{1}{2}}(y). \tag{2.13}$$

It will be shown below that the integral above converges uniformly in x and y when $\varkappa = 0$. Hence the kernel $F(B)(x, y)$ is continuous. Since the operator $F(B)$ is nonnegative, from the last relation and (2.10), after passing to the limit as $\varkappa \to 0$, it follows that

$$N(V) \leqslant \frac{1}{F(1)} \int_0^\infty \int_X W(x) \sum_{n=0}^{N} c_n p_n(t, x, x) dt \mu(dx). \tag{2.14}$$

Step 3. The Kac–Feynman formula allows us to write an "explicit" representation for the Schrödinger semigroup $e^{t(-H_0 - nW(x))}$ using the Markov process x_s associated to the unperturbed operator H_0. Namely, the solution of the parabolic problem

$$\frac{\partial u}{\partial t} = -H_0 u - nW(x)u, \quad t > 0, \quad u(0, x) = \varphi(x) \in C(X), \tag{2.15}$$

can be written in the form

$$u(t, x) = E_x e^{-n \int_0^t W(x_s) ds} \varphi(x_t).$$

Note that the finite-dimensional distributions of x_s (for $0 < t_1 < \ldots < t_m$, $\Gamma_1, \ldots, \Gamma_m \in \mathcal{B}(X)$) are given by the formula

$$P_x(x_{t_1} \in \Gamma_1, \ldots, x_{t_m} \in \Gamma_m) = \int_{\Gamma_1} \cdots \int_{\Gamma_m} p_0(t_1, x, x_1) \\ \times p_0(t_2 - t_1, x_1, x_2) \ldots p_0(t_m - t_{m-1}, x_{m-1}, x_m) \mu(dx_1) \ldots \mu(dx_m).$$

If $p_0(t, x, y) > 0$, then one can define the conditional process (bridge) $\widehat{b}_s = \widehat{b}_s^{x \to y, t}$, $s \in [0, t]$, which starts at x and ends at y. Its finite-dimensional distributions are

$$P_{x \to y}(\widehat{b}_{t_1} \in \Gamma_1, \ldots, \widehat{b}_{t_n} \in \Gamma_n) \\ = \frac{1}{p_0(t, x, y)} \int_{\Gamma_1} \cdots \int_{\Gamma_m} p_0(t_1, x, x_1) \ldots p_0(t_m - t_{m-1}, x_{m-1}, x_m) \\ \times p_0(t - t_m, x_m, y) \mu(dx_1) \ldots \mu(dx_m).$$

In particular, the bridge $\widehat{b}_s^{x \to x, t}$, $s \in [0, t]$, is defined since $p_0(t, x, x) > 0$ (cf. condition (a)).

Let $p = p_n(t, x, y)$ be the fundamental solution of the problem (2.15). Then $p_n(t, x, y)$ can be expressed in terms of the bridge $\widehat{b}_s = \widehat{b}_s^{x \to y, t}$, $s \in [0, t]$:

$$p_n(t, x, y) = p_0(t, x, y) E_{x \to y} e^{-n \int_0^t W(\widehat{b}_s) ds}. \tag{2.16}$$

One of the consequences of (2.16) is that

$$p_n(t, x, y) \leqslant p_0(t, x, y). \tag{2.17}$$

Another consequence of (2.16) is the uniform convergence of the integral in (2.13) (and in (2.14)). In fact, (2.12) implies that

$$\sum_{n=0}^{N} c_n e^{-n \int_0^t W(\widehat{b}_s) ds} \leqslant C \frac{t^m}{1 + t^m}.$$

Hence from (2.16) and (2.4) it follows that the integrand in (2.13) can be estimated from above by $C\pi(t)\frac{t^m}{1+t^m}$. Then the uniform convergence of the integral in (2.13) follows from (2.3).

Now, (2.14) and (2.16) imply

$$N(V) \leqslant \frac{1}{F(1)} \int_0^\infty \int_X W(x) p_0(t, x, x) E_{x \to x} [\sum_{n=0}^{N} c_n e^{-n \int_0^t W(\widehat{b}_s) ds}] \mu(dx) dt.$$

Here, $\widehat{b}_s = \widehat{b}_s^{x \to x, t}$.

Step 4. We would like to rewrite the last inequality in the form

$$N(V) \leqslant \frac{1}{F(1)} \int_0^\infty \int_X p_0(t,x,x) E_{x\to x}\Big[W(\widehat{b}_\tau) \sum_{n=0}^N c_n e^{-n\int_0^t W(\widehat{b}_s)ds}\Big] \mu(dx)dt \tag{2.18}$$

with an arbitrary $\tau \in [0,t]$. For that purpose, it is enough to show that

$$\int_X p_0(t,x,x) E_{x\to x}\Big[W(\widehat{b}_\tau) e^{-\int_0^t mW(\widehat{b}_s)ds}\Big] \mu(dx)$$
$$= \int_X p_0(t,x,x) W(x) E_{x\to x}\Big[e^{-\int_0^t mW(\widehat{b}_s)ds}\Big] \mu(dx). \tag{2.19}$$

The validity of (2.19) can be justified using the Markov property of $\widehat{b}_s$ and its symmetry (reversibility in time). We fix $\tau \in (0,t)$. Let $y = \widehat{b}_\tau$. We split $\widehat{b}_s$ into two bridges $\widehat{b}_u^{x\to y,\tau}$, $u \in [0,\tau]$, and $\widehat{b}_v^{y\to x,t}$, $v \in [\tau,t]$. The first bridge starts at x and ends at y, the second one starts at y and goes back to x. Using these bridges, one can represent the left hand side above as

$$\int_X \int_X W(y)[p_0(\tau,x,y)p_0(t-\tau,y,x) - p_m(\tau,x,y)p_m(t-\tau,y,x)]\mu(dx)\mu(dy)$$

$$= \int_X W(y)[p_0(t,y,y) - p_m(t,y,y)]\mu(dy),$$

which coincides with the right hand side of (2.19). This proves (2.18).

Step 5. We take the average of both sides of (2.18) with respect to $\tau \in [0,t]$. Thus, $N(V)$ does not exceed

$$\frac{1}{F(1)} \int_0^\infty \int_X \frac{p_0(t,x,x)}{t} E_{x\to x} \sum_0^N (c_m \int_0^t W(\widehat{b}_s)ds e^{-\int_0^t mW(\widehat{b}_s)ds}) \mu(dx)dt$$

$$= \frac{1}{F(1)} \int_0^\infty \int_X \frac{p_0(t,x,x)}{t} E_{x\to x}(uP(e^{-u}))\mu(dx)dt, \quad u = \int_0^t W(\widehat{b}_s)ds, \tag{2.20}$$

where P is the polynomial defined in (2.11) and (2.14).

Let P be such that

$$uP(e^{-u}) \leqslant G(u), \tag{2.21}$$

where G is defined in the statement of Theorem 2.1. Then one can replace $uP(e^{-u})$ in (2.20) by $G(u)$. Then the Jensen inequality implies that

$$G(\int_0^t W(\widehat{b}_s))ds = G(\frac{1}{t}\int_0^t tW(\widehat{b}_s))ds \leqslant \frac{1}{t}\int_0^t G(tW(\widehat{b}_s))ds.$$

This allows us to rewrite (2.20) in the form

$$N(V) \leqslant \frac{1}{F(1)} \int_0^\infty \int_X \frac{p_0(t,x,x)}{t} \frac{1}{t} \int_0^t E_{x\to x} G(tW(\widehat{b}_s)) ds \mu(dx) dt. \tag{2.22}$$

It is essential that one can use the exact formula for the distribution above:

$$E_{x\to x} G(tW(\widehat{b}_s)) = \int_X G(tW(z)) \frac{p_0(s,x,z)p_0(t-s,z,x)}{p_0(t,x,x)} \mu(dz).$$

From here and (2.22) it follows that $N(V)$ can be estimated from above by

$$\begin{aligned}
&\frac{1}{F(1)} \int_0^\infty \frac{1}{t^2} \int_0^t ds \int_X \int_X G(tW(z)) p_0(s,x,z) p_0(t-s,z,x) \mu(dx)\mu(dz) dt \\
&\quad = \frac{1}{F(1)} \int_0^\infty \frac{1}{t^2} \int_0^t ds \int_X \mu(dz) G(tW(z)) p_0(t,z,z) dt \\
&\quad = \frac{1}{F(1)} \int_0^\infty \frac{1}{t} \int_X G(tW(z)) p_0(t,z,z) \mu(dz) dt \\
&\quad \leqslant \frac{1}{F(1)} \int_0^\infty \frac{\pi(t)}{t} \int_X G(tW(z)) \mu(dz) dt,
\end{aligned} \tag{2.23}$$

where $F(1)$ is defined in (2.11).

Step 6. Now, we are going to specify the choice of the polynomial P which was used in the previous steps. It must be nonnegative and satisfy (2.3) and (2.21). The polynomial P will be determined by the choice of the function G. Note that it suffices to prove (2.5) for functions G which are linear at infinity. In fact, for arbitrary G, let $G_N \leqslant G$ be a continuous function which coincides with G when $z \leqslant N$ and is linear when $z \geqslant N$. For example, if G is smooth, G_N can be obtained if the graph of G for $z \geqslant N$ is replaced by the tangent line through the point $(N, G(N))$. Since $G_N \leqslant G$, the validity of (2.5) for G_N implies (2.5) with the function G in the integrand and $g(1)$ being replaced by $g_N(1)$. Passing to the limit as $N \to \infty$ in this inequality, one gets (2.5) since $g_N(1) \to g(1)$ as $N \to \infty$. Similar arguments allow us to assume that $G = 0$ in a neighborhood of the origin (The validity of (2.5) for $G_\varepsilon(z) = G(z-\varepsilon) \leqslant G(z)$ implies (2.5)). Now, consider $G^\varepsilon(z) = \max(G(z), y(\varepsilon, z))$, where $y(\varepsilon, z)) = z^{m+1}, z \leqslant \varepsilon$, $y(\varepsilon, z) = (m+1)(z-\varepsilon) + \varepsilon^{m+1}, z > \varepsilon$, with m defined in condition (b). We will show later that the right-hand side of (2.5) is finite for $G = G^\varepsilon$. Thus, if (2.5) is proved for $G = G^\varepsilon$, then passing to the limit as $\varepsilon \to 0$ one gets (2.5) for G. Hence we can assume that $G = az$ at infinity and $G = z^{m+1}$ in a neighborhood of the origin. Note that $a \neq 0$ since G is convex.

A special approximation of the function G by exponential polynomials will be used. Consider the function

$$H(z) = \frac{G(z)}{z(1-e^{-z})^m}, \quad z > 0.$$

It is continuous, nonnegative and has positive limits as $z \to 0$ and $z \to \infty$. Hence there is an exponential polynomial $p_\varepsilon(e^{-z})$ which approximates $H(z)$ from below, i.e.,

$$|H(z) - p_\varepsilon(e^{-z})| < \varepsilon, \quad 0 < p_\varepsilon(e^{-z}) \leqslant H(z) \leqslant 2p_\varepsilon(e^{-z}), \quad z > 0.$$

In order to find p_ε, one can change the variable $t = e^{-z}$ and reduce the problem to the standard Weierstrass theorem on the interval (0,1). If $P_\varepsilon(e^{-z}) = (1 - e^{-z})^m p_\varepsilon(e^{-z})$, then

$$\begin{gathered} |z^{-1}G(z) - P_\varepsilon(e^{-z})| < \varepsilon, \quad 0 < P_\varepsilon(e^{-z}) \leqslant z^{-1}G(z), \quad z > 0, \\ P_\varepsilon(e^{-z}) < Cz^m, \ z \to 0. \end{gathered} \tag{2.24}$$

We choose a polynomial P in (2.11) and (2.14) to be equal to P_ε. The last two relations in (2.24) show that $P = P_\varepsilon$ satisfies all the properties used to obtain (2.23). The function F in (2.23) is defined by (2.11) with $P = P_\varepsilon$, and therefore $F(1) = F_\varepsilon(1)$ depends on ε. From the first relation of (2.24) it follows that $F_\varepsilon(1) \to g(1)$ as $\varepsilon \to 0$. Thus, passing to the limit in (2.23) as $\varepsilon \to 0$ we complete the proof of the inequality (2.5) for $N(V)$.

Step 7. Now, we are going to show that the inequality (2.5) for $N(V)$ implies the validity of this inequality for $N_0(V)$ under the assumption that the integral (2.5) converges. We can assume that G is linear at infinity and $G(z) = z^{m+1}$ in a neighborhood of the origin (cf. step 6). (Note that the above properties of G imply easily the convergence of the integral (2.5) if $W \in C_{\text{com}}(X)$, but (2.5) must be proved without this additional restriction on W)

Let $\{\psi_i\}$, $1 \leqslant i \leqslant n$, be a basis for the null space of the operator H. We need to show that n is finite and $N(V) + n$ does not exceed the right-hand side of (2.5).

Let $V_k = V_k(x)$, $k = 1, 2, \ldots$, be arbitrary functions such that $V_k(x) = 1$ when $x \in X_k \in X$, $V_k(x) = 0$ when $x \notin X_k$, where the sets X_k have finite measures, $\mu(X_k) = c_k < \infty$, and

$$\int_X V_k(x)|\psi_j|^2(x)\mu(dx) = \int_{X_k} |\psi_j|^2(x)\mu(dx) > 0, \quad 1 \leqslant j \leqslant k.$$

Consider the operator

$$H_{\varepsilon,k} = -H_0 + V(x) - \varepsilon V_k(x) = -H_0 - W(x) - \varepsilon V_k(x).$$

This operators has at least $N(V) + k$ strictly negative eigenvalues. In fact, the Dirichlet form

$$\int_X [-(H_0\phi)\phi + (V - \varepsilon V_k)|\phi|^2]\mu(dx)$$

with $\varepsilon > 0$ is negative if ϕ is an eigenfunction of H with a negative eigenvalue or $\phi = \psi_j$, $j \leqslant k$. Therefore, there exist at least $N(V)+k$ linearly independent functions ϕ for which the Dirichlet form is negative. Now, from the inequality (2.5) for strictly negative eigenvalues of the operator $H_{\varepsilon,k}$ it follows that

$$N(V) + k \leqslant \frac{1}{g(1)} \int_0^\infty \frac{\pi(t)}{t} \int_X G(tW_\varepsilon(x))\mu(dx)dt, \quad W_\varepsilon = W + \varepsilon V_k. \quad (2.25)$$

Assume that the double integral in (2.25) converges when $\varepsilon = 1$. Then one can pass to the limit as $\varepsilon \to 0$ in (2.25) and get

$$N(V) + k \leqslant \frac{1}{g(1)} \int_0^\infty \frac{\pi(t)}{t} \int_X G(tW(x))\mu(dx)dt.$$

Here, k is arbitrary if $n = \infty$, and one can take $k = n$ if $n < \infty$. This proves (2.5). Hence it remains only to justify the convergence of the double integral in (2.25) with $\varepsilon = 1$ under the condition that the double integral in (2.5) converges.

The integrands in (2.25) and (2.5) coincide when $x \notin X_k$. Hence we only need to prove the convergence of the integral (2.25) with X replaced by X_k. We denote by I_k and J_k the double integral in (2.25) with X replaced by $X_k' = X_k \bigcap \{W(x) > 1\}$ and $X_k'' = X_k \bigcap \{W(x) < 1\}$ respectively. Since $W + V_k \leqslant 2W$, from the properties of G mentioned above it follows that

$$G(t(W + V_k)) \leqslant CG(tW), \quad x \in X_k'.$$

This implies that $I_k < \infty$. Since $\mu(X_k'') < \infty$ and

$$G(t(W + V_k)) \leqslant C\frac{t^{m+1}}{1 + t^{m+1}}, \quad x \in X_k'',$$

from (2.3) it follows that $J_k < \infty$. Hence (2.5) is proved.

Step 8. In order to prove (2.6), we note that

$$\begin{aligned}\sum_{i:E_i<0} |E_i|^\gamma &= \gamma \int_0^\infty E^{\gamma-1} N_E(V)dE \leqslant \gamma \int_0^\infty E^{\gamma-1} N_0(-(W - E)_+)dE \\ &\leqslant \frac{\gamma}{g(1)} \int_0^\infty E^{\gamma-1} \int_0^\infty \frac{\pi(t)}{t} \int_X G(t(W(x) - E)_+)\mu(dx)dtdE \\ &= \frac{\gamma}{g(1)} \int_0^\infty \frac{\pi(t)}{t} \int_X \int_0^W E^{\gamma-1} G(t(W(x) - E))dE\mu(dx)dt \\ &= \frac{\gamma}{g(1)} \int_0^\infty \frac{\pi(t)}{t} \int_X \int_0^1 u^{\gamma-1} W^\gamma(x) G(tW(x)(1 - u))du\mu(dx)dt.\end{aligned}$$

One can replace $G(tW(x)(1-u))$ by $G(tW(x))$ since G is monotonically increasing. This immediately implies (2.6). □

Theorem 2.2. *Let $H = H_0 + V(x)$, where H_0 is a selfadjoint, nonnegative operator on $L^2(X,\mathcal{B},\mu)$, the potential $V = V(x)$ is real valued, and condition* (a)–(c) *hold.*

If

$$\pi(t) \leqslant c/t^{\beta/2}, \quad t \to \infty; \quad \pi(t) \leqslant c/t^{\alpha/2}, \quad t \to 0 \tag{2.26}$$

for some $\beta > 2$ and $\alpha \geqslant 0$, then

$$N_0(V) \leqslant C(h)[\int_{X_h^-} W(x)^{\beta/2}\mu(dx) + \int_{X_h^+} bW(x)^{\max(\alpha/2,1)}\mu(dx)], \tag{2.27}$$

where $X_h^- = \{x : W(x) \leqslant h^{-1}\}$, $X_h^+ = \{x : W(x) > h^{-1}\}$, $b = 1$ if $\alpha \neq 2$, $b = ln(1 + W(x))$ if $\alpha = 2$.

In some cases, $\max(\alpha/2, 1)$ can be replaced by $\alpha/2$, as will be discussed in Section 3.

Proof of Theorem 2.2. We write (2.7) in the form $N_0(V) \leqslant I_- + I_+$, where $I_\mp$ correspond to integration in (2.7) over $X_h^\mp$ respectively.

Let $x \in X_h^-$, i.e., $W < h^{-1}$. Then the interior integral in (2.7) does not exceed

$$C(h)\int_{\frac{\sigma}{W}}^{\infty} t^{-\beta/2}dt = C(h)W^{(\beta/2)-1}. \tag{2.28}$$

Thus, I_- can be estimated by the first term on the right-hand side of (2.27). Similarly,

$$\begin{aligned} I_+ &\leqslant C(h)\int_{X_h^+} W\Big(\int_{\frac{\sigma}{W}}^{h} + \int_h^{\infty}\Big)\pi(t)dt \\ &\leqslant C(h)\int_{X_h^+} W\Big(\int_{\frac{\sigma}{W}}^{h} t^{-\alpha/2}dt + \int_h^{\infty} t^{-\beta/2}dt\Big)dx, \end{aligned}$$

which does not exceed the second term on the right-hand side of (2.27). □

Theorem 2.3. *Let $H = H_0 + V(x)$, where H_0 is a selfadjoint, nonnegative operator on $L^2(X,\mathcal{B},\mu)$, the potential $V = V(x)$ is real valued, and condition* (a)–(c) *hold.*

If

$$\pi(t) \leqslant ce^{-at^{\gamma}}, \quad t \to \infty; \quad \pi(t) \leqslant c/t^{\alpha/2}, \quad t \to 0 \tag{2.29}$$

for some $\gamma > 0$ and $\alpha \geqslant 0$, then for each $A > 0$,

$$N_0(V) \leqslant C(h,A)[\int_{X_h^-} e^{-AW(x)^{-\gamma}}\mu(dx) + \int_{X_h^+} bW(x)^{\max(\alpha/2,1)}\mu(dx)], \tag{2.30}$$

where X_h^-, X_h^+, b *are the same as in Theorem* 2.2.

Proof of Theorem 2.3. The proof is the same as that of Theorem 2.2. One only needs to replace (2.28) by the estimate

$$\begin{aligned} C(h)\int_{\frac{\sigma}{W}}^{\infty} e^{-at^{\gamma}}dt &= C(h)W^{-1}\int_{\sigma}^{\infty} e^{-a(\frac{\tau}{W})^{\gamma}}d\tau \\ &\leqslant C(h)W^{-1}e^{-\frac{a}{2}(\frac{\sigma}{W})^{\gamma}}\int_{\sigma}^{\infty} e^{-\frac{a}{2}(\frac{\tau}{W})^{\gamma}}d\tau \\ &\leqslant \Big[C(h)W^{-1}\int_{\sigma}^{\infty} e^{-\frac{a}{2}(h\tau)^{\gamma}}d\tau\Big]e^{-\frac{a}{2}(\frac{\sigma}{W})^{\gamma}} \end{aligned}$$

and note that σ can be chosen as large as we please. □

3 Low Local Dimension ($\alpha < 2$)

3.1 Operators on lattices and groups

It is easy to see that Theorems 2.3 and 2.2 are not exact if $\alpha \leqslant 2$. We are going to illustrate this fact now and provide a better result in the case $\alpha = 0$ which occurs, for example, when operators on lattices and discrete groups are considered. An important example with $\alpha = 1$ will be discussed in next subsection (operators on quantum graphs).

Let $X = \{x\}$ be a countable set, and let H_0 be a difference operator on $L^2(X)$ which is defined by

$$(H_0\psi)(x) = \sum_{y\in X} a(x,y)\psi(y), \tag{3.1}$$

where

$$a(x,x) > 0,\quad a(x,y) = a(y,x) \leqslant 0,\quad \sum_{y\in X} a(x,y) = 0.$$

A typical example of H_0 is the negative difference Laplacian on the lattice $X = Z^d$, i.e.,

$$(H_0\psi)(x) = -\Delta\psi = \sum_{y\in Z^d:|y-x|=1} [\psi(x) - \psi(y],\quad x \in Z^d, \tag{3.2}$$

We assume that $0 < a(x,x) \leqslant c_0 < \infty$. Then $\mathrm{Sp}H_0 \subset [0, 2c_0]$. The operator $-H_0$ defines the Markov chain $x(s)$ on X with continuous time $s \geqslant 0$ which spends exponential time with parameter $a(x,x)$ at each point $x \in X$ and then jumps to a point $y \in X$ with probability $r(x,y) = \frac{a(x,y)}{a(x,x)}$, $\sum_{y:y\neq x} r(x,y) = 1$.

The transition matrix $p(t,x,y) = P_x(x_t = y)$ is the fundamental solution of the parabolic problem

$$\frac{\partial p}{\partial t} + H_0 p = 0, \quad p(0,x,y) = \delta_y(x).$$

Let $\pi(t) = \sup_x p(t,x,x)$. Obviously, $\pi(t) \leqslant 1$ and $\pi(t) \to 1$ uniformly in x as $t \to 0$. The asymptotic behavior of $\pi(t)$ as $t \to \infty$ depends on the operator H_0 and can be more or less arbitrary.

Consider now the operator $H = H_0 - m\delta_y(x)$ with the potential supported on one point. The negative spectrum of H contains at most one eigenvalue (due to rank one perturbation arguments), and such an eigenvalue exists if $m \geqslant c_0$. The latter follows from the variational principle since

$$\langle H_0\delta_y,\, \delta_y\rangle - m\langle \delta_y,\, \delta_y\rangle \leqslant c_0 - m < 0.$$

However, Theorems 2.2 and 2.3 estimate the number of negative eigenvalues $N(V)$ of the operator H by Cm. Similarly, if

$$V = -\sum_{1\leqslant i\leqslant n} m_i \delta(x - x_i)$$

and $m_i \geqslant c_0$, then $N(V) = n$, but Theorems 2.2 and 2.3 give only that $N(V) \leqslant C\sum m_i$. The following statement provides a better result in the case under consideration than the theorems above. The meaning of the statement below is that we replace $\max(\alpha/2, 1) = 1$ in (2.27), (2.30) by $\alpha/2 = 0$. Let us also mention that these theorems cannot be strengthened in a similar way if $0 < \alpha \leqslant 2$ (cf. Example 3.1).

Theorem 3.1. *Let $H = H_0 + V(x)$, where H_0 is defined in* (3.1), *and let assumptions of Theorem* 2.1 *hold. Then for each $h > 0$*

$$N_0(V) \leqslant C(h)\Big[n(h) + \int_0^\infty \frac{\pi(t)}{t} \sum_{x\in X_h^-} G(tW(x))dt\Big], \quad n(h) = \#\{x \in X_h^+\}.$$

If, additionally, either (2.26) *or* (2.29) *is valid for $\pi(t)$ as $t \to \infty$, then for each $A > 0$*

$$N_0(V) \leqslant C(h)\Big[\sum_{x\in X_h^-} W(x)^{\frac{\beta}{2}} + n(h)\Big], \quad n(h) = \#\{x \in X_h^+\}, \tag{3.3}$$

$$N_0(V) \leqslant C(h,A)\Big[\sum_{x\in X_h^-} e^{-AW(x)^{-\gamma}} + n(h)\Big], \quad n(h) = \#\{x \in X_h^+\}$$

respectively.

Remark 3.1. The estimate (3.3) for $N(V)$ in the case $X = Z^d$ can be found in [25].

Proof. In order to prove the first inequality, we split the potential $V(x) = V_1(x) + V_2(x)$, where $V_2(x) = V(x)$ for $x \in X_h^+$, $V_2(x) = 0$ for $x \in X_h^-$. Now, for each $\varepsilon \in (0,1)$,

$$N_0(V) \leqslant N_0(\varepsilon^{-1}V_1) + N_0((1-\varepsilon)^{-1}V_2) = N_0(\varepsilon^{-1}V_1) + n(h). \tag{3.4}$$

It remains to apply Theorem 2.1 to the operator $-\Delta + \varepsilon^{-1}V_1$ and pass to the limit as $\varepsilon \to 1$. The next two inequalities follow from Theorems 2.2 and 2.3. □

3.2 Operators on quantum graphs

We consider a specific quantum graph Γ^d, the so-called Avron–Exner–Last graph. Its vertices are points of the lattice Z^d, and edges are all the segments of length one connecting neighboring vertices. Let $s \in [0,1]$ be the natural parameter on the edges (the distance from one of the end points of the edge). Consider the space D of smooth functions φ on edges of Γ^d with the following (Kirchhoff) boundary conditions at vertices: at each vertex φ is continuous and

$$\sum_{i=1}^{d} \varphi_i' = 0, \tag{3.5}$$

where φ_i' are the derivatives along the adjoint edges in the direction out of the vertex. The operator H_0 acts on functions $\varphi \in D$ as $-\frac{d^2}{ds^2}$. The closure of this operator in $L^2(\Gamma^d)$ is a selfadjoint operator with the spectrum $[0,\infty)$ (cf. [3])

Theorem 3.2. *The assumptions of Theorems* 2.1, 2.2 *hold for the operator H_0 introduced in this section with the constants α, β in Theorem* 2.2 *equal to* 1 *and d respectively.*

One can easily see that there is a Markov process with the generator $-H_0$, and condition (a) of Theorem 2.1 holds. In Appendix, we estimate the function p_0 in order to show that condition (b) holds and find constants α, β defined in Theorem 2.2. In fact, the same arguments can be used to verify condition (a) analytically.

As we discussed above, Theorem 2.2 is not exact if $\alpha \leqslant 2$. Theorem 3.1 provides a better result in the case $\alpha = 0$. The situation is more complicated if $\alpha = 1$. We illustrate it using the operator H_0 on the quantum graph Γ^d. We consider two specific classes of potentials. In one case, the inequality (2.27) is valid with $\max(\alpha/2, 1) = 1$ replaced by $\alpha/2 = 1/2$. However, the inequality (2.27) cannot be improved for potentials of the second type. The first class (regular potentials) consists of piecewise constant functions.

Theorem 3.3. *Let $d \geqslant 3$ and V be constant on each edge e_i of the graph: $V(x) = -v_i < 0$, $x \in e_i$. Then*

$$N_0(V) \leqslant c(h)\Big(\sum_{i:\ v_i \leqslant h^{-1}} v_i^{d/2} + \sum_{i:\ v_i > h^{-1}} \sqrt{v_i} \Big).$$

Proof. Put $V(x) = V_1(x) + V_2(x)$, where $V_1(x) = V(x)$ if $|V(x)| > h^{-1}$, $V_1(x) = 0$ if $|V(x)| \leqslant h^{-1}$. Then (cf. (3.4))

$$N_0(V) \leqslant N_0(2V_1) + N_0(2V_2).$$

One can estimate $N(V_1)$ from above (below) by imposing the Neumann (Dirichlet) boundary conditions at all vertices of Γ. This leads to the estimates

$$\sum_{i:\ v_i > h^{-1}} \frac{\sqrt{2v_i}}{\pi} \leqslant N_0(V) \leqslant \sum_{i:\ v_i > h^{-1}} \Big(\frac{\sqrt{2v_i}}{\pi} + 1 \Big) \leqslant c(h) \sum_{i:\ v_i > h^{-1}} \sqrt{v_i},$$

which, together with Theorem 2.2 applied to $N_0(2V_2)$, justifies the statement of the theorem. □

The same arguments allow one to get a more general result.

Theorem 3.4. *Let $d \geqslant 3$. Let Γ_-^d be the set of edges e_i of the graph Γ^d, where $W \leqslant h^{-1}$, Γ_+^d be the complementary set of edges, and let*

$$\frac{\sup_{x \in e_i} W(x)}{\min_{x \in e_i} W(x)} \leqslant k_0 = k_0(h), \quad x \in \Gamma_+^d,$$

where $W = V_-$. Then

$$N_0(V) \leqslant c(h, k_0)\Big(\int_{\Gamma_-^d} W(x)^{d/2} dx + \int_{\Gamma_+^d} \sqrt{W(x)} dx \Big).$$

Example 3.1. The next example shows that there are singular potentials on Γ^d for which $\max(\alpha/2, 1)$ in (2.27) cannot be replaced by any value less than one. Consider the potential

$$V(x) = -A \sum_{i=1}^{m} \delta(x - x_i),$$

where x_i are middle points of some edges, and $A > 4$. One can easily modify the example by considering δ -sequences instead of δ-functions (in order to get a smooth potential.) Then

$$\int_{\Gamma^d} W^{\sigma}(x) dx = 0$$

for any $\sigma < 1$, while $N(V) \geqslant m$. In fact, consider the Sturm–Liouville problem on the interval $[-1/2, 1/2]$:

$$-y'' - A\delta(x)y = \lambda y, \; y(-1/2) = y(1/2) = 0, \quad A > 4.$$

It has (a unique) negative eigenvalue which is the root of the equation $\tanh(\sqrt{-\lambda}/2) = 2\sqrt{-\lambda}/A$. The corresponding eigenfunction is

$$y = \sinh[\sqrt{-\lambda}(|x| + 1/2)].$$

The estimate $N(V) \geqslant m$ follows by imposing the Dirichlet boundary conditions on the vertices of Γ^d.

4 Anderson Model

4.1 Discrete case

Consider the classical Anderson Hamiltonian

$$H_0 = -\Delta + V(x, \omega)$$

on $L^2(Z^d)$ with random potential $V(x, \omega)$. Here,

$$\Delta\psi(x) = \sum_{x':|x'-x|=1} \psi(x') - 2d\psi(x).$$

We assume that random variables $V(x, \omega)$ on the probability space (Ω, F, P) have the Bernoulli structure, i.e., they are i.i.d. and $P\{V(\cdot) = 0\} = p > 0$, $P\{V(\cdot) = 1\} = q = 1 - p > 0$. The spectrum of H_0 is equal to (cf. [2])

$$\mathrm{Sp}(H_0) = \mathrm{Sp}(-\Delta) \oplus 1 = [0, 4d + 1].$$

Let us stress that $0 \in \mathrm{Sp}(H_0)$ due to the existence P-a.s. of arbitrarily large clearings in realizations of V, i.e., there are balls $B_n = \{x : |x - x_n| < r_n\}$ such that $V(x) = 0$, $x \in B_n$, and $r_n \to \infty$ as $n \to \infty$ (cf. the proof of the theorem below for details).

Let

$$H = H_0 - W(x), \quad W(x) \geqslant 0.$$

The operator H has discrete random spectrum on $(-\infty, 0]$ with possible accumulation point at $\lambda = 0$. Put $N_0(-W) = \#\{\lambda_i \leqslant 0\}$. Obviously, $N_0(-W)$ is random. Denote by E the expectation of a r.v., i.e.,

$$EN_0 = \int_\Omega N_0 P(d\omega).$$

Theorem 4.1. (a) *For each* $h > 0$ *and* $\gamma < \frac{d}{d+2}$,

$$EN_0(-W) \leqslant c_1(h)[\#\{x \in Z^d : W(x) \geqslant h^{-1}\}] + c_2(h,\gamma) \sum_{x:W(x)<h^{-1}} e^{-\frac{1}{W^{\gamma}(x)}}.$$

In particular, if

$$W(x) < \frac{C}{\log^{\sigma}|x|}, \quad |x| \to \infty,$$

with some $\sigma > \frac{d+2}{d}$, *then* $EN_0(-W) < \infty$, *i.e.*, $N_0(-W) < \infty$ *almost surely.*

(b) *If*

$$W(x) > \frac{C}{\log^{\sigma}|x|}, \ |x| \to \infty, \quad \text{and } \sigma < \frac{2}{d}, \tag{4.1}$$

then $N_0(-W) = \infty$ *a.s. (in particular,* $EN_0(-W) = \infty$*).*

Proof. Since $V \geqslant 0$, the kernel $p_0(t,x,y)$ of the semigroup $\exp(-tH_0) = \exp(t(\Delta - V))$ can be estimated by the kernel of $\exp(t\Delta)$, i.e., by the transition probability of the random walk with continuous time on Z^d. The diagonal part of this kernel $p_0(t,x,x,\omega)$ is a stationary field on Z^d. Due to the Donsker–Varadhan estimate (cf. [6, 7]),

$$Ep_0(t,x,x,\omega) = Ep_0(t,x,x,\omega) \overset{\log}{\sim} \exp(-c_d t^{\frac{d}{d+2}}), \quad t \to \infty,$$

i.e.,

$$\log Ep_0 \sim -c_d t^{\frac{d}{d+2}}, \quad t \to \infty.$$

On the rigorous level, the relations above must be understood as estimates from above and from below, and the upper estimate has the following form: for each $\delta > 0$,

$$Ep_0 \leqslant C(\delta) \exp(-c_d t^{\frac{d}{d+2} - \delta}), \quad t \to \infty. \tag{4.2}$$

Now, the first part of the theorem is a consequence of Theorems 2.1 and 2.3. In fact, from Remarks 2.2 and 2.3 and (4.2) it follows that

$$\begin{aligned} EN_0(V) &\leqslant \frac{1}{c(\sigma)} \int_X W(x) \int_{\frac{\sigma}{W(x)}}^{\infty} Ep_0(t,x,x,\omega) dt \mu(dx) \\ &\leqslant \frac{C(\delta)}{c(\sigma)} \int_X W(x) \int_{\frac{\sigma}{W(x)}}^{\infty} e^{-c_d t^{\frac{d}{d+2} - \delta}} dt \mu(dx). \end{aligned}$$

Then it only remains to repeat the arguments used to prove Theorem 2.3.

The proof of the second part is based on the following lemma which indicates the existence of large clearings at the distances which are not too large. We denote by $C(r)$ the cube in the lattice,

$$C(r) = \{x \in Z^d : |x_i| < r,\ 1 \leqslant i \leqslant d\}.$$

Let us divide Z^d into cubic layers $L_n = C(a^{n+1}) \backslash C(a^n)$ with some constant $a \geqslant 1$ which will be selected later. One can choose a set $\Gamma^{(n)} = \{z_i^{(n)} \in L_n\}$ in each layer L_n such that

$$|z_i^{(n)} - z_j^{(n)}| \geqslant 2n^{\frac{1}{d}} + 1, \qquad d(z_i^{(n)}, \partial L_n) > n^{\frac{1}{d}},$$

and

$$|\Gamma^{(n)}| \geqslant c\frac{(2a)^{n(d-1)}a^{n+1}}{(2n^{1/d})^d} \geqslant ca^{nd}, \qquad n \to \infty.$$

Let $C(n^{1/d}, i)$ be the cube $C(n^{1/d})$ with the center shifted to the point $z_i^{(n)}$. Obviously, cubes $C_{n^{1/d},i}$ do not intersect each other, $C(n^{1/d}, i) \subset L_n$ and $|C(n^{1/d}, i)| \leqslant c'n$.

Consider the following event A_n ={each cube $C(n^{1/d}, i) \subset L_n$ contains at least one point where $V(x) = 1$}. Obviously,

$$P(A_n) = (1-p^{|C(n^{1/d},i)|})^{|\Gamma^{(n)}|} \leqslant e^{-|\Gamma^{(n)}|p^{|C(n^{1/d},i)|}} \leqslant e^{-ca^{nd}c'p^n} = e^{-c(a^dp^{c'})^n}.$$

We choose a big enough, so that $a^dp^{c'} > 1$. Then $\sum P(A_n) < \infty$, and the Borel–Cantelli lemma implies that P-a.s. there exists $n_0(\omega)$ such that each layer L_n, $n \geqslant n_0(\omega)$, contains at least one empty cube $C(n^{1/d}, i)$, $i = i(n)$. Then from (4.1) it follows that

$$W(x) \geqslant \frac{C}{n^{\frac{2}{d}-\delta}} = \varepsilon_n, \qquad x \in C(n^{1/d}, i), \qquad i = i(n).$$

One can easily show that the operator $H = -\Delta - \varepsilon$ in a cube $C \subset Z^d$ with the Dirichlet boundary condition at ∂C has at least one negative eigenvalue if $|C|\varepsilon^{d/2}$ is big enough. Thus the operator H in $C(n^{1/d}, i(n))$ with the Dirichlet boundary condition has at least one eigenvalue if n is big enough, and therefore $N(-W) = \infty$. □

4.2 Continuous case

Theorem 4.1 is also valid for the Anderson operators in R^d. Let

$$H_0 = -\Delta + V(x, \omega)$$

on $L^2(R^d)$ with the random potential

$$V(x, \omega) = \sum_{n \in Z^d} \varepsilon_n I_{Q_n}(x),\ x \in R^d,\ n = (n_1, \ldots, n_d),$$

where $Q_n = \{x \in R^d : n_i \leqslant x_i < n_i + 1, \ i = 1, 2, \dots, d\}$ and ε_n are independent Bernoulli r.v. with $P\{\varepsilon_n = 0\} = p, \ P\{\varepsilon_n = 1\} = q = 1 - p$. Put

$$H = H_0 - W(x) = -\Delta + V(x, \omega) - W(x).$$

Theorem 4.2. (a) *If $d \geqslant 3$, then for each $h > 0$ and $\gamma < \frac{d}{d+2}$*

$$EN_0(-W) \leqslant c_1(h) \int_{W(x) \geqslant h^{-1}} W(x)^{d/2} dx + c_2(h, \gamma) \int_{W(x) < h^{-1}} e^{-\frac{1}{W^\gamma(x)}} dx.$$

In particular, if

$$W(x) < \frac{C}{\log^\sigma |x|}, \quad |x| \to \infty,$$

with some $\sigma > \frac{d+2}{d}$, then $EN_0(-W) < \infty$, i.e., $N_0(-W) < \infty$ almost surely.

(b) *If*

$$W(x) > \frac{C}{\log^\sigma |x|}, \ |x| \to \infty, \quad \text{and} \ \sigma < \frac{2}{d},$$

then $N_0(-W) = \infty$ a.s. (in particular, $EN_0(-W) = \infty$).

The proof of this theorem is identical to the proof of Theorem 4.1 with the only difference that now $p_0(t, 0, 0)$ is not bounded as $t \to 0$, but $p_0(t, 0, 0) \leqslant c/t^{d/2}$ as $t \to 0$.

5 Lobachevsky Plane, Processes with Independent Increments

5.1 Lobachevsky plane (*cf.* [8, 21])

We use the Poincaré upper half plane model, where $X = \{z = x + iy : y > 0\}$ and the (Riemannian) metric on X has the form

$$ds^2 = y^{-2}(dx^2 + dy^2). \tag{5.1}$$

The geodesic lines of this metric are circular arcs perpendicular to the real axis (half-circles whose origin is on the real axis) and straight vertical lines ending on the real axis. The group of transformations preserving ds^2 is $SL(2, R)$, i.e., the group of real valued 2×2 matrices with the determinant equal to one. For each $A = \begin{bmatrix} a & b \\ c & d \end{bmatrix} \in SL(2, R)$, the action $A(z)$ is defined by

$$A(z) = \frac{az + b}{cz + d}.$$

For each $z_0 \in X$, there is a one-parameter stationary subgroup which consists of A such that $Az_0 = z_0$. The Laplace–Beltrami operator Δ' (invariant with respect to $SL(2,R)$) is defined uniquely up to a constant factor, and is equal to

$$\Delta' = y^2 \Delta = y^2\Big(\frac{\partial^2}{\partial x^2} + \frac{\partial^2}{\partial y^2}\Big). \tag{5.2}$$

The operator $-\Delta'$ is selfadjoint with respect to the Riemannian measure

$$\mu(dz) = y^{-2} dxdy \tag{5.3}$$

and has absolutely continuous spectrum on $[1/4, \infty)$. In order to find the number $N'(V)$ of eigenvalues of the operator $-\Delta' + V(x)$ less than $1/4$, one can apply Theorem 2.1 to the operator $H_0 = -\Delta' - \frac{1}{4}I$.

One needs to know constants α, β in order to apply Theorem 2.2. It is shown in [12] that the fundamental solution for the parabolic equation $u_t = -\Delta' u$ has the following asymptotic behavior

$$p(t,0,0) \sim c_1/t, \quad t \to 0; \quad p(t,0,0) \sim c_2 e^{-t/4}/t^{3/2}, \quad t \to \infty.$$

Thus, $\alpha = 2$, $\beta = 3$ for the operator $H_0 = -\Delta' - \frac{1}{4}I$. A similar result for the Laplacian in the Hyperbolic space of the dimension $d \geqslant 3$ can be found in [25].

5.2 *Markov processes with independent increments (homogeneous pseudodifferential operators)*

We estimate $N_0(V)$ for shift invariant pseudodifferential operators H_0 associated with Markov processes with independent increments. Similar estimates were obtained in [5] for pseudodifferential operators under assumptions that the symbol $f(p)$ of the operator is monotone and nonnegative, and the parabolic semigroup e^{-tH_0} is positivity preserving. This class includes important cases of $f(p) = |p|^\alpha, \alpha < 2$ and $f(p) = \sqrt{p^2 + m^2} - m$. Note that necessary and sufficient conditions of the positivity of $p_0(t,x,x)$ are given by Levy–Khinchin formula. We omit the monotonicity condition. What is more important, the results are expressed in terms of the Levy measure responsible for the positivity of $p_0(t,x,x)$. This allows us to consider variety estimates with power and logarithmical decaying potentials.

Let H_0 be a pseudodifferential operator in $X = R^d$ of the form

$$H_0 u = F^{-1}\Phi(k) F u, \quad (Fu)(k) = \int_{R^d} u(x) e^{-i(x,k)} dx, \quad u \in S(R^d),$$

where the symbol $\Phi(k)$ of the operator H_0 has the following form

$$\Phi(k) = \int_{R^d} (1 - \cos(x,k))\nu(x)dx. \tag{5.4}$$

Here, $\mu(dx) = \nu(x)dx$ is an arbitrary measure (for simplicity we assumed that it has a density) such that

$$\int_{|x|>1} \nu(x)dx + \int_{|x|<1} |x|^2\nu(x)dx < \infty. \tag{5.5}$$

Assumption (5.4) is needed (and is sufficient) to construct a Markov process with the generator $L = -H_0$ below. However, we impose an additional restriction on the measure $\mu(dx)$ assuming that the density $\nu(x)$ has the following power asymptotics at zero and at infinity

$$\nu(x) \sim |x|^{-d-2+\rho}, \quad x \to 0, \quad \nu(x) \sim |x|^{-d-\delta}, \quad x \to \infty,$$

with some $\rho, \delta \in (0,2)$. Note that assumption (5.5) holds in this case. To be more rigorous, we assume that

$$\nu(x) = a\Big(\frac{x}{|x|}\Big)|x|^{-d-\rho}(1 + O(|x|^{\varepsilon})), \quad x \to 0, \tag{5.6}$$

$$\nu(x) = b\Big(\frac{x}{|x|}\Big)|x|^{-d-\delta}(1 + O(|x|^{-\varepsilon})), \quad x \to \infty, \tag{5.7}$$

where $a, b, \varepsilon > 0$. We also consider another special case where the asymptotic behavior of $\nu(x)$ at infinity is at logarithmical borderline for the convergence of the integral (5.5). Namely, we assume that (5.6) holds and

$$\nu(x) > C|x|^{-d}\log^{-\sigma}|x|, \quad x \to \infty, \quad \sigma > 1. \tag{5.8}$$

The solution of problem (2.1) is given by

$$p_0(t, x-y) = \frac{1}{(2\pi)^d}\int_{R^d} e^{-t\Phi(k)+i(x-y,k)}dk. \tag{5.9}$$

A special form of the pseudodifferential operator H_0 is chosen in order to guarantee that $p_0 \geqslant 0$. In fact, let x_s, $s > 0$, be a Markov process in R^d with symmetric independent increments. It means that for arbitrary $0 < s_1 < s_2 < \ldots$, the random variables $x_{s_1} - x_0, x_{s_2} - x_{s_1}, \ldots$ are independent and the distribution of $x_{t+s} - x_s$ is independent of s. The symmetry condition means that Law$(x_s - x_0)$ =Law$(x_0 - x_s)$, or $p(s,x,y) = p(s,y,x)$, where p is the transition density of the process. According to the Levy–Khinchin theorem (cf. [10]), the Fourier transform (characteristic function) of this distribution has the form

$$Ee^{i(k,x_{t+s}-x_s)} = e^{-t\Phi(k)},$$

with $\Phi(k)$ given by (5.4). Moreover, each measure (5.5) corresponds to some process. One can consider the family of processes $x_s^{(x_0)} = x_0 + x_s$, $s > 0$, with an arbitrary initial point x_0. The generator L of this family can be evaluated in the Fourier space. If $\varphi(x) \in S(R^d)$ and $\widehat{\varphi}(k) = F\varphi$, then

$$
\begin{aligned}
L\varphi(x) &= \lim_{t\to 0} \frac{E\varphi(x + x_t^{(0)}) - \varphi(x)}{t} \\
&= \lim_{t\to 0} \frac{1}{(2\pi)^d} \int_{R^d} \frac{Ee^{i(x+x_t^{(0)},k)} - e^{i(x,k)}}{t} \widehat{\varphi}(k)dk \\
&= \frac{-1}{(2\pi)^d} \int_{R^d} e^{i(x,k)} \Phi(k)\widehat{\varphi}(k)dk = -H_0\varphi.
\end{aligned}
$$

Thus, the function (5.9) is the transition density of some process, and therefore $p_0(t,x) \geqslant 0$, i.e., assumption (a) of Theorem 2.1 holds. Since the operator H_0 is translation invariant, assumption (b) also holds with $\pi(t) = p_0(t, 0)$. Hence Theorem 2.1 can be applied to study negative eigenvalues of the operator $H_0 + V(x)$ when the (Levy) measure νdx satisfies (5.5). If (5.6), (5.7) or (5.6), (5.8) hold, then Theorems 2.2 and 2.3 can be used. Namely, the following statement is valid.

Theorem 5.1. *If a measure νdx satisfies* (5.6) *and* (5.7), *then* (2.26) *is valid with $\beta = 2d/\delta$, $\alpha = 2d/\rho$.*

If a measure νdx satisfies (5.6) *and* (5.8), *then* (2.29) *is valid with $\gamma = 1/\sigma$, $\alpha = 2d/\rho$.*

Proof. Consider first the case where (5.6) and (5.7) hold. Let us prove that these relations imply the following behavior of $\Phi(k)$ at zero and at infinity

$$
\begin{aligned}
\Phi(k) &= f\Big(\frac{k}{|k|}\Big)|k|^{\delta}(1 + O(|k|^{\varepsilon_1})), \ k \to 0, \\
\Phi(k) &= g\Big(\frac{k}{|k|}\Big)|k|^{\rho}(1 + O(|k|^{-\varepsilon_1})), \ k \to \infty
\end{aligned}
\tag{5.10}
$$

with some $f, g, \varepsilon_1 > 0$. We write (5.4) in the form

$$
\Phi(k) = \int_{|x|<1} 2\sin^2(x,k))\nu(x)dx + \int_{|x|>1} 2\sin^2(x,k))\nu(x)dx = \Phi_1(k) + \Phi_2(k).
\tag{5.11}
$$

The term $\Phi_1(k)$ is analytic in k and is of order $O(|k|^2)$ as $k \to 0$. We represent the second term as

$$
\int_{R^d} 2\sin^2(x,k))b(\dot{x})|x|^{-d-\delta}dx - \int_{|x|<1} 2\sin^2(x,k))b(\dot{x})|x|^{-d-\delta}dx
$$
$$
+ \int_{|x|>1} 2\sin^2(x,k))h(x)dx,
$$

where $\dot{x} = x/|x|$ and

$$h(x) = \nu(x) - b(\dot{x})|x|^{-d-\delta}, \quad |h| \leqslant C|x|^{-d-\delta-\varepsilon}.$$

The middle term above is of order $O(|k|^2)$ as $k \to 0$. The first term above can be evaluated by substitution $x \to x/|k|$. It coincides with $f(\frac{k}{|k|})|k|^\delta$. One can reduce ε to guarantee that $\delta + \varepsilon < 2$. Then the last term can be estimated using the same substitution. This leads to the asymptotics (5.10) as $k \to 0$.

Now. let $|k| \to \infty$. Since $\Phi_2(k)$ is bounded uniformly in k, it remains to show that $\Phi_1(k)$ has the appropriate asymptotics as $|k| \to \infty$. We write $v(x)$ in the integrand of $\Phi_1(k)$ as follows:

$$v(x) = a(\dot{x})|x|^{-d-\rho} + g(x), \quad |g(x)| \leqslant C|x|^{-d-\rho+\varepsilon}.$$

Then

$$\Phi_1(k) = \int_{R^d} 2\sin^2(x,k))a(\dot{x})|x|^{-d-\rho}dx - \int_{|x|>1} 2\sin^2(x,k))a(\dot{x})|x|^{-d-\rho}dx$$
$$+ \int_{|x|<1} 2\sin^2(x,k))g(x)dx.$$

The middle term on the right hand side above is bounded uniformly in k. The substitution $x \to x/|k|$ justifies that the first term coincides with $g(\frac{k}{|k|})|k|^\rho$. The same substitution shows that the order of the last term is smaller if $\varepsilon < \rho$. This gives the second relation of (5.10), and therefore, (5.10) is proved.

Let us estimate $\pi(t)$ when (5.10) holds. From (5.9) it follows that

$$\pi(t) = \frac{1}{(2\pi)^d} \int_{|k|<1} e^{-t\Phi(k)}dk + O(e^{-\eta t}) \quad \text{as } t \to \infty, \quad \eta > 0. \tag{5.12}$$

Now, the substitution $k \to t^{-1/\delta}k$ leads to

$$\pi(t) \sim ct^{-d/\delta}, \quad t \to \infty, \quad c = \frac{1}{(2\pi)^d} \int_{R^d} e^{-g(\frac{k}{|k|})|k|^\delta} dk.$$

Hence the first relation in (2.26) holds with $\beta = 2d/\delta$. In order to estimate $\pi(t)$ as $t \to 0$, we put

$$\pi(t) = \frac{1}{(2\pi)^d} \int_{|k|>1} e^{-t\Phi(k)}dk + O(1), \quad t \to 0,$$

and make the substitution $k \to t^{-1/\rho}k$. This leads to

$$\pi(t) \sim ct^{-d/\rho}, \quad t \to 0, \quad c = \frac{1}{(2\pi)^d} \int_{R^d} e^{-f(\frac{k}{|k|})|k|^\rho} dk.$$

Hence the second relation in (2.26) holds with $\alpha = 2d/\rho$. The first statement of the theorem is proved.

Let us prove the second statement. If (5.6) and (5.8) hold, then

$$\Phi(k) \geqslant c\Big(\log\frac{1}{|k|}\Big)^{1-\sigma}, \quad k \to 0,$$
$$\Phi(k) = g\Big(\frac{k}{|k|}\Big)|k|^{\rho}(1+O(|k|^{-\varepsilon_1})), \quad k \to \infty. \tag{5.13}$$

In fact, only the integrability of $v(x)$ at infinity, but not (5.7), was used in the proof of the second relation of (5.10). Thus, the second relation of (5.13) is valid. Let us prove the first estimate. Let $\Omega_k = \{x : |k|^{-2} > |x| > |k|^{-1}\}$, $|k| < 1$. We have

$$\Phi(k) \geqslant \int_{\Omega_k} 2\sin^2(x,k))\nu(x)dx \geqslant C\int_{\Omega_k} \sin^2(x,k))|x|^{-d}\log^{-\sigma}|x|dx$$
$$\geqslant C(2\log\frac{1}{|k|})^{-\sigma}\int_{\Omega_k} \sin^2(x,k))|x|^{-d}dx, \quad |k| \to 0.$$

It remains to show that

$$\int_{\Omega_k} \sin^2(x,k))|x|^{-d}dx \sim \log\frac{1}{|k|}, \quad |k| \to 0. \tag{5.14}$$

After the substitution $x = y/|k|$, the last integral can be written in the form

$$\frac{1}{2}\int_{|k|^{-1}>|y|>1} |y|^{-d}dy - \frac{1}{2}\int_{|k|^{-1}>|y|>1} \cos(y,\dot{k}))|y|^{-d}dy.$$

This justifies (5.14) since the second term above converges as $|k| \to 0$. Hence (5.13) is proved.

Finally, we need to obtain (2.29). The estimation of $\pi(t)$ as $t \to 0$ remains the same as in the proof of the first statement of the theorem. To get the estimate as $t \to \infty$, we use (5.12) (with a smaller domain of integration) and (5.13). Then we obtain

$$\pi(t) \leqslant \frac{1}{(2\pi)^d}\int_{|k|<1/2} e^{-ct(\log\frac{1}{|k|})^{1-\sigma}}dk + O(e^{-\eta t}), \quad t \to \infty, \quad \eta > 0.$$

After integrating with respect to angle variables and substitution $\log\frac{1}{|k|} = z$, we get

$$\pi(t) \leqslant C\int_{\log 2}^{\infty} z^{d-1}e^{-z-ctz^{1-\sigma}}dz + O(e^{-\eta t}), \quad t \to \infty, \quad \eta > 0.$$

The asymptotic behavior of the last integral can be easily found using standard Laplace method, and the integral behaves as $C_1 t^{\frac{2d-1}{2d\sigma}} e^{-c_1 t^{\frac{1}{\sigma}}}$ when $t \to \infty$. This completes the proof of (2.29). □

6 Operators on Continuous and Discrete Groups

6.1 Free groups

Suppose that X is a group Γ with generators $a_1, a_2, \ldots, a_d$, inverse elements $a_{-1}, a_{-2}, \ldots, a_{-d}$, the unit element e, and with no relations between generators except $a_i a_{-i} = a_{-i} a_i = e$. The elements $g \in \Gamma$ are the shortest versions of the words $g = a_{i_1} \cdot \ldots \cdot a_{i_n}$ (with all factors e and $a_j a_{-j}$ being omitted). The metric on Γ is given by

$$d(g_1, g_2) = d(e, g_1^{-1} g_2) = m(g_1^{-1} g_2),$$

where $m(g)$ is the number of letters $a_{\pm i}$ in g. The measure μ on Γ is defined by $\mu(\{g\}) = 1$ for each $g \in \Gamma$. It is easy to see that $|\{g : d(e,g) = R\}| = 2d(2d-1)^{R-1}$, i.e., the group Γ has an exponential growth rate.

Define the operator Δ_Γ on $X = \Gamma$ by the formula

$$\Delta_\Gamma \psi(g) = \sum_{-d \leqslant i \leqslant d,\ i \neq 0} [\psi(g a_i) - \psi(g)]. \tag{6.1}$$

Obviously, the operator $-\Delta_\Gamma$ is bounded and nonnegative in $L^2(\Gamma, \mu)$. In fact, $||\Delta_\Gamma|| \leqslant 4d$. As it is easy to see, the operator Δ_Γ is left-invariant:

$$(\Delta_\Gamma \psi)(gx) = \Delta_\Gamma(\psi(gx)), \quad x \in \Gamma,$$

for each fixed $g \in \Gamma$. Thus, conditions (a), (b) hold for operator $-\Delta_\Gamma$. In order to apply Theorem 2.2, one also needs to find the parameters α and β.

Theorem 6.1. (a) *The spectrum of the operator $-\Delta_\Gamma$ is absolutely continuous and coincides with the interval $l_d = [\gamma, \gamma + 4\sqrt{2d-1}]$, $\gamma = 2d - 2\sqrt{2d-1} \geqslant 0$.*

(b) *The kernel of the parabolic semigroup $\pi_\Gamma(t) = (e^{t\Delta_\Gamma})(t, e, e)$ on the diagonal has the following asymptotic behavior at zero and infinity*

$$\pi_\Gamma(t) \to c_1, \quad t \to 0, \quad \pi_\Gamma(t) \sim c_2 \frac{e^{-\gamma t}}{t^{3/2}}, \quad t \to \infty. \tag{6.2}$$

Remark 6.1. Since the absolutely continuous spectrum of the operator $-\Delta_\Gamma$ is shifted (it starts from γ, not from zero), the natural question about the eigenvalues of the operator $-\Delta_\Gamma + V(g)$ is to estimate the number $N_\Gamma(V)$ of eigenvalues less than the threshold γ. Obviously, $N_\Gamma(V)$ coincides with the number $N(V)$ of the negative eigenvalues of the operator $H_0 + V(g)$, where $H_0 = -\Delta_\Gamma - \gamma I$. Hence one can apply Theorems 2.1, 3.1 to this operator. From (6.2) it follows that constants α, β for the operator $H_0 = -\Delta_\Gamma - \gamma I$ are equal to 0 and 3 respectively and

$$N_\Gamma(V) \leqslant c(h)\Big[n(h) + \sum_{g\in\Gamma: W(g)\leqslant h^{-1}} W(x)^{3/2}\Big],$$

$$n(h) = \#\{g \in \Gamma : W(g) > h^{-1}\}.$$

Proof of Theorem 6.1. Let us find the kernel $R_\lambda(g_1, g_2)$ of the resolvent $(\Delta_\Gamma - \lambda)^{-1}$. From the Γ-invariance it follows that $R_\lambda(g_1, g_2) = R_\lambda(e, g_1^{-1}g_2)$. Hence it is enough to determine $u_\lambda(g) = R_\lambda(e, g)$. This function satisfies the equation

$$\sum_{i\neq 0} u_\lambda(ga_i) - (2d+\lambda)u_\lambda(g) = -\delta_e(g), \tag{6.3}$$

where $\delta_e(g) = 1$ if $g = e$, $\delta_e(g) = 0$ if $g \neq e$. Since the equation above is preserved under permutations of the generators, the solution $u_\lambda(g)$ depends only on $m(g)$. Let $\psi_\lambda(m) = u_\lambda(g)$, $m = m(g)$. Obviously, if $g \neq e$, then $m(ga_i) = m(g) - 1$ for one of the elements a_i, $i \neq 0$, and $m(ga_i) = m(g) + 1$ for all other elements a_i, $i \neq 0$. Hence (6.3) implies

$$\begin{gathered} 2d\psi_\lambda(1) - (2d+\lambda)\psi_\lambda(0) = -1, \\ \psi_\lambda(m-1) + (2d-1)\psi_\lambda(m+1) - (2d+\lambda)\psi_\lambda(m) = 0, \quad m > 0. \end{gathered} \tag{6.4}$$

Two linearly independent solutions of these equations have the form $\psi_\lambda(m) = \nu_\pm^m$, where $\nu_\pm$ are the roots of the equation

$$\nu^{-1} + (2d-1)\nu - (2d+\lambda) = 0.$$

Thus

$$\nu_\pm = \frac{2d+\lambda \pm \sqrt{(2d+\lambda)^2 - 4(2d-1)}}{2(2d-1)}.$$

The interval l_d was singled out as the set of real λ such that the discriminant above is not positive. Since $\nu_+\nu_- = 1/(2d-1)$, we have

$$|\nu_\pm| = \frac{1}{\sqrt{2d-1}}, \ \lambda \in l_d; \quad |\nu_+| > \frac{1}{\sqrt{2d-1}}, \quad |\nu_-| < \frac{1}{\sqrt{2d-1}} \quad \text{for real } \lambda \notin l_d.$$

Now, if we take into account that the set $A_{m_0} = \{g \in \Gamma,\ m(g) = m_0\}$ has exactly $2d(2d-1)^{m_0-1}$ points, i.e., $\mu(A_{m_0}) = 2d(2d-1)^{m_0-1}$, we get that

$$\nu_-^{m(g)} \in L^2(\Gamma, \mu), \ \nu_+^{m(g)} \notin L^2(\Gamma, \mu) \text{ for real } \lambda \notin l_d, \tag{6.5}$$

and

$$\int_{\Gamma\cap\{g: m(g)\leqslant m_0\}} |\nu_\pm|^{2m(g)}\mu(dg) \sim m_0, \quad 0 \to \infty \quad \text{for } \lambda \notin l_d. \tag{6.6}$$

The relations (6.5) imply that $R\backslash l_d$ belongs to the resolvent set of the operator Δ_Γ and that $R_\lambda(e, g) = c\nu_-^{m(g)}$. The relation (6.6) implies that

l_d belongs to the absolutely continuous spectrum of the operator Δ_Γ with functions $(\nu_+^{m(g)} - \nu_-^{m(g)})$ being the eigenfunctions of the continuous spectrum. Hence statement (a) is justified.

Note that the constant c in the formula for $R_\lambda(e,g)$ can be found from (6.4). This gives

$$R_\lambda(e,g) = \frac{1}{(2d+\lambda) - 2d\nu_-}\nu_-^{m(g)}.$$

Thus

$$R_\lambda(e,e) = \frac{1}{(2d+\lambda) - 2d\nu_-}.$$

Hence for each $a > 0$

$$\pi_\Gamma(t) = \frac{1}{2\pi}\int_{a-i\infty}^{a+i\infty} e^{\lambda t} R_\lambda(e,e)d\lambda = \frac{1}{2\pi}\int_{a-i\infty}^{a+i\infty} e^{\lambda t}\frac{d\lambda}{(2d+\lambda) - 2d\nu_-}.$$

The integrand here is analytic with branching points at the ends of the segment l_d, and the contour of integration can be bent into the left half plane $\mathrm{Re}\lambda < 0$ and replaced by an arbitrary closed contour around l_d. This immediately implies the first relation of (6.2). The asymptotic behavior of the integral as $t \to \infty$ is defined by the singularity of the integrand at the point $-\gamma$ (the right end of l_d). Since the integrand there has the form

$$e^{\lambda t}[a + b\sqrt{\lambda+\gamma} + O(\lambda+\gamma)], \quad \lambda + \gamma \to 0,$$

this leads to the second relation of (6.2). □

6.2 General remark on left invariant diffusions on Lie groups

The examples below concern differential operators on the continuous and discrete noncommutative groups Γ (processes with independent increments considered in the previous section are examples of operators on the abelian groups R^d).

First, we consider the Heisenberg (nilpotent) group $\Gamma = H^3$ of the upper triangular matrices

$$g = \begin{bmatrix} 1 & x & z \\ 0 & 1 & y \\ 0 & 0 & 1 \end{bmatrix}, \quad (x,y,z) \in R^3, \tag{6.7}$$

with units on the diagonal, and its discrete subgroup ZH^3, $(x,y,z) \in Z^3$.

Then we study (solvable) group of the affine transformations of the real line: $x \to ax + b\,, a > 0$, which has the matrix representation:

$$\mathrm{Aff}\left(R^1\right) = \left\{ g = \begin{bmatrix} a & b \\ 0 & 1 \end{bmatrix},\ a > 0,\ (a,b) \in R^2 \right\},$$

and its subgroup generated by

$$\alpha_1 = \begin{bmatrix} e & e \\ 0 & 1 \end{bmatrix} \quad \text{and} \quad \alpha_2 = \begin{bmatrix} e & -e \\ 0 & 1 \end{bmatrix}$$

and their inverses

$$\alpha_{-1} = \begin{bmatrix} e^{-1} & -1 \\ 0 & 1 \end{bmatrix} \quad \text{and} \quad \alpha_{-2} = \begin{bmatrix} e^{-1} & 1 \\ 0 & 1 \end{bmatrix}.$$

There are two standard ways to construct the Laplacian on a Lie group. A usual differential-geometric approach starts with the Lie algebra $\mathfrak{A}\Gamma$ on Γ, which can be considered either as the algebra of the first order differential operators generated by the differentiations along the appropriate one-parameter subgroups of Γ, or simply as a tangent vector space $T\Gamma$ to Γ at the unit element I. The exponential mapping $\mathfrak{A}\Gamma \to \Gamma$ allows one to construct (at least locally) the general left invariant Laplacian $\triangle_\Gamma$ on Γ as the image of the differential operator $\sum_{ij} a_{ij} D_i D_j + \sum_i b_i D_i$ with constant coefficients on $\mathfrak{A}\Gamma$. The Riemannian metric ds^2 on Γ and the volume element dv can be defined now using the inverse matrix of the coefficients of the Laplacian $\triangle_\Gamma$. It is important to note that additional symmetry conditions are needed to determine $\triangle_\Gamma$ uniquely.

The central object in the probabilistic construction of the Laplacian (cf., for instance, McKean [14]) is the Brownian motion g_t on Γ. We impose the symmetry condition $g_t \overset{\text{law}}{=} g_t^{-1}$. Since $\mathfrak{A}\Gamma$ is a linear space, one can define the usual Brownian motion b_t on $\mathfrak{A}\Gamma$ with the generator $\sum_{ij} a_{ij} D_i D_j + \sum_i b_i D_i$. The symmetry condition holds if $(I + db_t) \overset{\text{law}}{=} (I + db_t)^{-1}$. The process g_t (diffusion on Γ) is given (formally) by the stochastic multiplicative integral

$$g_t = \prod_{s=0}^{t} (I + db_s),$$

or (more rigorously) by the Ito stochastic differential equation

$$dg_t = g_t db_t. \tag{6.8}$$

The Laplacian $\triangle_\Gamma$ is defined now as the generator of the diffusion

$$\triangle_\Gamma f(g) = \lim_{\Delta t \to 0} \frac{Ef(g(I + b_{\Delta t})) - f(g)}{\Delta t}, \quad f \in C^2(\Gamma). \tag{6.9}$$

The Riemannian metric form is defined as above (by the inverse matrix of the coefficients of the Laplacian).

We use the probabilistic approach to construct the Laplacian in the examples below since it allows us to easily incorporate the symmetry condition.

6.3 Heisenberg group

Consider the Heisenberg group $\Gamma = H^3$ of the upper triangular matrices (6.7) with units on the diagonal. We have

$$\mathfrak{A}\Gamma = \{A = \begin{bmatrix} 0 & \alpha & \gamma \\ 0 & 0 & \beta \\ 0 & 0 & 0 \end{bmatrix}, \quad (\alpha,\beta,\gamma) \in R^3\}, \qquad e^A = \begin{bmatrix} 1 & \alpha & \gamma + \frac{\alpha\beta}{2} \\ 0 & 1 & \beta \\ 0 & 0 & 1 \end{bmatrix}.$$

Thus, $A \to \exp(A)$ is a one-to-one mapping of $\mathfrak{A}\Gamma$ onto Γ. Consider the following Brownian motion on $\mathfrak{A}\Gamma$:

$$b_t = \begin{bmatrix} 0 & u_t & \sigma w_t \\ 0 & 0 & v_t \\ 0 & 0 & 0 \end{bmatrix},$$

where σ is a constant and u_t, v_t, w_t are (standard) independent Wiener processes. Then equation (6.8) has the form

$$dg_t = \begin{bmatrix} 0 & dx_t & dz_t \\ 0 & 0 & dy_t \\ 0 & 0 & 0 \end{bmatrix} = \begin{bmatrix} 1 & x_t & z_t \\ 0 & 1 & y_t \\ 0 & 0 & 1 \end{bmatrix} \begin{bmatrix} 0 & du_t & \sigma dw_t \\ 0 & 0 & dv_t \\ 0 & 0 & 0 \end{bmatrix},$$

which implies that

$$dx_t = du_t, \quad dy_t = dv_t, \quad dz_t = \sigma dw_t + x_t dv_t.$$

Under condition $g(0) = I$, we get

$$g_t = \begin{bmatrix} 1 & u_t & \sigma w_t + \int_0^t u_s dv_s \\ 0 & 1 & v_t \\ 0 & 0 & 1 \end{bmatrix}.$$

Let us note that the matrix

$$(g_t)^{-1} = \begin{bmatrix} 1 & -u_t & u_t v_t - \sigma w_t - \int_0^t u_s dv_s \\ 0 & 1 & -v_t \\ 0 & 0 & 1 \end{bmatrix} = \begin{bmatrix} 1 & -u_t & -\sigma w_t + \int_0^t v_s du_s \\ 0 & 1 & -v_t \\ 0 & 0 & 1 \end{bmatrix}$$

has the same law as g_t. Now, from (6.9) it follows that

$$(\Delta_\Gamma f)(x,y,z) = \frac{1}{2}[f_{xx} + f_{yy} + (\sigma^2 + x^2) f_{zz} + 2\sigma x f_{yz}].$$

The matrix of the left invariant Riemannian metric has the form

$$\begin{bmatrix} 1 & 0 & 0 \\ 0 & 1 & \sigma x \\ 0 & \sigma x & \sigma^2 + x^2 \end{bmatrix}^{-1} = \begin{bmatrix} 1 & 0 & 0 \\ 0 & \sigma^2 + x^2 & -\sigma x \\ 0 & -\sigma x & 1 \end{bmatrix},$$

i.e.,

$$ds^2 = dx^2 + (\sigma^2 + x^2)dy^2 + dz^2 - 2\sigma x dy dz, \qquad dV = dx dy dz.$$

Denote by $p_\sigma(t, x, y, z)$ the transition density for the process g_t (fundamental solution of the parabolic equation $u_t = \Delta_\Gamma u$). Let $\pi_\sigma(t) = p_\sigma(t, 0, 0, 0)$.

Theorem 6.2. *The function $\pi_\sigma(t)$ has the following asymptotic behavior at zero and infinity:*

$$\pi_\sigma(t) \sim \frac{c_0}{t^{3/2}}, \quad t \to 0; \quad \pi_\sigma(t) \sim \frac{c}{t^2}, \quad t \to \infty, \quad c = p_0(1, 0, 0), \tag{6.10}$$

i.e., Theorem 2.2 *holds for the operator $H = \Delta_\Gamma + V(x, y, z)$ with $\alpha = 3$, $\beta = 4$.*

Proof. Since H^3 is a three-dimensional manifold, the asymptotics at zero is obvious. Let us prove the second relation of (6.10). We start with the simple case of $\sigma = 0$. The operator Δ_Γ in this case is degenerate. However, the density $p_0(t, x, y, z)$ exists and can be found using Hörmander hypoellipticity theory or by direct calculations. In fact, the joint distribution of (x_t, y_t, z_t) is selfsimilar:

$$\left(\frac{u_t}{\sqrt{t}}, \frac{v_t}{\sqrt{t}}, \frac{\int_0^t u_s dv_s}{t}\right) = \left(u_1, v_1, \int_0^1 u_s dv_s\right),$$

i.e.,

$$p_0(t, x, y, z) = \frac{1}{t^2} p_0\left(1, \frac{x}{\sqrt{t}}, \frac{y}{\sqrt{t}}, \frac{z}{t}\right),$$

and therefore,

$$p_0(t, 0, 0, 0) = \frac{c}{t^2}, \quad c = p_0(1, 0, 0, 0).$$

Let $\sigma^2 > 0$. Then

$$p_\sigma(t, x, y, z) = \frac{1}{\sqrt{2\pi\sigma^2 t}} \int_{R^1} p_0(t, x, y, z_1) e^{-\frac{(z - z z_1)^2}{2\sigma^2 t}} dz_1.$$

After rescaling $\frac{x}{\sqrt{t}} \to x$, $\frac{y}{\sqrt{t}} \to y$, $\frac{z}{t} \to z$, we get

$$p_\sigma(t, x, y, z) = \frac{\sqrt{t}}{t^2 \sqrt{2\pi\sigma^2}} \int_{R^1} p_0(1, x, y, z_1) e^{-\frac{t(z - z z_1)^2}{2\sigma^2}} dz_1.$$

It follows that $p_\sigma(t, 0, 0, 0) \sim c/t^2$, $t \to \infty$, with $c = p_0(1, 0, 0, 0)$. $\square$

Theorem 6.2 can be proved for the group H^n of upper triangular $n \times n$ matrices with units on the diagonal. In this case,

$$\alpha = \dim H^n = \frac{n(n-1)}{2}, \quad \beta = (n-1)+2(n-2)+3(n-3)+\ldots = \frac{n(n^2-1)}{2}.$$

6.4 Heisenberg discrete group

Consider the Heisenberg discrete group $\Gamma = ZH^3$ of integer valued matrices of the form

$$g = \begin{pmatrix} 1 & x & y \\ 0 & 1 & z \\ 0 & 0 & 1 \end{pmatrix}, \quad x, y, z \in Z^1.$$

Consider the Markov process g_t on ZH^3 defined by the equation

$$g_{t+dt} = g_t \begin{pmatrix} 1 & d\xi_t & d\zeta_t \\ 0 & 1 & d\eta_t \\ 0 & 0 & 1 \end{pmatrix}, \tag{6.11}$$

where ξ_t, η_t, ζ_t are three independent Markov processes on Z^1 with generators

$$\Delta_1 \psi(n) = \psi(n+1) + \psi(n-1) - 2\psi(n), \qquad n \in Z^1.$$

Equation (6.11) can be solved using discretization of time. This gives

$$g_t = \begin{pmatrix} 1 & x_t & y_t \\ 0 & 1 & z_t \\ 0 & 0 & 1 \end{pmatrix} \begin{pmatrix} 1 & \xi_t & \zeta_t + \int_0^t \xi_s d\eta_s \\ 0 & 1 & \eta_t \\ 0 & 0 & 1 \end{pmatrix}.$$

The generator L of this process has the form (6.1) with

$$a_{\pm 1} = \begin{pmatrix} 1 & \pm 1 & 0 \\ 0 & 1 & 0 \\ 0 & 0 & 1 \end{pmatrix}, \quad a_{\pm 2} = \begin{pmatrix} 1 & 0 & 0 \\ 0 & 1 & \pm 1 \\ 0 & 0 & 1 \end{pmatrix}, \quad a_{\pm 3} = \begin{pmatrix} 1 & 0 & \pm 1 \\ 0 & 1 & 0 \\ 0 & 0 & 1 \end{pmatrix},$$

i.e.,

$$L = \Delta_\Gamma \psi(g) = \sum_{i=\pm 1, \pm 2, \pm 3} [\psi(g a_i) - \psi(g)]. \tag{6.12}$$

If $\psi = \psi(g)$ is considered as a function of $(x, y, z) \in Z^3$, then

$$\begin{aligned} L\psi(x,y,z) &= \psi(x+1,y,z) + \psi(x-1,y,z) \\ &+ \psi(x,y+1,z+x) + \psi(x,y-1,z-x) \\ &+ \psi(x,y,z+1) + \psi(x,y,z-1) - 6\psi(x,y,z). \end{aligned} \tag{6.13}$$

The analysis of the transition probability in this case is similar to the continuous case, and it leads to the following result

Theorem 6.3. *If g_t is the process on ZH^3 with the generator* (6.13), *then*

$$P\{g_t = I\} = P\{x_t = y_t = z_t = 0\} \sim \frac{c}{t^2}, \quad t \to \infty,$$

with c defined in (6.10). *In particular, Theorem* 3.1 *can be applied to the operator $H_0 = L$ with $\beta = 4$.*

This result is valid in a more general setting (cf. [13]). Consider three independent processes ξ_t, η_t, ζ_t, $t \geqslant 0$, on Z^1 with independent increments and such that

$$Ee^{ik\xi_t} = e^{-t(1-\sum_{i=1}^{\infty} p_i \cos ki)}, \quad \sum_{i=1}^{\infty} p_i = 1,$$

$$Ee^{ik\eta_t} = e^{-t(1-\sum_{i=1}^{\infty} q_i \cos ki)}, \quad \sum_{i=1}^{\infty} q_i = 1,$$

$$Ee^{ik\zeta_t} = e^{-t(1-\sum_{i=1}^{\infty} r_i \cos ki)}, \quad \sum_{i=1}^{\infty} r_i = 1.$$

Assume also that there exist $\alpha_1, \alpha_2, \alpha_3$ on the interval $(0, 2)$ such that

$$p_i \sim \frac{c_1}{i^{1+\alpha_1}}, \quad q_i \sim \frac{c_2}{i^{1+\alpha_2}}, \quad r_i \sim \frac{c_3}{i^{1+\alpha_3}}$$

as $i \to \infty$, i.e., distributions with characteristic functions

$$\sum_{i=1}^{\infty} p_i \cos ki, \quad \sum_{i=1}^{\infty} q_i \cos ki, \quad \sum_{i=1}^{\infty} r_i \cos ki$$

belong to the domain of attraction of the symmetric stable law with parameters α_1, α_2, α_3. Let g_t be the process on ZH^3 defined by (6.11). Then

$$P\{g_t = I\} \sim \frac{c}{t^\gamma}, \quad t \to \infty, \qquad \gamma = \max\left(\frac{2}{\alpha_1} + \frac{2}{\alpha_1}, \frac{1}{\alpha_3}\right).$$

6.5 Group Aff (R^1) *of affine transformations of the real line*

This group of transformations $x \to ax+b$, $a > 0$, has a matrix representation:

$$\Gamma = \mathrm{Aff}\,(R^1) = \left\{g = \begin{bmatrix} a & b \\ 0 & 1 \end{bmatrix}, \quad a > 0, \quad (a,b) \in R^2\right\}.$$

We start with the Lie algebra for Aff (R^1) :

$$\mathfrak{A}\Gamma = \left\{ \begin{bmatrix} \alpha\ \beta \\ 0\ 0 \end{bmatrix}, \quad (\alpha,\beta) \in R^2 \right\}.$$

Obviously, for arbitrary $A = \begin{bmatrix} \alpha\ \beta \\ 0\ 0 \end{bmatrix}$, one has

$$\exp(A) = \begin{bmatrix} e^{\alpha} & \beta\frac{e^{\alpha}-1}{\alpha} \\ 0 & 1 \end{bmatrix},$$

i.e., the exponential mapping of $\mathfrak{A}\Gamma$ coincides with the group Γ. Consider the diffusion

$$b_t = \begin{bmatrix} w_t + \alpha t & v_t \\ 0 & 0 \end{bmatrix}$$

on $\mathfrak{A}\Gamma$, where (w_t, v_t) are independent Wiener processes. Consider the matrix valued process $g_t = \begin{bmatrix} x_t & y_t \\ 0 & 1 \end{bmatrix}$, $g_0 = \begin{bmatrix} 1\ 0 \\ 0\ 1 \end{bmatrix}$, on Γ satisfying the equation

$$dg_t = g_t db_t = \begin{bmatrix} x_t & y_t \\ 0 & 1 \end{bmatrix} \begin{bmatrix} dw_t + \alpha dt & dv_t \\ 0 & 0 \end{bmatrix} = \begin{bmatrix} x_t\left(dw_t + \alpha dt\right) & x_t dv_t \\ 0 & 0 \end{bmatrix}.$$

This implies

$$\begin{aligned} dx_t &= x_t\left(dw_t + \alpha dt\right), \\ dy_t &= x_t dv_t, \end{aligned}$$

i.e.,

$$x_t = e^{w_t + (\alpha - \frac{1}{2})t}, \quad y_t = \int_0^t x_s dv_s$$

due to the Ito formula.

We impose the following symmetry condition:

$$(g_t)^{-1} \stackrel{\text{law}}{=} g_t, \tag{6.14}$$

It holds if $\alpha = \frac{1}{2}$. In fact,

$$g_t = \begin{bmatrix} e^{w_t} & \int_0^t e^{w_s} dv_s \\ 0 & 1 \end{bmatrix}, \quad g_t^{-1} = \begin{bmatrix} e^{-w_t} & -\int_0^t e^{w_s - w_t} dv_s \\ 0 & 1 \end{bmatrix}, \tag{6.15}$$

and (6.14) follows after the change of variables $s = t - \tau$ in the matrix g_t^{-1}. Then the generator of the process g_t has the form

$$\Delta_\Gamma f = \frac{x^2}{2}\left[\frac{\partial^2 f}{\partial x^2} + \frac{\partial^2 f}{\partial y^2}\right] + \frac{x}{2}\frac{\partial f}{\partial x}.$$

Theorem 6.4. *Operator $\triangle_\Gamma$ is selfadjoint with respect to the measure $x^{-1}dxdy$. The function $\pi(t) = p(t,0,0)$ has the following behavior at zero and infinity:*

$$\pi(t) \sim \frac{c_0}{t}, \quad t \to 0; \quad \pi(t) \sim \frac{C}{t^{3/2}}, \quad t \to \infty. \tag{6.16}$$

Remark 6.2. Let $H = \triangle_\Gamma + V$, where the negative part $W = V_-$ of the potential is bounded: $W \leqslant h^{-1}$. From (6.16) and Theorem 2.2 it follows that

$$N_0(V) \leqslant C(h) \int_{-\infty}^{\infty} \int_0^{\infty} \frac{W^{3/2}(x,y)}{x} dxdy.$$

Remark 6.3. The left-invariant Riemannian metric on Aff (R^1) is given by the inverse diffusion matrix of $\triangle_\Gamma$, i.e.,

$$d\xi^2 = x^{-2}\left(dx^2 + dy^2\right) \qquad \left(g = \begin{bmatrix} x & y \\ 0 & 1 \end{bmatrix}, \quad x > 0\right)$$

After the change $(x,y) \to (y,x)$, this formula coincides with the metric on the Lobachevsky plane (cf. the previous section). However, one cannot identity the Laplacian on Aff (R^1) and on the Lobachevsky plane L^2 since they are defined by different symmetry conditions. The plane L^2 has a three-dimensional group of transformations, and each point $z \in L^2$ has a one-parameter stationary subgroup. The Laplacian on the Lobachevsky plane was defined by the invariance with respect to this three-dimensional group of transformations. In the case $\Gamma =$ Aff (R^1), the group of transformations is two-dimensional. It acts as a left shift $g \to g_1 g$, $g_1, g \in \Gamma$, and the Laplacian is specified by the left invariance with respect to this two-dimensional group and the symmetry condition (6.14).

Proof. Since Γ is a two-dimensional manifold, the asymptotics of $\pi(t)$ at zero is obvious. One needs only to justify the asymptotics of $\pi(t)$ at infinity.

Let us find the density of

$$(x_t, y_t) = \left(e^{w_t}, \int_0^t e^{w_s} dv_s\right).$$

The second term, for a fixed realization of $w.$, has the Gaussian law with (conditional) variance

$$\sigma^2 = \int_0^t e^{2w_s} ds$$

and

$$P\{x_t \in 1+dx, y_t \in 0+dy\} = p(t,0,0)dxdy = \frac{1}{\sqrt{2\pi t}} E \frac{1}{\sqrt{2\pi \int_0^t e^{2\widehat{w}_s} ds}}. \tag{6.17}$$

Here, $\widehat{w}_s$, $s \in [0,t]$, is the Brownian bridge on $[0,t]$. The distribution of the exponential functional

$$A(t) = \int_0^t e^{2\widehat{w}_s} ds$$

and the joint distribution of $(A(t), w(t))$ were calculated in [26]. Together with (6.17), these easily imply the statement of the theorem. □

6.6 A relation between Markov processes and random walks on discrete groups

Let Γ be a discrete group generated by elements $a_1, \dots, a_d$, $a_{-1} = a_1^{-1}, \dots$, $a_{-d} = a_d^{-1}$, with some identities. Define the Laplacian on Γ by the formula

$$\Delta\psi(g) = \sum_{i=-d}^{d} \psi(ga_i) - 2d\psi(g), \quad g \in \Gamma.$$

Consider the Markov process g_t on Γ with continuous time and the generator Δ. Let $\widetilde{g}_k$, $k = 0,1,2,\dots$, be the Markov chain on Γ with discrete time (symmetric random walk) such that

$$P\{\widetilde{g}_0 = e\} = 1, \quad P\{\widetilde{g}_{n+1} = ga_i \mid \widetilde{g}_n = g\} = \frac{1}{2d}, \quad i = \pm 1, \pm 2, \dots, \pm d.$$

Then there is a relation between transition probability $p(t,e,g)$ of the Markov process g_t and the transition probability $P\{\widetilde{g}_k = g\}$ of the random walk. In particular, one can estimate $\pi(t) = p(t,e,e)$ for large t through $\widetilde{\pi}(2k) = P\{\widetilde{g}_{2k} = e\}$ under minimal assumptions on $\widetilde{\pi}(2k)$. For example, it is enough to assume that $\widetilde{\pi}(2k) = k^\gamma L(k)$, $\gamma \geqslant 0$, where $L(k)$ for large k can be extended as slowly varying monotonic function of continuous argument k. We are not going to provide a general statement of this type, but we restrict ourself to a specific situation needed in the next section. Note that we consider here only even arguments of $\widetilde{\pi}$ since $\widetilde{\pi}(2k+1) = 0$.

Theorem 6.5. *Let $\widetilde{\pi}(2n) \leqslant e^{-c_0(2n)^\alpha}$ as $n \to \infty$, $c_0 > 0$, $0 < \alpha < 1$. Then*

$$\pi(t) \leqslant e^{-c_0(2dt)^\alpha}, \quad t \geqslant t_0.$$

Proof. The number ν_t of jumps of the process g_t on the interval $(0,t)$ has Poisson distribution. At the moments of jumps, the process performs the symmetric random walk with discrete time and transition probabilities $P\{g \to ga_i\} = 1/2d$, $i = \pm 1, \pm 2, \dots, \pm d$. Thus (taking into account that $\widetilde{\pi}(2k+1) = 0$),

$$\pi(t) = p(t, e, e) = \sum_{n=0}^{\infty} \widetilde{\pi}(2n) P\{\nu_t = 2n\}.$$

Due to the exponential Chebyshev inequality

$$P\{|\nu_t - 2dt| \geqslant \varepsilon t\} \leqslant e^{-c\varepsilon^2 t}, \quad t \to \infty.$$

Secondly,

$$P\{\nu_t \text{ is even}\} = \frac{1}{2} + O(e^{-4dt}), \quad t \to \infty.$$

These relations imply that, for $t \to \infty$ and $\delta > 0$

$$\begin{aligned}
\pi(t) &= \sum_{n:|2n-2dt|<\varepsilon t} \widetilde{\pi}(2n) P\{\nu_t = 2n\} + O(e^{-c_0(2dt)^\alpha}) \\
&\leqslant \sum_{n:|2n-2dt|<\varepsilon t} e^{-c_0(2n)^\alpha} P\{\nu_t = 2n\} + O(e^{-c_0(2dt)^\alpha}) \\
&\leqslant (1+\delta) e^{-c_0(2dt)^\alpha} \sum_{n:|2n-2dt|<\varepsilon t} P\{\nu_t = 2n\} + O(e^{-c_0(2dt)^\alpha}) \\
&\leqslant \frac{1+\delta}{2} e^{-c_0(2dt)^\alpha} + O(e^{-c_0(2dt)^\alpha}).
\end{aligned}$$

The proof is complete □

6.7 Random walk on the discrete subgroup of Aff (R^1)

Let us consider the following two matrices:

$$\alpha_1 = \begin{bmatrix} e & e \\ 0 & 1 \end{bmatrix} \quad \text{and} \quad \alpha_2 = \begin{bmatrix} e & -e \\ 0 & 1 \end{bmatrix}$$

in Aff (R^1) and their inverses

$$\alpha_{-1} = \begin{bmatrix} e^{-1} & -1 \\ 0 & 1 \end{bmatrix} \quad \text{and} \quad \alpha_{-2} = \begin{bmatrix} e^{-1} & 1 \\ 0 & 1 \end{bmatrix}.$$

Let G be a subgroup of Aff (R^1) generated by $\alpha_{\pm 1}$and $\alpha_{\pm 2}$. Consider the random walk on G of the form

$$g_n = h_1 h_2 \dots h_n,$$

where one step random matrices h_i coincide with one of the matrices $\alpha_{\pm 1}$, $\alpha_{\pm 2}$ with probability $1/4$, i.e.,

$$h_i = \begin{bmatrix} e^{\varepsilon_i} & \delta_i \\ 0 & 1 \end{bmatrix},$$

where

$$\begin{aligned} P\{\varepsilon_i = 1,\ \delta_i = e\} &= P\{\varepsilon_i = 1,\ \delta_i = -e\} \\ &= P\{\varepsilon_i = -1,\ \delta_i = -1\} = P\{\varepsilon_i = -1,\ \delta_i = 1\} = 1/4. \end{aligned} \tag{6.18}$$

Let Δ_G be the Laplacian on G which corresponds to the generators $a_{\pm 1}, a_{\pm 2}$, i.e. (cf. (6.1), (6.12)),

$$L = \Delta_\Gamma \psi(g) = \sum_{i=\pm 1, \pm 2} [\psi(g a_i) - \psi(g)].$$

Theorem 6.6. (a) *The following estimate is valid for* $\widetilde{\pi}(2n)$:

$$\widetilde{\pi}(2n) \leqslant e^{-c_0 (2n)^{1/3}}, \quad n \to \infty, \quad c_0 > 0.$$

(b) *Theorem* 3.1 *can be applied to the operator* $H = \Delta_G + V(g)$ *with* $\gamma = 1/3$, *i.e.*,

$$N_0(V) \leqslant C(h, A)\Big[\sum_{g: V(g) \leqslant h^{-1}} e^{-AW(g)^{-1/3}} + n(h) \Big],$$

$$n(h) = \#\{g : W(g) > h^{-1}\}.$$

Proof. The random variables $(\varepsilon_i,\ \delta_i)$ are dependent, but (6.18) implies that $(\varepsilon_i,\ \widetilde{\delta}_i)$, where $\widetilde{\delta}_i = \mathrm{sgn}\delta_i$, are independent symmetric Bernoulli r.v. It is easy to see that

$$g_n = \begin{bmatrix} e^{S_n} & \sum_{k=1}^n \delta_k e^{S_{k-1}} \\ 0 & 1 \end{bmatrix},$$

where $S_0 = 1$, $S_k = \varepsilon_1 + \ldots + \varepsilon_k$, $k > 0$, is a symmetric random walk on Z^1. This formula is an obvious discrete analogue of (6.15). Our goal is to calculate the probability

$$\widetilde{\pi}(2n) = P\{g_{2n} = I\} = P\Big\{ S_{2n} = 0, \sum_{k=1}^{2n} \delta_k e^{S_{k-1}} = 0 \Big\}$$

$$= \binom{2n}{n} \frac{1}{2^{2n}} P\Big\{ \sum_{k=2}^{2n} \delta_k e^{\widehat{S}_{k-1}} = 0 \Big\} \sim \frac{1}{\sqrt{\pi n}} P\Big\{ \sum_{k=1}^{2n-1} \delta_{k+1} e^{\widehat{S}_k} = 0 \Big\}$$

as $n \to \infty$. Here, $\widehat{S}_k$, $k = 0, 1, \ldots, 2n$, is the discrete bridge, i.e., the random walk S_k under conditions $S_0 = S_{2n} = 0$.

Put $M_{2n} = \max_{k\leqslant 2n} \widehat{S}_k$, $m_{2n} = \min_{k\leqslant 2n} \widehat{S}_k$. Let Γ^+_{s-1}, Γ^-_s be the sets of moments of time k when the bridge $\widehat{S}_k$ changes value from $s-1$ to s or from s to $s-1$ respectively. Introduce local times $\tau^+_{s-1} = \mathrm{Card}\Gamma^+_{s-1}$ and $\tau^-_s = \mathrm{Card}\Gamma^-_s$, i.e., $\tau^+_{s-1} = \#(\text{jumps of } \widehat{S}_k \text{ from } s-1 \text{ to } s)$ and $\tau^-_s = \#(\text{jumps of } \widehat{S}_k \text{ from } s \text{ to } s-1)$. Note that $\delta_{k+1}e^{\widehat{S}_k} = \widetilde{\delta}_{k+1}e^s$ when $k \in \Gamma^+_{s-1} \cup \Gamma^-_s$, and therefore

$$\sum_{k=1}^{2n-1} \delta_{k+1}e^{\widehat{S}_k} = \sum_{s=m_{2n}+1}^{M_{2n}} e^s \sum_{j\in\Gamma^+_{s-1}\cup\Gamma^-_s} \widetilde{\delta}_j.$$

Since r.v. $\{\widetilde{\delta}_j\}$ are independent of the trajectory S_k and numbers e^s, $s = 0, \pm 1, \pm 2, \ldots$, are rationally independent, we have

$$\begin{aligned} P\{g_{2n} = I\} &\sim \frac{1}{\sqrt{\pi n}} E \prod_{s=m_{2n}+1}^{M_{2n}} \binom{2\tau^-_s}{\tau^-_s} \left(\frac{1}{2}\right)^{2\tau^-_s} \leqslant \frac{1}{\sqrt{\pi n}} \left(\frac{1}{2}\right)^{M_{2n}-m_{2n}} \\ &= \frac{1}{\sqrt{\pi n}} \left(\frac{1}{2}\right)^{M_{2n}-m_{2n}} \left[I_{M_{2n}-m_{2n}>\sqrt{2n}} + I_{M_{2n}-m_{2n}<\sqrt{2n}}\right] \\ &\leqslant \frac{1}{\sqrt{\pi n}} \left(\frac{1}{2}\right)^{\sqrt{2n}} + \sum_{r=1}^{\sqrt{2n}} (\frac{1}{2})^r P\{|S_k| \leqslant r,\ k = 1,2,\ldots,2n,\ S_{2n} = 0\} \\ &\leqslant e^{-c_1\sqrt{2n}} + \sum_{r=1}^{\sqrt{2n}} \left(\frac{1}{2}\right)^r P\{|S_k| \leqslant r,\ k = 1,2,\ldots,2n,\ \ S_{2n} = 0\}. \end{aligned}$$

The proof is complete. □

Lemma 6.1. $P\{|S_k| \leqslant r,\ k = 1,2,\ldots,2n,\ S_{2n} = 0\} \leqslant (\cos\frac{\pi}{2(r+1)})^{2n}$.

Proof. Introduce the operator

$$H_0\psi(x) = \frac{\psi(x+1) + \psi(x-1)}{2}$$

on the set $[-r,r] \in Z^1$ with the Dirichlet boundary conditions $\psi(r+1) = \psi(-r-1) = 0$. Then $\varphi(x) = \cos\frac{\pi x}{2(r+1)}$ is an eigenfunction of H_0 with the eigenvalue $\lambda_{0,r+1} = \cos\frac{\pi}{2(r+1)}$. Hence

$$H_0^{2n}\varphi(x) = \lambda_{0,r+1}^{2n}\varphi(x).$$

Let $p_r(k,x,z)$ be the transition probability of the random walk on $[-r,r] \in Z^1$ with the absorption at $\pm(r+1)$. Then

$$\sum_{|z|\leqslant r} p_r(2n,x,z)\varphi(z) = \lambda_{0,r+1}^{2n}\varphi(x).$$

Since $\varphi(z) \leqslant 1$, $\varphi(0) = 1$, the latter relation implies

$$\sum_{|z|\leqslant r} p_r(2n,x,z) \leqslant \lambda_{0,r+1}^{2n}.$$

Since S_k, $k = 0, 1, \ldots, 2n$, is the symmetric random walk on Z^1, we have

$$P\{|S_k| \leqslant r,\ k = 1,2,\ldots,2n,\ S_{2n} = 0\} = p_r(2n,0,0) \leqslant \lambda_{0,r+1}^{2n}.$$

A direct calculation shows that

$$\max_{r\leqslant\sqrt{2n}} \left(\frac{1}{2}\right)^r (\cos\frac{\pi}{2(r+1)})^{2n} \leqslant e^{-c(2n)^{1/3}},$$

with the maximum achieved at $r = r_0 \sim c_1(2n)^{1/3}$. Thus

$$P\{g_{2n} = I\} \leqslant \left(\frac{1}{2}\right)^{\sqrt{2n}} + \sqrt{2n}e^{-c_0(2n)^{1/3}} \leqslant e^{-\widetilde{c}_0(2n)^{1/3}}$$

for arbitrary $\widetilde{c}_0 < c_0$ and sufficiently large n. This proves the first statement of the theorem. Now, the second statement follows from Theorem 6.5. □

7 Appendix

Proof of Theorem 3.2. As it was mentioned after the statement of the theorem, it is enough to show the validity of condition (b) and evaluate α, β. Let

$$u_t = -H_0u,\ t > 0,\quad u|_{t=0} = f,$$

with a compactly supported f and

$$\varphi = \varphi(x,\lambda) = \int_0^\infty ue^{\lambda t}dt,\quad \mathrm{Re}\lambda \leqslant -a < 0,\quad x \in \Gamma^d.$$

Note that we replaced $-\lambda$ by λ in the Laplace transform above. It is convenient for future notations. Then φ satisfies the equation

$$(H_0 - \lambda)\varphi = f, \tag{7.1}$$

and u can be found using the inverse Laplace transform

$$u = \frac{1}{(2\pi)^d}\int_{-a-i\infty}^{-a+i\infty} \varphi e^{-\lambda t}d\lambda. \tag{7.2}$$

The spectrum of H_0 is $[0,\infty)$, and φ is analytic in λ when $\lambda \in C\backslash[0,\infty)$. We are going to study the properties of φ when $\lambda \to 0$ and $\lambda \to \infty$. Let $\psi(z) = \psi(z,\lambda)$, $z \in Z^d$, be the restriction of the function $\varphi(x,\lambda)$, $x \in \Gamma^d$, on the lattice Z^d. Let e be an arbitrary edge of Γ^d with end points $z_1, z_2 \in Z^d$ and parametrization from z_1 to z_2. By solving the boundary value problem on e, we can represent φ on e in the form

$$\begin{gathered}\varphi = \frac{\psi(z_1)\sin k(1-s) + \psi(z_2)\sin ks}{\sin k} + \varphi_{\text{par}},\\ \varphi_{\text{par}} = \int_0^1 G(s,t)f(t)dt,\end{gathered} \tag{7.3}$$

where $k = \sqrt{\lambda}$, $\operatorname{Im} k > 0$, and

$$G = \frac{1}{k\sin k}\begin{cases}\sin ks \sin k(1-t), & s < t,\\ \sin kt \sin k(1-s), & s \geqslant t.\end{cases}$$

Due to the invariance of H_0 with respect to translations and rotations in Z^d, it is enough to estimate $p_0(t,x,x)$ when x belongs to the edge e_0 with z_1 being the origin in Z^d and $z_2 = (1,0,\dots,0)$. Let f be supported on one edge e_0. Then (7.3) is still valid, but $\varphi_{\text{par}} = 0$ on all the edges except e_0. We substitute (7.3) into (3.5) and get the following equation for ψ :

$$(\varDelta - 2d\cos k)\psi(z) = \frac{1}{k}\int_0^1 \sin k(1-t)f(t)dt\delta_1 + \frac{1}{k}\int_0^1 \sin ktf(t)dt\delta_0, \quad z \in Z^d.$$

Here, $\varDelta$ is the lattice Laplacian defined in (3.2) and δ_0, δ_1 are functions on Z^d equal to one at z, y respectively, and vanish elsewhere. In particular, if f is the delta function at a point s of the edge e_0, then

$$(\varDelta - 2d\cos k)\psi = \frac{1}{k}\sin k(1-s)\delta_1 + \frac{1}{k}\sin ks\delta_0. \tag{7.4}$$

Let $R_\mu(z - z_0)$ be the kernel of the resolvent $(\varDelta - \mu)^{-1}$ of the lattice Laplacian. Then (7.4) implies that

$$\begin{gathered}\psi(z) = \frac{1}{\sqrt{\lambda}}\sin\sqrt{\lambda}sR_\mu(z) + \frac{1}{\sqrt{\lambda}}\sin\sqrt{\lambda}(1-s)R_\mu(z - z_2),\\ \mu = 2d\cos\sqrt{\lambda}.\end{gathered} \tag{7.5}$$

The function $R_\mu(z)$ has the form

$$R_\mu(z) = \int_T \frac{e^{i(\sigma,z)}d\sigma}{(\sum_{1\leqslant j\leqslant d} 2\cos\sigma_j) - \mu}, \quad T = [-\pi,\pi]^d.$$

Hence the function $\sin(\sqrt{\lambda}s)R_\mu(z)$, $s \in (0,1)$, $\mu = 2d\cos\sqrt{\lambda}$, decays exponentially as $|\mathrm{Im}\sqrt{\lambda}| \to \infty$. This allows one to change the contour of integration in (7.2), when $z \in Z^d$, and rewrite (7.2) in the form

$$u(z,t) = \frac{1}{(2\pi)^d}\int_l \psi_\lambda(z)e^{-\lambda t}d\lambda, \quad z \in Z^d, \tag{7.6}$$

where the contour l consists of the ray $\lambda = \rho e^{-i\pi/4}$, $\rho \in (\infty, 1)$, a smooth arc starting at $\lambda = e^{-\pi/4}$, ending at $\lambda = e^{\pi/4}$, and crossing the real axis at $\lambda = -a$, and the ray $\lambda = \rho e^{i\pi/4}$, $\rho \in (1,\infty)$. It is easy to see that $|\psi(z,\lambda)| \leqslant C/|\sqrt{\lambda}|$ as $\lambda \in l$ uniformly in s and $z \in Z^d$. This immediately implies that $|u(z,t)| \leqslant C/\sqrt{t}$. Now, from (7.3) it follows that the same estimate is valid for $p_0(t,x,x)$, $x \in e_0$, i.e., condition (b) holds, and $\alpha = 1$.

From (7.6) it also follows that the asymptotic behavior of u as $t \to \infty$ is determined by the asymptotic expansion of $\psi(z,\lambda)$ as $\lambda \to 0$, $\lambda \notin [0,\infty)$. Note that the spectrum of the difference Laplacian is $[-2d, 2d]$, and $\mu = 2d - d\lambda + O(\lambda^2)$ as $\lambda \to 0$. From here and the well known expansions of the resolvent of the difference Laplacian near the edge of the spectrum it follows that the first singular term in the asymptotic expansion of $R_\mu(z)$ as $\lambda \to 0$, $\lambda \notin [0,\infty)$, has the form

$$c_d\lambda^{d/2-1}(1+O(\lambda)), \quad d \text{ is odd},$$

$$c_d\lambda^{d/2-1}\ln\lambda(1+O(\lambda)), \quad d \text{ is even}.$$

Then (7.5) implies that a similar expansion is valid for $\psi(z,\lambda)$ with the main term independent of s and the remainder estimated uniformly in s. This allows one to replace l in (7.6) by the contour which consists of the rays $\arg\lambda = \pm\pi/4$. From here it follows that for each $z \in Z^d$ and uniformly in s,

$$u(z,t) \sim t^{-d/2}, \quad t \to \infty.$$

This and (7.3) imply the same behavior for $p_0(t,x,x)$, $x \in e_0$, i.e., $\beta = d$. □

Acknowledgment. The authors are very grateful to V. Konakov and O. Safronov for very useful discussions.

The work was supported in part by the NSF grant DMS-0706928.

References

1. Birman, M., Solomyak, M.: Estimates for the number of negative eigenvalues of the Schrödinger operator and its generalizations. Adv. Sov. Math. 7 (1991)
2. Carmona, R., Lacroix, J.: Spectral Theory of Random Schrödinger Operator. Birhauser, Basel etc. (1990)

3. Chen, K., Molchanov, S., Vainberg, B.: Localization on Avron–Exner–Last graphs: I. Local perturbations. Contemp. Math. **415**, 81–92 (2006)
4. Cwikel, M.: Weak type estimates for singular values and the number of bound states of Schrödinger operators. Ann. Math. **106**, 93–100 (1977)
5. Daubechies, I.: An uncertanty principle for fermions with generalized kinetic energy. Commun. Math. Phys. **90**, 511–520 (1983)
6. Donsker, M.D., Varadhan, S.R.S.: Asymptotic evaluation of the certain Markov process expectations for large time I.II. Commun. Pure Appl. Math. **28**, 1–47 (1975)
7. Donsker, M.D., Varadhan, S.R.S.: Asymptotics for the Wiener sausage. Commun. Pure Appl. Math. **28**, no. 4, 525–565 (1975)
8. Eisenhart, L. P.: Rimannian Geometry. Princeton Univ. Press, Princeton (1997)
9. Gaveau, B.: Principe de moindre action, propagation de la chaleur et estimees sous elliptiques sur certains groupes nilpotents. Acta Math. **139**, no. 1, 95–153 (1977)
10. Gikhman, I., Skorokhod, A.: Introduction to the Theory of Random Processes, Dover Publ. Inc., Mineola, NY (1996)
11. Guillotin-Plantard, N., Rene Schott: Dynamic random walks on Heisenberg groups. J. Theor. Probab. **19**, no. 2, 377–395 (2006)
12. Karpelevich, F.I., Tutubalin, V.N., Shur, M.G.: Limit theorems for the compositions of distributions in the Lobachevsky plane and space. Theor. Probab. Appl. **4**, 399–402 (1959)
13. Konakov, V., Menozzi, S., Molchanov, S.: [In preparation]
14. McKean, H.: Stochastic Integrals. Am. Math. Soc., Providence, RI (2005)
15. Lieb, E.: Bounds on the eigenvalues of the Laplace and Schrödinger operators. Bull. Am. Math. Soc. **82**, no. 5, 751–753 (1976)
16. Lieb, E.: The number of bound states of one-body Schrödinger operators and the Weyl problem. In: Geometry of the Laplace Operator (Proc. Sympos. Pure Math., Univ. Hawaii, Honolulu, Hawaii, 1979), pp. 241–252 (1980)
17. Lieb, E., Thirring, W.: Bound for the kinetic energy of fermions which proves the stability of matter. Phys. Rev. Lett. **35**, 687–689 (1975)
18. Lieb, E., Thirring, W.: Inequalities for the moments of the eigenvalues of the Schrödinger Hamiltonian and their relation to Sobolev inequalities. In: Studies in Mathematical Physics: Essays in Honor of Valentine Bargmann, pp. 269–303, Princeton Univ. Press, Princeton, (1976)
19. Maz'ya, V.: Analytic criteria in the qualitative spectral analysis of the Schrödinger operator. In: Spectral Theory and Mathematical Physics: a Festschrift in honor of Barry Simon's 60th birthday, 257–288. Am. Math. Soc., Providence, RI (2007)
20. Maz'ya, V., Shubin, M.: Discreteness of spectrum and positivity criteria for Schrödinger operators. Ann. Math. **162**, 1–24 (2005)
21. Rashevsky, P. K.: Riemannian Geometry and Tensor Analysis (Russian). "Nauka", Moscow (1967)
22. Reed, M., Simon, B.: Methods of Modern Mathematical Physics. 4. Acad. Press, N.Y. (1978)
23. Rozenblum, G.: Distribution of the discrete spectrum of singular differential operators (Russian). Dokl. Akad. Nauk SSSR **202** 1012–1015 (1972); English transl.: Sov. Math. Dokl. **13**, 245-249 (1972)

24. Rozenblum, G., Solomyak, M.: LR-estimate for the generators of positivity preserving and positively dominated semigroups (Russian). Algebra Anal. **9**, no. 6, 214–236 (1997); English transl.: St. Petersburg Math. J. **9**, no. 6, 1195–1211 (1998)
25. Rozenblum, G., Solomyak, M.: Counting Schrödinger boundstates: semiclassics and beyond. In: Maz'ya, V. (ed.), Sobolev Spaces in Mathematics. II: Applications in Analysis and Parrtial Differential Equations. Springer, New York; Tamara Rozhkovskaya Publisher, Novosibirsk. International Mathematical Series **9**, 329–354 (2009)
26. Yor, M.: On some exponential functionals of brownian motion. Adv. Appl. Prob. **24**, 509–531 (1992)

Estimates for the Counting Function of the Laplace Operator on Domains with Rough Boundaries

Yuri Netrusov and Yuri Safarov

Abstract We present explicit estimates for the remainder in the Weyl formula for the Laplace operator on a domain Ω, which involve only the most basic characteristics of Ω and hold under minimal assumptions about the boundary $\partial\Omega$.

This is a survey of results obtained by the authors in the last few years. Most of them were proved or implicitly stated in our papers [10, 11, 12]; we give precise references or outline proofs wherever it is possible. The results announced in Subsection 5.2 are new.

Let $\Omega \subset \mathbb{R}^n$ be an open bounded domain in $\mathbb{R}^n$, and let $-\Delta_{\rm B}$ be the Laplacian on Ω subject to the Dirichlet (B = D) or Neumann (B = N) boundary condition. Further on, we use the subscript B in the cases where the corresponding statement refers to (or result holds for) both the Dirichlet and Neumann Laplacian. Let $N_{\rm B}(\Omega,\lambda)$ be the number of eigenvalues of $\Delta_{\rm B}$ lying below λ^2. If the number of these eigenvalues is infinite or $-\Delta_{\rm B}$ has essential spectrum below λ^2, then we define $N_{\rm N}(\Omega,\lambda) := +\infty$. Let

$$R_{\rm B}(\Omega,\lambda) := N_{\rm B}(\Omega,\lambda) - (2\pi)^{-n}\,\omega_n\,|\Omega|\,\lambda^n\,,$$

where ω_n is the volume of the n-dimensional unit ball and $|\Omega|$ denotes the volume of Ω. According to the Weyl formula, $R_{\rm B}(\Omega,\lambda) = o(\lambda^n)$ as $\lambda \to +\infty$. If B = D, then this is true for every bounded domain [4]. If B = N, then the Weyl formula holds only for domains with sufficiently regular boundaries. In

Yuri Netrusov
Department of Mathematics, University of Bristol, University Walk, Bristol BS8 1TW, UK
e-mail: y.netrusov@bristol.ac.uk

Yuri Safarov
Department of Mathematics, King's College London, Strand, London WC2R 2LS, UK
e-mail: yuri.safarov@kcl.ac.uk

A. Laptev (ed.), *Around the Research of Vladimir Maz'ya III: Analysis and Applications*, International Mathematical Series 13, DOI 10.1007/978-1-4419-1345-6_9,

the general case, R_{N} may well grow faster than λ^n; moreover, the Neumann Laplacian on a bounded domain may have a nonempty essential spectrum (see, for instance, Remark 6.1 or [6]). The necessary and sufficient conditions for the absence of the essential spectrum in terms of capacities were obtained by Maz'ya [8].

The aim of this paper is to present estimates for $R_{\mathrm{B}}(\Omega,\lambda)$, which involve only the most basic characteristics of Ω and constants depending only on the dimension n. The estimate from below (1.2) for $R_{\mathrm{B}}(\Omega,\lambda)$ and the estimate from above (4.1) for $R_{\mathrm{D}}(\Omega,\lambda)$ hold for all bounded domains. The upper bound (4.2) for $R_{\mathrm{N}}(\Omega,\lambda)$ is obtained for domains Ω of class C, i.e., under the following assumption:

- every point $x\in\partial\Omega$ has a neighborhood U_x such that $\Omega\cap U_x$ coincides (in a suitable coordinate system) with the subgraph of a continuous function f_x.

If all the functions f_x satisfy the Hölder condition of order α, one says that Ω belongs to the class C^α. For domains $\Omega\in C^\alpha$ with $\alpha\in(0,1)$ our estimates $R_{\mathrm{D}}(\Omega,\lambda)=O(\lambda^{n-\alpha})$ and $R_{\mathrm{N}}(\Omega,\lambda)=O(\lambda^{(n-1)/\alpha})$ are order sharp in the scale C^α as $\lambda\to\infty$. The latter estimate implies that the Weyl formula holds for the Neumann Laplacian whenever $\alpha>1-\frac{1}{n}$. If $\alpha\leqslant 1-\frac{1}{n}$, then there exist domains in which the Weyl formula for $N_{\mathrm{N}}(\Omega,\lambda)$ fails (see Remark 4.2 for details or [11] for more advanced results).

For domains of class C^∞ our methods only give the known remainder estimate $R_{\mathrm{B}}(\Omega,\lambda)=O(\lambda^{n-1}\log\lambda)$. To obtain the order sharp estimate $O(\lambda^{n-1})$, one has to use more sophisticated techniques. The most advanced results in this direction were obtained in [7], where the estimate $R_{\mathrm{B}}(\Omega,\lambda)=O(\lambda^{n-1})$ was established for domains which belong to a slightly better class than C^1.

Throughout the paper, we use the following notation.

$d(x)$ is the Euclidean distance from the point $x\in\Omega$ to the boundary $\partial\Omega$;
$\Omega^{\mathrm{b}}_\delta:=\{x\in\Omega\,|\,d(x)\leqslant\delta\}$ is the internal closed δ-neighborhood of $\partial\Omega$;
$\Omega^{\mathrm{i}}_\delta:=\Omega\setminus\Omega^{\mathrm{b}}_\delta$ is the interior part of Ω.

1 Lower Bounds

Denote by $\Pi_{\mathrm{B}}(\lambda)$ the spectral projection of the operator $-\Delta_{\mathrm{B}}$ corresponding to the interval $[0,\lambda^2)$. Let $e_{\mathrm{B}}(x,y;\lambda)$ be its integral kernel (the so-called *spectral function*). It is well known that $e_{\mathrm{B}}(x,y;\lambda)$ is an infinitely differentiable function on $\Omega\times\Omega$ for each fixed λ and that $e_{\mathrm{B}}(x,x;\lambda)$ is a nondecreasing polynomially bounded function of λ for each fixed $x\in\Omega$.

By the spectral theorem, the cosine Fourier transform of $\frac{\mathrm{d}}{\mathrm{d}\lambda}e_{\mathrm{B}}(x,y;\lambda)$ coincides with the fundamental solution $u_{\mathrm{B}}(x,y;t)$ of the wave equation in Ω. On the other hand, due to the finite speed of propagation, $u_{\mathrm{B}}(x,x;t)$ is equal to $u_0(x,x;t)$ whenever $t\in(-d(x),d(x))$, where $u_0(x,y;t)$ is the fundamental solution of the wave equation in $\mathbb{R}^n$. By a direct calculation, $u_0(x,x;t)$ is independent of x and coincides with the cosine Fourier transform of the function $n\,(2\pi)^{-n}\omega_n\,\lambda_+^{n-1}$. Applying the Fourier Tauberian theorem proved in [12], we obtain

$$|e_{\rm B}(x,x;\lambda)-(2\pi)^{-n}\omega_n\,\lambda^n| \leqslant \frac{2n(n+2)^2\,(2\pi)^{-n}\omega_n}{d(x)}\left(\lambda+\frac{(n+2)\,\sqrt[n+2]{3}}{d(x)}\right)^{n-1} \tag{1.1}$$

for all $x\in\Omega$ and $\lambda>0$ [12, Corollary 3.1]. Since

$$N_{\rm B}(\Omega,\lambda)=\int_\Omega e_{\rm B}(x,x;\lambda)\,{\rm d}x \geqslant \int_{\Omega^{\rm i}_\delta} e_{\rm B}(x,x;\lambda)\,{\rm d}x$$

for all $\delta>0$, integrating (1.1) over $\Omega^{\rm i}_{\lambda^{-1}}$, we arrive at

$$R_{\rm B}(\lambda,\Omega)\geqslant -2n(n+2)^2\,(2\pi)^{-n}\omega_n\bigl(1+(n+2)\,\sqrt[n+2]{3}\bigr)^{n-1}\,\lambda^{n-1}\int_{\Omega^{\rm i}_{\lambda^{-1}}}\frac{{\rm d}x}{d(x)}\,.$$

Estimating constants and taking into account the obvious inequality

$$\int_{\Omega^{\rm i}_\delta}\frac{{\rm d}x}{d(x)}=\int_\delta^\infty s^{-1}\,{\rm d}(|\Omega^{\rm b}_s|)\leqslant\int_0^{\delta^{-1}}|\Omega^{\rm b}_{t^{-1}}|\,{\rm d}t\,,$$

we see that

$$\begin{aligned} R_{\rm B}(\lambda,\Omega) &\geqslant -\,C_{n,1}\,\lambda^{n-1}\int_{\lambda^{-1}}^\infty s^{-1}\,{\rm d}(|\Omega^{\rm b}_s|)\\ &\geqslant -\,C_{n,1}\,\lambda^{n-1}\int_0^\lambda |\Omega^{\rm b}_{t^{-1}}|\,{\rm d}t \end{aligned}\tag{1.2}$$

for all $\lambda>0$, where $C_{n,1}:=\dfrac{2\,(n+2)^{n+1}}{\pi^{n/2}\,\Gamma(n/2)}$ and Γ is the gamma-function.

2 Variational Formulas

In order to obtain upper bounds for $R_{\rm B}(\lambda,\Omega)$, we need to estimate the contribution of $\Omega^{\rm b}_\delta$. For the Neumann Laplacian

$$\int_{\Omega^{\rm b}_\delta} e_{\rm N}(x,x;\lambda)\,{\rm d}x$$

may well not be polynomially bounded, even if $\Omega \in C$. In this case, the Fourier Tauberian theorems are not applicable. Instead, we use the variational technique.

The idea is to represent Ω as the union of relatively simple domains and estimate the counting function for each of these domains. Then upper bounds for $N_{\mathrm{B}}(\lambda, \Omega)$ are obtained with the use of the following two lemmas.

Let $N_{\mathrm{N,D}}(\widetilde{\Omega}, \Upsilon, \lambda)$ be the counting function of the Laplacian on $\widetilde{\Omega}$ with Dirichlet boundary condition on $\Upsilon \subset \partial\widetilde{\Omega}$ and Neumann boundary condition on $\partial\widetilde{\Omega} \setminus \Upsilon$.

Lemma 2.1. If $\{\Omega_i\}$ is a countable family of disjoint open sets $\Omega_i \subset \Omega$ such that $|\Omega| = |\cup_i \Omega_i|$, then

$$\sum_i N_{\mathrm{D}}(\Omega_i, \lambda) \leqslant N_{\mathrm{D}}(\Omega, \lambda) \leqslant N_{\mathrm{N}}(\Omega, \lambda) \leqslant \sum_i N_{\mathrm{N}}(\Omega_i, \lambda)$$

and

$$N_{\mathrm{N}}(\Omega, \lambda) \geqslant \sum_i N_{\mathrm{N,D}}(\Omega_i, \partial\Omega_i \setminus \partial\Omega, \lambda).$$

Proof. It is an elementary consequence of the Rayleigh–Ritz formula. □

Given a collection of sets $\{\Omega_j\}$, let us denote by $\aleph\{\Omega_j\}$ the multiplicity of the covering $\{\Omega_j\}$, i.e., the maximal number of the sets Ω_j containing a common element.

Lemma 2.2. Let $\{\Omega_j\}$ be a countable family of open sets $\Omega_j \subset \Omega$ such that $|\Omega| = |\cup_j \Omega_j|$, and let $\aleph\{\Omega_j\} \leqslant \varkappa < +\infty$. If $\Upsilon \subset \partial\Omega$ and $\Upsilon_j := \partial\Omega_j \bigcap \Upsilon$, then

$$N_{\mathrm{N,D}}(\Omega, \Upsilon, \varkappa^{-1/2}\lambda) \leqslant \sum_j N_{\mathrm{N,D}}(\Omega_j, \Upsilon_j, \lambda).$$

Proof. See [11, Lemma 2.2]. □

Remark 2.1. Lemmas 2.1 and 2.2 remain valid for more general differential operators. This allows one to extend our results to some classes of higher order operators [11].

3 Partitions of Ω

The following theorem is due to H. Whitney.

Theorem 3.1. There exists a countable family $\{Q_{i,m}\}_{m\in\mathcal{M}_i,\, i\in\mathcal{I}}$ of mutually disjoint open n-dimensional cubes $Q_{i,m}$ with edges of length 2^{-i} such that

$$\overline{\Omega} = \bigcup_{i \in \mathcal{I}} \bigcup_{m \in \mathcal{M}_i} \overline{Q_{i,m}} \quad \text{and} \quad Q_{i,m} \subset \left(\Omega^{\mathrm{b}}_{4\delta_i} \setminus \Omega^{\mathrm{b}}_{\delta_i}\right)$$

where $\delta_i := \sqrt{n}\,2^{-i}$, $\mathcal{I}$ is a subset of $\mathbb{Z}$, and $\mathcal{M}_i$ are some finite index sets.

Proof. See, for example, [13, Chapter VI]. □

Lemma 3.1. For every $\delta > 0$ there exists a finite family of disjoint open sets $\{M_k\}$ such that

(i) each set M_k coincides with the intersection of Ω and an open n-dimensional cube with edges of length δ;
(ii) $\Omega^{\mathrm{b}}_{\delta_0} \subset \bigcup_k \overline{M_k} \subset \Omega^{\mathrm{b}}_{\delta_1} \bigcup \partial\Omega$, where $\delta_0 := \delta/\sqrt{n}$ and $\delta_1 := \sqrt{n}\,\delta + \delta/\sqrt{n}$.

Proof. Consider an arbitrary covering of $\mathbb{R}^n$ by cubes with disjoint interiors of size δ and select the cubes which have nonempty intersections with Ω. □

Theorem 3.1 and Lemma 3.1 imply that Ω can be represented (modulo a set of measure zero) as the union of Whitney cubes and the subsets M_k lying in cubes of size δ. This is sufficient to estimate $R_{\mathrm{D}}(\lambda, \Omega)$. However, the condition (i) of Lemma 3.1 does not imply any estimates for $N_{\mathrm{N}}(\lambda, M_k)$. In order to obtain an upper bound for $R_{\mathrm{N}}(\lambda, \Omega)$, one has to consider a more sophisticated partition of Ω.

If Ω' is an open $(d-1)$-dimensional set and f is a continuous real-valued function on the closure $\overline{\Omega'}$, let

- $G_{f,b}(\Omega') := \{x \in \mathbb{R}^n \mid b < x_d < f(x'),\ x' \in \Omega'\}$, where b is a constant such that $\inf f > b$;
- $\operatorname{Osc}(f, \Omega') := \sup\limits_{x' \in \Omega'} f(x') - \inf\limits_{x' \in \Omega'} f(x')$;
- $\mathcal{V}_\delta(f, \Omega')$ be the maximal number of disjoint $(n-1)$-dimensional cubes $Q'_i \subset \Omega'$ such that $\operatorname{Osc}(f, Q'_i) \geqslant \delta$ for each i.

If $n = 2$, then, roughly speaking, $\mathcal{V}_\delta(f, \Omega')$ coincides with the maximal number of oscillations of f which are not smaller than δ. Further on,

- $\mathbf{V}(\delta)$ is the class of domains V which are represented in a suitable coordinate system in the form $V = G_{f,b}(Q')$, where Q' is an $(n-1)$-dimensional cube with edges of length not greater than δ, $f : \overline{Q'} \mapsto \mathbb{R}$ is a continuous function, $b = \inf f - \delta$, and $\operatorname{Osc}(f, Q') \leqslant \delta/2$;
- $\mathbf{P}(\delta)$ is the set of n-dimensional rectangles such that the length of the maximal edge does not exceed δ.

Assume that $\Omega \in C$. Then there is a finite collection of domains $\Omega_l \subset \Omega$ such that $\Omega_l = G_{f_l, b_l}(Q'_l) \in \mathbf{V}(\delta_l)$ with some $\delta_l > 0$ and $\partial\Omega \subset \bigcup_{l \in \mathcal{L}} \overline{\Omega_l}$. Let us fix such a collection, and set

- n_Ω is the number of the sets Ω_l;

- $\mathcal{V}_\delta(\Omega) := \max\{1, \mathcal{V}_\delta(f_1, Q_1'), \mathcal{V}_\delta(f_2, Q_2'), \ldots\}$;
- δ_Ω is the largest positive number such that $\Omega^{\mathrm{b}}_{\delta_\Omega} \subset \bigcup_{l\in\mathcal{L}} \Omega_l$, $\delta_\Omega \leqslant \operatorname{diam} Q_l'$, and $2\delta_\Omega \leqslant \inf f_l - b_l$ for all l.

Theorem 3.2. Let $\Omega \in C$. Then for each $\delta \in (0, \delta_\Omega]$ there exist finite families of sets $\{P_j\}$ and $\{V_k\}$ satisfying the following conditions:

(i) $P_j \in \mathbf{P}(\delta)$ and $V_k \in \mathbf{V}(\delta)$;

(ii) $\aleph\{P_j\} \leqslant 4^n\, n_\Omega$ and $\aleph\{V_k\} \leqslant 4^{n-1}\, n_\Omega$;

(iii) $\Omega^{\mathrm{b}}_{\delta_0} \subset \cup_{j,k} \left(\overline{P_j} \bigcup \overline{V_k}\right) \subset \Omega^{\mathrm{b}}_{\delta_1}$, where $\delta_0 := \delta/\sqrt{n}$ and $\delta_1 := \sqrt{n}\,\delta + \delta/\sqrt{n}$;

(iv) $\#\{V_k\} \leqslant 2^{3(n-1)} \left(3^{n-1}\, \mathcal{V}_{\delta/2}(\Omega) + n_\Omega\, \delta^{-n}\, |\Omega^{\mathrm{b}}_{\delta_1}|\right)$ and

$$\#\{P_j\} \leqslant 2^{3n-1} 3^{n-1}\, \delta^{-1} \int_{(2\operatorname{diam}\Omega)^{-1}}^{4/\delta} t^{-2}\, \mathcal{V}_{t^{-1}}(\Omega)\, \mathrm{d}t \;+\; 2^{3n}\, n^{n/2}\, n_\Omega\, \delta^{-n}\, |\Omega^{\mathrm{b}}_{\delta_1}|\,.$$

Proof. The theorem follows from [11, Corollary 3.8]. □

4 Upper Bounds

The counting functions of the Laplacian on Whitney cubes can be evaluated explicitly. For other domains introduced in the previous section the counting functions are estimated as follows.

Lemma 4.1. (i) If $P \in \mathbf{P}(\delta)$, then $N_{\mathrm{N}}(P, \lambda) = 1$ for all $\lambda \leqslant \pi\delta^{-1}$.

(ii) If $V \in \mathbf{V}(\delta)$, then $N_{\mathrm{N}}(V, \lambda) = 1$ for all $\lambda \leqslant (1 + 2\pi^{-2})^{-1/2}\delta^{-1}$.

(iii) If M is a subset of an n-dimensional cube Q with edges of length δ and $\Upsilon := \partial M \bigcap Q$, then

$$N_{\mathrm{N,D}}(M, \Upsilon, \lambda) = 0 \quad \textit{for all } \lambda \leqslant (2^{-1} - 2^{-1}\delta^{-n}|M|)^{1/2}\, \pi\delta^{-1}$$

and

$$N_{\mathrm{N,D}}(M, \Upsilon, \lambda) \leqslant 1 \quad \text{for all } \lambda \leqslant \pi\delta^{-1}.$$

Proof. See [11, Lemma 2.6]. □

Remark 4.1. The first result in Lemma 4.1(iii) is very rough. Much more precise results in terms of capacities were obtained in [9, Chapter 10, Section 1].

Applying Theorem 3.1 and Lemmas 2.1, 2.2, 3.1, 4.1 and putting $\delta = C\lambda^{-1}$ with an appropriate constant C, we obtain

$$R_{\mathrm{D}}(\Omega,\lambda) \leqslant 2^{7n}\, n^{2n}\, \lambda^{n-1} \int_0^{\lambda} |\Omega_{t^{-1}}^{\mathrm{b}}|\, \mathrm{d}t \quad \forall \lambda > 0\,. \tag{4.1}$$

Similarly, if $\Omega \in C$, then Theorems 3.1, 3.2 and Lemmas 2.1, 2.2, 4.1 imply

$$\begin{aligned} R_{\mathrm{N}}(\Omega,\lambda) \leqslant{}& 2^{7n}\, n_\Omega^{1/2} \lambda \int_{(2\,\mathrm{diam}\Omega)^{-1}}^{C_\Omega \lambda} t^{-2}\, \mathcal{V}_{t^{-1}}(\Omega)\, \mathrm{d}t \\ &+ 2^{8n}\, n^{2n}\, n_\Omega\, \lambda^{n-1} \int_0^{C_\Omega \lambda} |\Omega_{t^{-1}}^{\mathrm{b}}|\, \mathrm{d}t \end{aligned} \tag{4.2}$$

for all $\lambda \geqslant \delta_\Omega^{-1}$, where $C_\Omega := 2^{n+3}\, n_\Omega^{1/2}$ (see [11] for details). Note that

$$|\Omega_{t^{-1}}^{\mathrm{b}}| \leqslant 2^{2n-2}\, 3^n\, n_\Omega\, (\mathrm{diam}\Omega)^{d-1} t^{-1} + 2^{3n-3}\, 3^{2n}\, t^{-n}\, \mathcal{V}_{t^{-1}}(\Omega)$$

for all $t > 0$ [11, Lemma 4.3]. Therefore, (4.2) implies the estimate

$$R_{\mathrm{N}}(\Omega,\lambda) \leqslant C'_\Omega\, \lambda^{n-1} \left(\log \lambda + \int_{(2\,\mathrm{diam}\Omega)^{-1}}^{C'_\Omega \lambda} t^{-n}\, \mathcal{V}_{t^{-1}}(\Omega)\, \mathrm{d}t \right) \tag{4.3}$$

with a constant C'_Ω depending on Ω.

Remark 4.2. Assume that Ω belongs to the Hölder class C^α for some $\alpha \in (0,1)$. Then, by [11, Lemma 4.5], there are constants C'_1 and C'_2 such that

$$\mathcal{V}_{t^{-1}}(\Omega) \leqslant C'_1\, t^{(n-1)/\alpha} + C'_2\,.$$

Now, (1.2) and (4.2) imply that

$$R_{\mathrm{N}}(\Omega,\lambda) = O\big(\lambda^{(n-1)/\alpha}\big)\,, \qquad \lambda \to \infty\,.$$

This estimate is order sharp. More precisely, for each $\alpha \in (0,1)$ there exists a domain Ω with C^α-boundary such that $R_{\mathrm{N}}(\Omega,\lambda) \geqslant c\,\lambda^{(n-1)/\alpha}$ for all sufficiently large λ, where c is a positive constant [11, Theorem 1.10]. The inequalities (1.2) and (4.1) imply the well known estimate

$$R_{\mathrm{D}}(\Omega,\lambda) = O\left(\lambda^{n-\alpha}\right)\,, \qquad \lambda \to \infty\,.$$

It is obvious that $(n-1)/\alpha > n - \alpha$. Moreover, if $\alpha < 1 - n^{-1}$, then $(n-1)/\alpha > n$, which means that $R_{\mathrm{N}}(\Omega,\lambda)$ may grow faster than λ^n as $\lambda \to \infty$.

Remark 4.3. In a number of papers, estimates for $R_{\mathrm{D}}(\Omega,\lambda)$ were obtained in terms of the so-called upper Minkowski dimension and the corresponding Minkowski content of the boundary (see, for instance, [2, 3] or [5]). Our formulas (1.2) and (4.1) are universal and imply the known estimates.

5 Planar Domains

In the two-dimensional case, it is much easier to construct partitions of a domain Ω, since the intersection of Ω with any straight line consists of disjoint open intervals. This allows one to refine the above results. Throughout this section, we assume that $\Omega \subset \mathbb{R}^2$.

5.1 The Neumann Laplacian

Consider the domain

$$\Omega = G_\varphi := \{(x,y) \in \mathbb{R}^2 \mid 0 < x < 1, -1 < y < \varphi(x)\}, \tag{5.1}$$

where $\varphi : (0,1) \mapsto [0,+\infty]$ is a lower semicontinuous function such that $|G_\varphi| < \infty$ (this implies, in particular, that φ is finite almost everywhere). Note that Ω does not have to be bounded; the results of this subsection hold for unbounded domains of the form (5.1).

For each fixed $s > 0$ the intersection of G_φ with the horizontal line $\{y = s\}$ coincides with a countable collection of open intervals. Let us consider the open set $E(\varphi, s)$ obtained by projecting these intervals onto the horizontal axis $\{y = 0\}$,

$$E(\varphi,s) = \{x \in (0,1) \mid (x,s) \in G_\varphi\} = \bigcup_{j \in \Gamma(\varphi,s)} I_j\,,$$

where I_j are the corresponding open disjoint subintervals of $(0,1)$ and $\Gamma(\varphi,s)$ is an index set. It is obvious that $E(\varphi,s_2) \subset E(\varphi,s_1)$ whenever $s_2 > s_1$.

It turns out that the spectral properties of the Neumann Laplacian on G_φ are closely related to the following function, describing geometric properties of G_φ. Given $t \in \mathbb{R}_+$, let us denote

$$n(\varphi,t) = \sum_{k=1}^{+\infty} \#\left\{j \in \Gamma(\varphi,kt) \mid \mu(I_j) < 2\,\mu\big(I_j \bigcap E(\varphi,kt+t)\big)\right\},$$

where $\mu(\cdot)$ is the one-dimensional measure of the corresponding set. Note that $n(\varphi,t)$ may well be $+\infty$.

Recall that the first eigenvalue of the Neumann Laplacian is equal to zero and the corresponding eigenfunction is constant. If the rest of the spectrum is separated from 0 and lies in the interval $[\nu^2,\infty)$, then we have the so-called Poincaré inequality

$$\inf_{c\in\mathbb{R}} \|u-c\|^2_{L_2(\Omega)} \leqslant \nu^{-2}\,\|\nabla u\|^2_{L_2(\Omega)} \quad \forall\, u \in W^{2,1}(\Omega)\,,$$

where $W^{2,1}(\Omega)$ is the Sobolev space.

Theorem 5.1. The Poincaré inequality holds in $\Omega = G_\varphi$ if and only if there exists $t > 0$ such that $n(\varphi, t) = 0$. Moreover, there is a constant $C \geqslant 1$ independent of φ such that

$$C^{-1}(t_0 + 1) \leqslant \nu^{-2} \leqslant C\,(t_0 + 1),$$

where $t_0 := \inf\{t > 0 \mid n(\varphi, t) = 0\}$ and ν^{-2} is the best possible constant in the Poincaré inequality.

Proof. See [10, Theorem 1.2]. □

Theorem 5.2. The spectrum of Neumann Laplacian on G_φ is discrete if and only if $n(\varphi, t) < +\infty$ for all $t > 0$.

Proof. See [10, Corollary 1.4]. □

Theorem 5.3. Let $\Psi : [1, +\infty) \mapsto (0, +\infty)$ be a function such that

$$C^{-1} s^a \leqslant \frac{\Psi(s\,t)}{\Psi(t)} \leqslant C\, s^b \quad \forall s, t \geqslant 1,$$

where $a > 1$, $b \geqslant a$ and $C \geqslant 1$ are some constants. Then the following two conditions are equivalent.

(i) There exist constants $C_1 \geqslant 1$ and $\lambda_* > 0$ such that

$$C_1^{-1}\,\Psi(\lambda) \leqslant R_{\rm N}(G_\varphi, \lambda) \leqslant C_1\,\Psi(\lambda) \quad \forall \lambda \geqslant \lambda_*.$$

(ii) There exist constants $C_2 \geqslant 1$ and $t_* > 0$ such that

$$C_2^{-1}\,\Psi(t) \leqslant n(\varphi, t^{-1}) \leqslant C_2\,\Psi(t) \quad \forall t \geqslant t_*.$$

Proof. See [10, Theorem 1.6]. □

5.2 The Dirichlet Laplacian

Berry [1] conjectured that the Weyl formula for the Dirichlet Laplacian on a domain with rough boundary might contain a second asymptotic term depending on the fractal dimension of the boundary. This problems was investigated by a number of mathematicians and physicists and was discussed in many papers (see, for instance, [2, 5] and the references therein). To the best of our knowledge, positive results were obtained only for some special classes of domains (such as domains with model cusps and disconnected selfsimilar fractals). The following theorem justifies the conjecture for planar domains of class C.

Theorem 5.4. Let Ω be a planar domain of class C such that

$$|\Omega_\delta^{\mathrm{b}}| = C_1\delta^{\alpha_1} + \cdots + C_m\delta^{\alpha_m} + o(\delta^\beta), \qquad \delta \to 0,$$

where C_j, α_i and β are real constants such that $0 < \alpha_1 < \alpha_2 < \cdots < \alpha_m \leqslant \beta < 1$ and $\beta < (1+\alpha_1)/2$. Then

$$R_{\mathrm{D}}(\Omega, \lambda) = \tau_{\alpha_1} C_1 \lambda^{2-\alpha_1} + \cdots + \tau_{\alpha_m} C_m \lambda^{2-\alpha_m} + o(\lambda^{2-\beta}), \qquad \lambda \to \infty,$$

where τ_{α_j} is a constant depending only on α_j for each $j = 1, \ldots, m$.

Recall that the interior Minkowski content of order α of a planar domain Ω is defined as

$$M_\alpha^{\mathrm{int}}(\Omega) := c(\alpha) \lim_{\delta\to 0} \delta^{\alpha-2}\, |\Omega_\delta^{\mathrm{b}}| \tag{5.2}$$

provided that the limit exists. Here, $\alpha \in (0,2)$ and $c(\alpha)$ is a normalizing constant. Theorem 5.4 with $m = 1$ and $\alpha_1 = \beta = \alpha$ immediately implies the following assertion.

Corollary 5.1. If Ω is a planar domain of class C and $0 < M_\alpha^{\mathrm{int}}(\Omega) < +\infty$ for some $\alpha \in (1,2)$, then

$$\lim_{\lambda\to+\infty} R_{\mathrm{D}}(\Omega,\lambda)/\lambda^{2-\alpha} = \tau_\alpha\, M_\alpha^{\mathrm{int}}(\Omega),$$

where τ_α is a constant depending only on α.

The proof of Theorem 5.4 consists of two parts, geometric and analytic. The first part uses the technique developed in [10] and the following lemma about partitions of planar domains $\Omega \in C$.

Lemma 5.1. For every planar domain $\Omega \in C$ there exists a finite collection of open connected disjoint subsets $\Omega_i \subset \Omega$ and a set D such that

(i) $\Omega \subset ((\cup_i \Omega_i) \cup D) \subset \overline{\Omega}$;

(ii) D coincides with the union of a finite collection of closed line segments;

(iii) each set Ω_i is either a Lipschitz domain or is obtained from a domain given by (5.1) with a continuous function φ_i by translation, rotation and dilation.

The second, analytic part of the proof involves investigation of some one-dimensional integral operators.

6 Concluding Remarks and Open Problems

Remark 6.1. It is not clear how to obtain upper bounds for $N_{\mathrm{N}}(\Omega,\lambda)$ for general domains Ω. It is not just a technical problem; for instance, the Neumann Laplacian on the relatively simple planar domain Ω obtained from the square $(0,2)\times(0,2)$ by removing the line segments $\frac{1}{n}\times(0,1)$, $n=1,2,3\ldots$, has a nonempty essential spectrum.

Remark 6.2. It may be possible to extend and/or refine our results, using a combination of our variational approach with the technique developed by Ivrii [7].

Remark 6.3. There are strong reasons to believe that Theorem 5.4 cannot be extended to higher dimensions.

Finally, we draw reader's attention to the following open problems.

Problem 6.1. By Lemma 2.2, $N_{\mathrm{N}}(\Omega,\varkappa^{-1/2}\lambda)\leqslant\sum_j N_{\mathrm{N}}(\Omega_j,\lambda)$ for any finite family $\{\Omega_j\}$ of open sets $\Omega_j\subset\Omega$ such that $|\Omega|=|\cup_j\Omega_j|$ and $\aleph\{\Omega_j\}\leqslant\varkappa<+\infty$. It is possible that the better estimate

$$N_{\mathrm{N}}(\Omega,\lambda)\leqslant\sum_j N_{\mathrm{N}}(\Omega_j,\lambda)$$

holds. This conjecture looks plausible and is equivalent to the following statement: if $\Omega_1\subset\Omega$, $\Omega_2\subset\Omega$ and $\Omega\subset\Omega_1\bigcup\Omega_2$, then

$$N_{\mathrm{N}}(\Omega_1,\lambda)+N_{\mathrm{N}}(\Omega_2,\lambda)\geqslant N_{\mathrm{N}}(\Omega,\lambda).$$

Problem 6.2. It would be interesting to know whether the converse statement to Corollary 5.1 is true. Namely, assume that Ω is a planar domain of class C such that

$$R_{\mathrm{D}}(\Omega,\lambda)=C\,\lambda^{2-\alpha}+o(\lambda^{2-\alpha}),\qquad \lambda\to\infty,$$

with some constant C. Does this imply that the limit (5.2) exists and finite?

Problem 6.3. Is it possible to improve the estimate $R_{\mathrm{B}}(\Omega,\lambda)=O(\lambda^{n-1}\log\lambda)$ for Lipschitz domains? The variational methods are applicable to all domains Ω of class C but do not allow one to remove the $\log\lambda$, whereas Ivrii's technique gives the best possible result $R_{\mathrm{B}}(\Omega,\lambda)=O(\lambda^{n-1})$ but works only for Ω which are "logarithmically" better than Lipschitz domains.

References

1. Berry, M. V.: Some geometric aspects of wave motion: wavefront dislocations, diffraction catastrophes, diffractals. Geometry of the Laplace Operator. Proc. Sympos. Pure Math. **36**, 13–38 (1980)
2. Brossard, J., Carmona, R.: Can one hear the dimension of a fractal? Commun. Math. Phys. **104**, 103–122 (1986)
3. van den Berg, M., Lianantonakis, M.: Asymptotics for the spectrum of the Dirichlet Laplacian on horn-shaped regions. Indiana Univ. Math. J. **50**, 299–333 (2001)
4. Birman, M.S., Solomyak, M.Z.: The principal term of spectral asymptotics for "non-smooth" elliptic problems (Russian). Funkt. Anal. Pril. **4:4**, 1–13 (1970); English transl.: Funct. Anal. Appl. **4** (1971)
5. Fleckinger-Pellé, J., Vassiliev, D.: An example of a two-term asymptotics for the "counting function" of a fractal drum. Trans. Am. Math. Soc. **337:1**, 99–116 (1993)
6. Hempel, R., Seco, L., Simon, B.: The essential spectrum of Neumann Laplacians on some bounded singular domains. J. Funct. Anal. **102**, 448–483 (1991)
7. Ivrii, V.: Sharp spectral asymptotics for operators with irregular coefficients. II. Domains with boundaries and degenerations. Commun. Partial Differ. Equ. **28**, 103–128 (2003)
8. Maz'ya, V.G.: On Neumann's problem for domains with irregular boundaries (Russian). Sib. Mat. Zh. **9**, 1322-1350 (1968); English transl.: Sib. Math. J. **9**, 990–1012 (1968)
9. Maz'ya, V.G.: Sobolev Spaces. Springer, Berlin etc. (1985)
10. Netrusov, Y.: Sharp remainder estimates in the Weyl formula for the Neumann Laplacian on a class of planar regions. J. Funct. Anal. **250**, 21–41 (2007)
11. Netrusov, Y., Safarov, Y.: Weyl asymptotic formula for the Laplacian on domains with rough boundaries. Commun. Math. Phys. **253**, 481–509 (2005)
12. Safarov, Y.: Fourier Tauberian Theorems and applications. J. Funct. Anal. **185**, 111–128 (2001)
13. Stein, E.: Singular Integrals and Differentiability Properties of Functions. Princeton Univ. Press, Princeton (1970)

$W^{2,p}$-Theory of the Poincaré Problem

Dian K. Palagachev

Abstract We present some recent results regarding the $W^{2,p}$-theory of a degenerate oblique derivative problem for second order uniformly elliptic operators. The boundary operator is prescribed in terms of directional derivative with respect to a vector field $\boldsymbol{\ell}$ which is tangent to $\partial\Omega$ at the points of a nonempty set $\mathcal{E} \subset \partial\Omega$. Sufficient conditions are given ensuring existence, uniqueness and regularity of solutions in the L^p-Sobolev scales. Moreover, we show that the problem considered is of Fredholm type with index zero.

1 Introduction

The general aim of this survey is to present some recent results regarding the $W^{2,p}$-theory of the degenerate oblique derivative problem for second order elliptic equations known as the *Poincaré problem.* Precisely, let $\Omega \subset \mathbb{R}^n$, $n \geqslant 3$, be a bounded domain with smooth enough boundary $\partial\Omega$ for which $\boldsymbol{\nu}(x) = (\nu_1(x), \ldots, \nu_n(x))$ is the unit *outward* normal at the point $x \in \partial\Omega$. Given a unit vector field $\boldsymbol{\ell}(x) = \big(\ell^1(x), \ldots, \ell^n(x)\big)$ on $\partial\Omega$, we decompose it into a sum of tangential and normal components,

$$\boldsymbol{\ell}(x) = \boldsymbol{\tau}(x) + \gamma(x)\boldsymbol{\nu}(x) \quad \forall\, x \in \partial\Omega,$$

where $\boldsymbol{\tau}(x)$, $\boldsymbol{\tau}\colon \partial\Omega \to \mathbb{R}^n$, is the projection of $\boldsymbol{\ell}(x)$ on the tangent hyperplane to $\partial\Omega$ at the point $x \in \partial\Omega$, and $\gamma\colon \partial\Omega \to \mathbb{R}$ denotes the Euclidean inner product $\gamma(x) := \boldsymbol{\ell}(x) \cdot \boldsymbol{\nu}(x)$. Actually, the set of zeroes of the function $\gamma(x)$,

$$\mathcal{E} := \big\{x \in \partial\Omega\colon\ \gamma(x) = 0\big\}$$

Dian K. Palagachev
Department of Mathematics, Technical University of Bari, Via E. Orabona 4, 70125 Bari, Italy
e-mail: palaga@poliba.it

A. Laptev (ed.), *Around the Research of Vladimir Maz'ya III: Analysis and Applications*, International Mathematical Series 13, DOI 10.1007/978-1-4419-1345-6_10,

is the subset of the boundary where the field $\boldsymbol{\ell}$ becomes tangent to it.

Given a second order uniformly elliptic operator $\mathcal{L}$ with measurable in Ω coefficients, our goal is to study the oblique derivative problem

$$\begin{aligned} \mathcal{L}u &:= a^{ij}(x)D_{ij}u + b^i(x)D_iu + c(x)u = f(x) \quad \text{a.e. } \Omega, \\ \mathcal{B}u &:= \partial u/\partial \boldsymbol{\ell} + \sigma(x)u = \varphi(x) \qquad\qquad\qquad \text{on } \partial\Omega \end{aligned} \tag{$\mathcal{P}$}$$

in the *degenerate* case where $\mathcal{E} \not\equiv \varnothing$. We consider $(\mathcal{P})$ within the framework of the Sobolev scales $W^{2,p}(\Omega)$ for *each* value of $p > 1$, assuming as low as possible regularity of the coefficients of $\mathcal{L}$ and $\mathcal{B}$.

Tangential problems like $(\mathcal{P})$ arise naturally in mathematical models of determining the gravitational fields of the celestial bodies. Actually, it was Poincaré the first to arrive at a problem of that type in his studies on tides [29]. We refer the reader to [14], where a model of $(\mathcal{P})$ governs diffraction of ocean waves by islands, to [11] for the applications in scattering of large objects, and to [39] and [5] which link $(\mathcal{P})$ to the gas dynamics problems. The theory of stochastic processes is another area where $(\mathcal{P})$ models real phenomena (cf. [35]). Now the operator $\mathcal{L}$ describes analytically a strong Markov process with continuous paths in Ω, such as the Brownian motion, $\partial u/\partial \boldsymbol{\ell}$ corresponds to reflection along $\boldsymbol{\ell}$ on $\partial\Omega \backslash \mathcal{E}$ and to diffusion at the points of $\mathcal{E}$, while the term σu is usually associated to absorption phenomena.

It is well known from the general theory of PDEs that $(\mathcal{P})$ is *not* an elliptic boundary value problem. Actually, reduction of $(\mathcal{P})$ to a pseudodifferential equation on $\partial\Omega$ leads to a problem of principal type which is no more elliptic (cf. [13, 3]). A necessary and sufficient condition for $(\mathcal{P})$ to be a *regular elliptic* boundary value problem is that the couple of operators $(\mathcal{L}, \mathcal{B})$ satisfies the Shapiro–Lopatinskij complementary condition (cf. [1]) which means that $\boldsymbol{\ell}$ must be transversal to $\partial\Omega$ when $n \geqslant 3$ and $|\boldsymbol{\ell}| \neq 0$ as $n = 2$. If $\boldsymbol{\ell}$ becomes *tangent* to $\partial\Omega$, then $(\mathcal{P})$ is a *degenerate* problem and new effects occur in contrast to the regular case. The qualitative properties of $(\mathcal{P})$ depend strongly on the behavior of $\boldsymbol{\ell}$ near the set of tangency $\mathcal{E}$ and precisely on the way the normal component $\gamma\boldsymbol{\nu}$ of $\boldsymbol{\ell}$ changes or no its orientation with respect to $\partial\Omega$ along the trajectories of $\boldsymbol{\ell}$ when these cross $\mathcal{E}$. The general results were obtained by Hörmander [12], Egorov and Kondrat'ev [4], Eskin [6], Maz'ya [17], Maz'ya and Paneyah [18], Melin and Sjöstrand [19], Paneyah [26, 27] and precise details can be found in the monographs by Popivanov and Palagachev [33] and Paneyah [28], and in the survey of Popivanov [30]. All these studies were carried out within the framework of the Sobolev classes $H^s (\equiv H^{s,2})$ assuming C^∞-smooth data and this naturally involved pseudodifferential techniques and Hilbert space approach.

The simplest case occurs when $\gamma = \boldsymbol{\ell} \cdot \boldsymbol{\nu}$, even if vanishing on $\mathcal{E}$, conserves its sign on $\partial\Omega$ (Fig. 1 (a)). According to the physical interpretation of $(\mathcal{P})$ in the theory of Brownian motion (cf. [2, 33]) this is the case of *neutral* vector field $\boldsymbol{\ell}$ and the Poincaré problem $(\mathcal{P})$ is well-posed (in the sense of Hadamard) in the *appropriate* functional classes (cf. [12, 4]). Assume now

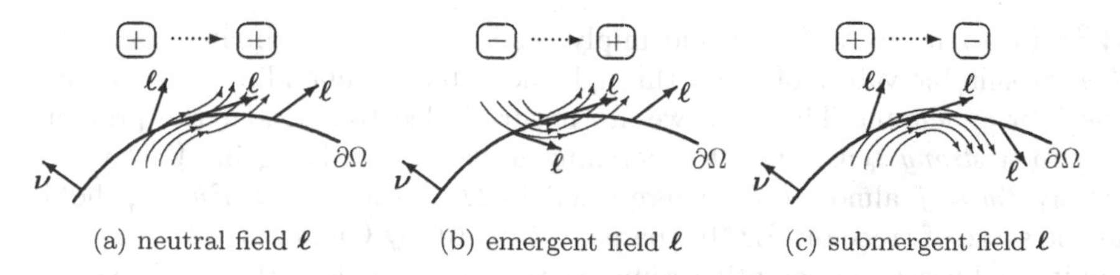

(a) neutral field $\boldsymbol{\ell}$ (b) emergent field $\boldsymbol{\ell}$ (c) submergent field $\boldsymbol{\ell}$

Fig. 1 The three types of the vector field $\boldsymbol{\ell}$.

that γ changes its sign from "−" to "+" in the positive direction along the $\boldsymbol{\tau}$-integral curves when these cross $\mathcal{E}$ (cf. Fig. 1 (b)). Then the field $\boldsymbol{\ell}$ is of *emergent* type, the tangency set $\mathcal{E}$ is called *attracting* and the problem $(\mathcal{P})$ has a *kernel of infinite dimension.* As is shown by Egorov and Kondrat'ev [4] in the case $\operatorname{codim}_{\partial\Omega}\mathcal{E} = 1$, we have to modify $(\mathcal{P})$ by prescribing a Dirichlet type boundary condition on $\mathcal{E}$ in order to get a well-posed problem. Finally, let γ changes its sign from "+" to "−" in the positive direction along the $\boldsymbol{\tau}$-integral curves (Fig 1 (c)). Now $\boldsymbol{\ell}$ is of *submergent* type and $\mathcal{E}$ corresponds to a *repellent* manifold. The problem $(\mathcal{P})$ has *infinite-dimensional cokernel* ([12]) and Maz'ya and Paneyah [18] were the first to propose a relevant modification of $(\mathcal{P})$ by violating the boundary condition at the points of $\mathcal{E}$, ($\operatorname{codim}_{\partial\Omega}\mathcal{E} = 1$). As consequence, a Fredholm problem arises, but the restriction $u|_{\partial\Omega}$ has a finite smooth jump at $\mathcal{E}$ which could be determined by the data of $(\mathcal{P})$.

In contrast to the case of regular oblique derivative problems when the solution gains two derivatives from f and one derivative from φ, the new feature of the Poincaré problem $(\mathcal{P})$ is that its solution *loses regularity* from the data f and φ no matter what type the vector field $\boldsymbol{\ell}$ is. That loss of smoothness is precisely characterized by the *subelliptic* estimates derived for the solutions of tangential problems (cf. Hörmander [12], Maz'ya and Paneyah [18], Winzell [37], Guan [8], Guan and Sawyer [9, 10]) and it depends on the *order of contact* k between $\boldsymbol{\ell}$ and $\partial\Omega$. Loosely speaking, the solution u of $(\mathcal{P})$ gains $2 - k/(k+1)$ derivatives from f and $1 - k/(k+1)$ derivatives from φ.

Regarding the topological structure of the set $\mathcal{E}$, it was assumed initially to be a submanifold of $\partial\Omega$ of codimension one. Melin and Sjöstrand [19] and Paneyah [26, 27] were the first to consider $(\mathcal{P})$ in the general situation when $\mathcal{E}$ is a subset of $\partial\Omega$ of positive $(n-1)$-dimensional Hausdorff measure which is *nontrapping* for the $\boldsymbol{\ell}$-trajectories, i.e., each integral curve of $\boldsymbol{\ell}$ through a point of $\mathcal{E}$ leaves it in a *finite time* in both directions. The results were extended by Winzell [36, 38] to the settings of Hölder's classes assuming $C^{1,\alpha}$-regularity of the coefficients of $\mathcal{L}$. Let us note that $\operatorname{meas}_{\partial\Omega}\mathcal{E} > 0$ automatically implies an *infinite* order of contact between $\boldsymbol{\ell}$ and $\partial\Omega$, and therefore gain of one derivative from f and zero derivative from φ for the solutions to $(\mathcal{P})$.

When dealing with nonlinear Poincaré problems, however, we have to dispose of precise information on the linear problem $(\mathcal{P})$ with coefficients less regular than C^∞ (cf. [20, 31, 33, 15, 34, 32]). Indeed, *a priori* estimates in

$W^{2,p}$ for solutions to $(\mathcal{P})$ would imply easily pointwise estimates for u and Du for suitable values of $p > 1$ through the Sobolev imbedding theorem and the Morrey lemma. This way, we are naturally led to consider the problem $(\mathcal{P})$ in a *strong* sense, i.e., to searching for solutions lying in $W^{2,p}$ which satisfy $\mathcal{L}u = f$ almost everywhere (a.e.) in Ω and for which $\mathcal{B}u = \varphi$ holds in the sense of trace on $\partial\Omega$. In the papers [9, 10] by Guan and Sawyer, solvability and precise subelliptic estimates were obtained for $(\mathcal{P})$ in $H^{s,p}$-spaces ($\equiv W^{s,p}$ for integer s!). However, the pseudodifferential technique involved in [9] requires C^{∞}-smooth coefficients, while in [10] the coefficients of $\mathcal{L}$ are $C^{0,\alpha}$-smooth, but the field $\boldsymbol{\ell}$ is supposed to be of finite type, and this automatically excludes the possibility to deal with sets $\mathcal{E}$ of *positive* surface measure.

In this survey, we present some recent results regarding the $W^{2,p}$-theory of the *neutral* and the *emergent* Poincaré problem $(\mathcal{P})$ for *arbitrary* $p \in (1,\infty)$. Our general goal is to built a relevant solvability theory weakening both the Winzell assumptions on the $C^{1,\alpha}$-regularity of the coefficients of $\mathcal{L}$ and these of Guan and Sawyer on the *finite type* of $\boldsymbol{\ell}$. Indeed, the loss of smoothness for the degenerate problems already mentioned requires some more regularity of the data near the set of tangency $\mathcal{E}$. Precisely, we suppose essential boundedness near $\mathcal{E}$ for the $\boldsymbol{\ell}$-directional derivatives of the coefficients of $\mathcal{L}$, whereas discontinuity controlled in VMO is allowed for a^{ij} away from $\mathcal{E}$, and b^i, $c \in L^{\infty}$ there. Later on, $\boldsymbol{\ell}$ is a Lipschitz vector field on $\partial\Omega$ with Lipschitz continuous first derivatives near $\mathcal{E}$, with *no restrictions* imposed on its order of contact with $\partial\Omega$. Regarding the structure of the tangency set $\mathcal{E}$, it could be a quite general subset of $\partial\Omega$ of *positive* $(n-1)$-dimensional Hausdorff measure, which is subordinated *only* to the *nontrapping* condition that all trajectories of $\boldsymbol{\ell}$ through the points of $\mathcal{E}$ are nonclosed and leave $\mathcal{E}$ in a finite time in both directions.

We start with *a priori* estimate for the $W^{2,p}(\Omega)$-solutions to $(\mathcal{P})$ with *arbitrary* $p \in (1,\infty)$. The technique adopted in the *neutral* case is based on a dynamical system approach employing the fact that $\partial u/\partial\boldsymbol{\ell}$ is a local solution near $\mathcal{E}$ of a Dirichlet problem with right-hand side depending on the solution u itself. Application of the L^p-estimates for such problems leads to an estimate for the $W^{2,p}$-norms of u over a family of subdomains which, starting *away* from $\mathcal{E}$, evolve along the $\boldsymbol{\ell}$-trajectories and exhaust a sort of their tubular neighborhoods. The desired *a priori* bound follows through suitable iteration with respect to the curvilinear parameter on the trajectories of $\boldsymbol{\ell}$.

For what concerns the *emergent* Poincaré problem, it has a kernel of *infinite dimension,* as mentioned above, and in order to get a well-posed problem one has to prescribe the values of u over a *codimension-one* submanifold $\mathcal{E}_0 \subset \mathcal{E}$ of $\partial\Omega$ such that $\boldsymbol{\ell}$ is *transversal* to $\mathcal{E}_0$. This way, we are led to consider the modified Poincaré problem

$$\begin{aligned} &\mathcal{L}u = f(x) && \text{a.e. } \Omega, \\ &\mathcal{B}u = \varphi(x) \quad \text{on } \partial\Omega, \quad u = \mu(x) && \text{on } \mathcal{E}_0 \end{aligned} \qquad (\mathcal{MP})$$

instead of $(\mathcal{P})$. To derive the $W^{2,p}(\Omega)$-estimates for the solutions of $(\mathcal{MP})$, we make use of the fact that on the parts of $\partial\Omega$ *away* from $\mathcal{E}_0$, where γ is nonnegative/nonpositive, $(\mathcal{MP})$ behaves as a *neutral* problem and therefore the $W^{2,p}(\Omega)$-estimates from the *neutral* case are available. For what concerns the layer near $\mathcal{E}_0$, $\boldsymbol{\ell}$ is transversal to $\mathcal{E}_0$ and it turns out that each point x close to $\mathcal{E}_0$ could be reached from a unique point $x' \in \mathcal{E}_0$ through an integral curve of the field $\boldsymbol{\ell}$. An integration of $\partial u/\partial \boldsymbol{\ell}$ along that curve expresses $u(x)$ in terms of $u(x')$ and integral of $\partial u/\partial \boldsymbol{\ell}$ over the intermediate arc connecting x' and x. The supplementary condition $u|_{\mathcal{E}_0} = \mu$ provides a $W^{2,p}(\mathcal{E}_0)$-estimate for the restriction $u|_{\mathcal{E}_0}$ which solves a uniformly elliptic Dirichlet problem over the manifold $\mathcal{E}_0$, while $\partial u/\partial \boldsymbol{\ell}$ is a local solution to a Dirichlet-type problem with right-hand side depending on u, and it estimates as in the *neutral* case.

Regarding the *transversality* of $\boldsymbol{\ell}$ to $\mathcal{E}_0$ in the *emergent* case $(\mathcal{MP})$, let us note the very deep paper of Maz'ya [17], where the field $\boldsymbol{\ell}$ can be tangent also to $\mathcal{E}_0$ at the points of lower-dimensional submanifolds. Taking account of the degenerate character of $(\mathcal{P})$, the author proposed an adequate weak formulation of $(\mathcal{P})$ and even solved a local-analysis-problem posed by V.I. Arnold. Maz'ya's technique, however, seems inapplicable to our situation because of its strong dependence on the C^∞ structure of $(\mathcal{P})$.

Another advantage of the dynamical system approach used in deriving the *a priori* bounds is the *improving-of-integrability* property obtained for the solutions of $(\mathcal{P})$ and $(\mathcal{MP})$. Roughly speaking, it means that the Poincaré problem, even if a *degenerate* one, behaves as an *elliptic* problem for what concerns the degree p of integrability. In other words, if $p \in [q, \infty)$ and $u \in W^{2,q}(\Omega)$ is a strong solution to $(\mathcal{P})$ or $(\mathcal{MP})$ with $f \in L^p(\Omega)$ and $\partial f/\partial \boldsymbol{\ell} \in L^p$ near $\mathcal{E}$, $\varphi \in W^{1-1/p,p}(\partial\Omega)$ and $\varphi \in W^{2-1/p,p}$ near $\mathcal{E}$, and $\mu \in W^{2-1/p,p}(\mathcal{E}_0)$ in the case of $(\mathcal{MP})$, then $u \in W^{2,p}(\Omega)$.

The importance of the improving-of-integrability property is indispensable in deriving unicity of the $W^{2,p}(\Omega)$-solutions to both the *neutral* and *emergent* Poincaré problem when $c(x) \leqslant 0$ a.e. in Ω. In fact, the difference u of any two such solutions solves the homogeneous problem and therefore $u \in W^{2,q}(\Omega)$ for *all* $q > 1$. In particular, $u \in W^{2,n}(\Omega)$ whence $u \equiv 0$ by means of the Aleksandrov maximum principle. The uniqueness established leads to refined *a priori* estimates for the solutions to the Poincaré problem and turns out to be a sufficient condition ensuring strong solvability in $W^{2,p}(\Omega)$ for *each* $p > 1$ when $c(x) \leqslant 0$ a.e. in Ω. Moreover, invoking the Riesz–Schauder theory, it follows that $(\mathcal{P})$ and $(\mathcal{MP})$ are problems of Fredholm type with index zero.

We present here mainly the results from [22] and [25], where the Poincaré problem is studied in $W^{2,p}(\Omega)$ for each value of $p \in (1, \infty)$ and general, *nontrapping* structure of the tangency set $\mathcal{E}$. The reader is referred to [16], where the *neutral* problem $(\mathcal{P})$ is studied in $W^{2,p}(\Omega)$, $p > n$, by the method of *elliptic regularization,* and to [21] for a constructive approach to the *emergent* problem $(\mathcal{MP})$ in the case $\mathcal{E} \equiv \mathcal{E}_0$, i.e., when the whole set of tangency is a codimension-one submanifold of $\partial\Omega$.

2 A Priori Estimates

Hereafter, we adopt the standard summation convention on repeated indices and $D_i := \partial/\partial x_i$, $D_{ij} := \partial^2/\partial x_i \partial x_j$. The class of functions with Lipschitz continuous k-th order derivatives is denoted by $C^{k,1}$, $W^{k,p}$ stands for the Sobolev space of functions with L^p-summable weak derivatives up to order $k \in \mathbb{N}$ and normed by $\|\cdot\|_{W^{k,p}}$, while $W^{s,p}(\partial\Omega)$ with $s > 0$ noninteger, $p \in (1, +\infty)$, is the fractional order Sobolev space. The Sarason class of functions with vanishing mean oscillation is denoted as usual by $VMO(\Omega)$. We use the standard parametrization $t \mapsto \psi_{\boldsymbol{L}}(t, x)$ for the *trajectory* (*maximal integral curve*) of a given vector field $\boldsymbol{L}$ passing through the point x, i.e., $\partial_t \psi_{\boldsymbol{L}}(t, x) = \boldsymbol{L} \circ \psi_{\boldsymbol{L}}(t, x)$ and $\psi_{\boldsymbol{L}}(0, x) = x$.

We fix hereafter $\Sigma \subset \overline{\Omega}$ to be a *closed* neighborhood of $\mathcal{E}$ in $\overline{\Omega}$ and assume:

- *Uniform ellipticity of the operator $\mathcal{L}$:* there is a constant $\lambda > 0$ such that

$$\lambda^{-1}|\xi|^2 \leqslant a^{ij}(x)\xi_i\xi_j \leqslant \lambda|\xi|^2, \quad a^{ij}(x) = a^{ji}(x) \tag{2.1}$$

  for almost all (a.a.) $x \in \Omega$ and all $\xi \in \mathbb{R}^n$;
- *Regularity of the data:*

$$\begin{aligned} &a^{ij} \in VMO(\Omega) \cap C^{0,1}(\Sigma), \quad b^i, c \in L^\infty(\Omega) \cap C^{0,1}(\Sigma), \\ &\ell^i, \sigma \in C^{0,1}(\partial\Omega) \cap C^{1,1}(\partial\Omega \cap \Sigma), \\ &\partial\Omega \in C^{1,1}, \ \partial\Omega \cap \Sigma \in C^{2,1}; \end{aligned} \tag{2.2}$$

- *Non-trapping condition:*

$$\begin{aligned} &\text{the arcs of the } \boldsymbol{\tau}\text{-trajectories lying in } \mathcal{E} \\ &\text{are all nonclosed and of finite length,} \end{aligned} \tag{2.3}$$

  i.e., the integral curves of the field $\boldsymbol{\ell}$, which coincide with these of $\boldsymbol{\tau}$ on $\mathcal{E}$, leave the set of tangency $\mathcal{E}$ in a *finite time* in both directions.

Regarding the behavior of the vector field $\boldsymbol{\ell}$ near the set of tangency $\mathcal{E}$, we suppose either

- *Neutral type of the vector field $\boldsymbol{\ell}$:*

$$\gamma(x) = \boldsymbol{\ell}(x) \cdot \boldsymbol{\nu}(x) \geqslant 0 \quad \forall x \in \partial\Omega \tag{2.4}$$

  which simply means that $\boldsymbol{\ell}(x)$ is either tangent to $\partial\Omega$ or is directed outwards Ω at each point $x \in \partial\Omega$;

or

- *Emergent type of the field $\boldsymbol{\ell}$:*

$$\begin{aligned}
&\partial\Omega = \partial\Omega^+ \cup \partial\Omega^- \text{ with closed } \partial\Omega^{\pm} := \{x \in \partial\Omega\colon\ \gamma(x) \gtreqless 0\},\\
&\mathcal{E}_0 := \partial\Omega^+ \cap \partial\Omega^- \subset \mathcal{E}, \quad \operatorname{codim}_{\partial\Omega}\mathcal{E}_0 = 1, \mathcal{E}_0 \in C^{2,1},\\
&\boldsymbol{\ell}(x) \text{ is strictly transversal to } \mathcal{E}_0 \text{ and}\\
&\text{points from } \partial\Omega^- \text{ into } \partial\Omega^+ \text{ for each } x \in \mathcal{E}_0.
\end{aligned} \tag{2.5}$$

Actually, $\partial\Omega^+$ is the set of all boundary points x, where the field $\boldsymbol{\ell}(x)$ points *outwards* Ω or is tangent to $\partial\Omega$, while $\boldsymbol{\ell}$ is tangent or directed *inward* Ω on $\partial\Omega^-$, and the tangency set $\mathcal{E}$ may contain points both from $\partial\Omega^+$ and $\partial\Omega^-$. In other words, given a point $x \in \mathcal{E}$ we dispose of a large amount of freedom for what concerns its assignment to $\partial\Omega^+$ or $\partial\Omega^-$. In this sense, the definitions of $\partial\Omega^+$ and $\partial\Omega^-$ are rather flexible and have to respect *only* the requirement that the field $\boldsymbol{\ell}$ is *transversal* to their common boundary $\mathcal{E}_0$.

An important outgrowth of the nontrapping condition (2.3) is the *length boundedness* of the $\boldsymbol{\tau}$-integral curves contained in $\mathcal{E}$. In facts, it follows from (2.3), the compactness of $\mathcal{E}$ and the semi-continuity properties of the lengths of the $\boldsymbol{\tau}$-maximal integral curves that *there is a finite upper bound* $\varkappa_0$ *for the arclengths of the* $\boldsymbol{\tau}$*-trajectories lying in* $\mathcal{E}$ (cf. [38, Proposition 3.1] and [33, Proposition 3.2.4] for details).

In what follows, we will use a suitable extension of the field $\boldsymbol{\ell}$ in a neighborhood of $\partial\Omega$. For each $x \in \mathbb{R}^n$ and close enough to $\partial\Omega$ set $d(x) = \operatorname{dist}(x, \partial\Omega)$ for the distance function and define $\Gamma := \{x \in \mathbb{R}^n\colon\ d(x) \leqslant d_0\}$ with $d_0 > 0$ sufficiently small (note that Γ contains also points exterior to Ω!). Thus, to each $x \in \Gamma$ there corresponds a unique $y(x) \in \partial\Omega$ closest to x, $d(x) = |x - y(x)|$, $\big(d(x)\big)^2 \in C^{1,1}(\Gamma)$, and $y(x) \in C^{0,1}(\Gamma)$ while $y(x) \in C^{1,1}(\Gamma \cap \Sigma)$ (cf. [7, Chapter 14]). We define

$$\boldsymbol{L}(x) := \begin{cases} \dfrac{\boldsymbol{\ell}(y(x)) + (d(x))^2\boldsymbol{\nu}(y(x))}{|\boldsymbol{\ell}(y(x)) + (d(x))^2\boldsymbol{\nu}(y(x))|} & \forall x \in \Gamma \quad \text{if } \boldsymbol{\ell} \text{ is neutral},\\[2ex] \boldsymbol{\ell}(y(x)) & \forall x \in \Gamma \quad \text{if } \boldsymbol{\ell} \text{ is emergent}. \end{cases}$$

It is clear that the extended field $\boldsymbol{L}$ preserves the regularity properties of $\boldsymbol{\ell}$. Moreover, $\boldsymbol{L}$ is *strictly transversal* to $\partial(\Omega \setminus \Gamma)$ in the *neutral* case and each point of Γ can be reached from $\partial(\Omega\setminus\Gamma)\cap\Omega$ through an $\boldsymbol{L}$-trajectory of length at most $\varkappa = \varkappa(\varkappa_0) = \text{const} > 0$. Regarding the *emergent* case, the extended field $\boldsymbol{L}$ is *strictly transversal* to the $C^{1,1}$-smooth, $(n-1)$-dimensional manifold

$$\mathcal{N} := \{x \in \Omega\colon\ y(x) \in \mathcal{E}_0\}$$

formed by the *inward* normals to $\partial\Omega$ at the points of $\mathcal{E}_0$.

In order to state our main results, we need to introduce special functional spaces which take into account the higher regularity of the data of the Poincaré problem near the tangency set $\mathcal{E}$. For any $p \in (1, \infty)$ define the Banach spaces

$$\mathcal{F}^p(\Omega,\Sigma) := \{f \in L^p(\Omega):\ \partial f/\partial \mathbf{L} \in L^p(\Sigma)\}$$

normed by $\|f\|_{\mathcal{F}^p(\Omega,\Sigma)} := \|f\|_{L^p(\Omega)} + \|\partial f/\partial \mathbf{L}\|_{L^p(\Sigma)}$ and

$$\Phi^p(\partial\Omega,\Sigma) := \left\{\varphi \in W^{1-1/p,p}(\partial\Omega):\ \varphi \in W^{2-1/p,p}(\partial\Omega \cap \Sigma)\right\}$$

equipped with the norm

$$\|\varphi\|_{\Phi^p(\partial\Omega,\Sigma)} := \|\varphi\|_{W^{1-1/p,p}(\partial\Omega)} + \|\varphi\|_{W^{2-1/p,p}(\partial\Omega\cap\Sigma)}.$$

We use also the space

$$\mathcal{W}^{2,p}(\Omega,\Sigma) := \{u \in W^{2,p}(\Omega):\ \partial u/\partial \mathbf{L} \in W^{2,p}(\Sigma)\}$$

with the norm $\|u\|_{\mathcal{W}^{2,p}(\Omega,\Sigma)} := \|u\|_{W^{2,p}(\Omega)} + \|\partial u/\partial \mathbf{L}\|_{W^{2,p}(\Sigma)}$.

Hereafter, when referring to the Poincaré problem we mean $(\mathcal{P})$ in case of *neutral* field $\boldsymbol{\ell}$ and the modified problem $(\mathcal{MP})$ if $\boldsymbol{\ell}$ is of *emergent* type. The following is our first result asserting *improving-of-integrability* and providing an L^p-*a priori estimate* for the strong solutions to the Poincaré problem.

Theorem 2.1. *Suppose* (2.1)–(2.3) *together with* (2.4) *if $\boldsymbol{\ell}$ is of neutral type and* (2.5) *if $\boldsymbol{\ell}$ is emergent. Let $u \in W^{2,q}(\Omega)$ be a strong solution to $(\mathcal{P})$ in the neutral case, and to $(\mathcal{MP})$ in the emergent one with $f \in \mathcal{F}^p(\Omega,\Sigma)$, $\varphi \in \Phi^p(\partial\Omega,\Sigma)$, $\mu \in W^{2-1/p,p}(\mathcal{E}_0)$, and $1 < q \leqslant p < \infty$.*

Then $u \in W^{2,p}(\Omega)$ and there is a constant C, depending on the data of the problem only, such that

$$\|u\|_{W^{2,p}(\Omega)} \leqslant \begin{cases} CK(u,f,\varphi) & \text{if } \boldsymbol{\ell} \text{ is neutral,} \\ C\left(K(u,f,\varphi) + \|\mu\|_{W^{2-1/p,p}(\mathcal{E}_0)}\right) & \text{if } \boldsymbol{\ell} \text{ is emergent,} \end{cases} \tag{2.6}$$

where $K(u,f,\varphi) := \|u\|_{L^p(\Omega)} + \|f\|_{\mathcal{F}^p(\Omega,\Sigma)} + \|\varphi\|_{\Phi^p(\partial\Omega,\Sigma)}$.

The complete proofs of Theorem 2.1 are given in [22] and [25] for the neutral and the emergent case respectively, while [21, 23] and [24] contain simplified versions in the situation when the whole set of tangency $\mathcal{E}$ is a codimension-one submanifold of $\partial\Omega$. We restrict ourselves here to give only a general idea of the proofs.

Let $\Sigma' \subset \Sigma'' \subset \Sigma$ be closed neighborhoods of $\mathcal{E}$ in $\overline{\Omega}$ with Σ'' so "narrow" that $\Sigma'' \subset \Gamma$. Taking into account (2.1), (2.2) and the fact that $\gamma(x) \neq 0$ for all $x \in \partial\Omega \setminus \Sigma'$, we get that both $(\mathcal{P})$ and $(\mathcal{MP})$ are *regular* oblique derivative problems in $\Omega\setminus\Sigma'$ for the uniformly elliptic operator $\mathcal{L}$ with VMO principal coefficients, and therefore the L^p-theory of such problems (cf. [15, Theorem 2.3.1]) yields $u \in W^{2,p}(\Omega \setminus \Sigma')$ and

$$\|u\|_{W^{2,p}(\Omega\setminus\Sigma')} \leqslant C\left(\|u\|_{L^p(\Omega)} + \|f\|_{L^p(\Omega)} + \|\varphi\|_{W^{1-1/p,p}(\partial\Omega)}\right). \tag{2.7}$$

Remark 2.1. Let us note that the directional derivative $\partial u/\partial \boldsymbol{L}$ of any $W^{2,q}$-solution to $(\mathcal{P})$ is a $W^{2,q}(\Sigma)$-function. In fact, $u \in W^{2,q}$ implies $\partial u/\partial \boldsymbol{L} \in W^{1,q}(\Sigma)$ and taking the difference quotients in $\boldsymbol{L}$-direction, the regularity theory of the generalized solutions to uniformly elliptic equations (cf. [7, Chapter 8 and Lemma 7.24]) yields $\partial u/\partial \boldsymbol{L} \in W^{2,q}(\Sigma)$. Moreover, $\partial u/\partial \boldsymbol{L}$ is a strong local solution to the Dirichlet problem

$$\begin{cases} \mathcal{L}\left(\partial u/\partial \boldsymbol{L}\right) = \partial u/\partial \boldsymbol{L} + 2a^{ij} D_j L^k D_{ki} u + (a^{ij} D_{ij} L^k + b^i D_i L^k) D_k u \\ \qquad\qquad - \left(\partial a^{ij}/\partial \boldsymbol{L}\right) D_{ij} u - \left(\partial b^i/\partial \boldsymbol{L}\right) D_i u - \left(\partial c/\partial \boldsymbol{L}\right) u \quad \text{a.e. } \Sigma, \\ \partial u/\partial \boldsymbol{L} = \varphi - \sigma u \qquad\qquad\qquad\qquad\qquad\qquad\qquad \text{on } \partial\Omega \cap \Sigma \end{cases}$$

with $\boldsymbol{L}(x) = (L^1(x), \ldots, L^n(x)) \in C^{1,1}(\Sigma)$. This way, once having $u \in W^{2,p}(\Omega)$ and the estimate (2.6), the L^p-theory of the uniformly elliptic equations (cf. [1], [7, Chapter 9]) gives

$$\|\partial u/\partial \boldsymbol{L}\|_{W^{2,p}(\widetilde{\Sigma})} \leqslant C' \left(\|u\|_{L^p(\Omega)} + \|f\|_{\mathcal{F}^p(\Omega,\Sigma)} + \|\varphi\|_{\Phi^p(\partial\Omega,\Sigma)}\right)$$

for any *closed* neighborhood $\widetilde{\Sigma}$ of $\mathcal{E}$ in $\overline{\Omega}$, $\widetilde{\Sigma} \subset \Sigma$, where the constant C' depends on $\operatorname{dist}(\widetilde{\Sigma}, \Omega \setminus \Sigma)$ in addition. In other words, if a strong solution u to the Poincaré problem belongs to $W^{2,p}(\Omega)$, then automatically $u \in \mathcal{W}^{2,p}(\Omega, \Sigma)$ provided $f \in \mathcal{F}^p(\Omega, \Sigma)$ and $\varphi \in \Phi^p(\partial\Omega, \Sigma)$.

Let us concentrate on the *neutral* case now and consider the problem $(\mathcal{P})$. Without loss of generality, we may assume $\sigma \equiv 0$ in $(\mathcal{P})$. In fact, according to (2.3) for each $x \in \Gamma \cap \overline{\Omega}$ there is a unique $\alpha(x) \in C^{0,1}(\Gamma \cap \overline{\Omega}) \cap C^{1,1}(\Sigma)$ such that $\boldsymbol{\psi}_{\boldsymbol{L}}(-\alpha(x), x) \in \partial(\Gamma \setminus \Omega) \cap \Omega$. Suppose that $\sigma(x)$ is extended in $\Gamma \cap \overline{\Omega}$ such that $\sigma \in C^{0,1}(\Gamma \cap \overline{\Omega}) \cap C^{1,1}(\Sigma)$ and define

$$\widetilde{\sigma}(x) := \int_0^{\alpha(x)} \sigma \circ \boldsymbol{\psi}_{\boldsymbol{L}}(t - \alpha(x), x))dt.$$

Since $(\partial\widetilde{\sigma}/\partial \boldsymbol{L})(x) = \sigma(x)$ for all $x \in \Gamma \cap \overline{\Omega}$ and $\widetilde{\sigma}$, $\partial\widetilde{\sigma}/\partial \boldsymbol{L} \in C^{1,1}(\Sigma)$, it is clear that $\widetilde{u}(x) := u(x)e^{\widetilde{\sigma}(x)} \in W^{2,q}(\Sigma) \cap W^{2,p}(\Sigma \setminus \Sigma')$ solves in Σ a problem like $(\mathcal{P})$ with $\sigma \equiv 0$.

Let $x_0 \in \mathcal{E}$ be arbitrary. We set $t \mapsto \boldsymbol{\psi}_{\boldsymbol{L}}(t, x_0)$ for the parametrization of the $\boldsymbol{L}$-trajectory through x_0. According to (2.3) that trajectory will leave $\mathcal{E}$ in both directions for a finite time. Thus, there exist $t^- < 0 < t^+$ such that $\boldsymbol{\psi}_{\boldsymbol{L}}(t^-, x_0) \in \Sigma'' \setminus \Sigma'$, $\boldsymbol{\psi}_{\boldsymbol{L}}(t^+, x_0) \in \mathbb{R}^n \setminus \overline{\Omega}$. Let $\mathcal{H}$ be the $(n-1)$-dimensional hyperplane through x_0 and orthogonal to $\boldsymbol{L}(x_0)$. We set

$$B_r(x_0) := \{x \in \mathcal{H}: \quad |x - x_0| < r\}$$

with radius $r > 0$ to be specified later. The Picard inequality

$$|\boldsymbol{\psi}_{\boldsymbol{L}}(t, x') - \boldsymbol{\psi}_{\boldsymbol{L}}(t, x'')| \leqslant e^{t\|\boldsymbol{L}\|_{C^1(\Sigma)}}|x' - x''|$$

implies that, if r is small enough, then the $\boldsymbol{L}$-flow

$$B'_r(x_0) := \boldsymbol{\psi_L}(t^-, B_r(x_0)) := \{\boldsymbol{\psi_L}(t^-, y)\colon \quad y \in B_r(x_0)\}$$

of $B_r(x_0)$ along the $\boldsymbol{L}$-trajectories at time t^- is *fully contained* in $\Sigma'' \setminus \Sigma'$ and thus $B'_r(x_0) \cap \mathcal{E} = \varnothing$. The set

$$\Theta_r := \{\boldsymbol{\psi_L}(t, x')\colon\ x' \in B'_r(x_0),\ t \in (0, t^+ - t^-)\}$$

is a sort of tubular n-dimensional neighborhood of the $\boldsymbol{L}$-trajectory through x_0 of base $B'_r(x_0)$, and we define

$$\mathcal{T}_r := \Theta_r \cap \Omega,$$

noting that $\mathcal{T}_r \subset \Sigma''$ if $r > 0$ is small enough. The boundary $\partial\mathcal{T}_r$ consists of the "base" $B'_r(x_0)$ and the "lateral" components $\partial_1\mathcal{T}_r := \partial\mathcal{T}_r \cap \partial\Omega$ and $\partial_2\mathcal{T}_r := (\partial\mathcal{T}_r \cap \Omega) \setminus B'_r(x_0)$. Take $\chi\colon \mathcal{H} \to \mathbb{R}^+$ to be a C^∞ cut-off function with support contained in $B_{3r/4}(x_0)$ and which equals 1 in $B_{r/2}(x_0)$. Extend χ to $\mathbb{R}^n$ as a constant along the $\boldsymbol{L}$-trajectory through $y \in \mathcal{H}$ and define

$$U(x) := \chi(x)u(x).$$

It is clear that $U \in W^{2,q}(\Sigma)$ is a strong solution to the problem

$$\begin{cases} \mathcal{L}U = F(x) := \chi f + 2a^{ij}D_j\chi D_i u + (a^{ij}D_{ij}\chi + b^i D_i\chi)u & \text{a.e. } \mathcal{T}_r, \\ \partial U/\partial\boldsymbol{L} = \Phi := \begin{cases} \chi\varphi & \text{on } \partial_1\mathcal{T}_r, \\ 0 & \text{near } \partial_2\mathcal{T}_r, \\ \chi\partial u/\partial\boldsymbol{L} & \text{on } B'_r(x_0) \subset \Sigma'' \setminus \Sigma'. \end{cases} \end{cases}$$

We have $u \in W^{2,q}(\Sigma)$ whence $Du \in L^{nq/(n-q)}$ if $q < n$ and $Du \in L^s$ for all $s > 1$ when $q \geqslant n$. Therefore, $F \in L^{p'}(\Sigma)$ with $p' := \min\{p, nq/(n-q)\}$ if $q < n$, and $p' = p$ otherwise. Moreover, from Remark 2.1 it follows that $\partial F/\partial\boldsymbol{L} \in L^{p'}(\Sigma'')$. Similarly, $\partial u/\partial\boldsymbol{L} \in W^{2,p}(\Sigma'' \setminus \Sigma')$ by (2.7), whence $\Phi \in W^{2-1/p,p}(\partial\mathcal{T}_r)$. Therefore, (2.1), (2.2), $\mathcal{T}_r \subset \Sigma''$, and Remark 2.1 give that

$$V(x) := \partial U/\partial\boldsymbol{L}$$

is a $W^{2,q}(\mathcal{T}_r)$-solution of the Dirichlet problem

$$\begin{cases} \mathcal{L}V = \partial F/\partial\boldsymbol{L} + 2a^{ij}D_jL^kD_{ik}U + \left(a^{ij}D_{ij}L^k + b^iD_iL^k\right)D_kU \\ \qquad - \left(\partial a^{ij}/\partial\boldsymbol{L}\right)D_{ij}U - \left(\partial b^i/\partial\boldsymbol{L}\right)D_iU - (\partial c/\partial\boldsymbol{L})\,U \text{ a.e. } \mathcal{T}_r, \\ V = \Phi \qquad\qquad\qquad\qquad\qquad\qquad\qquad\qquad \text{on } \partial\mathcal{T}_r. \end{cases} \tag{2.8}$$

At this point change the variables passing from $x \in \Theta_r$ into (x', ξ), where $x' = \boldsymbol{\psi_L}(-\xi(x), x) \in B'_r(x_0)$ and $\xi\colon\ \Theta_r \to (0, t^+ - t^-)$, $\xi(x) \in C^{1,1}(\Theta_r)$. The field $\boldsymbol{L}$ is transversal to $B'_r(x_0) = \psi_{\boldsymbol{L}}(t^-, B_r(x_0))$ and therefore $x \mapsto (x', \xi)$ is a $C^{1,1}$-diffeomorphism. Moreover, $\partial/\partial \boldsymbol{L}$ rewrites into $\partial/\partial\xi$, $\psi_{\boldsymbol{L}}(t, x') = (x', t)$ and $V(x', \xi) = \partial U(x', \xi)/\partial\xi$ as $(x', \xi) \in \mathcal{T}_r$ (cf. [28, Proposition 1.3]). The function $V(x', \xi)$ is absolutely continuous in ξ for a.a. $x' \in B'_r(x_0)$ (after redefining it on a set of zero measure, if necessary) and therefore

$$U(x', \xi) = U(x', 0) + \int_0^\xi V(x', t)dt \quad \text{for a.a. } (x', \xi) \in \mathcal{T}_r. \tag{2.9}$$

The point $(x', 0) \in B'_r(x_0)$ lies in $\Sigma'' \setminus \Sigma'$ and $U(x', 0)$ is a $W^{2,p}$-function there by (2.7), the Fubini theorem and [21, Remark 2.1]. Taking the derivatives of (2.9) up to the second order and substituting them into the right-hand side of (2.8), rewrites it as

$$\begin{cases} \mathcal{L}'V = F_1(x', \xi) + \displaystyle\int_0^\xi \mathcal{D}_2(\xi)V(x', t)dt & \text{a.e. } \mathcal{T}_r, \\ V = \Phi & \text{on } \partial\mathcal{T}_r. \end{cases} \tag{2.10}$$

Here, $\mathcal{L}'$ is the operator $\mathcal{L}$ in terms of $(x', \xi) = (x'_1, \dots, x'_{n-1}, \xi)$,

$$F_1(x', \xi) := \partial F/\partial \boldsymbol{L} + \mathcal{D}_1 V(x', \xi) + \mathcal{D}'_1 U(x', \xi) + \mathcal{D}'_2 U(x', 0),$$

$$\mathcal{D}_2(\xi)V(x', t) := \sum_{i,j=1}^{n-1} A^{ij}(x', \xi) D_{x'_i x'_j} V(x', t), \quad A^{ij} \in L^\infty,$$

and $\mathcal{D}_1, \mathcal{D}'_1, \mathcal{D}'_2$ are linear differential operators with L^∞-coefficients, $\operatorname{ord}\mathcal{D}_1 = \operatorname{ord}\mathcal{D}'_1 = 1$, $\operatorname{ord}\mathcal{D}'_2 = 2$. We have $\partial F/\partial\boldsymbol{L} \in L^{p'}(\Sigma'')$, $U(x', 0) \in W^{2,p}(B'_r(x_0))$, $U, V \in W^{2,q}(\Sigma'')$ and therefore the Sobolev imbedding theorem implies $F_1 \in L^{p'}(\mathcal{T}_r)$. For what concerns the second order operator $\mathcal{D}_2(\xi)$ it has a quite rough characteristic form that is neither symmetric nor sign-definite. Anyway, the improving-of-integrability property holds for (2.10) thanking to the particular structure of $\mathcal{T}_r$ as union of $\boldsymbol{L}$-trajectories through $B'_r(x_0)$. In fact, it turns out that if $V \in W^{2,p'}$ on a subset of $\mathcal{T}_r$ with $\xi < T$, then V remains a $W^{2,p'}$-function in a larger subset with $\xi < T + r$ provided r is small enough. Precisely, setting

$$\mathcal{P}_{r,T} := \{(x', \xi) \in \mathcal{T}_r : \quad \xi < T\} \quad \forall T \in (0, t^+ - t^-),$$

it follows that for fixed $r > 0$, $\{\mathcal{P}_{r,T}\}_{T \geqslant 0}$ is a nondecreasing family of domains exhausting $\mathcal{T}_r$. Moreover, $\mathcal{P}_{r,T} \subset \Sigma'' \setminus \Sigma'$ for small values of T and $V \in W^{2,p'}$ here by means of Remark 2.1. Suppose that $V \in W^{2,p'}(\mathcal{P}_{r,T})$. Let $(x', \xi) \in \mathcal{P}_{r,T+r}$. Rewriting the integral in (2.10) as

$$\int_0^T \mathcal{D}_2(\xi)V(x', t)dt + \int_T^\xi \mathcal{D}_2(\xi)V(x', t)dt,$$

the first term lies in $W^{2,p'}$ because the operator $\mathcal{D}_2(\xi)$ acts in the x'-variables only, while $\xi \in (T, T+r)$ in the second one and contraction mapping arguments show (cf. [22, Proposition 3.6]) that the solution V of (2.10) belongs to $W^{2,p'}(\mathcal{P}_{r,T+r})$ if r is less than some positive r_0 depending on the point $x_0 \in \mathcal{E}$ and which is the same for all other points of $\mathcal{E}$ lying on the same $\boldsymbol{L}$-trajectory as x_0. The improving of integrability of U in $\mathcal{T}_r$ then follows from (2.7) and (2.9) by means of iteration with respect of the curvilinear parameter ξ along the $\boldsymbol{L}$-trajectory and standard bootstrapping (cf. [22, Proposition 3.7]). Finally, selecting a finite set $\{\mathcal{T}_r^j\}_{j=1}^N$ of neighborhoods each of the type $\mathcal{T}_r$ above and such that the closure $\left(\bigcup_{j=1}^N \mathcal{T}_{r/2}^j\right) \subset \Sigma''$ is a closed neighborhood of $\mathcal{E}$ in $\overline{\Omega}$, we get

$$\|u\|_{W^{2,p}(\Sigma'')} \leqslant C\big(\|u\|_{L^p(\Omega)} + \|f\|_{\mathcal{F}^p(\Omega,\Sigma)} + \|\varphi\|_{\Phi^p(\partial\Omega,\Sigma)}\big)$$

and the desired estimate (2.6) in the *neutral* case follows from (2.7).

Remark 2.2. If the improving-of-integrability holds on a set $S \subset \overline{\Omega}$, then it is guaranteed, on the base of (2.9), on the whole flow $\psi_{\boldsymbol{L}}(S) := \{\psi_{\boldsymbol{L}}(t,x) \in \overline{\Omega}\colon x \in S,\ t \in \mathbb{R}\}$ of S along the $\boldsymbol{L}$-trajectories. Precisely, $u \in W^{2,p}(S)$ implies $u \in W^{2,p}(\psi_{\boldsymbol{L}}(S))$ and

$$\|u\|_{W^{2,p}(\psi_{\boldsymbol{L}}(S'))} \leqslant C\big(K(u,f,\varphi) + \|\partial u/\partial \boldsymbol{L}\|_{W^{1,p}(\psi_{\boldsymbol{L}}(S))}\big) \quad \forall S' \subsetneq S.$$

Consider now the *emergent* modified Poincaré problem $(\mathcal{MP})$ and assume $\sigma \equiv 0$ which always could be obtained by suitable change of variables as above. The first step in proving Theorem 2.1 in the *emergent* case is to establish it in a small enough closed neighborhood $\Sigma_0 \subset \overline{\Omega}$ of the manifold $\mathcal{E}_0$. Take $x_0 \in \mathcal{E}_0$ to be an arbitrary point and assume that the normal manifold $\mathcal{N}$ has the local representation $\{x_n = 0\}$ near x_0. We set $\mathcal{H}$ for the $(n-1)$-dimensional hyperplane through x_0 and containing locally $\mathcal{N}$, and define

$$B'_r := \{x' \in \mathcal{H}\colon\ |x'| < r\}, \quad \Omega'_r := \Omega \cap B'_r,$$

with $r > 0$ under control. Let $\mathcal{T}'_r$ be the flow of Ω'_r along the $\boldsymbol{L}$-trajectories in both directions,

$$\mathcal{T}'_r := \{\psi_{\boldsymbol{L}}(t,x') \in \Omega\colon\ x' \in \Omega'_r,\ t \in \mathbb{R}\}.$$

From (2.3) and (2.5) it follows that $\mathcal{T}'_r$ is a *connected* set fully contained in Σ if $r > 0$ is small enough. Let $\partial_1\mathcal{T}'_r := \partial\mathcal{T}'_r \cap \partial\Omega$ and $\partial_2\mathcal{T}'_r := \partial\mathcal{T}'_r \cap \Omega$. Introduce new coordinates in $\mathcal{T}'_r$ passing from $x \in \mathcal{T}'_r$ into (x',ξ), where $x' \in \Omega'_r$, $x' = \psi_{\boldsymbol{L}}(-\xi(x),x)$ and $\xi(x)\colon\ \mathcal{T}'_r \to \mathbb{R}$, $\xi(x) \in C^{1,1}(\mathcal{T}'_r)$, is the signed arclength of the $\boldsymbol{L}$-trajectory joining x with its projection x' on Ω'_r. The field $\boldsymbol{L}$ is transversal to $\mathcal{N}$ and therefore the map $x \mapsto (x',\xi)$ is a $C^{1,1}$-diffeomorphism. Take a cut-off function $\chi \in C_0^\infty(B'_{3r/4})$, $\chi(x') = 1$ in $x' \in B'_{r/2}$, and extend it in $\mathcal{T}'_r$ as a constant along each $\boldsymbol{L}$-trajectory through Ω'_r.

We have as above that $U(x) := \chi(x)u(x) \in W^{2,q}(\mathcal{T}'_r)$, $\partial U/\partial \boldsymbol{L} \equiv \partial U/\partial \xi$ and U solves the problem

$$\begin{cases} \mathcal{L}U = F(x) := \chi f + 2a^{ij}D_j\chi D_i u + (a^{ij}D_{ij}\chi + b^i D_i\chi)u & \text{a.e. } \mathcal{T}'_r, \\ \partial U/\partial \boldsymbol{L} = \Phi(x) := \chi\varphi = \begin{cases} \chi\varphi & \text{on } \partial_1\mathcal{T}'_r, \\ 0 & \text{near } \partial_2\mathcal{T}'_r, \end{cases} & U = \chi\mu \quad \text{on } \mathcal{N}\cap\partial\mathcal{T}'_r. \end{cases}$$

We have $F \in L^{p'}(\mathcal{T}'_r)$ as before, where p' is the Sobolev conjugate of p, and Remark 2.1 implies $\partial F/\partial \boldsymbol{L} \in L^{p'}(\mathcal{T}'_r)$ and $\Phi \in W^{2-1/p,p}(\partial\mathcal{T}'_r)$. Therefore, from (2.2) and Remark 2.1 it follows that $\partial U/\partial \boldsymbol{L}$ is a $W^{2,q}$-solution of the Dirichlet problem (2.8) in $\mathcal{T}'_r$. This way,

$$U(x) = U \circ \psi_{\boldsymbol{L}}(-\xi(x), x) + \int_0^{\xi(x)} \frac{\partial U}{\partial \boldsymbol{L}} \circ \psi_{\boldsymbol{L}}(t - \xi(x), x)dt \quad \text{for a.a. } x \in \mathcal{T}'_r,$$

i.e.,

$$U(x', \xi) = U(x', 0) + \int_0^{\xi} \frac{\partial U}{\partial t}(x', t)dt \quad \text{for a.a. } (x', \xi) \in \mathcal{T}'_r \tag{2.11}$$

with $x' \in \Omega'_r \subset \mathcal{N}$. The higher integrability of U in $\mathcal{T}'_r$ and thus of u will follow from the improving-of-summability property of $U(x', 0)$ and $\partial U/\partial t$. For, let $\mathcal{L}''$ be the action of the differential operator $\mathcal{L}$ on the functions defined in $\mathcal{T}'_r$ and which are constant on almost every trajectory of $\boldsymbol{L}$ through Ω'_r. Thus,

$$\begin{aligned} (\mathcal{L}''U)(x', 0) \equiv\, & a^{ij}(x', 0)D_{x'_i x'_j}U(x', 0) + b^i(x', 0)D_{x'_i}U(x', 0) + c(x', 0)U(x', 0) \\ =\, & F'(x') := F(x', 0) - a^{in}(x', 0)D_{x'_i}(D_\xi U)(x', 0) \\ & - a^{nn}(x', 0)(D_\xi(D_\xi U))(x', 0) - b^n(x', 0)(D_\xi U)(x', 0) \\ =\, & \widetilde{F}(x') - \widetilde{a}^{in}(x')(\widetilde{D_{x'_i}(D_\xi U)})(x') \\ & - \widetilde{a}^{nn}(x')(\widetilde{D_\xi(D_\xi U)})(x') - \widetilde{b}^n(x')(\widetilde{D_\xi U})(x'), \end{aligned}$$

where the summation in i and j runs from 1 to $n-1$ and the "tilde" over a function stands for its trace value on Ω'_r taken along the $\boldsymbol{L}$-trajectories (cf. [21, Remark 2.1]). One has $F,\ \partial F/\partial \boldsymbol{L} \in L^{p'}(\mathcal{T}'_r)$ and [21, Remark 2.1] gives $\widetilde{F}(x') = F(x', 0) \in L^{p'}(\Omega'_r)$. Further on, $D_\xi U \in W^{2,q}(\mathcal{T}'_r)$ and therefore $\widetilde{D_\xi U}, \widetilde{D_{x'}(D_\xi U)}, \widetilde{D_\xi(D_\xi U)} \in L^s(\Omega'_r)$ with $s = (n-1)q/(n-q)$ if $q < n$ and any $s > 1$ when $q \geqslant n$. This way, $F'(x') \in L^{p''}(\Omega'_r)$, where $p'' = \min\{p', s\} = \min\{p, (n-1)q/(n-q)\}$ if $q < n$, and $p'' = p$ otherwise. Since the restriction $\widetilde{U}(x') := U(x', 0)$ is a $W^{2,q}(\Omega'_r)$-solution to the Dirichlet problem

$$\mathcal{L}''\widetilde{U} = F' \quad \text{a.e. } \Omega'_r, \qquad \widetilde{U} = \chi\mu \quad \text{on } \partial\Omega'_r$$

with $F' \in L^{p''}(\Omega'_r)$, $\chi\mu \in W^{2-1/p,p}(\partial\Omega'_r)$, we get $\widetilde{U} \in W^{2,p''}(\Omega'_r)$ with $p'' \leqslant p'$ and $p \geqslant p'' > q$.

To get $D_\xi U \in W^{2,p''}$ near $\mathcal{E}_0$, and therefore to conclude $U \in W^{2,p''}$ on the base of (2.11) and $U(x',0) = \widetilde{U}(x') \in W^{2,p''}(\Omega'_r)$, we compute the second order derivatives of U from (2.11) and substitute them into the right-hand side of (2.8), obtaining this way that $V(x',\xi) := D_\xi U(x',\xi)$ solves in $\mathcal{T}'_r$ a Dirichlet problem like (2.10). Starting from this point and repeating the procedure in the *neutral* case above, it follows $V \in W^{2,p''}(\mathcal{T}'_r)$ if $r > 0$ is sufficiently small, and this yields $U \in W^{2,p''}(\mathcal{T}'_r)$ on the base of $U \in W^{2,p''}(\Omega'_r)$ and (2.11), whence $u \in W^{2,p''}(\mathcal{T}'_{r/2})$. Interpolating between p'', p' and p, and covering the compact $\mathcal{E}_0$ by the union of finite number of sets like $\mathcal{T}'_{r/2}$, we get (cf. [25, Proposition 2.5]) that there exists a closed neighborhood $\Sigma_0 \subset \overline{\Omega}$ of $\mathcal{E}_0$, $\Sigma_0 \subset \Sigma'$, such that $u \in W^{2,p}(\Sigma_0)$ and

$$\|u\|_{W^{2,p}(\Sigma_0)} \leqslant C\big(K(u,f,\varphi) + \|\mu\|_{W^{2-1/p,p}(\mathcal{E}_0)} + \|u\|_{W^{1,p}(\Sigma')} + \|\partial u/\partial \boldsymbol{L}\|_{W^{1,p}(\Sigma')}\big).$$

Now, from Remark 2.2 it follows that $u \in W^{2,p}(\psi_{\boldsymbol{L}}(\Sigma_0))$, while the vector field $\boldsymbol{\ell}$ behaves like normal one on the set $\mathcal{E} \setminus \psi_{\boldsymbol{L}}(\Sigma_0)$, and therefore each point of $\Sigma'' \setminus \psi_{\boldsymbol{L}}(\Sigma_0)$ can be reached from a *strictly interior* for Ω subset, where we dispose of (2.7). Thus, $u \in W^{2,p}(\Sigma'' \setminus \psi_{\boldsymbol{L}}(\Sigma_0))$ as consequence of (2.7) and Remark 2.2. The desired estimate (2.6) follows by interpolation of the $W^{2,p}$-norms of u and making use of Remark 2.1.

3 Uniqueness, Strong Solvability, and Fredholmness

We start with a maximum principle which is crucial in establishing that the Poincaré problem is of Fredholm type (cf. [22, Lemma 4.1] and [25, Lemma 3.1]).

Lemma 3.1. *Assume that* (2.1), (2.2), *and* (2.3) *hold.*

Neutral case. *Let $\boldsymbol{\ell}$ be of neutral type* (2.4) *and suppose that $c(x) \leqslant 0$ a.e. Ω, $\sigma(x) \geqslant 0$ on $\partial\Omega$ and either $c \not\equiv 0$ or $\sigma \not\equiv 0$. Let $u \in W^{2,n}_{\mathrm{loc}}(\Omega) \cap C^1(\overline{\Omega})$ satisfy*

$$\begin{aligned} a^{ij}(x)D_{ij}u + b^i(x)D_iu + c(x)u &\geqslant 0 \quad \text{a.e. } \Omega, \\ \partial u/\partial \boldsymbol{\ell} + \sigma(x)u &\leqslant 0 \quad \text{on } \partial\Omega. \end{aligned}$$

Then $u(x) \leqslant 0$ in $\overline{\Omega}$.

Emergent case. *Suppose that $\boldsymbol{\ell}$ is of emergent type* (2.5). *Let $c(x) \leqslant 0$ a.e. Ω, $\sigma(x) \gtrless 0$ on $\partial\Omega^\pm$, and either $c \not\equiv 0$ or $\sigma \not\equiv 0$. Assume that $u \in W^{2,n}_{\mathrm{loc}}(\Omega) \cap C^1(\overline{\Omega})$ satisfies*

$$a^{ij}(x)D_{ij}u + b^i(x)D_iu + c(x)u \geqslant 0 \qquad a.e.\ \Omega,$$
$$\partial u/\partial \boldsymbol{\ell} + \sigma(x)u \lesseqgtr 0 \quad on\ \partial\Omega^{\pm}, \quad u \leqslant 0 \quad on\ \mathcal{E}_0,$$

Then $u(x) \leqslant 0$ *in* $\overline{\Omega}$.

The proof is carried out arguing by contradiction. Assuming that $u(x)$ attains positive values in Ω and setting $u(x_0) = \max_{\overline{\Omega}} u(x) = M > 0$, it follows from $u \in W^{2,n}_{\text{loc}}(\Omega)$ and the strong Aleksandrov maximum principle [7, Theorem 9.6] that $x_0 \in \partial\Omega$, unless $u =$ const, which contradicts the boundary condition. Later on, $\boldsymbol{\ell}$ is transversal to $\partial\Omega$ on $\partial\Omega \setminus \mathcal{E}$ and therefore $u \in C^1(\overline{\Omega})$ and the Hopf boundary point lemma yield $x_0 \in \mathcal{E}$. The field $\boldsymbol{\ell}(x_0)$ is tangent to $\partial\Omega$ now, $\partial u/\partial\boldsymbol{\ell}(x_0) = 0$ and another contradiction with the boundary condition follows if $\sigma(x_0) \neq 0$. Otherwise, $\sigma(x_0) = 0$ and the maximum M of u propagates along the $\boldsymbol{\tau}$-trajectory through x_0 up to a point x_0' at which either $\sigma(x_0') \neq 0$, or $x_0' \in \partial\Omega \setminus \mathcal{E}$ (thanks to the nontrapping condition (2.3)), or $x_0' \in \mathcal{E}_0$ if $\boldsymbol{\ell}$ is emergent. In all the cases, a contradiction with the boundary condition follows and this gives the claim.

Lemma 3.1 provides sufficient conditions ensuring *unicity* for the Poincaré problem in $W^{2,p}(\Omega)$ for *each* $p > 1$. In fact, let $u, v \in W^{2,p}(\Omega)$ be two solutions of $(\mathcal{P})$ or $(\mathcal{MP})$ respectively. Their difference $w \in W^{2,p}(\Omega)$ solves the corresponding *homogeneous* Poincaré problem, and therefore $w \in W^{2,q}(\Omega)$ for all $q > 1$ as it follows from Theorem 2.1. In particular, $w \in W^{2,n}(\Omega)$ and Lemma 3.1 yields $w \equiv 0$ in $\overline{\Omega}$. In other words, the following assertion holds.

Corollary 3.1 (uniqueness). *Under the assumptions of Lemma* 3.1, *the Poincaré problem may have at most one solution in* $W^{2,p}(\Omega)$, $p > 1$.

Another outgrowth of Lemma 3.1 is the possibility to refine considerably the *a priori* estimate (2.6) by dropping out the L^p-norm of the solution from the right-hand side. Precisely (cf. [22, Theorem 4.4] and [25, Theorem 3.4]), the following assertion holds.

Theorem 3.1 (refined a priori estimate). *Under the assumptions of Lemma* 3.1, *let* $p \in (1, \infty)$, *and let* $u \in W^{2,p}(\Omega)$ *be a strong solution to the neutral problem* $(\mathcal{P})$ *with* $f \in \mathcal{F}^p(\Omega, \Sigma)$ *and* $\varphi \in \Phi^p(\partial\Omega, \Sigma)$ *or to the emergent problem* $(\mathcal{MP})$ *with* $f \in \mathcal{F}^p(\Omega, \Sigma)$, $\varphi \in \Phi^p(\partial\Omega, \Sigma)$ *and* $\mu \in W^{2-1/p,p}(\mathcal{E}_0)$. *Then there exists an absolute constant* C *such that*

$$\|u\|_{W^{2,p}(\Omega)} \leqslant \begin{cases} C\big(\|f\|_{\mathcal{F}^p(\Omega,\Sigma)} + \|\varphi\|_{\Phi^p(\partial\Omega,\Sigma)}\big) & \text{if } \boldsymbol{\ell} \text{ is neutral,} \\ C\big(\|f\|_{\mathcal{F}^p(\Omega,\Sigma)} + \|\varphi\|_{\Phi^p(\partial\Omega,\Sigma)} & \\ \qquad\qquad + \|\mu\|_{W^{2-1/p,p}(\mathcal{E}_0)}\big) & \text{if } \boldsymbol{\ell} \text{ is emergent.} \end{cases} \tag{3.1}$$

To get (3.1) we argue by contradiction. In fact, assuming that (3.1) is false, there is a sequence $\{u_k\}_{k\in\mathbb{N}} \in W^{2,p}(\Omega)$ such that

$$\|u_k\|_{L^p(\Omega)} = 1, \quad \lim_{k\to\infty} \|\mathcal{L}u_k\|_{\mathcal{F}^p(\Omega,\Sigma)} = \lim_{k\to\infty} \|\mathcal{B}u_k\|_{\Phi^p(\partial\Omega,\Sigma)} = 0$$

and $\lim_{k\to\infty} \|u_k\|_{W^{2-1/p,p}(\mathcal{E}_0)} = 0$ in case $\boldsymbol{\ell}$ is of emergent type. The estimate (2.6) implies boundedness of $\|u_k\|_{W^{2,p}(\Omega)}$ and thus a subsequence $\{u_{k'}\}_{k'\in\mathbb{N}}$ is *weakly convergent* to $u \in W^{2,p}(\Omega)$. It follows that $\mathcal{L}u = 0$ a.e. Ω, $\mathcal{B}u = 0$ on $\partial\Omega$ and $u = 0$ on $\mathcal{E}_0$ in the case of $(\mathcal{MP})$, and thus $u = 0$ in $\overline{\Omega}$ by means of Corollary 3.1 and contrary to $\|u\|_{L^p(\Omega)} = 1$.

With the refined *a priori* estimate (3.1) at hand, it is standard to derive solvability of $(\mathcal{P})$ and $(\mathcal{MP})$ in $W^{2,p}(\Omega)$. In fact, for each $k \in \mathbb{N}$ consider the approximating problems

$$\mathcal{L}_k u_k = f_k \text{ a.e. } \Omega, \quad \mathcal{B}_k u_k = \varphi_k \text{ on } \partial\Omega \tag{$\mathcal{P}_k$}$$

if $\boldsymbol{\ell}$ is *neutral*, and

$$\mathcal{L}_k u_k = f_k \text{ a.e. } \Omega, \quad \mathcal{B}_k u_k = \varphi_k \text{ on } \partial\Omega, \quad u_k = \mu_k \text{ on } \mathcal{E}_0 \tag{$\mathcal{MP}_k$}$$

if $\boldsymbol{\ell}$ is *emergent*. Here, $\mathcal{L}_k$ and $\mathcal{B}_k$ are second order elliptic operator and first order boundary operator respectively, both with $C^\infty(\overline{\Omega})$-coefficients which approximate the coefficients of $\mathcal{L}$ or $\mathcal{B}$ in the corresponding functional spaces according to (2.2). The approximating vector fields $\boldsymbol{\ell}_k$, $\lim_{k\to\infty} \boldsymbol{\ell}_k = \boldsymbol{\ell}$ in $C^{0,1}(\partial\Omega)\cap C^{1,1}(\partial\Omega\cap\Sigma)$, satisfy the *nontrapping* condition (2.3) and (2.4) in the case of $(\mathcal{P}_k)$ or (2.5) when dealing with $(\mathcal{MP}_k)$. Further on, the sequences $\{f_k\}_{k\in\mathbb{N}} \in C^\infty(\overline{\Omega})$, $\{\varphi_k\}_{k\in\mathbb{N}} \in C^\infty(\partial\Omega)$ and $\{\mu_k\}_{k\in\mathbb{N}} \in C^\infty(\mathcal{E}_0)$ approximate f, φ and μ as $k \to \infty$ in $\mathcal{F}^p(\Omega,\Sigma)$, $\Phi^p(\partial\Omega,\Sigma)$ and $W^{2-1/p,p}(\mathcal{E}_0)$ respectively. Under the hypotheses of Lemma 3.1, each of the problems $(\mathcal{P}_k)$ and $(\mathcal{MP}_k)$ has unique *classical* solution $u_k \in C^{2,\alpha}(\overline{\Omega})$, $\alpha \in (0,1)$, by virtue of [38, Theorem 1] for $(\mathcal{P}_k)$ and from [38, Theorem 2] for $(\mathcal{MP}_k)$. The sequence $\{u_k\}_{k\in\mathbb{N}}$ is bounded in $W^{2,p}(\Omega)$ as it follows from (3.1), whence a subsequence converges weakly to $u \in W^{2,p}(\Omega)$. Passage to the limit as $k \to \infty$ in $(\mathcal{P}_k)$ and $(\mathcal{MP}_k)$ implies strong solvability of $(\mathcal{P})$ and $(\mathcal{MP})$ respectively. The reader is referred to [22, Theorem 5.1] and [25, Theorem 3.5] for alternative proofs of the following assertion.

Theorem 3.2 (strong solvability). *Under the hypotheses of Lemma* 3.1, *the Poincaré problems* $(\mathcal{P})$ *and* $(\mathcal{MP})$ *are uniquely solvable in* $W^{2,p}(\Omega)$ *for arbitrary* $p \in (1,\infty)$ *and any* $f \in \mathcal{F}^p(\Omega,\Sigma)$, $\varphi \in \Phi^p(\partial\Omega,\Sigma)$ *and* $\mu \in W^{2-1/p,p}(\mathcal{E}_0)$.

We are in a position now to prove that the Poincaré problem, even if a degenerate one, is a problem of Fredholm type with *zero index*. For this goal, take a $p \in (1,\infty)$ and define the *kernel* and the *range* of the *neutral* problem $(\mathcal{P})$ by the settings

$$\mathcal{K}_p^{(\mathcal{P})} := \big\{u \in \mathcal{W}^{2,p}(\Omega,\Sigma)\colon\ \mathcal{L}u = 0 \text{ a.e. } \Omega,\ \mathcal{B}u = 0 \text{ on } \partial\Omega\big\},$$
$$\mathcal{R}_p^{(\mathcal{P})} := \mathcal{F}^p(\Omega,\Sigma) \times \Phi^p(\partial\Omega,\Sigma),$$

and these of the *emergent* problem $(\mathcal{MP})$ by

$$\mathcal{K}_p^{(\mathcal{MP})} := \{u \in \mathcal{W}^{2,p}(\Omega, \Sigma)\colon\ \mathcal{L}u = 0 \text{ a.e. } \Omega,\ \mathcal{B}u = 0 \text{ on } \partial\Omega,\ u = 0 \text{ on } \mathcal{E}_0\},$$
$$\mathcal{R}_p^{(\mathcal{MP})} := \mathcal{F}^p(\Omega, \Sigma) \times \Phi^p(\partial\Omega, \Sigma) \times W^{2-1/p,p}(\mathcal{E}_0).$$

Theorem 3.3 (Fredholmness). *Assume that* (2.1), (2.2), *and* (2.3) *hold and consider the problem* $(\mathcal{P})$ *when the field* $\boldsymbol{\ell}$ *is of neutral type* (2.4) *or the problem* $(\mathcal{MP})$ *if* $\boldsymbol{\ell}$ *is a field of emergent type* (2.5).

For each $p \in (1, \infty)$ *there exists a closed subspaces* $\widetilde{\mathcal{R}}_p^{(\mathcal{P})}$ *and* $\widetilde{\mathcal{R}}_p^{(\mathcal{MP})}$ *of finite codimensions in* $\mathcal{R}_p^{(\mathcal{P})}$ *and* $\mathcal{R}_p^{(\mathcal{MP})}$ *respectively, and such that the problems* $(\mathcal{P})$ *and* $(\mathcal{MP})$ *are solvable in* $W^{2,p}(\Omega)$ *for each* $(f, \varphi) \in \widetilde{\mathcal{R}}_p^{(\mathcal{P})}$ *and each* $(f, \varphi, \mu) \in \widetilde{\mathcal{R}}_p^{(\mathcal{MP})}$ *respectively, and the corresponding solutions satisfy* (2.6). *Moreover,*

$$\operatorname{codim} \widetilde{\mathcal{R}}_p^{(\mathcal{P})} = \dim \mathcal{K}_p^{(\mathcal{P})}, \quad \operatorname{codim} \widetilde{\mathcal{R}}_p^{(\mathcal{MP})} = \dim \mathcal{K}_p^{(\mathcal{MP})}.$$

In the particular case where the hypotheses of Lemma 3.1 are verified, the kernels $\mathcal{K}_p^{(\mathcal{P})}$ *and* $\mathcal{K}_p^{(\mathcal{MP})}$ *are trivial,* $\widetilde{\mathcal{R}}_p^{(\mathcal{P})} \equiv \mathcal{R}_p^{(\mathcal{P})}$ *and* $\widetilde{\mathcal{R}}_p^{(\mathcal{MP})} \equiv \mathcal{R}_p^{(\mathcal{MP})}$, *and therefore* $(\mathcal{P})$ *and* $(\mathcal{MP})$ *are uniquely solvable for arbitrary data* $(f, \varphi) \in \mathcal{R}_p^{(\mathcal{P})}$ *and* $(f, \varphi, \mu) \in \widetilde{\mathcal{R}}_p^{(\mathcal{MP})}$ *respectively, with solutions satisfying* (3.1).

Proof. Subtracting a suitable function $v \in \mathcal{W}^{2,p}(\Omega, \Sigma)$, without loss of generality we may suppose $\sigma \equiv 0$ and consider the Poincaré problems $(\mathcal{P})$ and $(\mathcal{MP})$ with *homogeneous* boundary data (cf. the proofs of Theorem 5.2 in [22] and Theorem 3.6 in [25]). Thus, define the linear space

$$\mathfrak{W}_p := \begin{cases} \{u \in \mathcal{W}^{2,p}(\Omega, \Sigma)\colon \partial u/\partial \boldsymbol{\ell} = 0 \text{ on } \partial\Omega\} & \text{if } \boldsymbol{\ell} \text{ is neutral,} \\ \{u \in \mathcal{W}^{2,p}(\Omega, \Sigma)\colon \partial u/\partial \boldsymbol{\ell} = 0 \text{ on } \partial\Omega,\ u = 0 \text{ on } \mathcal{E}_0\} & \text{if } \boldsymbol{\ell} \text{ is emergent} \end{cases}$$

and take a constant $\Lambda > \|c\|_{L^\infty(\Omega)}$. From Theorem 3.2 it follows that the uniformly elliptic operator

$$\mathcal{L}_\Lambda := \mathcal{L} - \Lambda\colon\ \mathfrak{W}_p \to \mathcal{F}^p(\Omega, \Sigma)$$

is *invertible,* while the estimate (3.1) ensures that the inverse $\mathcal{L}_\Lambda^{-1}$ maps bounded sets of $\mathcal{F}^p(\Omega, \Sigma)$ into bounded sets in $\mathfrak{W}_p$. Since $\mathfrak{W}_p$ is compactly imbedded in $\mathcal{F}^p(\Omega, \Sigma)$ by the Rellich–Kondrachov theorem, this means that

$$\mathcal{L}_\Lambda^{-1}\colon\ \mathcal{F}^p(\Omega, \Sigma) \to \mathcal{F}^p(\Omega, \Sigma)$$

is a *compact operator.* Setting Id for the identity operator from $\mathcal{F}^p(\Omega, \Sigma)$ into itself, the equation $\mathcal{L}u = f$ a.e. Ω is equivalent to $\mathcal{L}_\Lambda u + \Lambda u = f$, i.e., to

$$\left(\mathrm{Id} + \Lambda \mathcal{L}_\Lambda^{-1}\right) u = \mathcal{L}_\Lambda^{-1} f$$

and the claim follows from the Riesz–Schauder theory. □

Actually, Theorem 3.3 provides a Fredholm alternative for the Poincaré problem.

Corollary 3.2. *Under the assumptions of Theorem 3.3, let $p > 1$ be arbitrary. Then, either*

- *The homogeneous Poincaré problem ($(\mathcal{P})$ or $(\mathcal{MP})$) admits only the trivial solution and then the inhomogeneous problem is uniquely solvable in $W^{2,p}(\Omega)$ for arbitrary data $(f, \varphi) \in \mathcal{F}^p(\Omega, \Sigma) \times \Phi^p(\partial\Omega, \Sigma)$ in the case of $(\mathcal{P})$ and arbitrary $(f, \varphi, \mu) \in \mathcal{F}^p(\Omega, \Sigma) \times \Phi^p(\partial\Omega, \Sigma) \times W^{2-1/p,p}(\mathcal{E}_0)$ in case of $(\mathcal{MP})$,*

or

- *The homogeneous Poincaré problem has nontrivial solutions which span a subspace of $W^{2,p}(\Omega)$ of finite dimension $k > 0$. Then the inhomogeneous problem $(\mathcal{P})$ (respectively, $(\mathcal{MP})$) is solvable only for those data $(f, \varphi) \in \mathcal{F}^p(\Omega, \Sigma) \times \Phi^p(\partial\Omega, \Sigma)$ (respectively, $(f, \varphi, \mu) \in \mathcal{F}^p(\Omega, \Sigma) \times \Phi^p(\partial\Omega, \Sigma) \times W^{2-1/p,p}(\mathcal{E}_0)$) which satisfy k complementary conditions.*

References

1. Agmon, S., Douglis, A, Nirenberg, L.: Estimates near the boundary for solutions of elliptic partial differential equations satisfying general boundary conditions I. Commun. Pure Appl. Math. **12**, 623–727 (1959); II, ibid. **17**, 35–92 (1964)
2. Borrelli, R.L.: The singular, second order oblique derivative problem. J. Math. Mech. **16**, 51–81 (1966)
3. Egorov, Y.V.: Linear Differential Equations of Principal Type. Contemporary Soviet Mathematics, New York (1986)
4. Egorov, Y.V., Kondrat'ev, V.: The oblique derivative problem. Math. USSR Sb. **7**, 139–169 (1969)
5. Elling, V.; Liu, T.-P.: Supersonic flow onto a solid wedge. Commun. Pure Appl. Math. **61**, 1347–1448 (2008)
6. Eskin, G.I.: Degenerate elliptic pseudodifferential equations of principal type. Math. USSR Sb. **11**, 539–582 (1971)
7. Gilbarg, D., Trudinger, N.S.: Elliptic Partial Differential Equations of Second Order. 2nd ed., Springer, Berlin et. (1983)
8. Guan, P.: Hölder regularity of subelliptic pseudodifferential operators. Duke Math. J. **60**, 563–598 (1990)
9. Guan, P., Sawyer, E.: Regularity estimates for the oblique derivative problem. Ann. Math. **137**, 1–70 (1993)
10. Guan, P., Sawyer, E.: Regularity estimates for the oblique derivative problem on nonsmooth domains I. Chine. Ann. Math., Ser. B **16**, 1–26 (1995); II, ibid. **17**, 1–34 (1996)
11. Holmes, J.J.: Theoretical development of laboratory techniques for magnetic measurement of large objects. IEEE Trans. Magnetics **37**, 3790–3797 (2001)

12. Hörmander, L.: Pseudodifferential operators and non-elliptic boundary value problems. Ann. Math. **83**, 129–209 (1966)
13. Hörmander, L.: The Analysis of Linear Partial Differential Operators III: Pseudo-Differential Operators; IV: Fourier Integral Operators. Springer, Berlin (1985)
14. Martin, P.A.: On the diffraction of Poincaré waves. Math. Methods Appl. Sci. **24**, 913–925 (2001)
15. Maugeri, A., Palagachev, D.K., Softova, L.G.: Elliptic and Parabolic Equations with Discontinuous Coefficients. Math. Res. **109**, Wiley–VCH, Berlin (2000)
16. Maugeri, A., Palagachev, D.K., Vitanza, C.: A singular boundary value problem for uniformly elliptic operators. J. Math. Anal. Appl. **263**, 33–48 (2001)
17. Maz'ya, V.G.: The degenerate problem with an oblique derivative (Russian). Mat. Sb., N. Ser. **87(129)**, 417–454 (1972); English transl.: Math. USSR Sb. **16**, 429–469 (1972)
18. Maz'ya, V.G.: Paneyah, B.P.: Degenerate elliptic pseudo-differential operators and the problem with oblique derivative (Russian). Tr. Moskov. Mat. Ob. **31**, 237–295 (1974)
19. Melin, A., Sjöstrand, J.: Fourier integral operators with complex phase functions and parametrix for an interior boundary value problem. Commun. Partial Differ. Equ. **1**, 313–400 (1976)
20. Palagachev, D.K.: The tangential oblique derivative problem for second order quasilinear parabolic operators. Commun. Partial Differ. Equ. **17**, 867–903 (1992)
21. Palagachev, D.K.: The Poincaré problem in L^p-Sobolev spaces I: Codimension one degeneracy. J. Funct. Anal. **229**, 121–142 (2005)
22. Palagachev, D.K.: Neutral Poincaré Problem in L^p-Sobolev Spaces: Regularity and Fredholmness. Int. Math. Res. Notices **2006**, Article ID 87540 (2006)
23. Palagachev, D.K.: $W^{2,p}$-a priori estimates for the neutral Poincaré problem. J. Nonlinear Convex Anal. **7**, 499–513 (2006)
24. Palagachev, D.K.: $W^{2,p}$-a priori estimates for the emergent Poincaré problem. J. Glob. Optim. **40**, 305–318 (2008)
25. Palagachev, D.K.: The Poincaré problem in L^p-Sobolev spaces II: Full dimension degeneracy. Commun. Partial Differ. Equ. **33**, 209–234 (2008)
26. Paneyah, B.P.: On a problem with oblique derivative. Sov. Math. Dokl. **19**, 1568–1572 (1978)
27. Paneyah, B.P.: On the theory of solvability of the oblique derivative problem. Math. USSR Sb. **42**, 197–235 (1982)
28. Paneyah, B.P.: The Oblique Derivative Problem. The Poincaré Problem. Math. Topics **17**, Wiley–VCH, Berlin (2000)
29. Poincaré, H.: Lecons de Méchanique Céleste, Tome III, Théorie de Marées, Gauthiers–Villars, Paris (1910)
30. Popivanov, P.R.: On the tangential oblique derivative problem — methods, results, open problems. Gil, Juan (ed.) et al., Aspects of boundary problems in analysis and geometry. Birkhaüser, Basel. Operator Theory: Advances and Applications **151**, pp. 430–471 (2004)
31. Popivanov, P.R., Kutev, N.D.: The tangential oblique derivative problem for nonlinear elliptic equations. Commun. Partial Differ. Equ. **14**, 413–428 (1989)
32. Popivanov, P.R., Kutev, N.D.: Viscosity solutions to the degenerate oblique derivative problem for fully nonlinear elliptic equations. Math. Nachr. **278**, 888–903 (2005)

33. Popivanov, P.R., Palagachev, D.K.: The Degenerate Oblique Derivative Problem for Elliptic and Parabolic Equations. Math. Res. **93**, Akademie-Verlag, Berlin (1997)
34. Softova, L.G.: $W_p^{2,1}$-solvability for the parabolic Poincaré problem. Commun. Partial Differ. Equ. **29**, 1783–1789 (2004)
35. Taira, K.: Semigroups, Boundary Value Problems and Markov Processes. Springer Monographs in Mathematics, Springer, Berlin (2004)
36. Winzell, B.: The oblique derivative problem I. Math. Ann. **229**, 267–278 (1977)
37. Winzell, B.: Sub-elliptic estimates for the oblique derivative problem. Math. Scand. **43**, 169–176 (1978)
38. Winzell, B.: A boundary value problem with an oblique derivative. Commun. Partial Differ. Equ. **6**, 305–328 (1981)
39. Zheng, Y: A global solution to a two-dimensional Riemann problem involving shocks as free boundaries. Acta Math. Appl. Sin. Engl. Ser. **19**, 559–572 (2003)

Weighted Inequalities for Integral and Supremum Operators

Luboš Pick

Abstract We survey results on weighted inequalities for integral and supremum operators with particular emphasize on certain recent developments. We discuss various, mostly recent, results concerning several topics that are in one way or another connected with the Hardy integral operator. It is my great pleasure and honor to dedicate this paper to Professor Vladimir Maz'ya, a true classic of the field, whose astonishing mathematical achievements have increased considerably the beauty of this part of mathematical analysis and therefore inspired and attracted many other mathematicians.

1 Prologue

The *Hardy integral operator*

$$Hf(t) := \int_0^t f(s)\,ds, \tag{1.1}$$

together with its many various modifications, where $t \in (0, \infty)$ and f is a nonnegative locally-integrable function on $(0, \infty)$, plays a central role in several branches of analysis and its applications. It becomes of a particular interest when functional-analytic methods are applied to finding solutions of partial differential equations.

In this survey, we concentrate on its properties such as boundedness or compactness on various function spaces, with a particular emphasize on weighted spaces. We further study the action of the Hardy operator restricted

Luboš Pick
Faculty of Mathematics and Physics, Department of Mathematical Analysis, Charles University, Sokolovská 83, 186 75 Praha 8, Czech Republic
e-mail: pick@karlin.mff.cuni.cz

A. Laptev (ed.), *Around the Research of Vladimir Maz'ya III: Analysis and Applications*, International Mathematical Series 13, DOI 10.1007/978-1-4419-1345-6_11,

to the cone of monotone functions and finally we treat the analogous operators where the integration is replaced by taking a supremum.

Such knowledge is useful since it has been known that various problems involving functions of several variables can be reduced to a one-dimensional task, often quite more manageable than the original question, in which the operator H plays a key role. As a particular instance we can recall the reduction theorem which asserts that Sobolev type embedding

$$W^m X(\Omega) \hookrightarrow Y(\Omega)$$

where Ω is an open bounded subset of $\mathbb{R}^n$, $1 \leqslant m \leqslant n-1$, X, Y are rearrangement-invariant Banach function spaces, and $W^1 X$ is the Sobolev space built upon X, is equivalent to the boundedness of the operator

$$H_{\frac{m}{n}} f(t) := \int_t^1 s^{\frac{m}{n}-1} f(s)\, ds$$

from $\overline{X}(0,1)$ to $\overline{Y}(0,1)$, where $\overline{X}$, $\overline{Y}$ are the representation spaces of X, Y respectively (cf. the exact definitions below). This result was proved in [28, 19, 47].

Reduction techniques using the Hardy operator have been successfully applied for example also to the dimension-independent logarithmic Gaussian–Sobolev inequalities ([22]) or to various problems dealing with traces of functions in a Sobolev space (cf., for example, [21] or [23]).

Recent results containing (among other topics) applications of Hardy inequalities to optimal Sobolev embeddings were mentioned in our previous survey paper [74].

Here, we have a different goal. We will be interested in results concerning the Hardy operator itself rather than its applications.

The paper is divided into three sections. In the first one, we give a brief outline of the history of boundedness and compactness of the Hardy operator in weighted function spaces. We concentrate on weighted Lebesgue spaces and on the action of the Hardy operator from a weighted Banach function space into L^∞ and (weighted) BMO. The second section is devoted to the weighted Hardy type inequalities restricted to the cone of nonincreasing functions. This subject is in the focus of many authors especially since 1990's. We in particular survey the necessary and sufficient conditions of all possible mutual embeddings between classical Lorentz spaces of types Λ, Γ and S. This theory has been completed only very recently. We also present a few applications of these results. In the last section, we treat the so-called supremum operators, or Hardy type operators involving suprema, a topic whose interest is rapidly growing in recent years and whose applications don't seem to stop appearing in various most unpredictable mathematical problems. Again, we concentrate mainly on weighted inequalities for Lebesgue spaces.

The reader interested in a more detailed history of the Hardy inequality is referred to the wonderful book [55] from which we borrow heavily. Plenty of results which would have deserved to be mentioned here are omitted because of the limited space. Several related topics have been left out for the sake of brevity; this concerns, for example, weak classical Lorentz spaces (cf. [15, 24, 35]), abstract duality and reduction principles ([33]), applications to Sobolev embeddings (many papers mentioned throughout), Hardy type operators involving suprema over the interval $(0,t)$ (we restrict ourselves to intervals (t,∞)), reduction theorems for supremum operators ([36]) and many more.

As usual, throughout the paper, by $A\lesssim B$ and $A\gtrsim B$ we mean that $A \leqslant CB$ and $CA \geqslant B$ respectively, where C is a positive constant independent of appropriate quantities involved in the expressions A and B. We also write $A \approx B$ when both $A\lesssim B$ and $A\gtrsim B$ are satisfied.

2 Hardy Operator

Almost a century ago, in 1915, G.H. Hardy proved an estimate for the arithmetic means, namely, he showed that there exists a positive constant, C, such that, for every sequence $\{a_n\}$ of real numbers one has

$$\sum_{n=1}^{\infty}\left|\frac{1}{n}\sum_{k=1}^{n}a_k\right|^2 \leqslant C\sum_{n=1}^{\infty}|a_n|^2.$$

The inequality has been then considered in a more general context, namely with the power 2 replaced by a general $p \in (1,\infty)$, and, finally, in the celebrated 1925 Hardy's paper [41], the integral version

$$\int_0^\infty\left(\frac{1}{t}\int_0^t f(s)\,ds\right)^p dt \leqslant \left(\frac{p}{p-1}\right)^p\int_0^\infty f(t)^p\,dt \tag{2.1}$$

was proved for every function $f \geqslant 0$ on $(0,\infty)$.

The estimate (2.1) can be considered as an inequality involving the *Hardy average operator*

$$Af(t) := \frac{1}{t}\int_0^t f(s)\,ds, \tag{2.2}$$

which is a simple modification of the classical Hardy operator (1.1).

The principal analytic feature of Hardy type operators is how they act on function spaces. Such information is indispensable when functional-theoretical methods are applied to the solution of the partial differential equations.

For example, (2.1) can be interpreted as the boundedness of the average operator A on every Lebesgue space $L^p(0,\infty)$ for $1 < p \leqslant \infty$. It can be also interpreted in a little different way: given a *weight function*, or *weight* (i.e., a nonnegative measurable function w on $(0,\infty)$), we can introduce a *weighted Lebesgue space* $L^p_w(0,\infty)$ as a set of measurable functions f on $(0,\infty)$ such that $\int_0^\infty |f(x)|^p w(x)\,dx < \infty$, and endowed with the norm

$$\|f\|_{L^p_w(0,\infty)} := \left(\int_0^\infty |f(x)|^p w(x)\,dx\right)^{\frac{1}{p}}, \qquad 1 \leqslant p < \infty.$$

Then (2.1) can be interpreted as the boundedness of the classical Hardy operator H from $L^p(0,\infty)$ into $L^p_w(0,\infty)$, where $w(t) = t^{-p}$.

The above-described double-interpretation principle of a Hardy type inequality in terms of Hardy type integral operators is possible in Lebesgue spaces because of the homogeneity of their norms, but in more general function spaces does not necessarily work. As it has been known for decades, although Lebesgue spaces still play a primary role in the theory and applications of function spaces, there are other types of function spaces which are also of interest and importance.

In general, one can study properties of the *weighted Hardy operator*

$$H_{u,v}f(t) := u(t)\int_0^t f(s)v(s)\,ds, \tag{2.3}$$

where u and v are *weights*, i.e., positive measurable functions on $(0,\infty)$. The question of boundedness of the operator $H_{u,v}$ from a Lebesgue space $L^p(0,\infty)$ to another one, $L^q(0,\infty)$, transforms via a simple change of variables to the equivalent question, for which weights w and v and for which parameters $p, q \in (0,\infty)$, there is a positive constant, C, possibly dependent on w, v, p, q, but not on f, such that

$$\left(\int_0^\infty \left(\int_0^t f(s)\,ds\right)^q w(t)\,dt\right)^{\frac{1}{q}} \leqslant C\left(\int_0^\infty f(t)^p v(t)\,dt\right)^{\frac{1}{p}}. \tag{2.4}$$

We recall that, by (2.4), this is true for $p = q \in (1,\infty)$, $w(t) = \frac{1}{t}$ and $v \equiv 1$. It is a routine matter to generalize this inequality to more general, but still power, weights. The question how to verify (2.4) given two general weights v, w and a couple of possibly different parameters p, q is a much deeper and more difficult task.

There exists a vast amount of literature dedicated to the study of Hardy type inequalities, including several monographs. We do not have any ambition in this note to give a thorough and comprehensive set of references or historical survey. We instead refer the reader to the book by Kufner, Maligranda and Persson [55], where the history of the Hardy inequality is given in full detail. We quote here only those references which have a direct connection to the particular inequalities we intend to treat here.

According to the (rather surprising) information revealed in [55], quoted there to a personal communication with Professor Vladimir Maz'ya, an apparently first treatment of the weighted Hardy inequality (2.1) was carried out by Kac and Krein in the paper [44], which is almost neglected in literature. Here, the inequality was characterized in the case $p = q = 2$ and $v \equiv 1$ by the condition

$$\sup_{t\in(0,\infty)} t \int_t^\infty w(s)\,ds < \infty. \tag{2.5}$$

A systematic investigation of the weighted Hardy inequality was then done by many authors. For instance, in 1950's and 1960's plenty of results were obtained by Beesack (for details and references cf. [55]). The next major step was the characterization of (2.4) for $p = q$ and a pair of general weights. This was obtained around 1969-1972 by several authors independently, including Artola, Tomaselli, Talenti, Boyd–Erdos and Muckenhoupt, of which some published their results and some did not. The result reads as follows. (Here and throughout, given $p \in (1,\infty)$, we denote, as usual, $p' = \frac{p}{p-1}$.)

Theorem 2.1. *The necessary and sufficient condition for* (2.4) *when* $p = q$, $1 < p < \infty$, *is*

$$\sup_{t\in(0,\infty)} \left(\int_t^\infty w(s)\,ds\right)^{1/p} \left(\int_0^t v(s)^{1-p'}\,ds\right)^{1/p'} < \infty. \tag{2.6}$$

Since a particularly nice and easy direct proof was given by Muckenhoupt in [69], (2.6) is often called the *Muckenhoupt condition.*

In 1970's, an extensive research was dedicated to the extension of Theorem 2.1 to the case where the parameters p and q are different. While this is almost immediate when $p \leqslant q$, in the opposite case the characterization of (2.4) is much more difficult. This diversity is important and interesting and we return to it later in connection with other topics.

Early steps in searching necessary and sufficient conditions for (2.4) when $p < q$ can be traced back to 1930 when Hardy and Littlewood [42] found it for power weights. Various partial results were then obtained by many authors including Walsh, Boyd–Erdos, Juberg, Riemenschneider, Muckenhoupt and many more. Around 1978–1979, several authors obtained, or at least announced, a complete characterization. Among these one should name Maz'ya–Rozin, Kokilashvili, and Bradley. Later still, in 1982, Andersen and Muckenhoupt presented a different proof and characterized also corresponding weak type inequalities. In particular, Bradley [10] gave a complete characterization of (2.4) for $1 \leqslant p \leqslant q$ with a nice and simple proof in the spirit of Muckenhoupt. The result reads as follows.

Theorem 2.2. *Let $1 \leqslant p \leqslant q < \infty$ and let v, w be weights on $(0,\infty)$. Then there exists a constant C such that the inequality (2.4) holds for every $f \geqslant 0$ on $(0,\infty)$ if and only if*

$$\sup_{t>0} \left(\int_t^\infty w(s)\, ds \right)^{1/q} \left(\int_0^t v(s)^{1-p'}\, ds \right)^{1/p'} < \infty \tag{2.7}$$

when $p > 1$, and

$$\sup_{t>0} \left(\int_t^\infty w(s)\, ds \right)^{1/q} \operatorname*{ess\,sup}_{s\in(0,t)} \frac{1}{v(s)} < \infty \tag{2.8}$$

when $p = 1$.

The Bradley's paper became quite well-known, and the case $1 \leqslant p \leqslant q < \infty$ has been often called the *Bradley case.* However, since the condition (2.7) is in some sense just a double-parameter version of (2.6), we refer to it as the *Muckenhoupt condition.*

Bradley moreover made a simple, but interesting observation that (2.7) is *necessary* for (2.4) for *any* values of p and q. The fact that it is *not sufficient* when $p > q$ was since noticed by several authors, certainly by Maz'ya–Rozin and also by Sawyer. In 1984, Sawyer [78] gave the following interesting characterization of (2.4).

Theorem 2.3. *Let $0 < p < \infty$, $1 \leqslant q < \infty$ and define r by*

$$\frac{1}{r} = \frac{1}{p} - \frac{1}{q}. \tag{2.9}$$

Let v, w be weights on $(0,\infty)$. Then there exists a constant C such that the inequality (2.4) holds for every $f \geqslant 0$ on $(0,\infty)$ if and only if

$$\sup_{\{x_k\}} \left\{ \sum_{k\in\mathbb{Z}} \left(\int_{x_k}^{x_{k+1}} w(x)\, dx \right)^{r/q} \left(\int_{x_{k-1}}^{x_k} v(x)^{1-p'}\, dx \right)^{r/p'} \right\}^{1/r} < \infty, \tag{2.10}$$

where the supremum is taken over all positive increasing sequences $\{x_k\} \subset (0,\infty)$.

Sawyer's paper contained more general results concerning Lorentz spaces and also a more general weighted Hardy operator. His condition is formulated in what we call *discretized* form. The supremum is extended over all possible double-infinite sequences, sometimes called *discretization sequences* or *covering sequences* of points in $(0,\infty)$. It has proved to be of a great significance from the theoretical point of view (and we see more of it later on), however in practice, discretized conditions are for a given pair of weights v, w in general not easy to verify. From this point of view the necessary and sufficient condition for (2.4) in the case $1 \leqslant q < p < \infty$ obtained by Maz'ya and Rozin in late 1970's (cf. [66, pp. 72–76] and [67, pp.45–48]) is more manageable. The result reads as follows.

Theorem 2.4. *Let $1 \leqslant q < p < \infty$ and let r be defined by (2.9). Let v, w be weights on $(0,\infty)$. Then there exists a constant C such that the inequality (2.4) holds for every $f \geqslant 0$ on $(0,\infty)$ if and only if*

$$\left(\int_0^\infty \left(\int_x^\infty w(t)\,dt\right)^{r/q} \left(\int_0^x v(t)^{1-p'}\,dt\right)^{r/q'} v(x)^{1-p'}\,dx\right)^{1/r} < \infty \quad (2.11)$$

The condition (2.11) is currently often called a *Maz'ya type condition* and the weighted Hardy type inequality with parameters satisfying $1 \leqslant q < p < \infty$ is usually in literature referred to as the *Maz'ya case* (in comparison to the above-mentioned *Bradley case* in which the inequality between p and q is converse). Furthermore, the Maz'ya type conditions are sometimes called of *integral* type in comparison of the above-mentioned discretized ones. To find a direct proof of the equivalence between conditions of Maz'ya and of Sawyer is an interesting exercise. It should be noted that, when $p > q$, the Muckenhoupt condition is *strictly weaker* than any of these, i.e., the implication (2.11)⇒(2.7) cannot be reversed.

In 1980's, one of the cases of parameters that still remained open, namely, $0 < q < 1$ and $p > 1$, was studied by Sinnamon, who proved in [80] that, in that situation, again, the Maz'ya condition (2.11) is necessary and sufficient for (2.4). In fact, in the case $q < 1$, one has q' negative, so it is better to replace (2.11) by

$$\left(\int_0^\infty \left(\int_0^x v(t)^{1-p'}\,dt\right)^{r/p'} \left(\int_x^\infty w(t)\,dt\right)^{r/p} w(x)\,dx\right)^{1/r} < \infty, \quad (2.12)$$

whose equivalence to (2.11) is obtained through integration by parts.

Sinnamon used new methods based on the Halperin level function, a tool that, as we will see, has been used later by several other authors, too.

The case $0 < q < 1$ and $p = 1$ was solved by Sinnamon and Stepanov in [81].

Theorem 2.5. *Let $0 < q < 1$, $p = 1$ and let v, w be weights on $(0,\infty)$. Then there exists a constant C such that the inequality (2.4) holds for every $f \geqslant 0$ on $(0,\infty)$ if and only if*

$$\left(\int_0^\infty \left(\operatorname*{ess\,inf}_{0<t<x} w(t)\right)^{\frac{q}{q-1}} \left(\int_x^\infty v(t)\,dt\right)^{\frac{q}{1-q}} v(x)\,dx\right)^{\frac{1-q}{q}} < \infty. \quad (2.13)$$

Finally, the case $0 < p < 1$ is not of interest since (2.4) can never hold in such situation in reasonable circumstances (this was noticed by several authors, including Rudin, Maligranda, Lai, and Prokhorov–Stepanov, the reader interested in details is kindly referred to [55]).

The action of Hardy type operators on various function spaces was extensively studied by many authors. Particular results devoted to Orlicz spaces, Lorentz spaces and more general structures have been obtained.

All the above-mentioned function spaces and many more can be hidden under a common roof of the so-called Banach function spaces. Sometimes it is useful to work in this rather general setup. We now give some basic preliminary material from the theory of Banach function spaces. Here, a standard general reference is [7].

Here and throughout, χ_E denotes the characteristic function of a set E.

Definition 2.6. Let v be a weight on $(0,\infty)$. We say that a normed linear space (X,v) is a *Banach function space* if

1. the norm is defined for all measurable functions f, and $\|f\|_{(X,v)} < \infty$ if and only if $f \in (X,v)$;
2. the norm in (X,v) of f is always equal to the norm of $|f|$;
3. if $0 \leqslant f_n \nearrow f$ almost everywhere, then $\|f_n\|_{(X,v)} \nearrow \|f\|_{(X,v)}$;
4. if $v(E) < \infty$, then $\|\chi_E\|_{(X,v)} < \infty$;
5. for every $E \subset (0,\infty)$ of finite measure there exists a constant C_E such that
$$\int_E f(x)v(x)\,dx \leqslant C_E\|f\|_{(X,v)}.$$

Definition 2.7. Let v be a weight on $(0,\infty)$ and let (X,v) be a Banach function space. Then the set

$$(X',v) := \left\{f;\ \int_0^\infty f(x)g(x)v(x)\,dx < \infty \quad \forall\, g \in (X,v)\right\}$$

is called the *associate space of* (X,v).

The space (X',v) is equipped with the norm

$$\|f\|_{(X',v)} := \sup\left\{\int_0^\infty f(x)g(x)v(x)\,dx;\ \|g\|_{(X,v)} \leqslant 1\right\}.$$

The spaces (X,v) and (X',v) are Banach spaces and it is always true that $(X'',v) = (X,v)$. Moreover, the Hölder inequality

$$\int_0^\infty f(x)g(x)v(x)\,dx \leqslant \|f\|_{(X,v)}\|g\|_{(X',v)}$$

holds, and is saturated in the sense that for any ε and any function $f \in (X,v)$ there exists a function $g \in (X',v)$ such that $\|g\|_{(X',v)} \leqslant 1$ and

$$\|f\|_{(X,v)} \geqslant \int_0^\infty f(x)g(x)v(x)\,dx - \varepsilon.$$

A very important notion in the theory of Banach function spaces is the absolute continuity of a norm. This property is particularly important when compactness of operators is in question, as we will see later.

Definition 2.8. Let $\{E_n\}_{n\in\mathbb{N}}$ be a sequence of subsets of $(0,\infty)$. We say that E_n tends to an empty set, written $E_n \to \varnothing$, if the characteristic functions χ_{E_n} of E_n converge in n pointwise almost everywhere to zero. If the sequence is moreover monotone in the sense that $E_{n+1} \subset E_n$ for every $n \in \mathbb{N}$, then we write $E_n \searrow \varnothing$.

It should be noted that E_n in Definition 2.8 are not required to have finite measure.

Definition 2.9. Let (X,v) be a Banach function space on $(0,\infty)$. We say that a function $f \in (X,v)$ has an *absolutely continuous norm* in (X,v) if $\|f\chi_{E_n}\| \to 0$ whenever $E_n \to \varnothing$. The set of all functions in (X,v) of absolutely continuous norm is denoted by $(X,v)_a$. If $(X,v) = (X,v)_a$, then the space (X,v) itself is said to have *absolutely continuous norm.* If K is a subset of (X,v), we say that K has *uniformly absolutely continuous norms* in (X,v) if, for every sequence $E_n \to \varnothing$,

$$\lim_{n\to\infty} \sup_{f\in K} \|f\chi_{E_n}\|_{(X,v)} = 0.$$

Example 2.10. If $1 \leqslant p < \infty$, the weighted Lebesgue space $L^p_v(0,\infty)$ has absolutely continuous norm. On the other hand, L^∞_a contains only the zero function.

In connection with the investigation of the compactness properties of the Hardy operator which we quote later, a slightly modified property that the absolute continuity was introduced in [56].

Definition 2.11. Let (X,v) be a Banach function space on $(0,\infty)$. We say that a function $f \in (X,v)$ has a *continuous norm* in (X,v) if for any $a \in [0,\infty)$, $b \in (0,\infty]$ and for any sequences $\{x_n\}$, $\{y_n\}$ such that either $x_n \searrow a$ or $y_n \nearrow b$ we have

$$\lim_{n\to\infty} \|f\chi_{(a,x_n)}\|_{(X,v)} = \lim_{n\to\infty} \|f\chi_{(y_n,b)}\|_{(X,v)} = 0.$$

While absolute continuity of a norm is a classical notion in the theory of Banach function spaces, and its connection to compactness properties of integral operators has been widely known for many decades (cf., for example, [63]), the pointwise continuity of norm was, as far as we know, first introduced in [56]. As we will see, it is a key tool to balance compactness theorems in situations where attempts to give necessary conditions in terms of absolute continuity of norm have failed. For a similar purpose it was used in [27].

Obviously, $(X,v)_a \subset (X,v)_c \subset (X,v)$. In [56], the question whether any of these two inclusions can be proper in general, was left open. This question turned out to be quite deep, but only a little later Lang and Nekvinda in [58] proved that, actually, both these inclusions can be proper (which means in particular that the notion of continuity of norm is strictly weaker than that of the absolute continuity of norm). Later still, in [59], together with Rákosník, they constructed a wild enough space (X,v) in which every function has continuous norm, but only the zero function has absolutely continuous norm. Their ingenious constructions were based on properties of the Cantor set.

The action of a Hardy operator on general weighted Banach function spaces was studied by Berezhnoi [8]. He used the concept of the ℓ-convexity and ℓ-concavity of function spaces, treated in detail, for example, by Lindenstrauss and Tzafriri in [61]. His condition is a discretized one, in some sense related to the capacitary conditions treated before by Maz'ya.

We now concentrate on one of the possible extreme cases of the Hardy inequality, namely, we investigate what happens when the target space is near L^∞ (endowed with the usual norm $\|f\|_{L^\infty} := \operatorname{ess\,sup}_{x\in(0,\infty)} |f(x)|$). The boundedness of the Hardy operator from a weighted Banach function space into L^∞ was treated in [56], but the simple result can be traced in other works, too. It reads as follows.

Theorem 2.12. *Let (X,v) be a weighted Banach function space on $(0,\infty)$. Then there exists a constant $C>0$ such that for all $f\in(X,v)$*

$$\|Hf\|_{L^\infty} \leqslant C\|f\|_{(X,v)} \tag{2.14}$$

if and only if the function v^{-1} belongs to the space (X',v), i.e.,

$$\left\|\frac{1}{v}\right\|_{(X',v)} < \infty. \tag{2.15}$$

Moreover, the best constant C in (2.14) equals $\|v^{-1}\|_{(X',v)}$.

The proof of Theorem 2.12 is in fact nothing deeper than a simple exercise, but it illustrates nicely the situation in the sense that the decisive role here is played by certain properties of the function v^{-1} as an element of the associate space to (X,v). These facts proved useful when more difficult tasks were treated.

In many situations, L^∞ is not a convenient space for various reasons. This happens, for example, in harmonic analysis or Fourier analysis where many important operators such as the Hilbert transform or singular integrals in general do not have satisfactory properties near L^∞ or in the interpolation theory where suitable replacements for L^∞ as an endpoint function space have to be found. The role of such a replacement is in some circumstances successfully played by the space BMO. We here concentrate on the characterization of the boundedness of the Hardy operator into a weighted version of the BMO space.

We first need some notation and definitions.

Let w be a weight on $(0,\infty)$, let f be a measurable function on $(0,\infty)$ and let E be a measurable subset of $(0,\infty)$. We then denote by $f_{w,E}$ the *weighted integral mean* of f over E, i.e.,

$$f_{w,E} := \frac{1}{w(E)} \int_E f(x)w(x)\,dx,$$

if the integral makes sense, where, as usual, we write

$$w(E) := \int_E w(x)\,dx.$$

Definition 2.13. Let w be a weight on $(0,\infty)$. The space (BMO, w) is the set of all measurable functions f on $(0,\infty)$ such that

$$\|f\| := \sup_{[a,b]\subset(0,\infty)} \frac{1}{b-a} \int_a^b \left|f(x) - f_{w,(a,b)}\right| w(x)\,dx < \infty. \tag{2.16}$$

Now, we are interested in the question when the Hardy operator H is bounded from (X,v) into (BMO, w). This is a much more difficult task than the analogous one with the target space equal to L^∞. This problem was studied first in [56] and later still in [60]. The key task toward the characterization was to find a suitable replacement for the function v^{-1} whose role was described in Theorem 2.12. It turned out that in the case of weighted BMO, this role is played not by a single function, but by a whole bunch of functions parametrized by a system of subintervals of $(0,\infty)$.

Definition 2.14. Let $0 < a < b < \infty$. We define

$$G_{(a,b)}(x) := \frac{w(a,x)w(x,b)}{(b-a)w(a,b)v(x)} \chi_{(a,b)}(x). \tag{2.17}$$

With this notation, we can present our characterization.

Theorem 2.15. *Let (X,v) be a weighted Banach function space on $(0,\infty)$ and let w be another weight on $(0,\infty)$. Then there exists a constant $C>0$ such that for all $f \in (X,v)$*

$$\|Hf\|_{(\mathrm{BMO},w)} \leqslant C\|f\|_{(X,v)} \tag{2.18}$$

if and only if

$$\sup_{[a,b]\subset(0,\infty)} \left\|G_{(a,b)}\right\|_{(X',v)} < \infty. \tag{2.19}$$

In the case of nonweighted BMO (i.e., $w \equiv 1$), the condition (2.19) is of course replaced by

$$\sup_{[a,b]\subset(0,\infty)} \left\| \frac{(x-a)(b-x))}{(b-a)^2 v(x)} \chi_{(a,b)}(x) \right\|_{(X',v)} < \infty. \tag{2.20}$$

It should be noted that the nonweighted BMO is a strictly larger function space that L^∞. For certain cases of the parameters involved it may happen that the conditions (2.20) and (2.15) coincide; this happens, for example, when $X = L^1$. However, in general, (2.20) is strictly weaker than (2.15). A simple exercise shows that such an example is obtained by putting $v(x) := x^{p-1}$ and $X = L^p$, $1 < p < \infty$.

Aside from the boundedness, the most important property of the Hardy operator from the point of view of applications, is its compactness on various function spaces. Again, it has been studied by many authors and a vast literature is available on this topic. We start here with a summary of results for $1 < p, q < \infty$. Partial steps have been done by Kac–Krein, Stuart, Juberg, Riemenschneider, Maz'ya, Opic–Kufner, Mynbaev–Otelbaev, Stepanov, and others. An interesting feature about the following theorem is how it reveals one very important difference between the Bradley case and the Maz'ya case.

Theorem 2.16. (i) *Let $1 < p \leqslant q < \infty$. Then the operator H, defined by* (1.1), *is compact from $L^p_v(0,\infty)$ into $L^q_w(0,\infty)$ if and only if*

$$\lim_{x\to 0+} \left(\int_x^\infty w(s)\,ds \right)^{1/q} \left(\int_0^x v(s)^{1-p'}\,ds \right)^{1/p'} = 0 \tag{2.21}$$

and

$$\lim_{x\to \infty} \left(\int_x^\infty w(s)\,ds \right)^{1/q} \left(\int_0^x v(s)^{1-p'}\,ds \right)^{1/p'} = 0. \tag{2.22}$$

(ii) *Let $1 < q < p < \infty$. Then the operator H, defined by* (1.1), *is compact from $L^p_v(0,\infty)$ into $L^q_w(0,\infty)$ if and only if it is bounded, i.e., if and only if the Maz'ya condition* (2.11) *holds.*

So it turns out that, in the Bradley case, compactness is born from boundedness by forcing the endpoint limits to be zero, while, in the Maz'ya case, there is no difference between boundedness and compactness, boundedness already entails compactness.

In case a space ℓ of sequences can be found such that the above-mentioned ℓ-convexity/concavity relation between the space (X,v) and (Y,w) takes place, a discretization characterization of the compactness of T from (X,v) to (Y,w) was found by Berezhnoi [8].

We now survey some results from [27]. We work with four possibly different weights, v, w, φ, and ψ and with the Hardy operator T, defined as

$$Tf(x) := \varphi(x) \int_0^x f(t)\psi(t)v(t)\,dt.$$

Of course, our use of a double weight ψv rather than a single one comes purely from a typographical convenience. We also need the following modification of the operator T: Given an interval $I \subset (0,\infty)$, we denote

$$T_I f(x) := \chi_I(x)\varphi(x) \int_0^x \chi_I(t) f(t)\psi(t)v(t)\,dt.$$

We are interested in the action of the operator T from one weighted Banach function space, (X,v), into another one (Y,w). First, one easily gets a necessary condition in the spirit of the Muckenhoupt condition (2.6) in the following way:

Assume

$$T : (X,v) \to (Y,w),$$

i.e., there exists a constant C such that, for every $f \geqslant 0$,

$$\|Tf\|_{(Y,w)} \leqslant C\|f\|_{(X,v)}.$$

Thus, given any $R > 0$, we have

$$\begin{aligned} C \geqslant C\|f\|_{(X,v)} \geqslant \|Tf\|_{(Y,w)} &\geqslant \|Tf\chi_{(R,\infty)}\|_{(Y,w)} \\ &\geqslant \int_0^R f(t)\psi(t)v(t)\,dt \|\varphi\chi_{(R,\infty)}\|_{(Y,w)}. \end{aligned}$$

Taking now the supremum over all such f and R and using Definition 2.7, we obtain

$$\sup_{R\in(0,\infty)} \|\varphi\chi_{(R,\infty)}\|_{(Y,w)} \|\psi\chi_{(0,R)}\|_{(X',v)} < \infty.$$

Since a similar reasoning works for T_I in place of T, we finally get

Proposition 2.17. *Let I be an arbitrary subinterval of $(0,\infty)$. Suppose that*

$$T_I : (X,v) \to (Y,w). \tag{2.23}$$

Then,

$$\sup_{R\in I} \|\varphi\chi_I\chi_{(R,\infty)}\|_{(Y,w)} \|\psi\chi_I\chi_{(0,R)}\|_{(X',v)} < \infty. \tag{2.24}$$

Note that in the particular case $I = (0,\infty)$, $X = L^p$ and $Y = L^q$, the condition (2.24) coincides with the Muckenhoupt condition (2.6). In consistency of this observation, we call (2.24) a *Muckenhoupt condition* even in the general setting of weighted Banach function spaces.

Proposition 2.17 shows that the Muckenhoupt condition is always necessary for the boundedness of the operator T from (X,v) into (Y,w). This is just a more general version of the above-mentioned observation of Bradley.

We recall however, that, in the case of weighted Lebesgue spaces, the Muckenhoupt condition is sufficient for $T : (X, v) \to (Y, w)$ when $1 \leqslant p \leqslant q < \infty$, but it is not sufficient when $1 < q < p < \infty$. Thus, naturally, the collection of all pairs of Banach function spaces $((X, v), (Y, w))$ can be divided into two subclasses; one containing those pairs for which (2.24) is sufficient for $T_I : (X, v) \to (Y, w)$ and those for which it is not. This led us in [27] to the following definition.

Definition 2.18. We say that a pair of weighted Banach function spaces $((X, v), (Y, w))$ belongs to the *Muckenhoupt category* $\mathfrak{M}(\varphi, \psi)$, if for any interval $I \subset (0, \infty)$ the condition (2.24) implies (2.23) and the estimate

$$\begin{aligned} C^{-1}\|T_I\|_{(X,v)\to(Y,w)} &\leqslant \sup_{R\in I} \|\varphi\chi_I\chi_{(R,\infty)}\|_{(Y,w)}\|\psi\chi_I\chi_{(0,R)}\|_{(X',v)} \\ &\leqslant C\|T_I\|_{(X,v)\to(Y,w)} \end{aligned}$$

holds with some constant C independent of v, w and I.

For pairs of weighted Banach function spaces belonging to $\mathfrak{M}(\varphi, \psi)$, the compactness of T from (X, v) into (Y, w) can be characterized. We survey the results in the following theorem.

Theorem 2.19. *Let v, w, φ, ψ be weights on $(0, \infty)$ and let (X, v) and (Y, w) be two weighted Banach function spaces on $(0, \infty)$. Suppose that the pair $((X, v), (Y, w))$ belongs to the Muckenhoupt category $\mathfrak{M}(\varphi, \psi)$.*

Then $T_{\varphi,\psi}$ is compact from (X, v) into (Y, w) if and only if the following two conditions are satisfied:

(i) *both conditions*

$$\lim_{a\to 0+} \sup_{R\in(0,a)} \|\varphi\chi_{(R,a)}\|_{(Y,w)}\|\psi\chi_I\chi_{(0,R)}\|_{(X',v)} = 0 \tag{2.25}$$

and

$$\lim_{b\to\infty} \sup_{R\in(b,\infty)} \|\varphi\chi_{(R,\infty)}\|_{(Y,w)}\|\psi\chi_I\chi_{(b,R)}\|_{(X',v)} = 0; \tag{2.26}$$

(ii) *the functions φ and ψ have continuous norms at every point $\alpha \in (0, \infty)$.*

To finish this section, we once again return to the properties of the Hardy operator in the cases where the target space is L^∞ or BMO, but this time we investigate its compactness.

The following theorem was proved in [56].

Theorem 2.20. *Let (X, v) be a weighted Banach function space on $(0, \infty)$. Then the Hardy operator H is compact from (X, v) into L^∞ if and only if the function v^{-1} has continuous norm in (X', v).*

Replacing L^∞ by the (nonweighted) BMO space leads to the following result, which was proved in [60] (in the earlier work [56] a slightly different and not so nice characterizing condition was given).

Theorem 2.21. *Let (X, v) be a weighted Banach function space on $(0, \infty)$. Let $K \subset (X, v)$ be the collection of all functions $G_I(x)$, defined by*

$$G_I(x) := \chi_I(x) \frac{(b-x)(x-a)}{(b-a)^2} \frac{1}{v(x)},$$

where $I = [a, b] \subset (0, \infty)$. Then, the Hardy operator H is compact from (X, v) into L^∞ if and only if K has uniformly continuous norms in (X', v).

An interesting question, which was treated in [60], is how big is the gap between L^∞ and the (bigger) space BMO when measured by the action of the Hardy operator from weighted Banach function spaces. Considering the four statements, namely

(i) H is bounded from $X(v)$ into L^∞;
(ii) H is bounded from $X(v)$ into BMO;
(iii) H is compact from $X(v)$ into L^∞;
(iv) H is compact from $X(v)$ into BMO,

we can ask about all possible general implications among them. Now, we obviously have the trivial ones,

$$\text{(iii)} \Rightarrow \text{(i)} \Rightarrow \text{(ii)} \qquad \text{and} \qquad \text{(iii)} \Rightarrow \text{(iv)} \Rightarrow \text{(ii)},$$

but the question is whether there are any other implications valid. In [60] it was shown that this is not the case. We now survey the corresponding counterexamples.

Example 2.22. With no loss of generality we work here on $(0, \frac{1}{4})$ rather than on $(0, \infty)$. Let $1 < p < \infty$, and define for $n = 2, 3, \ldots$

$$\alpha_n := 2^{-n} e^{-1/n}, \qquad \beta_n = 2^{-n}, \qquad J_n := (\alpha_n, \beta_n), \qquad R_n := (\beta_{n+1}, \alpha_n),$$

and

$$v(x) = \sum_{n=2}^{\infty} \left(x^{p-1} \chi_{J_n}(x) + \chi_{R_n}(x) \right), \; x \in \left(0, \frac{1}{4}\right).$$

Put $(X, v) = L^p_v$. Then (iv) is true and (i) is false.

For the next example, we need the *Lorentz space* $L^{p,1}(v)$, defined as the collection of all functions f such that

$$\|f\|_{L^{p,1}(v)} := \int_0^\infty \lambda v\left(\{|f| > \lambda\}\right)^{\frac{1}{p}-1} d\lambda$$

is finite.

Example 2.23. Let $1 < p < \infty$, $v(x) = x^{p-1}$ for $x \in (0,\infty)$, and let $(X, v) = L^{p,1}(v)$. Then (i) is true and (iv) is false.

The above two examples demonstrate the interesting fact that there is no relation between (i) and (iv). Next, we concentrate on the fact that none of the implications above can be reversed.

Example 2.24. Let $1 < p < \infty$, $v(x) = x^{p-1}$ for $x \in (0,\infty)$, and let $(X, v) = L^p(v)$. Then (ii) is true, but both (iv) and (i) are false.

For the final counterexample we need to introduce a new function space. Let Z be defined as the set of all functions f on $(0,1)$ such that

$$\|f\|_Z := \sup_{0<\alpha\leqslant 1} \int_0^1 |f(x)|\, \alpha x^{\alpha-1}\, dx$$

is finite. It is not difficult to verify that Z is a Banach function space (non-weighted, i.e., with $w \equiv 1$) and that it does not have absolutely continuous norm. Some more interesting properties of this space are derived in [60].

Example 2.25. Similarly to Example 2.22, with no loss of generality we work here on $(0,1)$ rather than on $(0,\infty)$. Let $(X, v) = Z'$, the associate space of Z. Then both (iv) and (i) are true but (iii) is false.

Since Example 2.25 in fact works with Z' rather than Z, it would be of interest to give a reasonable direct characterization of Z. Some attempts have been made in unpublished manuscripts, but unfortunately none was correct. So as far as we know, this problem is still open.

3 Hardy Operator on Monotone Functions

A very special part of the investigation of properties of the Hardy operator is how it acts on functions that are a-priori assumed to be monotone on $(0,\infty)$. The real boom of this topic started in 1990, but particular results in this direction can be found in earlier works, for example of Bennett, Renaud or Burenkov. More details on the history of the subject can be found, again, in [55].

First of all, we need some background on the nonincreasing rearrangement of a function and on related function spaces.

We denote by $\mathfrak{M}(0,\infty)$ the class of real-valued measurable functions on $(0,\infty)$ and by $\mathfrak{M}_+(0,\infty)$ the class of nonnegative functions in $\mathfrak{M}(0,\infty)$. Given $f \in \mathfrak{M}(0,\infty)$, its *nonincreasing rearrangement* is defined by

$$f^*(t) = \inf\{\lambda > 0;\ |\{x \in (0,\infty);\ |f(x)| > \lambda\}| \leqslant t\}, \quad t \in [0,\infty).$$

We also define the *maximal nonincreasing rearrangement* of f by

$$f^{**}(t) = t^{-1} \int_0^t f^*(s)\, ds, \quad t \in (0, \infty).$$

It should be noted that the operation $f \mapsto f^*$ is not subadditive. This fact causes a lot of problems when function spaces are defined in terms of f^*. For two functions f and g, one only has the estimate

$$(f+g)^*(s+t) \leqslant f^*(s) + g^*(t), \qquad s, t \in (0, \infty).$$

On the other hand, the operation $f \mapsto f^{**}$ is more friendly from this point of view in the sense that

$$(f+g)^{**}(t) \leqslant f^{**}(t) + g^{**}(t), \qquad t \in (0, \infty).$$

Definition 3.1. A Banach function space X of functions defined on $(0, \infty)$, equipped with the norm $\|\cdot\|_X$, is said to be *rearrangement-invariant* if it satisfies, aside from the axioms (1)–(5) of Definition 2.6, the relation

$$\|f\|_X = \|g\|_X \text{ whenever } f^* = g^*.$$

A basic tool for working with rearrangement-invariant spaces is the *Hardy–Littlewood–Pólya* (HLP) *principle*, treated in [7, Chapter 2, Theorem 4.6]). It asserts that $f^{**}(t) \leqslant g^{**}(t)$ for every $t \in (0, \infty)$ implies $\|f\|_X \leqslant \|g\|_X$ for every r.i. space X.

The *inequality of Hardy and Littlewood* states that

$$\int_0^\infty |f(x)g(x)|\, dx \leqslant \int_0^\infty f^*(t)g^*(t)\, dt, \qquad f, g \in \mathfrak{M}(0, \infty). \tag{3.1}$$

R.i. spaces can be in general defined for functions acting on a general σ-finite measure space Ω, not necessarily on $(0, \infty)$ (cf. the applications to Sobolev embeddings that are mentioned in the Section 1). In such cases, for every r.i. space $X(\Omega)$, there exists a unique r.i. space $\overline{X}(0, \infty)$ on $(0, \infty)$, satisfying $\|f\|_{X(\Omega)} = \|f^*\|_{\overline{X}(0,\infty)}$. Such a space, endowed with the norm

$$\|f\|_{\overline{X}(0,\infty)} = \sup_{\|g\|_{X(\Omega)} \leqslant 1} \int_0^\infty f^*(t)g^*(t)\, dt,$$

is called the *representation space* of $X(\Omega)$.

Let X be an r.i. space. Then, the function $\varphi_X : [0, \infty) \to [0, \infty)$ given by

$$\varphi_X(t) = \begin{cases} \|\chi_{(0,t)}\|_{\overline{X}} & \text{for } t \in (0, \infty), \\ 0 & \text{for } t = 0, \end{cases}$$

is called the *fundamental function* of X. For every r.i. space X, its fundamental function φ_X is *quasiconcave* on $[0,\infty)$, i.e., it is nondecreasing on $[0,\infty)$, $\varphi_X(0)=0$, and $\frac{\varphi_X(t)}{t}$ is nonincreasing on $(0,\infty)$. Moreover, one has

$$\varphi_X(t)\varphi_{X'}(t)=t, \qquad \text{for } t\in[0,\infty).$$

For a comprehensive treatment of r.i. spaces we refer the reader to [7].

In 1990, two fundamental papers appeared. The first one was due to Ariño and Muckenhoupt [2] who studied the action of the Hardy–Littlewood maximal operator, defined at $f\in L^1_{\text{loc}}(\mathbb{R}^n)$ by

$$(Mf)(x)=\sup_{Q\ni x}|Q|^{-1}\int_Q|f(y)|\,dy, \qquad x\in\mathbb{R}^n,$$

(where the supremum is extended over all cubes $Q\subset\mathbb{R}^n$ with sides parallel to the coordinate axes and $|E|$ denotes the n-dimensional Lebesgue measure of $E\subset\mathbb{R}^n$), on the co-called *classical Lorentz spaces*. The second key paper of 1990, due to Sawyer [79], established among other things an important duality principle for weighted Lebesgue spaces restricted to the cone of nonincreasing functions. Since then, many authors were attracted to this topic and, again, a vast amount of results was obtained. We now concentrate on the relations between various types of classical Lorentz spaces.

Definition 3.2. Let $p\in(0,\infty)$ and let v be a weight. Then the *classical Lorentz space* $\Lambda^p(v)$ is defined as

$$\Lambda^p(v)=\left\{f\in\mathfrak{M}(0,\infty);\, \|f\|_{\Lambda^p(v)}:=\left(\int_0^\infty f^*(t)^p v(t)\,dt\right)^{1/p}<\infty\right\}.$$

The spaces $\Lambda^p(v)$ were introduced by Lorentz in 1951 in [62]. For appropriate values of p and for an appropriate weight v this space is a rearrangement-invariant Banach function space. However, $\|\cdot\|_{\Lambda^p(v)}$ is not always a norm (consider, for example, the cases where $p\in(0,1)$).

The difficulties concerning the norm properties of the functional $\|\cdot\|_{\Lambda^p(v)}$ (even in cases when $p\geqslant 1$) are primarily caused by the fact that the operation $f\mapsto f^*$ is not subadditive, so the triangle inequality is not guaranteed in general. This obstacle has been sometimes overcome by replacing f^* by f^{**}. Spaces whose norms involve f^{**} appeared in early 1960's, for example, in Calderón's paper [11], but they can be traced also to the works of Hunt, Peetre, O'Neil, and others. In 1990, a major step in this direction was taken by Sawyer in his already mentioned paper [79] in which the new spaces $\Gamma^p(v)$ are introduced.

Definition 3.3. Let $p\in(0,\infty)$ and let v be a weight. Then the *classical Lorentz space* $\Gamma^p(v)$ is defined as

$$\Gamma^p(v) = \left\{ f \in \mathfrak{M}(0,\infty); \|f\|_{\Gamma^p(v)} := \left(\int_0^\infty f^{**}(t)^p v(t)\, dt \right)^{1/p} < \infty \right\}.$$

The spaces $\Gamma^p(v)$ proved later to be quite useful in many branches of functional analysis, for example, in interpolation theory or in applications to Sobolev embeddings.

The question when the space $\Lambda^p(v)$ is a Banach space, was studied by several authors. Lorentz [62] proved that, for $p \geqslant 1$, $\|f\|_{\Lambda^p(w)}$ is a norm if and only if w is nonincreasing. The class of weights for which $\|f\|_{\Lambda^p(w)}$ is merely *equivalent* to a Banach norm is however considerably larger. In fact it consists of all those weights w which, for some C and all $t > 0$, satisfy

$$t^p \int_t^\infty x^{-p} w(x)\, dx \leqslant C \int_0^t w(x)\, dx \quad \text{when } p \in (1,\infty) \tag{3.2}$$

([79, Theorem 4], cf. also [2]), or

$$\frac{1}{t} \int_0^t w(x)\, dx \leqslant \frac{C}{s} \int_0^s w(x)\, dx \qquad \text{for } 0 < s \leqslant t \quad \text{when } p = 1 \tag{3.3}$$

([12, Theorem 2.3]). It has been also proved that, for $p \in [1,\infty)$, the space $\Lambda^p(w)$ is equivalent to a Banach space if and only if the Hardy–Littlewood maximal operator acts boundedly from $\Lambda^p(w)$ to a weak version of $\Lambda^p(w)$ ([17, Theorem 3.3], [18, Theorem 3.9], [12]). Furthermore, for $p > 1$, this is equivalent to $\Lambda^p(w) = \Gamma^p(w)$ ([79]). We write that a weight w *satisfies the condition* B_p, written $w \in B_p$, if (3.2) is true.

In [40, Theorem 1.1] (the proper publication of this paper appeared much later as [82]), cf. also [16, Corollary 2.2], [46, p. 6], it was observed that the functional $\|f\|_{\Lambda^p(w)}$, $0 < p \leqslant \infty$, does not have to be a quasinorm. It was shown that it is a quasinorm if and only if the function $W(t) := \int_0^t w(s)\, ds$ satisfies the Δ_2-*condition* i.e.,

$$W(2t) \leqslant CW(t) \qquad \text{for some } C > 1 \text{ and all } t \in (0,\infty). \tag{3.4}$$

In [24], an even perhaps more exotic behavior of classical Lorentz spaces was observed: it can happen that the sum of two functions in the "space" is not in the space. Obviously $f \in \Lambda^p(w)$ implies that $\lambda f \in \Lambda^p(w)$ for every $\lambda \in \mathbb{R}$. However in general, perhaps surprisingly, $\Lambda^p(w)$ is not a linear space. We survey the main results of [24]. We start with a proposition which illustrates well the problem.

Proposition 3.4. *Suppose that w and v are nonnegative measurable functions on $(0,\infty)$ and a is a positive number such that $w(t) = 0$ for a.e. $t \in (0,a)$ and $w(t) = v(t)$ for all $t \geqslant a$, and the functions $V(t) = \int_0^t v(s)ds$ and $W(t) = \int_0^t w(s)ds$ are finite for all $t > 0$.*

Let $f : (0,\infty) \to \mathbb{R}$ be a measurable function. Then the following are equivalent:

(i) $f \in \Lambda^p(w)$.

(ii) $\min\{\lambda, |f|\} \in \Lambda^p(v)$ *for some positive number* λ.

(iii) $\min\{\lambda, |f|\} \in \Lambda^p(v)$ *for every positive number* λ.

With the help of Proposition 3.4, we can formulate the characterization of the situations in which the space $\Lambda^p(v)$ is linear.

Theorem 3.5. *The following are equivalent:*

(i) $\Lambda^p(w)$ *is not a linear space.*

(ii) *There exists a sequence of positive numbers* t_n *tending either to* 0 *or to* ∞, *such that* $W(2t_n) > 2^n W(t_n)$ *for all* $n \in \mathbb{N}$.

(iii) *There exists a sequence of positive numbers* t_n *such that* $W(2t_n) > 2^n W(t_n) > 0$ *for all* $n \in \mathbb{N}$.

Theorem 3.5 has an interesting corollary.

Corollary 3.6. *The following are equivalent:*

(i) $\Lambda^p(w)$ *is a linear space.*

(ii) *There exist positive constants* α, β *and* C *such that* $W(2t) \leqslant CW(t)$ *for all* $t \leqslant \alpha$ *and all* $t \geqslant \beta$.

(iii) *There exists a constant* C' *such that one of the following two conditions hold: Either*

(iii-A)

$$W(2t) \leqslant C'W(t) \tag{3.5}$$

for all $t > 0$, *or*

(iii-B) $W(t) = 0$ *on some interval* $(0, a)$ *and* (3.5) *holds for all* t *in some interval* (b, ∞).

Example 3.7. We will find a measurable function $w : (0,\infty) \to [0,\infty)$ such that the two functions $W(x) = \int_0^x w(t)dt$ and $\Phi(x) = \int_x^\infty t^{-p} w(t)dt$ are both finite for all $x > 0$, but the set $\Lambda^p(w) = \left\{f : \int_0^\infty f^*(t)^p w(t)dt < \infty\right\}$ is not a linear space.

These conditions on W and Φ are apparently necessary and sufficient to ensure that the space $\Gamma^p(w)$ is nontrivial so it seems relevant to impose them here.

Initially the w which we construct can assume the value 0. But, as we will see, it is easy to modify this to an example where w is strictly positive.

Here is the construction:

Let us first define a sequence of positive numbers w_n recursively by setting $w_1 = 1$ and

$$w_n = (2^n - 1)(w_1 + w_2 + \dots + w_{n-1}) \tag{3.6}$$

for all $n > 1$. Then we define a second sequence of positive numbers s_n recursively by setting $s_1 = 2$ and, for each $n > 1$,

$$s_n = \max\left\{2s_{n-1}, w_n^{1/p}2^{n/p} + 1\right\}. \tag{3.7}$$

The function $w : (0, \infty) \to [0, \infty)$ is defined by $w = \sum_{n=1}^{\infty} w_n \chi_{(s_n-1,s_n]}$. It follows from (3.7) that $w_n/(s_n - 1)^p \leqslant 2^{-n}$ and so

$$\Phi(0) = \int_0^{\infty} t^{-p} w(t)dt = \sum_{n=1}^{\infty} w_n \int_{s_n-1}^{s_n} t^{-p}dt \leqslant \sum_{n=1}^{\infty} w_n(s_n - 1)^{-p} \leqslant 1.$$

Obviously we also have $\int_0^x w(t)dt < \infty$ for each $x > 0$. Thus w satisfies the conditions mentioned above which ensure that $\Gamma^p(w)$ is nontrivial.

We claim that for this particular choice of w, the set $\Lambda^p(w)$ is not a linear space.

To show this let us first observe that, by (3.7) we have $s_{n-1} \leqslant s_n/2 \leqslant s_n - 1$ and so

$$\int_0^{s_{n-1}} w(t)dt = \int_0^{s_n/2} w(t)dt = w_1 + w_2 + \dots + w_{n-1}$$

It follows, using (3.6), that

$$\begin{aligned} W(s_n) &= \int_0^{s_n} w(t)dt = w_1 + w_2 + \dots + w_{n-1} + w_n \\ &= 2^n(w_1 + w_2 + \dots + w_{n-1}) = 2^n \int_0^{s_n/2} w(t)dt = 2^n W(s_n/2). \end{aligned}$$

Thus the sequence $\{t_n\}$ given by $t_n = s_n/2$ tends to ∞ and satisfies

$$W(2t_n) = 2^n W(t_n).$$

In other words, w satisfies condition (iii) of Theorem 3.5. Consequently, $\Lambda^p(w)$ is not a linear space.

To get another less exotic example, i.e. where the weight function is strictly positive, we simply replace w by $w + u$ where u is any strictly positive function for which the above function f satisfies $\int_0^{\infty} f(t)^p u(t)dt < \infty$ and $\int_t^{\infty} s^{-p}(w(s) + u(s))\, ds < \infty$.

Example 3.8. Let $w = \chi_{[1,\infty)}$. Then $\Lambda^p(w)$ is not quasinormable, but is a linear space.

As is clear from the above examples, the analysis of the structure of the classical Lorentz space $\Lambda^p(v)$ requires some knowledge of the boundedness of the Hardy operator H restricted to the monotone functions, i.e., the existence of a constant C such that, for every f,

$$\left(\int_0^{\infty} \left(\frac{1}{t}\int_0^{\infty} f^*(s)\, ds\right)^q w(t)\, dt\right)^{\frac{1}{q}} \leqslant C\left(\int_0^{\infty} f^*(s)^p v(t)\, dt\right)^{\frac{1}{p}}. \tag{3.8}$$

Moreover, since, due to the Riesz–Wiener estimate from above and the Stein–Herz estimate from below, the Hardy–Littlewood maximal operator M satisfies

$$C^{-1}(Mf)^*(t) \leqslant f^{**}(t) \leqslant C(Mf)^*(t)$$

with some C independent of all f and $t \in (0,\infty)$, we conclude that the inequality (3.8) is equivalent to saying that M is bounded from $\Lambda^p(v)$ to $\Lambda^q(w)$.

This was first observed by Ariño and Muckenhoupt in [2] who noticed that the boundedness of the Hardy–Littlewood maximal operator on a classical Lorentz space $\Lambda^p(v)$ is equivalent to (3.8) (with $p = q$ and $v = w$), which can be considered as the boundedness of the operator H on the weighted Lebesgue space $L^p(w)$, but *restricted to nonincreasing functions.* Ariño and Muckenhoupt showed that the set of weights for which the *restricted* inequality holds is considerably wider than that of those weights, for which the unrestricted Hardy inequality is true. Even though some results in this direction have been available before (due to Boyd, Maz'ya, Hunt, Calderón, O'Neil, Peetre, Krein–Petunin–Semenov, Renaud, Bennett, Maligranda, and others), the real avalanche of papers on this topic started only after Ariño and Muckenhoupt's paper appeared in 1990.

We now quote the fundamental duality result of Sawyer [79] which, again, illustrates the difference between restricted and unrestricted inequalities. We use the symbol $f\downarrow$ to indicate that f is nonicreasing on $(0,\infty)$.

Theorem 3.9. *Let $1 < p < \infty$ and let v be a weight on $(0,\infty)$ and let g be a nonnegative function on $(0,\infty)$. Then*

$$\sup_{f\geqslant 0, f\downarrow} \frac{\int_0^\infty f(x)g(x)\,dx}{\|f\|_{L^p_v}}$$
$$\approx \left(\int_0^\infty \left(\int_0^x g(t)\,dt\right)^{p'} \frac{v(x)}{V(x)^{p'}}\,dx\right)^{\frac{1}{p'}} + \frac{\int_0^\infty g(x)\,dx}{\left(\int_0^\infty v(x)\,dx\right)^{\frac{1}{p}}}. \tag{3.9}$$

It is instructive to compare this result with the classical (nonrestricted) duality relation

$$\sup_{f\geqslant 0} \frac{\int_0^\infty f(x)g(x)\,dx}{\|f\|_{L^p_v}} \approx \left(\int_0^\infty g(x)^{p'} v(x)^{1-p'}\,dx\right)^{\frac{1}{p'}}.$$

Even in the case where the second summand in (3.9) vanishes, the right-hand sides are essentially different.

One of the major applications of the Sawyer's result is the following characterization of an embedding between two classical Lorentz spaces.

Theorem 3.10. *Let $1 < p \leqslant q < \infty$ and let v and w be weights on $(0,\infty)$. Then the embedding*

$$\Lambda^p(v) \hookrightarrow \Lambda^q(w)$$

holds if and only if there exists a constant C such that for every $t > 0$

$$\left(\int_0^t w(x)\,dx\right)^{\frac{1}{q}} \leqslant C\left(\int_0^t v(x)\,dx\right)^{\frac{1}{p}}. \tag{3.10}$$

In 1993, Stepanov [83] discovered the very important *reduction principle*, based on the observation that a restricted weighted inequality can be often reduced to a nonrestricted one (which might be easier to handle or for which criteria might be known) by considering only those nonincreasing functions f^* on $(0,\infty)$ which allow the representation in the form $f^*(t) = \int_t^\infty h(s)\,ds$ for some $h \geqslant 0$ on $(0,\infty)$. Of course not every nonincreasing function is of this form, but it is very often the case that a given weighted inequality can be reduced to such functions without loss of information. This can be proved by a standard approximation argument, which is carried out in detail, for example, in [14, Proof of Proposition 7.2]. The *Stepanov trick* brought into the play a new nice simple proof of the Sawyer's duality result and also allowed to extend the range of parameters in Theorem 3.10.

Since 1990's, a great effort has been spent by many authors to obtain a characterization of all possible embeddings between the spaces $\Lambda^p(v)$ and $\Gamma^p(w)$. Important contributions were made by Neugebauer, Andersen, Stepanov, Raynaud, Lai, Carozza, Carro–García del Amo–Soria, Gol'dman, Carro–Soria, Gol'dman–Heinig–Stepanov, Braverman, Heinig–Stepanov, Myasnikov–Persson–Stepanov, Bergh–Burenkov–Persson, Burenkov–Gol'dman, Heinig–Maligranda, Sinnamon–Stepanov, Maligranda, Montgomery-Smith, Kamińska–Maligranda, Carro–Raposo–Soria, and others. A survey describing the state of the topic by the end of 1990's was given in [15]. However, the situation has considerably changed since then. Some of the cases that either had not been known at all then or had been characterized by some unsatisfactory conditions were obtained recently. In fact, we can safely say that, today, the problem of characterizing all possible mutual embeddings between all possible types of classical Lorentz spaces has been settled completely.

In [38], a new approach based on discretization techniques of [71] and [37] was applied to classical Lorentz spaces in order to obtain necessary and sufficient conditions on parameters $p, q \in (0,\infty)$ and weights v, w such that the embedding $\Gamma^p(v) \hookrightarrow \Gamma^q(w)$ or the embedding $\Gamma^p(v) \hookrightarrow \Lambda^1(w)$ hold. The former embedding is useful in interpolation theory while a standard argument applied to the latter provides a characterization of the associate space to $\Gamma^q(w)$. The results of [38] meant a considerable step ahead. However, to verify the conditions formulated through the discretizing sequences is almost impossible. After [38] was published many authors tried to obtain more manageable conditions (expressed, if possible, in an integral form – compare the Maz'ya conditions for the weighted Hardy inequality that was treated in Section 2).

Another technique which was used with certain success, is based on the level function of Halperin. It was apparently first used by Sinnamon. In connection with embeddings between weak versions of classical Lorentz spaces, it was applied in [15].

In [34], a breakthrough was made into the conditions obtained by discretization. A new method, called *antidiscretization* was developed, and with the help of it, integral type conditions were obtained in cases of embeddings for which till then only discretized conditions have been known. The antidiscretization method is in some sense based on the blocking technique of K. Grosse–Erdmann [39].

Given weights v, w, we throughout write, as above,

$$V(t) := \int_0^t v(s)\,ds \qquad \text{and} \qquad W(t) := \int_0^t w(s)\,ds.$$

We start with the embeddings of type $\Lambda \hookrightarrow \Lambda$. The full characterization is given by the following theorem.

Theorem 3.11. *Let v, w be weights on $(0,\infty)$ and let $p, q \in (0,\infty)$.*

(i) *If $0 < p \leqslant q < \infty$, then the embedding*

$$\Lambda^p(v) \hookrightarrow \Lambda^q(w) \tag{3.11}$$

holds if and only if

$$\sup_{t>0} W^{1/q}(t) V^{-1/p}(t) < \infty. \tag{3.12}$$

(ii) *Let $0 < q < p < \infty$ and let r be given by* (2.9). *Then the embedding* (3.11) *holds if and only if*

$$\begin{aligned} &\left(\int_0^\infty \left(\frac{W(t)}{V(t)} \right)^{\frac{r}{p}} w(t)\,dt \right)^{\frac{1}{r}} \\ &= \left(\frac{q}{r} \frac{W(\infty)^{r/q}}{V(\infty)^{r/p}} + \frac{q}{p} \int_0^\infty \left(\frac{W(t)}{V(t)} \right)^{r/q} v(t)\,dt \right)^{1/r} < \infty. \end{aligned} \tag{3.13}$$

The proofs can be found in [79, Remark (i), p. 148] for $1 < p,\ q < \infty$ and in [83, Proposition 1] for all values $0 < p, q < \infty$. Part (i) also follows from a more general result in [16, Corollary 2.7].

We now turn to embeddings of type $\Lambda \hookrightarrow \Gamma$. We note that such an embedding is equivalent to the boundedness of the Hardy–Littlewood maximal operator from the space $\Lambda^p(v)$ to $\Lambda^q(w)$.

Theorem 3.12. *Let v, w be weights on $(0,\infty)$ and let $p, q \in (0,\infty)$. In the case $q < p$, we define r by* (2.9).

(i) *If* $1 < p \leqslant q < \infty$, *then the embedding*

$$\Lambda^p(v) \hookrightarrow \Gamma^q(w) \tag{3.14}$$

holds if and only if (3.12) *is satisfied and*

$$\sup_{t>0} \left(\int_t^\infty \frac{w(s)}{s^q}\, ds \right)^{1/q} \left(\int_0^t \frac{v(s) s^{p'}}{V^{p'}(s)}\, ds \right)^{1/p'} < \infty. \tag{3.15}$$

(ii) *Let* $0 < p \leqslant 1$ *and* $0 < p \leqslant q < \infty$. *Then* (3.14) *holds if and only if* (3.12) *is satisfied and*

$$\sup_{t>0} t \left(\int_t^\infty \frac{w(s)}{s^q}\, ds \right)^{1/q} V^{-1/p}(t) < \infty. \tag{3.16}$$

(iii) *Let* $1 < p < \infty$, $0 < q < p < \infty$ *and* $q \neq 1$. *Then* (3.14) *holds if and only if* (3.13) *is satisfied and*

$$\left(\int_0^\infty \left[\left(\int_t^\infty \frac{w(s)}{s^q}\, ds \right)^{1/q} \left(\int_0^t \frac{v(s) s^{p'}}{V^{p'}(s)}\, ds \right)^{(q-1)/q} \right]^r \frac{v(t) t^{p'}}{V^{p'}(t)}\, dt \right)^{1/r} \tag{3.17}$$

$$\approx \left(\int_0^\infty \left[\left(\int_t^\infty \frac{w(s)}{s^q}\, ds \right)^{1/p} \left(\int_0^t \frac{v(s) s^{p'}}{V^{p'}(s)}\, ds \right)^{1/p'} \right]^r \frac{w(t)}{t^q}\, dt \right)^{1/r} < \infty.$$

(iv) *Let* $1 = q < p < \infty$. *Then* (3.14) *holds if and only if* (3.13) *is satisfied and*

$$\left(\int_0^\infty \left(\frac{W(t) + t \int_t^\infty \frac{w(s)}{s}\, ds}{V(t)} \right)^{p'-1} \int_t^\infty \frac{w(s)}{s}\, ds\, dt \right)^{1/p'} \tag{3.18}$$

$$\approx \frac{W(\infty)}{V^{1/p}(\infty)} + \left(\int_0^\infty \left(\frac{W(t) + t \int_t^\infty \frac{w(s)}{s}\, ds}{V(t)} \right)^{p'} v(t)\, dt \right)^{1/p'} < \infty.$$

(v) *Let* $0 < q < p = 1$. *Then* (3.14) *holds if and only if* (3.13) *is satisfied and*

$$\left(\int_0^\infty \left(\int_t^\infty \frac{w(s)}{s^q}\, ds \right)^{q/(1-q)} \left(\operatorname*{ess\,inf}_{0<s<t} \frac{V(s)}{s} \right)^{q/(q-1)} \frac{w(t)}{t^q}\, dt \right)^{(1-q)/q}. \tag{3.19}$$

(vi) *Let* $0 < q < p < 1$. *Then* (3.14) *holds if and only if* (3.13) *is satisfied and*

$$\left(\int_0^\infty \sup_{0<s\leqslant t} \frac{s^r}{V(s)^{\frac{r}{p}}} \left(\int_t^\infty \frac{w(s)}{s^q}\, ds\right)^{\frac{r}{p}} \frac{w(t)}{t^q}\, dt\right)^{\frac{1}{r}} < \infty. \tag{3.20}$$

The results of Theorem 3.12 are scattered in literature. The case (i) is due to Sawyer [79, Theorem 2]. The case (ii) was obtained independently by Carro and Soria in [17, Proposition 2.6b], and also by Stepanov in [83, Theorem 3b]; for a particular case see also [57, Theorem 2.2]. The case (iii) can be found in [79, Theorem 2] for $1 < q < p < \infty$ and in [83, Theorem 3a] for $0 < q < 1 < p < \infty$. The case (iv) follows by a simple reduction of a $\Gamma^1(w)$ space to a $\Lambda^1(u)$ space with an appropriate new weight u (an exercise with the Fubini theorem), and then by Theorem 3.11, some more details can be found on [15]. The case (v) was established by Sinnamon and Stepanov in [81, Theorem 4.1]. The remaining case (vi) has been resisting for quite long time, and, naturally, it is the most recent result. The condition presented here was obtained in [14] in connection with the investigation of weighted inequalities involving two Hardy operators. Earlier results such as [83, Proposition 2] or [37, Theorem 6], see also the survey in [15, Theorem 4.1], had different conditions, but only sufficient.

We now turn our attention to the embeddings of type $\Gamma \hookrightarrow \Lambda$. Note that such an embedding can be considered in some sense as a reverse inequality for the Hardy–Littlewood maximal operator. We also note that for this case of embeddings, the state of known results dramatically changed since the survey given in [15], where only some of the cases of parameters are mentioned and in one of them a discretized condition is given. The new results were enabled by the anti-discretization technique developed in [34]. We here follow the approach of [34]; in particular, we give the result in a more general situation, involving one more weight. More precisely, we characterize inequalities of type

$$\left(\int_0^\infty f^*(t)^q w(t)\, dt\right)^{\frac{1}{q}} \leqslant C \left(\int_0^\infty f_u^{**}(t)^p v(t)\, dt\right)^{\frac{1}{p}} \tag{3.21}$$

for every f, where $f_u^{**}(t) = \frac{1}{U(t)} \int_0^t f^*(s) u(s)\, ds$, u, v, w are weights, $U(t) = \int_0^t u(s)\, ds$, and $p, q \in (0, \infty)$. The embedding

$$\Gamma^p(v) \hookrightarrow \Lambda^q(w) \tag{3.22}$$

can be then easily obtained from the theorem by putting $u \equiv 1$ and hence $U(t) = t$.

Theorem 3.13. *Let u, v, w be weights on $(0, \infty)$ and let $p, q \in (0, \infty)$. In the case $q < p$, we define r by (2.9).*

(i) *If $0 < p \leqslant q < \infty$ and $1 \leqslant q < \infty$, then (3.21) holds for some $C > 0$ and all f if and only if*

$$A(1) = \sup_{t\in(0,\infty)} \frac{W(t)^{\frac{1}{q}}}{\left(V(t) + U(t)^p \int_t^\infty U(s)^{-p} v(s)\, ds\right)^{\frac{1}{p}}} < \infty.$$

(ii) *If* $1 \leqslant q < p < \infty$, *then* (3.21) *holds for some* $C > 0$ *and all* f *if and only if*

$$A(2) = \left(\int_0^\infty \frac{U(t)^r [\sup_{y\in[t,\infty)} U(y)^{-r} W(y)^{\frac{r}{q}}] V(t) \int_t^\infty U(s)^{-p} v(s)\, ds}{(V(t) + U(t)^p} \right.$$

$$\left. \times \int_t^\infty U(s)^{-p} v(s)\, ds)^{\frac{r}{p}+2} d(U^p(t))\right)^{\frac{1}{r}} < \infty.$$

(iii) *If* $0 < p \leqslant q < 1$, *then* (3.21) *holds for some* $C > 0$ *and all* f *if and only if*

$$A(3) = \sup_{t\in(0,\infty)} \frac{W(t)^{\frac{1}{q}} + U(t) \left(\int_t^\infty W(s)^{\frac{q}{1-q}} w(s) U(s)^{-\frac{q}{1-q}}\, ds\right)^{\frac{1-q}{q}}}{\left(V(t) + U(t)^p \int_t^\infty U(s)^{-p} v(s)\, ds\right)^{\frac{1}{p}}} < \infty.$$

(iv) *If* $0 < q < 1$ *and* $0 < q < p$, *then* (3.21) *holds for some* $C > 0$ *and all* f *if and only if*

$$A(4) = \left(\int_0^\infty \frac{\left[W(t)^{\frac{1}{1-q}} + U(t)^{\frac{q}{1-q}} \int_t^\infty W(s)^{\frac{q}{1-q}} w(s) U(s)^{-\frac{q}{1-q}}\, ds\right]^{\frac{r(1-q)}{q}-1}}{\left(V(t) + U(t)^p \int_t^\infty U(s)^{-p} v(s)\, ds\right)^{\frac{r}{p}}} \right.$$

$$\left. \times W(t)^{\frac{q}{1-q}} w(t)\, dt\right)^{\frac{1}{r}} < \infty.$$

Moreover, $A(4) \approx A(5)$, *where*

$$A(5) = \left(\int_0^\infty \frac{[W(t)^{\frac{1}{1-q}} + U(t)^{\frac{q}{1-q}} \int_t^\infty W(s)^{\frac{q}{1-q}} w(s) U(s)^{-\frac{q}{1-q}}\, ds]^{\frac{r(1-q)}{q}}}{(V(t) + U(t)^p \int_t^\infty U(s)^{-p} v(s)\, ds)^{\frac{r}{p}+2}} \right.$$

$$\left. \times V(t) \int_t^\infty U(s)^{-p} v(s)\, ds\, d(U^p(t))\right)^{\frac{1}{r}}.$$

Again, there are many papers that contributed to the results in Theorem 3.13. For example, with $u \equiv 1$, it was proved in [70, Theorem 3.2] for $1 \leqslant p = q < \infty$, in [57, Theorem 2.1] for $1 \leqslant p \leqslant q < \infty$ and in [84, p. 473] for $0 < p \leqslant q < \infty$, $1 \leqslant q < \infty$. The proof with the best constant can be found in [43, Theorem 3.2 (a) and (c)] or [4], where a multidimensional case is treated. Certain reduction to Theorem 3.11 is possible, again, when $p = 1$, details can be found in [15]. When w itself is nonincreasing, then a discretized condition in the case $q = 1 < p < \infty$ is given in [38, Theorem 2.2a], and the same result is, for general w, obtained in [15, Proposition 2.12].

It remains to treat the case of embeddings of type $\Gamma \hookrightarrow \Gamma$. Such embeddings were studied by Gol'dman, Heinig and Stepanov [38], where a complete characterization was given. However, in the case $0 < q < p < \infty$, the conditions were in a discrete form. Integral type conditions were established in [34] using the antidiscretization technique. We survey the results in the following theorem. We follow the approach from [34], in particular, we study a more general inequality

$$\left(\int_0^\infty f_u^{**}(t)^q w(t)\,dt\right)^{\frac{1}{q}} \leqslant C\left(\int_0^\infty f_u^{**}(t)^p v(t)\,dt\right)^{\frac{1}{p}}, \tag{3.23}$$

where $p, q \in (0,\infty)$ and u, v, w are weights.

Theorem 3.14. *Let $p, q \in (0,\infty)$ and let u, v, w be weights. In cases when $q < p$ we define r by (2.9).*

(i) *Let $0 < p \leqslant q < \infty$. Then (3.23) holds if and only if*

$$A(6) = \sup_{t\in(0,\infty)} \frac{\left(W(t) + U(t)^q \int_t^\infty U(s)^{-q} w(s)\,ds\right)^{\frac{1}{q}}}{\left(V(t) + U(t)^p \int_t^\infty U(s)^{-p} v(s)\,ds\right)^{\frac{1}{p}}} < \infty.$$

(ii) *Let $0 < q < p < \infty$. Then (3.23) holds if and only if*

$$A(7) = \left(\int_0^\infty \frac{\left(W(t) + U(t)^q \int_t^\infty U(y)^{-q} w(y)\,dy\right)^{\frac{r}{q}}}{\left(V(t) + U(t)^p \int_t^\infty U(s)^{-p} v(s)\,ds\right)^{\frac{r}{p}+2}} \times V(t) \int_t^\infty U(s)^{-p} v(s)\,ds\,d(U^p)(t)\right)^{\frac{1}{r}} < \infty.$$

Moreover, $A(7) \approx A(8)$, where

$$A(8) = \left(\int_0^\infty \frac{\left(W(t) + U(t)^q \int_t^\infty U(s)^{-q} w(s)\,ds\right)^{\frac{r}{q}-1} w(t)}{\left(V(t) + U(t)^p \int_t^\infty U(s)^{-p} v(s)\,ds\right)^{\frac{r}{p}}}\,dt\right)^{\frac{1}{r}} < \infty.$$

As already mentioned above, among many applications of the characterization of embeddings between classical Lorentz spaces an important part is played by the characterization of associate spaces. To know an associate space to a given function space is often very important (for example, plenty of applications to optimal Sobolev embeddings are known, see our previous survey [74] and the references therein). It follows easily from the very definition of an associate space and the Hardy–Littlewood–Pólya principle that once we know an embedding into the space $\Lambda^1(w)$, then, applying the resulting characterizing quantity expressed in terms of w to a general function g, we obtain the desired associate space. We now apply the results above to obtain a characterization of associate spaces of classical Lorentz spaces. The results follow immediately from the above theorems.

We first treat the spaces of type Λ.

Theorem 3.15. *Let $p \in (0,\infty)$ and let v be a weight.*

(i) *Let $0 < p \leqslant 1$, then*

$$\|g\|_{\Lambda^p(v)'} \approx \sup_{t>0} g^{**}(t) \frac{t}{V^{1/p}(t)}.$$

(ii) *Let $1 < p < \infty$, then*

$$\|g\|_{\Lambda^p(v)'} \approx \left(\int_0^\infty (g^{**}(t))^{p'} \frac{t^{p'} v(t)}{V^{p'}(t)} \, dt \right)^{1/p'} + V^{-1/p}(\infty) \int_0^\infty g^*(t) \, dt.$$

And now, we consider the spaces

$$\Gamma_u^p(v) = \left\{ f \in \mathfrak{M}(0,\infty); \|f\|_{\Gamma_u^p(v)} := \left(\int_0^\infty (f_u^{**}(t))^p v(t) \, dt \right)^{1/p} < \infty \right\},$$

where u, v are weights and $p \in (0,\infty)$. For such spaces, we have the following result from [34].

Theorem 3.16. *Let u, v be positive weights on $[0,\infty)$ and let $p \in (0,\infty)$.*

(i) *Let $0 < p \leqslant 1$, then*

$$\|g\|_{\Gamma_u^p(v)'} \approx \sup_{t \in (0,\infty)} \frac{\int_0^t g^*(s) \, ds}{\left(V(t) + U(t)^p \int_t^\infty U(s)^{-p} v(s) \, ds\right)^{\frac{1}{p}}}.$$

(ii) *Let $1 < p < \infty$, then*

$$\|g\|_{\Gamma_u^p(v)'} \approx \left(\int_0^\infty \frac{\left[\sup_{y \in [t,\infty)} U(y)^{-p'} \left(\int_0^y g^*(\tau) \, d\tau\right)^{p'}\right]}{\left(V(t) + U(t)^p \int_t^\infty U(s)^{-p} v(s) \, ds\right)^{p'+1}} \right. \\ \left. \times U(t)^{p'} V(t) \int_t^\infty U(s)^{-p} v(s) \, ds \, d(U^p)(t) \right)^{\frac{1}{p'}}.$$

A curious consequence of the characterization of the embeddings between classical Lorentz spaces, which is of certain independent interest, was obtained in [15, Corollary 8.2]. We first need to define endpoint spaces.

Definition 3.17. Let φ be a concave nondecreasing function on $[0,\infty)$ such that $\varphi(t) = 0$ if and only if $t = 0$, and $\varphi(t)/t$ is nonincreasing on $(0,\infty)$. The *endpoint Lorentz space* Λ_φ and the *endpoint Marcinkiewicz space* M_φ are defined as the sets of functions f from $\mathfrak{M}(0,\infty)$ such that

$$\|f\|_{\Lambda_\varphi} = \int_0^\infty f^*(t)\, d\varphi(t) < \infty, \qquad \|f\|_{M_\varphi} = \sup_{t>0} f^{**}(t)\varphi(t) < \infty$$

respectively.

It was shown in [15, Corollary 8.2] that when φ is a fundamental function of an r.i. space X and the endpoint Lorentz space Λ_φ coincides with the endpoint Marcinkiewicz space M_φ, then, necessarily, X is equivalent to one of the spaces L^1, L^∞, $L^1 \cap L^\infty$ or $L^1 + L^\infty$. It is clear that when X is one of these four spaces, then the endpoint spaces coincide, but the converse implication is interesting. The details read as follows.

Theorem 3.18. *Let φ be a concave nondecreasing function on $[0,\infty)$ such that $\varphi(t) = 0$ if and only if $t = 0$, and $\varphi(t)/t$ is nonincreasing on $(0,\infty)$. Then $\Lambda_\varphi = M_\varphi$ if and only if φ is on $(0,\infty)$ equivalent to one of the following four functions: t, 1, $\max\{1,t\}$ or $\min\{1,t\}$.*

Recently, the two types of classical Lorentz spaces (Λ and Γ) have been enriched on one more type of a function space, in some sense related to the classical Lorentz ones. The primary role of this type of a space is to measure oscillation of functions after rearrangements have been taken. In other words, the spaces in question work with the quantity $f^{**} - f^*$ rather than just f^* (involved in Λ type spaces) or f^{**} (involved in Γ type spaces).

Definition 3.19. Let $p \in (0,\infty)$ and let v be a weight. Then the space $S^p(v)$ is defined as

$$S^p(v) = \left\{ f \in \mathfrak{M}(0,\infty);\, \|f\|_{S^p(v)} := \left(\int_0^\infty (f^{**}(t) - f^*(t))^p v(t)\, dt \right)^{1/p} < \infty \right\}.$$

The spaces of this type have long and interesting history.

In 1981, in order to obtain a Marcinkiewicz- type interpolation theorem for operators that are unbounded on L^∞, Bennett, De Vore, and Sharpley in [6], introduced a new rearrangement-invariant space consisting of those measurable functions for which $f^{**} - f^*$ is bounded, which plays the role of certain "weak-L^∞" in the sense that it contains L^∞ and possesses appropriate interpolation properties. Since, $f^{**} - f^*$ can be seen as some kind of measure of the oscillation of f^*, they proved that weak-$L^\infty(Q)$, where Q is a cube in $\mathbb{R}^n$, is, in fact, the rearrangement–invariant hull of $BMO(Q)$.

The quantity $f^{**} - f^*$ has an intimate connection to Besov type spaces that measure smoothness of functions by means of moduli of continuity. A pioneering result in measuring smoothness by means of rearrangements is the Pólya–Szegö inequality which states that the L^p-norm of the gradient of the symmetric rearrangement of a given function f does not exceed that of f. Further results were obtained by several authors in connection with a problem posed by Ul'yanov [86]. Various estimates, mostly in the setting of Lebesgue spaces, were obtained, for example, by Ul'yanov [86], Kolyada [50], Kolyada and Lerner [53], Storozhenko [85], and others. An excellent survey was provided by Kolyada [52]; some results in this direction can be also found in an earlier book by Kudryavtsev and Nikol'skii [54].

In [3], Bagby and Kurtz replaced the "good λ" inequality

$$\mu(\{x : Tf(x) > 2\lambda \text{ and } Mf(x) \leqslant \varepsilon\lambda\}) \leqslant C(\varepsilon)\mu(\{x : Tf(x) > \lambda\})$$

(here T is a maximal Calderón–Zygmund singular operator, M is the classical Hardy–Littlewood maximal function and $d\mu = w(x)\,dx$, with w in the Muckenhoupt's weight class A^∞), by the rearrangement inequality

$$(Tf)_w^{**}(t) - (Tf)_w^{*}(t) \leqslant c(Mf)_w^{**}(t). \tag{3.24}$$

This estimate, which is intimately connected to yet another one, later called the "*better* λ" inequality, contains more information than the good–λ inequality. In particular, it does not involve the parameter ε which would have to vary. The absence of ε enables one to obtain the best possible weighted L^p bounds and, via extrapolation, also sharp exponential integrability estimates.

Inequality (3.24) was the starting point used by Boza and Martín (cf. [9]) in order to obtain sufficient conditions for the boundedness of a maximal Calderón–Zygmund singular integral operators between classical Lorentz spaces.

In [31], Franciosi, motivated by the fact that functions belonging to $L^\infty \cap VMO$, but not necessarily continuous, have recently been considered in very important regularity problems connected with elliptic differential equations and systems with rough coefficients, obtained boundedness and VMO results for a function f integrable on cube $Q \subset \mathbb{R}^n$, such that

$$\int_0^{|Q|} (f^{**}(t) - f^*(t))\frac{dt}{t} < \infty.$$

In [64] (cf. also [28]) an elementary proof based on the Maz'ya truncation trick of the sharp Sobolev embedding

$$W_0^{1,n}(\Omega) \subset BW_n(\Omega) \tag{3.25}$$

was shown where Ω is an open subset of $\mathbb{R}^n$ with $|\Omega| = 1$ and $BW_n(\Omega)$ is the Maz'ya–Hansson–Brézis–Wainger space defined by the condition

$$\int_0^1 \Big(\frac{f^*(t)}{\log(e/t)}\Big)^n \frac{dt}{t} < \infty,$$

by showing that

$$W_0^{1,n}(\Omega) \subset W_n \subset BW_n(\Omega) \tag{3.26}$$

where

$$W_n = \Big\{f : \Big(\int_0^1 (f^*(\tfrac{t}{2}) - f^*(t))^n \frac{dt}{t}\Big)^{1/n} < \infty\Big\}.$$

In [5], Bastero, Milman and Ruiz, using a natural extension of $L_{(p,q)}(\Omega)$ spaces, introduced the spaces

$$L(\infty,q)(\Omega)=\{f: t^{-1/q}(f^{**}(t)-f^{*}(t))\in L_q(\Omega)\},$$

and showed that in fact $W_n = L(\infty,n)$. Moreover, using certain version of the Pólya-Szegö symmetrization principle, they proved

$$f^{**}(t)-f^{*}(t)\leqslant c_n(\nabla f)^{**}(t)t^{1/n}, \tag{3.27}$$

which readily implies the first embedding of (3.26), and thus the proof of (3.25) is reduced to an embedding result for rearrangement-invariant spaces. We note that (3.27) was in various forms obtained by several authors. The first one who proved it was apparently Kolyada [51, Lemma 5.1], who showed in fact a seemingly weaker, but equivalent estimate with the left-hand side replaced by $t^{-\frac{1}{n}}(f^{*}(t)-f^{*}(2t))$ – see also a direct proof in his survey [52, Lemma 3.1], and, independently, Alvino, Trombetti and Lions [1, (A7), p. 219]. Another detailed proof, in a more general setting of metric spaces, was given by Kališ and Milman in [45].

In [68], Milman and Pustylnik extended (3.26) to the case $k>1$, by showing that

$$W_0^{k,n}(\Omega)\subset L(\infty,n/k)(\Omega).$$

Here, the sets of functions with finite quantities $\|f^{**}(t)-f^{*}(t)\|$ appear for a large system of norms in a natural context of optimal Sobolev embeddings. They extend the results of [28] to the setup of rearrangement-invariant spaces. The fractional case of the above embedding has been considered in [65].

It should be noticed that in the study function spaces defined in terms of the functional $f^{**}-f^{*}$ certain care must be exercised. In particular, this functional vanishes on constant functions and, moreover, the operation $f\to f^{**}-f^{*}$ is not subadditive. Therefore, quantities involving $f^{**}-f^{*}$ do not have norm properties, which makes the study of the corresponding function spaces difficult.

In [13], the basic functional properties of spaces $S^p(v)$ were studied. Various conditions were given for $S^p(v)$ to have a lattice property, to be normable, and to be linear. We omit the details here, since they are quite complicated. In [14], a sequel to [13], relations of the spaces $S^p(v)$ to classical Lorentz spaces were studied.

We now consider general relations between the spaces of types Γ, Λ and S. Having the theory of embeddings of the spaces of type Λ and Γ complete, it is time for the spaces of type S to come into the play.

First, it is not difficult to prove that

$$\Gamma^p(v)=\Lambda^p(v)\cap S^p(v). \tag{3.28}$$

We recall that, following [2], we say that a weight v on $(0, \infty)$ belongs to the class B_p if

$$\int_t^\infty \frac{v(s)}{s^p}\,ds \lesssim \frac{1}{t^p}\int_0^t v(s)\,ds, \qquad t \in (0, \infty).$$

We say that a weight v on $(0, \infty)$ belongs to the reverse class B_p, written $w \in RB_p$ (cf. [2, 13]) if

$$\frac{1}{t^p}\int_0^t v(s)\,ds \lesssim \int_t^\infty \frac{v(s)}{s^p}\,ds, \qquad t \in (0, \infty).$$

From this, one immediately obtains

Corollary 3.20. *Let $0 < p < \infty$ and let v be a weight on $(0, \infty)$. Then, the following statements are equivalent:*

(i)
$$\Lambda^p(v) \hookrightarrow S^p(v);$$

(ii)
$$\Lambda^p(v) = \Gamma^p(v);$$

(iii)
$$v \in B_p.$$

For $p \geqslant 1$, we have the following simple characterization.

Corollary 3.21. *Let $1 \leqslant p < \infty$ and let v be a weight on $(0, \infty)$. Then, the following statements are equivalent:*

(i)
$$S^p(v) \hookrightarrow \Lambda^p(v);$$

(ii)
$$S^p(v) = \Gamma^p(v);$$

(iii)
$$v \in RB_p.$$

We now investigate embeddings of type $S \hookrightarrow S$, i.e., we intend to characterize

$$S^p(v) \hookrightarrow S^q(w). \tag{3.29}$$

We denote

$$\widetilde{v}_p(t) := v\left(\frac{1}{t}\right) t^{p-2} \quad \text{and} \quad \widetilde{w}_q(t) := w\left(\frac{1}{t}\right) t^{q-2}. \tag{3.30}$$

With this notation, it is easy to observe that embeddings of S type spaces are equivalent to those of Λ type spaces with an appropriate change of weights.

Theorem 3.22. *Let $0 < p, q < \infty$ and let v, w be weights on $(0,\infty)$. Then,* (3.29) *is equivalent to*

$$\Lambda^p(\widetilde{v}_p) \hookrightarrow \Lambda^q(\widetilde{w}_q), \tag{3.31}$$

with $\widetilde{v}_p$ and $\widetilde{w}_q$ from (3.30).

Now, we can present the full characterization.

Theorem 3.23. (i) *If $0 < p \leqslant q < \infty$, then the embedding* (3.29) *holds if and only if*

$$\sup_{t\in(0,\infty)} \frac{\left(\int_t^\infty s^{-q} w(s)\, ds\right)^{\frac{1}{q}}}{\left(\int_t^\infty s^{-p} v(s)\, ds\right)^{\frac{1}{p}}} < \infty. \tag{3.32}$$

(ii) *If $0 < q < p < \infty$, and r is defined by* (2.9), *then* (3.29) *holds if and only if*

$$\left(\int_0^\infty \left[\frac{\int_t^\infty s^{-q} w(s)\, ds}{\int_t^\infty s^{-p} v(s)\, ds}\right]^{\frac{r}{p}} \frac{w(t)}{t^q}\, dt\right)^{\frac{1}{r}} < \infty. \tag{3.33}$$

We can turn to relations between spaces of type S and Γ. We begin with a slight restriction. As mentioned above, the functional $f^{**}(t) - f^*(t)$ is zero when f is constant on $(0,\infty)$. Hence constant functions always belong to any $S^p(v)$ whereas they do not necessarily belong to analogous spaces of type Λ and Γ. For this reason, we study the appropriate embeddings restricted to the set

$$\mathbb{A} = \left\{ f \in \mathfrak{M}_+(0,\infty);\ \lim_{t\to\infty} f^*(t) = 0 \right\}.$$

We denote this restriction by writing, for example,

$$S^p(v) \hookrightarrow \Lambda^q(w), \qquad f \in \mathbb{A},$$

meaning

$$\left(\int_0^\infty f^*(t)^q w(t)\, dt\right)^{\frac{1}{q}} \leqslant C \left(\int_0^\infty \left(f^{**}(t) - f^*(t)\right)^p v(t)\, dt\right)^{\frac{1}{p}} \quad \forall\, f \in \mathbb{A},$$

and so on.

The first reduction is, again, obtained via a simple change of variables.

Theorem 3.24. *Let $0 < p, q < \infty$ and let v, w be weights on $(0,\infty)$. Then the embedding*

$$S^p(v) \hookrightarrow \Gamma^q(w), \quad f \in \mathbb{A}, \tag{3.34}$$

is equivalent to

$$\Lambda^p(\widetilde{v}_p)\hookrightarrow\Gamma^q(\widetilde{w}_q). \tag{3.35}$$

Similarly, the embedding

$$\Gamma^p(v)\hookrightarrow S^q(w), \quad f\in\mathbb{A}, \tag{3.36}$$

is equivalent to

$$\Gamma^p(\widetilde{v}_p)\hookrightarrow\Lambda^q(\widetilde{w}_q). \tag{3.37}$$

Now, the main result reads as follows.

Theorem 3.25. *Let $0<p,q<\infty$ and let v,w be weights on $(0,\infty)$. In the case $q<p$, we define r by (2.9).*

(i) *If $1<p\leqslant q<\infty$, then the embedding (3.34) holds if and only if (3.32) holds and*

$$\sup_{t\in(0,\infty)} W(t)^{\frac{1}{q}}\left(\int_t^\infty \frac{v(s)}{\left(s^p\int_{\frac{1}{s}}^\infty v(y)y^{-p}\,dy\right)^{p'}}\,ds\right)^{\frac{1}{p'}}<\infty. \tag{3.38}$$

(ii) *If $0<p<1$ and $0<p\leqslant q<\infty$, then (3.34) holds if and only if (3.32) holds and*

$$\sup_{t\in(0,\infty)}\frac{W(t)^{\frac{1}{q}}}{t\left(\int_t^\infty v(s)s^{-p}\,ds\right)^{\frac{1}{p}}}. \tag{3.39}$$

(iii) *If $1<p<\infty$, $0<q<p<\infty$ and $q\neq 1$, then (3.34) holds if and only if (3.33) holds and*

$$\left(\int_0^\infty \frac{W(t)^{\frac{r}{q}}\left(\int_t^\infty \frac{v(s)}{\left(s^p\int_s^\infty \frac{v(y)}{y^p}\,dy\right)^{p'}}\,ds\right)^{\frac{r(q-1)}{q}}}{\left(t^p\int_s^\infty \frac{v(y)}{y^p}\,dy\right)^{p'}}\,v(t)\,dt\right)^{\frac{1}{r}}<\infty. \tag{3.40}$$

(iv) *If $1=q<p<\infty$, then (3.34) holds if and only if (3.33) holds and*

$$\frac{\int_0^\infty \frac{w(s)}{s}\,ds}{\left(\int_0^\infty \frac{v(s)}{s^p}\,ds\right)^{\frac{1}{p}}}+\left(\int_0^\infty\left[\frac{\frac{W(t)}{t}+\int_t^\infty \frac{w(s)}{s}\,ds}{\int_t^\infty \frac{v(s)}{s^p}\,ds}\right]^{p'}\frac{v(t)}{t^p}\,dt\right)^{\frac{1}{p'}}<\infty. \tag{3.41}$$

(v) *If $0<q<p\leqslant 1$, then (3.34) holds if and only if*

$$\left(\int_0^\infty \frac{\left(\int_t^\infty \frac{w(s)}{s^q}\,ds\right)^{\frac{r}{p}}}{\left(\int_t^\infty \frac{v(s)}{s^p}\,ds\right)^{\frac{r}{p}}}\frac{w(t)}{t^q}\,dt\right)^{\frac{1}{r}}<\infty \tag{3.42}$$

and

$$\left(\int_0^\infty W(t)^{\frac{r}{p}} \operatorname*{ess\,sup}_{t\leqslant s<\infty} \frac{1}{s^r\left(\int_s^\infty \frac{v(y)}{y^p}\,dy\right)^{\frac{r}{p}}}\frac{w(t)}{t^q}\,dt\right)^{\frac{1}{r}}<\infty. \tag{3.43}$$

The converse embedding is covered by the following theorem.

Theorem 3.26. *Let $0<p,q<\infty$ and let v,w be weights on $(0,\infty)$. In the case $q<p$, we define r by (2.9).*

(i) *If $0<p\leqslant q<\infty$ and $1\leqslant q<\infty$, then the embedding (3.36) holds if and only if*

$$\sup_{t\in(0,\infty)} \frac{\left(\int_t^\infty \frac{w(s)}{s^q}\,ds\right)^{\frac{1}{q}}}{\left(\frac{V(t)}{t^p}+\int_t^\infty \frac{v(s)}{s^p}\,ds\right)^{\frac{1}{p}}}<\infty. \tag{3.44}$$

(ii) *If $1\leqslant q<p<\infty$, then (3.36) holds if and only if*

$$\left(\int_0^\infty \frac{\sup_{s\in(0,t)} s^r\left(\int_s^\infty \frac{w(y)}{y^q}\,dy\right)^{\frac{r}{q}}}{t^{r+p+1}\left(\frac{V(t)}{t^p}+\int_t^\infty \frac{v(s)}{s^p}\,ds\right)^{\frac{r}{p}+2}}V(t)\int_t^\infty \frac{v(s)}{s^p}\,ds\,dt\right)^{\frac{1}{r}}<\infty. \tag{3.45}$$

(iii) *If $0<p\leqslant q<1$, then (3.36) holds if and only if*

$$\sup_{t\in(0,\infty)} \frac{\left(\int_t^\infty \frac{w(s)}{s^q}\,ds\right)^{\frac{1}{q}}+\frac{1}{t}\left(\int_0^t\left[s\int_s^\infty \frac{w(y)}{y^q}\,dy\right]^{\frac{q}{1-q}}\frac{w(s)}{s^q}\,ds\right)^{\frac{1-q}{q}}}{\left(\frac{V(t)}{t^p}+\int_t^\infty \frac{v(s)}{s^p}\,ds\right)^{\frac{1}{p}}}<\infty. \tag{3.46}$$

(iv) *If $0<q<1$ and $0<q<p$, then (3.36) holds if and only if*

$$\Bigg(\frac{[(\int_t^\infty \frac{w(s)}{s^q}\,ds)^{\frac{1}{1-q}}+\frac{1}{t^{\frac{q}{1-q}}}\int_0^t(s\int_s^\infty \frac{w(y)}{y^q}\,dy)^{\frac{q}{1-q}}\frac{w(s)}{s^q}\,ds]^{\frac{r(1-q)}{q}-1}}{(\frac{V(t)}{t^p}+\int_t^\infty \frac{v(s)}{s^p}\,ds)^{\frac{1}{p}}} \times\left(\int_t^\infty \frac{w(s)}{s^q}\,ds\right)^{\frac{q}{1-q}}\frac{w(t)}{t^q}\,dt\Bigg)^{\frac{1}{r}}. \tag{3.47}$$

The most interesting and also the most difficult task is the characterization of mutual embeddings between the spaces S and the spaces Λ. We first reduce the problem to a modified one.

We observe that the embeddings $S^p(v)\hookrightarrow\Lambda^q(w)$ and $\Lambda^p(v)\hookrightarrow S^q(w)$ (restricted to $f\in\mathbb{A}$) are interchangeable (with an appropriate change of weights) and therefore it is enough to investigate only one of them.

Proposition 3.27. *Let $0 < p, q < \infty$ and let v, w be weights on $(0,\infty)$. Then the embedding*

$$S^p(v) \hookrightarrow \Lambda^q(w), \quad f \in \mathbb{A}, \tag{3.48}$$

is equivalent to the embedding

$$\Lambda^p(\widetilde{v}_p) \hookrightarrow S^q(\widetilde{w}_q), \quad f \in \mathbb{A}, \tag{3.49}$$

where $\widetilde{v}_p$ and $\widetilde{w}_q$ are from (3.30).

The next step of our analysis is the use of the Stepanov trick to reduce the restricted inequality to a nonrestricted one. An interesting fact is that its application to the functional $f^{**} - f^*$ entails one to investigate weighted inequalities involving two Hardy operators.

Proposition 3.28. *Let $0 < p, q < \infty$ and let v, w be weights on $(0,\infty)$. Then the embedding*

$$\Lambda^p(v) \hookrightarrow S^q(w), \quad f \in \mathbb{A}, \tag{3.50}$$

holds if and only if the inequality

$$\left(\int_0^\infty \left(\frac{1}{t}\int_0^t f(s)\,ds\right)^q w(t)\,dt\right)^{\frac{1}{q}} \leqslant C\left(\int_0^\infty \left(\int_t^\infty \frac{f(s)}{s}\,ds\right)^p v(t)\,dt\right)^{\frac{1}{p}} \tag{3.51}$$

is satisfied for every nonnegative function h.

So, what remains, is just to give a characterization of the two-operator Hardy inequality (3.51). That is done in the following theorem.

Theorem 3.29. *Let $0 < p \leqslant \infty$, $1 \leqslant q \leqslant \infty$. In the case $q < p$, we define r by* (2.9). *Let v, w be weights on $(0,\infty)$. Then* (3.51) *holds if and only if one of the following conditions holds:*

(i) $1 < p \leqslant q < \infty$ *and*

$$\sup_{0<t<\infty} \left(\int_t^\infty \frac{w(s)}{s^q}\,ds\right)^{\frac{1}{q}} \left(\int_0^t \frac{s^{p'} v(s)}{V(s)^{p'}}\,ds + \frac{t^{p'}}{V(t)^{p'-1}}\right)^{\frac{1}{p'}} < \infty;$$

(ii) $1 < p < \infty$, $q = \infty$ *and*

$$\sup_{0<t<\infty} \left(\int_0^t \frac{s^{p'} v(s)}{V(s)^{p'}}\,ds + \frac{t^{p'}}{V(t)^{p'-1}}\right)^{\frac{1}{p'}} \operatorname*{ess\,sup}_{t\leqslant s<\infty} \frac{w(s)}{s} < \infty;$$

(iii) $p = q = \infty$ *and*

$$\sup_{0<t<\infty} \left(\frac{t}{\operatorname{ess\,sup}_{0<s\leqslant t} v(s)} + \int_0^t s\,d\left(\frac{-1}{\operatorname{ess\,sup}_{0<y\leqslant s} v(y)}\right)\right) \operatorname*{ess\,sup}_{t\leqslant s<\infty} \frac{w(s)}{s} < \infty;$$

(iv) $1 < q < p < \infty$,

$$\left(\int_0^\infty \left(\sup_{0<s\leqslant t} s^q \int_s^\infty \frac{w(y)}{y^q}\,dy\right)^{\frac{r}{q}} \frac{v(t)}{V(t)^{\frac{r}{q}}}\,dt\right)^{\frac{1}{r}} < \infty,$$

$$\frac{\sup_{0<t<\infty} t\left(\int_t^\infty \frac{w(s)}{s^q}\,ds\right)^{\frac{1}{q}}}{V(\infty)^{\frac{1}{p}}} < \infty$$

and

$$\left(\int_0^\infty \left(\int_0^t \frac{s^{p'}v(s)}{V(s)^{p'}}\,ds\right)^{\frac{r}{p'}} \frac{t^{p'}v(t)}{V(t)^{p'}} \left(\int_t^\infty \frac{w(s)}{s^q}\,ds\right)^{\frac{r}{q}} dt\right)^{\frac{1}{r}} < \infty;$$

(v) $p = \infty$, $1 < q < \infty$,

$$\int_0^\infty \Big(\int_0^t s\,d\Big(\frac{-1}{\operatorname{ess\,sup}_{0<y\leqslant s} v(y)}\Big)\Big)^q \times t\Big(\int_t^\infty \frac{w(s)}{s^q}\,ds\Big)^{\frac{1}{q}} d\Big(\frac{-1}{\operatorname{ess\,sup}_{0<s\leqslant t} v(s)}\Big) < \infty,$$

$$\frac{\operatorname{ess\,sup}_{0<s<\infty} w(s)}{\operatorname{ess\,sup}_{0<s<\infty} v(s)} < \infty$$

and

$$\int_0^\infty \Big(\int_t^\infty d\Big(\frac{-1}{\operatorname{ess\,sup}_{0<y\leqslant s} v(y)}\Big)\Big)^{q-1} \times \sup_{0<s\leqslant t} s\Big(\int_s^\infty \frac{w(y)}{y^q}\,dy\Big)^{\frac{1}{q}} d\Big(\frac{-1}{\operatorname{ess\,sup}_{0<s\leqslant t} v(s)}\Big) < \infty;$$

(vi) $0 < p \leqslant 1 < q < \infty$ *and*

$$\sup_{0<t<\infty} \frac{t}{V(t)^{\frac{1}{p}}} \left(\int_t^\infty \frac{w(s)}{s^q}\,ds\right)^{\frac{1}{q}} < \infty;$$

(vii) $0 < p \leqslant 1$, $q = \infty$ *and*

$$\sup_{0<t<\infty} \left(\sup_{0<s\leqslant t} \frac{s}{V(s)^{\frac{1}{p}}}\right) \operatorname*{ess\,sup}_{t\leqslant s<\infty} \frac{w(s)}{s} < \infty;$$

(viii) $q = 1 < p < \infty$,

$$\left(\int_0^\infty \left(\sup_{0<s\leqslant t} s \int_s^\infty \frac{w(y)}{y}\,dy\right)^{p'} \frac{v(t)}{V(t)^{p'}}\,dt\right)^{\frac{1}{p'}} < \infty$$

and

$$\frac{\sup_{0<t<\infty} t \int_t^\infty \frac{w(s)}{s}\,ds}{V(\infty)^{\frac{1}{p}}} < \infty;$$

(ix) $0 < p \leqslant 1 = q$ *and*

$$\sup_{0<t<\infty} \frac{t}{V(t)^{\frac{1}{p}}} \int_t^\infty \frac{w(s)}{s}\,ds < \infty;$$

(x) $p = \infty$, $q = 1$,

$$\int_0^\infty \sup_{0<s\leqslant t} s \int_s^\infty \frac{w(s)}{s}\,d\left(\frac{-1}{\operatorname{ess\,sup}_{0<s\leqslant t} v(s)}\right) < \infty$$

and

$$\frac{\sup_{0<t<\infty} t \int_t^\infty \frac{w(s)}{s}\,ds}{\operatorname{ess\,sup}_{0<s<\infty} v(s)} < \infty;$$

(xi) $p = 1$, $0 < q < 1$ *and*

$$\left(\int_0^\infty \left(\int_t^\infty \frac{w(s)}{s^q}\right)^{\frac{q}{1-q}} \left[\operatorname*{ess\,sup}_{0<s\leqslant t} \frac{1}{(Pv)(s)}\right]^{\frac{q}{1-q}} \frac{w(t)}{t^q}\,dt\right)^{\frac{q}{1-q}} < \infty.$$

And, of course, the following result is an immediate consequence of the theorems above.

Corollary 3.30. *Let* $0 < p < \infty$, $1 \leqslant q < \infty$, *and let* v, w *be weights on* $(0, \infty)$. *Then, the embedding*

$$\Lambda^p(v) \hookrightarrow S^q(w), \quad f \in \mathbb{A},$$

holds if and only if one of the conditions (i)–(xi) *of* Theorem 3.29 *holds.*

Let us finally note that the inequality (3.51) is clearly of independent interest, and it will have many other important consequences apart from the one presented here. It has not been studied in the required generality; apparently the only result available in literature seems to be its characterization when $1 < p = q < \infty$ and $v = w$ due to Neugebauer [70].

4 Hardy Type Operators Involving Suprema

In the last section, we study one quite special class of operators, related in some sense to the Hardy type operators, but different in another one. The operators we have in mind involve the operation of taking a pointwise (essential) supremum where Hardy type operators involve integration.

Recall that, in [2], Ariño and Muckenhoupt characterized when the Hardy–Littlewood maximal operator M is bounded on the classical Lorentz space $\Lambda^p(w)$ when $p \in [1, \infty)$. In other words, they characterized the class of weights w for which the inequality

$$\int_0^\infty (Mf)^*(t)^p w(t)\, dt \lesssim \int_0^\infty f^*(t)^p w(t)\, dt$$

holds for every locally integrable f on $\mathbb{R}^n$. The approach of [2] has two main ingredients. The first of them is the well-known above-mentioned two-sided estimate due to Riesz, Wiener, Stein and Herz (cf. [7, Chapter 3, Theorem 3.8])

$$(Mf)^*(t) \approx f^{**}(t), \quad f \in L^1_{\text{loc}}(\mathbb{R}^n), \quad t \in (0, \infty). \tag{4.1}$$

The second key ingredient is a characterization of the boundedness of A on the cone $\mathfrak{M}^+(0, \infty; \downarrow)$ in $L^p_w(0, \infty)$. An analogous problem for the *fractional maximal operator* in place of the Hardy–Littlewood maximal operator was studied in [20]. The fractional maximal operator, M_γ, $\gamma \in (0, n)$, is defined at $f \in L^1_{\text{loc}}(\mathbb{R}^n)$ by

$$M_\gamma f(x) = \sup_{Q \ni x} |Q|^{\frac{\gamma}{n} - 1} \int_Q |f(y)|\, dy, \qquad x \in \mathbb{R}^n,$$

where the supremum is extended over all cubes $Q \subset \mathbb{R}^n$ with sides parallel to the coordinate axes. It was shown in [20, Theorem 1.1] that

$$(M_\gamma f)^*(t) \lesssim \sup_{t \leqslant s < \infty} s^{\frac{\gamma}{n}} f^{**}(s) \tag{4.2}$$

for every $f \in L^1_{\text{loc}}(\mathbb{R}^n)$ and $t \in (0, \infty)$. This estimate is not two-sided as (4.1), but it is sharp in the following sense: for every $\varphi \in \mathfrak{M}^+(0, \infty; \downarrow)$ there exists a function f on $\mathbb{R}^n$ such that $f^* = \varphi$ a.e. on $(0, \infty)$ and

$$(M_\gamma f)^*(t) \gtrsim \sup_{t \leqslant s < \infty} s^{\frac{\gamma}{n}} f^{**}(s), \quad t \in (0, \infty). \tag{4.3}$$

Consequently, the role of f^{**} in the argument of Ariño and Muckenhoupt [2] is in the case of fractional maximal operator taken over by the expression on the right-hand sides of (4.2) and (4.3). Thus, in order to characterize boundedness of the fractional maximal operator M_γ between classical Lorentz spaces it is necessary and sufficient to characterize the validity of the weighted inequality

$$\left(\int_0^\infty \left(\sup_{t\leqslant\tau<\infty} \tau^{\frac{\gamma}{n}-1}\int_0^\tau \varphi(s)\,ds\right)^q w(t)\,dt\right)^{\frac{1}{q}} \lesssim \left(\int_0^\infty \varphi(t)^p v(t)\,dt\right)^{\frac{1}{p}}$$

for all $\varphi \in \mathfrak{M}^+(0,\infty;\downarrow)$. This last estimate can be interpreted as a restricted weighted inequality for the operator T_γ, defined by

$$T_\gamma g(t) = \sup_{t\leqslant s<\infty} s^{\frac{\gamma}{n}-1}\int_0^s g(y)\,dy, \qquad g \in \mathfrak{M}_+(0,\infty), \quad t\in(0,\infty). \tag{4.4}$$

The operator T_γ is a typical example of what we call *a Hardy type operator involving suprema.* Rather interestingly, such operators have been recently encountered in various research projects. They have been found indispensable in the search for optimal pairs of rearrangement-invariant norms for which a Sobolev type inequality holds ([47]). They constitute a very useful tool for characterization of the associate norm of an operator-induced norm, which naturally appears as an optimal domain norm in a Sobolev embedding ([75, 76]). Supremum operators are also very useful in limiting interpolation theory as can be seen from their appearance, for example, in [26, 30, 25, 77].

The role of supremum operators varies with the given problem, but very often it is used first to make some quantity monotone and then, when it has done its job, it is peeled off thanks to its boundedness on certain function spaces. A typical example is [47], where optimal Sobolev embeddings are studied, and a key part is played by the quantity $t^\alpha g^{**}(t)$, with some positive α. Now, of course, t^α is nondecreasing and $g^{**}(t)$ nonincreasing, and since g is arbitrary, it is not clear who wins. On the other hand, in order to handle certain important norm-like estimates, one needs to work with the quantity $\left[s^\alpha g^{**}(s)\right]^*(t)$, which might be quite difficult to evaluate in general. Here, a supremum operator comes at a great help. Namely, the operator

$$Tg(t) := t^{-\alpha}\sup_{t\leqslant s\leqslant 1} s^\alpha g^*(s), \qquad t\in(0,1),$$

has the two key properties: an obvious pointwise estimate

$$g^* \leqslant Tg$$

and the important monotonicity property

$$t^\alpha Tg(t)\downarrow \qquad \text{for any } g,$$

implying of course that

$$\left[s^\alpha Tg(s)\right]^*(t) = t^\alpha Tg(t).$$

By doing this, one obstacle is over. Of course, we have to pay for it by having one more operator around, but if certain its reasonable boundedness properties are known, it might be peeled off from the whole business at an appropriate place. In [47, 48, 49, 22] and more, this idea works beautifully. The peeling-off step is illustrated nicely by [47, Theorem 3.13] and [22, Lemma 4.5]. Some more details and further information on this topic and its applications can be found in [75].

Now the question is when the supremum operators are bounded on function spaces. For Lebesgue spaces, a thorough comprehensive study of weighted inequalities for supremum operators was carried out in [32], particular results for the case when $1 < p \leqslant q < \infty$ were established in the earlier work [20] (for more general operators and for extended range of p and q cf. [72, Theorem 2.10]). Some further extensions and applications can be found in [29].

We concentrate on the main results now. We follow [32].

Definition 4.1. Let u be a continuous weight on $(0,\infty)$. We define the operator R_u at $\varphi \in \mathfrak{M}^+(0,\infty;\downarrow)$ by

$$R_u\varphi(t) = \sup_{t\leqslant\tau<\infty} u(\tau)\varphi(\tau), \quad t \in (0,\infty).$$

Let b be a weight and let $B(t) = \int_0^t b(s)\,ds$, $t \in (0,\infty)$. Assume that b is such that $0 < B(t) < \infty$ for every $t \in (0,\infty)$. The operator $T_{u,b}$ is defined at $g \in \mathfrak{M}_+(0,\infty)$ by

$$T_{u,b}g(t) = \sup_{t\leqslant\tau<\infty} \frac{u(\tau)}{B(\tau)} \int_0^\tau g(s)b(s)\,ds, \quad t \in (0,\infty).$$

We also make use of the weighted Hardy operator

$$P_{u,b}g(t) = \frac{u(t)}{B(t)} \int_0^t g(s)b(s)\,ds, \quad t \in (0,\infty).$$

We start with a simple lemma which enables us to restrict our considerations to special weights u, namely to those for which $\frac{u}{B}$ is nonincreasing. For this purpose, we put

$$\overline{u}(t) = B(t) \sup_{t\leqslant\tau<\infty} \frac{u(\tau)}{B(\tau)}, \quad t \in (0,\infty).$$

Note that $\frac{\overline{u}}{B}$ is nonincreasing on $(0,\infty)$. In the second part of the lemma, we develop a principle which (under certain assumptions) reduces the inequalities involving $T_{u,b}$ to those involving R_u and $P_{u,b}$, which are considerably more manageable operators. This idea was first used in [20].

Lemma 4.2. *Let u and b be as in* Definition 4.1.

(i) *For every $g \in \mathfrak{M}_+(0,\infty)$ and $t \in (0,\infty)$,*

$$T_{u,b}g(t) = T_{\overline{u},b}g(t).$$

(ii) *Assume that*

$$\sup_{0<t<\infty} \frac{u(t)}{B(t)} \int_0^t \frac{b(s)}{u(s)}\, ds < \infty. \tag{4.5}$$

Then, for all $\varphi \in \mathfrak{M}^+(0,\infty;\downarrow)$,

$$T_{u,b}\varphi(t) \approx R_u\varphi(t) + P_{\overline{u},b}\varphi(t), \quad t \in (0,\infty).$$

We are particularly interested in the situation when (4.5) is valid with u replaced by $\overline{u}$. It is easy to show that this is equivalent to

$$\sup_{0<t<\infty} \frac{u(t)}{B(t)} \int_0^t \frac{b(s)}{\overline{u}(s)}\, ds < \infty. \tag{4.6}$$

We now treat restricted weighted inequalities, i.e., those involving only nonincreasing functions.

Let $0 < p, q < \infty$ and let u be a continuous weight. Our first main goal is to give a characterization of weights v and w such that the inequality

$$\left(\int_0^\infty R_u\varphi(t)^q w(t)\, dt \right)^{\frac{1}{q}} \lesssim \left(\int_0^\infty \varphi(t)^p v(t)\, dt \right)^{\frac{1}{p}} \tag{4.7}$$

holds for all $\varphi \in \mathfrak{M}^+(0,\infty;\downarrow)$. We also present a simpler characterization in this case provided that u is equivalent to a nondecreasing function on $(0,\infty)$. We are led to considering such a special case by the important examples when $u(t) = t^\alpha$ with $\alpha \in (0,1)$ (cf. [20, 47] etc.), $u(t) = t^\alpha |\log t|^\beta$ with $\alpha \in (0,1)$ and $\beta \in \mathbb{R}$ (cf. [72, 29]), and $u(t) = \frac{1}{\sqrt{1+\log \frac{1}{t}}}$ that comes out in the study of optimal dimension-independent Gaussian–Sobolev embeddings in [22].

In the following theorem, case (ii), we have only a discretized condition. We will return to this problem later. Now, we have to introduce the notion of a covering sequence, familiar to us from the Sawyer's result quoted in Section 1.

Definition 4.3. Let $\{x_k\}_{k=-\infty}^{\infty}$ be an increasing sequence in $(0,\infty)$ such that $\lim_{k\to-\infty} x_k = 0$ and $\lim_{k\to\infty} x_k = \infty$. Then we say that $\{x_k\}$ is a *covering sequence*. We also admit increasing sequences $\{x_k\}_{k=J}^{k=K}$, where either $J \in \mathbb{Z}$ and $x_J = 0$, or $K \in \mathbb{Z}$ and $x_K = \infty$, or both.

Theorem 4.4. *Let* $0 < p, q < \infty$ *and let* u *be a continuous weight. Let* v *and* w *be weights such that* $0 < \int_0^x v(t)\, dt < \infty$ *and* $0 < \int_0^x w(t)\, dt < \infty$ *for every* $x \in (0,\infty)$. *In the case* $q < p$, *we define* r *by* (2.9).

(i) *Let* $0 < p \leqslant q < \infty$. *Then* (4.7) *is satisfied for all* $\varphi \in \mathfrak{M}^+(0,\infty;\downarrow)$ *if and only if*

$$\left(\int_0^x \Big[\sup_{t\leqslant\tau\leqslant x} u(\tau) \Big]^q w(t)\, dt \right)^{\frac{1}{q}} \lesssim \left(\int_0^x v(t)\, dt \right)^{\frac{1}{p}} \quad \forall\, x \in (0,\infty). \tag{4.8}$$

(ii) *Let* $0 < q < p < \infty$. *Then* (4.7) *is satisfied for all* $\varphi \in \mathfrak{M}^+(0,\infty;\downarrow)$ *if and only if*

$$\sup_{\{x_k\}} \sum_k \left(\int_{x_k}^{x_{k+1}} [\sup_{t \leqslant \tau \leqslant x_{k+1}} u(\tau)]^q w(t)\, dt \right)^{\frac{r}{q}} \left(\int_0^{x_{k+1}} v(t)\, dt \right)^{-\frac{r}{p}} < \infty, \quad (4.9)$$

where the supremum is taken over all covering sequences $\{x_k\}$.

The proof of the above theorem is quite involved and contains certain fine real analysis.

Remark 4.5. Suppose that u is equivalent to a nondecreasing function on $(0,\infty)$. Then (4.8) reduces to

$$u(x) \left(\int_0^x w(t)\, dt \right)^{\frac{1}{q}} \lesssim \left(\int_0^x v(t)\, dt \right)^{\frac{1}{p}} \quad \forall\, x \in (0,\infty).$$

As a particular case of this result, we recover [20, Lemma 3.1]. Analogously, (4.9) is simplified to

$$\sup_{\{x_k\}} \sum_k u(x_{k+1})^r \left(\int_{x_k}^{x_{k+1}} w(t)\, dt \right)^{\frac{r}{q}} \left(\int_0^{x_{k+1}} v(t)\, dt \right)^{-\frac{r}{p}} < \infty, \quad (4.10)$$

Moreover, the next theorem shows that in this case (4.9) can be replaced by an integral condition in the spirit of [73, Theorem 1.15].

Theorem 4.6. *Let* $0 < q < p < \infty$ *and let* u, v, w *and* r *be as in* Theorem 4.4. *Moreover, assume that* u *is equivalent to a nondecreasing function on* $(0,\infty)$. *Then the inequality* (4.7) *holds if and only if*

$$\int_0^\infty \left(\int_0^t w(s)\, ds \right)^{\frac{r}{p}} [\sup_{t \leqslant \tau < \infty} u(\tau) \left(\int_0^\tau v(s)\, ds \right)^{-\frac{1}{p}}]^r w(t)\, dt < \infty. \quad (4.11)$$

Our next aim is to characterize the validity of the inequality

$$\left(\int_0^\infty [(T_{u,b}\varphi)(t)]^q\, w(t)\, dt \right)^{\frac{1}{q}} \lesssim \left(\int_0^\infty [\varphi(t)]^p\, v(t)\, dt \right)^{\frac{1}{p}} \quad (4.12)$$

for all $\varphi \in \mathfrak{M}^+(0,\infty;\downarrow)$. To this end one can combine Lemma 4.2, Theorem 4.4 and the known results on weighted inequalities for monotone functions due to Sawyer [79], Stepanov [83], and others. For example, when (4.5) is true with u replaced by $\overline{u}$ (i.e., when (4.6) is satisfied), then we obtain the following theorem.

Theorem 4.7. *Let* $0 < p, q < \infty$. *In the case* $q < p$, *we define* r *by* (2.9). *Let* b *be a weight such that* $0 < B(t) < \infty$ *for every* $t \in (0,\infty)$, *where*

$B(t) = \int_0^t b(s)\,ds$. *Let* u, v *and* w *be as in* Theorem 4.4 *and assume that* (4.6) *is satisfied.*

(i) *Let* $1 < p \leqslant q < \infty$. *Then* (4.12) *holds on* $\mathfrak{M}^+(0,\infty;\downarrow)$ *if and only if*

$$\left(\int_0^x \bigl[\sup_{t\leqslant\tau\leqslant x} \overline{u}(\tau)\bigr]^q w(t)\,dt\right)^{\frac{1}{q}} \lesssim \left(\int_0^x v(t)\,dt\right)^{\frac{1}{p}} \quad \forall\, x \in (0,\infty), \tag{4.13}$$

and

$$\sup_{x>0}\left(\int_x^\infty \left(\frac{\overline{u}(t)}{B(t)}\right)^q w(t)\,dt\right)^{\frac{1}{q}} \left(\int_0^x \left(\frac{B(t)}{V(t)}\right)^{p'} v(t)\,dt\right)^{\frac{1}{p'}} < \infty. \tag{4.14}$$

(ii) *Let* $0 < p \leqslant q < \infty$ *and* $0 < p \leqslant 1$. *Then* (4.12) *holds on* $\mathfrak{M}^+(0,\infty;\downarrow)$ *if and only if* (4.13) *is satisfied and*

$$B(x)\left(\int_x^\infty \left(\frac{\overline{u}(t)}{B(t)}\right)^q w(t)\,dt\right)^{\frac{1}{q}} \lesssim \left(\int_0^x v(t)\,dt\right)^{\frac{1}{p}} \quad \forall\, x \in (0,\infty). \tag{4.15}$$

(iii) *Let* $1 < p < \infty$, $0 < q < p < \infty$ *and* $q \neq 1$. *Then* (4.12) *holds on* $\mathfrak{M}^+(0,\infty;\downarrow)$ *if and only if*

$$\sup_{\{x_k\}} \sum_k \left(\int_{x_k}^{x_{k+1}} \bigl[\sup_{t\leqslant\tau\leqslant x_{k+1}} \overline{u}(\tau)\bigr]^q w(t)\,dt\right)^{\frac{r}{q}} \left(\int_0^{x_{k+1}} v(t)\,dt\right)^{-\frac{r}{p}} < \infty, \tag{4.16}$$

where the supremum is taken over all covering sequences, and

$$\int_0^\infty \left(\int_t^\infty \left(\frac{\overline{u}(s)}{B(s)}\right)^q w(s)\,ds\right)^{\frac{r}{q}} \left(\int_0^t \left(\frac{B(s)}{V(s)}\right)^{p'} v(s)\,ds\right)^{\frac{r}{q'}} \times \left(\frac{B(t)}{V(t)}\right)^{p'} v(t)\,dt < \infty. \tag{4.17}$$

(iv) *Let* $1 < p < \infty$ *and* $q = 1$. *Then* (4.12) *holds on* $\mathfrak{M}^+(0,\infty;\downarrow)$ *if and only if* (4.16) *is satisfied and*

$$\left(\int_0^\infty \left(\int_0^t \overline{u}(s)w(s)\,ds + B(t)\int_t^\infty \frac{\overline{u}(s)}{B(s)} w(s)\,ds\right)^{p'} \frac{v(t)}{[V(t)]^{p'}}\,dt\right)^{\frac{1}{p'}} < \infty. \tag{4.18}$$

(v) *Let* $0 < q < p \leqslant 1$. *Then* (4.12) *holds on* $\mathfrak{M}^+(0,\infty;\downarrow)$ *if and only if* (4.16) *is satisfied and*

$$\left(\int_0^\infty \left(\operatorname*{ess\,sup}_{0<s\leqslant t} \frac{B(s)}{[V(s)]^{\frac{1}{p}}}\right)^r \left(\int_t^\infty \left(\frac{\overline{u}(s)}{B(s)}\right)^q w(s)\,ds\right)^{\frac{r}{p}} \times \left(\frac{\overline{u}(t)}{B(t)}\right)^q w(t)\,dt\right)^{\frac{1}{r}} < \infty. \tag{4.19}$$

Remark 4.8. As in Theorem 3.14, in the case where $\overline{u}$ is equivalent to a nondecreasing function on $(0,\infty)$, the condition (4.16) can be replaced by the integral condition

$$\int_0^\infty \left(\int_0^t w(s)\,ds\right)^{\frac{r}{p}} \Big[\sup_{t\leqslant\tau<\infty} \overline{u}(\tau) \left(\int_0^\tau v(s)\,ds\right)^{-\frac{1}{p}}\Big]^r w(t)\,dt < \infty.$$

Remark 4.9. Theorem 4.7 with slightly modified assumptions can also be proved using a reduction theorem from [33]. Note that Theorem 6.2 in [33] states that, for $1<p<\infty$ and $0<q<\infty$, the inequality (4.12) holds for all $\varphi \in \mathfrak{M}^+(0,\infty;\downarrow)$ if and only if (4.7) holds for all $\varphi \in \mathfrak{M}^+(0,\infty;\downarrow)$ and simultaneously the estimate

$$\left(\int_0^\infty T_{u,b}g(t)^q w(t)\,dt\right)^{\frac{1}{q}} \lesssim \left(\int_0^\infty g(t)^p \left(\frac{V(t)}{B(t)}\right)^p v(t)^{1-p}\,dt\right)^{\frac{1}{p}}$$

is satisfied for all $g \in \mathfrak{M}_+(0,\infty)$. This motivates us to study when the inequality

$$\left(\int_0^\infty T_{u,b}g(t)^q w(t)\,dt\right)^{\frac{1}{q}} \lesssim \left(\int_0^\infty g(t)^p v(t)\,dt\right)^{\frac{1}{p}} \tag{4.20}$$

holds for all $g \in \mathfrak{M}_+(0,\infty)$. This will be done next.

We study the same weighted inequalities as above, but without the restriction to nonincreasing functions. In particular, we wish to establish when the inequality (4.20) holds for all $g \in \mathfrak{M}_+(0,\infty)$.

Recall that if $p \in (1,\infty)$, then the conjugate number p' is given by $p' = \frac{p}{p-1}$. We also use the following notation.

Notation. For a given weight v, $0 \leqslant \alpha < \beta \leqslant \infty$ and $1 \leqslant p < \infty$, we denote

$$\sigma_p(\alpha,\beta) = \begin{cases} \left(\int_\alpha^\beta v(t)^{1-p'}\,dt\right)^{\frac{1}{p'}} & \text{when } 1<p<\infty \\ \operatorname*{ess\,sup}_{\alpha<t<\beta} \frac{1}{v(t)} & \text{when } p=1. \end{cases}$$

We first consider the particular case $b \equiv 1$. Then we write $T_u := T_{u,b}$ and, since $B(t)=t$, $t\in(0,\infty)$, we have

$$T_u g(t) = \sup_{t \leqslant s < \infty} \frac{u(s)}{s} \int_0^s g(y)\, dy.$$

We characterize the validity of the inequality

$$\left(\int_0^\infty T_u g(t)^q w(t)\, dt \right)^{\frac{1}{q}} \lesssim \left(\int_0^\infty g(t)^p v(t)\, dt \right)^{\frac{1}{p}} \tag{4.21}$$

on the set $\mathfrak{M}_+(0, \infty)$.

Theorem 4.10. *Assume that* $1 \leqslant p < \infty$ *and* $0 < q < \infty$. *In the case* $q < p$, *we define* r *by* (2.9). *Let* u, v *and* w *be as in* Theorem 4.4.

(i) *Let* $p \leqslant q$. *Then* (4.21) *holds on* $\mathfrak{M}_+(0, \infty)$ *if and only if*

$$\sup_{x>0} \left(\left(\frac{\overline{u}(x)}{x} \right)^q \int_0^x w(t)\, dt + \int_x^\infty \left(\frac{\overline{u}(t)}{t} \right)^q w(t)\, dt \right)^{\frac{1}{q}} \sigma_p(0, x) < \infty. \tag{4.22}$$

(ii) *Let* $q < p$. *Then* (4.21) *holds on* $\mathfrak{M}_+(0, \infty)$ *if and only if*

$$\sup_{\{x_k\}} \left(\sum_k \left(\int_{x_{k-1}}^{x_{k+1}} \min\left\{ \frac{\overline{u}(x_k)}{x_k}, \frac{\overline{u}(t)}{t} \right\}^q w(t)\, dt \right)^{\frac{r}{q}} [\sigma_p(x_{k-1}, x_k)]^r \right)^{\frac{1}{r}} < \infty, \tag{4.23}$$

where the supremum is taken over all covering sequences $\{x_k\}$.

Our next aim is to obtain an analogous result involving a general operator $T_{u,b}$.

Theorem 4.11. *Assume that* $1 \leqslant p < \infty$ *and* $0 < q < \infty$. *In the case* $q < p$, *we define* r *by* (2.9). *Let* b *be a weight such that* $b(t) > 0$ *for a.e.* $t \in (0, \infty)$ *and* $B(t) := \int_0^t b(s)\, ds < \infty$ *for all* $t \in (0, \infty)$. *Let* u, v *and* w *be as in* Theorem 4.4. *For* $0 \leqslant \alpha < \beta \leqslant \infty$, *we denote*

$$\sigma_{p,b}(\alpha, \beta) = \begin{cases} \left(\int_\alpha^\beta v(t)^{1-p'} b(t)^{p'}\, dt \right)^{\frac{1}{p'}} & \text{when } 1 < p < \infty, \\ \operatorname*{ess\,sup}_{\alpha < t < \beta} \frac{b(t)}{v(t)} & \text{when } p = 1. \end{cases}$$

(i) *Let* $p \leqslant q$. *Then* (4.20) *holds on* $\mathfrak{M}_+(0, \infty)$ *if and only if*

$$\sup_{x>0} \left(\left(\frac{\overline{u}(x)}{B(x)} \right)^q \int_0^x w(t)\, dt + \int_x^\infty \left(\frac{\overline{u}(t)}{B(t)} \right)^q w(t)\, dt \right)^{\frac{1}{q}} \sigma_{p,b}(0, x) < \infty.$$

(ii) *Let* $q < p$. *Then* (4.20) *holds on* $\mathfrak{M}_+(0, \infty)$ *if and only if*

$$\sup_{\{x_k\}} \left(\sum_k \left(\int_{x_{k-1}}^{x_{k+1}} \min \left\{ \frac{\overline{u}(x_k)}{B(x_k)}, \frac{\overline{u}(t)}{B(t)} \right\}^q w(t)\, dt \right)^{\frac{r}{q}} [\sigma_{p,b}(x_{k-1}, x_k)]^r \right)^{\frac{1}{r}} < \infty,$$

where the supremum is taken over all covering sequences $\{x_k\}$.

We have also an alternative integral criterion that characterizes (4.21).

Theorem 4.12. *Let* u, v *and* w *be as in* Theorem 4.10. *Let* $1 \leqslant p < \infty$ *and* $0 < q < p$. *In the case* $q < p$, *we define* r *by* (2.9). *Then the inequality* (4.21) *holds for all* $g \in \mathfrak{M}_+(0, \infty)$ *if and only if*

$$\left(\int_0^\infty \left(\int_t^\infty \left(\frac{\overline{u}(s)}{s} \right)^q w(s)\, ds \right)^{\frac{r}{p}} \left(\frac{\overline{u}(t)}{t} \right)^q \sigma_p(0,t)^r w(t)\, dt \right)^{\frac{1}{r}} < \infty \quad (4.24)$$

and

$$\left(\int_0^\infty \left(\int_0^t w(s)\, ds \right)^{\frac{r}{p}} \Big[\sup_{t \leqslant \tau < \infty} \frac{\overline{u}(\tau)}{\tau} \sigma_p(0,\tau) \Big]^r w(t)\, dt \right)^{\frac{1}{r}} < \infty. \quad (4.25)$$

Acknowledgment. This research was supported by the research project MSM 0021620839 of the Czech Ministry of Education.

References

1. Alvino, A., Trombetti, G., Lions, P.L.: On optimization problems with prescribed rearrangements. Nonlinear Anal. **13**, 185–220 (1989)
2. Ariño, M., Muckenhoupt, B.: Maximal functions on classical Lorentz spaces and Hardy's inequality with weights for nonincreasing functions. Trans. Am. Math. Soc. **320**, 727–735 (1990)
3. Bagby, R.J., Kurtz, D.S.: A rearrangement good-λ inequality. Trans. Am. Math. Soc. **293**, 71–81 (1986)
4. Barza, S., Persson, L.-E., Soria, J.: Sharp weighted multidimensional integral inequalities for monotone functions. Math. Nachr. **210**, 43–58 (2000)
5. Bastero, J., Milman, M., Ruiz, F.: A note in $L(\infty, q)$ spaces and Sobolev embeddings. Indiana Univ. Math. J. **52**, 1215–1230 (2003)
6. Bennett, C., De Vore, R., Sharpley, R.: Weak L^∞ and BMO. Ann. Math. **113**, 601–611 (1981)
7. Bennett, C., Sharpley, R.: Interpolation of Operators. Academic Press, Boston (1988)
8. Berezhnoi, E.I.: Weighted inequalities of Hardy type in general ideal spaces. Sov. Math. Dokl. **43**, 492–495 (1991)
9. Boza, S., Martín, J.: Equivalent expressions for norms in classical Lorentz spaces. Forum Math. **17**, 361–373 (2005)

10. Bradley, J.S.: Hardy inequalities with mixed norms. Canad. Math. Bull. **21**, 405–408 (1978)
11. Calderón, A.P.: Intermediate spaces and interpolation, the complex method. Studia Math. **24**, 113–190 (1964)
12. Carro, M.J., García del Amo, A., Soria, J.: Weak type weights and normable Lorentz spaces. Proc. Am. Math. Soc. **124**, 849–857 (1996)
13. Carro, M., Gogatishvili, A., Martín, J., Pick, L.: Functional properties of rearrangement invariant spaces defined in terms of oscillations. J. Funct. Anal. **229**, no. 2, 375–404 (2005)
14. Carro, M., Gogatishvili, A., Martín, J., Pick, L.: Weighted inequalities involving two Hardy operators with applications to embeddings of function spaces. J. Operator Theory **59**, no. 2, 101–124 (2008)
15. Carro, M., Pick, L., Soria, J., Stepanov, V.D.: On embeddings between classical Lorentz spaces. Math. Ineq. Appl. **4**, 397–428 (2001)
16. Carro, M.J., Soria, J.: Weighted Lorentz spaces and the Hardy operator. J. Funct. Anal. **112**, 480–494 (1993)
17. Carro, M.J., Soria, J.: Boundedness of some integral operators. Canad. J. Math. **45**, 1155–1166 (1993)
18. Carro, M.J., Soria, J.: The Hardy-Littlewood maximal function and weighted Lorentz spaces. J. London Math. Soc. **55**, 146–158 (1997)
19. Cianchi, A.: Symmetrization and second-order Sobolev inequalities. Ann. Mat. Pura Appl. **183**, 45–77 (2004)
20. Cianchi, A., Kerman, R., Opic, B., Pick, L.: A sharp rearrangement inequality for fractional maximal operator. Studia Math. **138**, 277–284 (2000)
21. Cianchi, A., Kerman, R., Pick, L.: Boundary trace inequalities and rearrangements. J. Anal. Math. **105**, 241–265 (2008)
22. Cianchi, A., Pick, L.: Optimal Gaussian Sobolev embeddings. J. Funct. Anal. **256**, 3588–3642 (2009)
23. Cianchi, A., Pick, L.: An optimal endpoint trace embedding. Ann. Inst. Fourier (Grenoble) [To appear]
24. Cwikel, M., Kamińska, A., Maligranda, L., Pick, L.: Are generalized Lorentz "spaces" really spaces? Proc. Am. Math. Soc. **132**, 3615–3625 (2004)
25. Cwikel, M., Pustylnik, E.: Weak type interpolation near "endpoint" spaces. J. Funct. Anal. **171**, 235–277 (1999)
26. Doktorskii, R.Ya.: Reiteration relations of the real interpolation method. Sov. Math. Dokl. **44**, 665–669 (1992)
27. Edmunds, D.E., Gurka, P., Pick, L.: Compactness of Hardy type integral operators in weighted Banach function spaces. Studia Math. **109**, 73–90 (1994)
28. Edmunds, D.E., Kerman, R., Pick, L.: Optimal Sobolev imbeddings involving rearrangement-invariant quasinorms. J. Funct. Anal. **170**, 307–355 (2000)
29. Edmunds, D.E., Opic, B.: Boundedness of fractional maximal operators between classical and weak type Lorentz spaces. Disser. Math. **410**, 1–50 (2002)
30. Evans, W.D., Opic, B.: Real interpolation with logarithmic functors and reiteration. Canad. J. Math. **52**, 920–960 (2000)
31. Franciosi, M.: A condition implying boundedness and VMO for a function f. Studia Math. **123**, 109–116 (1997)
32. Gogatishvili, A., Opic, B., Pick, L.: Weighted inequalities for Hardy type operators involving suprema. Collect. Math. **57**, no. 3, 227–255 (2006)
33. Gogatishvili, A., Pick, L.: Duality principles and reduction theorems. Math. Inequal. Appl. **3**, 539–558 (2000)

34. Gogatishvili, A., Pick, L.: Discretization and anti-discretization of rearrangement-invariant norms. Publ. Mat. **47**, 311–358 (2003)
35. Gogatishvili, A., Pick, L.: Embeddings and duality theorems for weak classical Lorentz spaces. Canad. Math. Bull. **49**, no. 1, 82–95 (2006)
36. Gogatishvili, A., Pick, L.: A reduction theorem for supremum operators. J. Comput. Appl. Math. **208**, 270–279 (2007)
37. Gol'dman, M.L.: On integral inequalities on the set of functions with some properties of monotonicity. In: Function spaces, Differential Operators and Nonlinear Analysis **133**, 274–279 (1993)
38. Gol'dman, M.L., Heinig, H.P., Stepanov, V.D.: On the principle of duality in Lorentz spaces. Canad. J. Math. **48**, 959–979 (1996)
39. Grosse-Erdmann, K.-G.: The Blocking Technique, Weighted Mean Operators and Hardy's Inequality. Lect. Notes Math. 1679, Springer, Berlin etc. (1998)
40. Haaker, A.: On the conjugate space of Lorentz space. Technical Report, Lund, 1–23 (1970)
41. Hardy, G.H.: Notes on some points in the integral calculus, LX. An inequality between integrals. Messenger Math. **54**, 150–156 (1925)
42. Hardy, G.H., Littlewood, J.E.: Notes on the theory of series. XII: On certain inequalities connected with the calculus of variations. J. London Math. Soc. **5**, 34–39 (1930)
43. Heinig, H.P., Maligranda, L.: Weighted inequalities for monotone and concave functions. Studia Math. **116**, 133–165 (1995)
44. Kac, I.S., Krein, M.G.: Criteria for discreteness of the spectrum of a singular string (Russian). Izv. Vyss. Uchebn. Zav. Mat. **2**, 136–153 (1958)
45. Kališ, J., Milman, M.: Symmetrization and sharp Sobolev inequalities in metric spaces. arXiv: 0806.0052, subject: Math.FA, Math.AP
46. Kamińska, A., Maligranda, L.: Order convexity and concavity in Lorentz spaces with arbitrary weight. Research Report, Lulera University of Technology **4**, 1–21 (1999)
47. Kerman, R., Pick, L.: Optimal Sobolev imbeddings. Forum Math. **18**, no. 4, 535–570 (2006)
48. Kerman, R., Pick, L.: Optimal Sobolev imbedding spaces. Studia Math. **192**, no. 3, 195–217 (2009)
49. Kerman, R., Pick, L.: Compactness of Sobolev imbeddings involving rearrangement-invariant norms. Studia Math. **186**, no. 2, 127–160 (2008)
50. Kolyada, V.I.: Estimates of rearrangements and embedding theorems. Math. USSR-Sb. **64**, no. 1, 1–21 (1989)
51. Kolyada, V.I.: Rearrangement of functions and embedding theorems. Russ. Math. Surv. **44**, 73–118 (1989)
52. Kolyada, V.I.: On embedding theorems. In: J. Rákosník (ed.) Nonlinear Analysis, Function Spaces and Applications 8, Proceedings of the Spring School held in Prague, Czech Republic, May 30-June 6, 2006, pp. 35–94. Institute of Mathematics, Academy of Sciences of the Czech Republic, Prague (2007)
53. Kolyada, V.I., Lerner, A.K.: On limiting embeddings of Besov spaces. Studia Math. **171**, no. 1, 1–13 (2005)
54. Kudryavtsev, L.D., Nikolskii, S.M.: Spaces of differentiable functions of several variables and imbedding theorems. In: Analysis III, Encyclopaedia Math. Sci. **26**, pp. 1–140. Springer, Berlin (1991)
55. Kufner, A., Maligranda, L., Persson, L.-E.: The Hardy Inequality. About its history and some related results. Pilsen, Vydavatelský Servis (2007)

56. Lai, Q., Pick, L.: The Hardy operator, L_∞, and BMO. J. London Math. Soc. **48**, 167–177 (1993)
57. Lai, S.: Weighted norm inequalities for general operators on monotone functions. Trans. Am. Math. Soc. **340**, 811–836 (1993)
58. Lang, J., Nekvinda, A.: A difference between the continuous norm and the absolutely continuous norm in Banach function spaces. Czechoslovak Math. J. **47**, no. 122, 221–232 (1997)
59. Lang, J., Nekvinda, A., Rákosník, J.: Continuous norms and absolutely continuous norms in Banach function spaces are not the same. Real Anal. Exch. **26**, no. 1, 345–364 (2000/2001)
60. Lang, J., Pick, L.: The Hardy operator and the gap between L_∞ and BMO. J. London Math. Soc. **57**, 196–208 (1998)
61. Lindenstrauss, J., Tzafriri, L.: Classical Banach Spaces I and II. Springer, Berlin (1997)
62. Lorentz, L.L.: On the theory of spaces Λ. Pacific J. Math **1**, 411–429 (1951)
63. Luxemburg, W.A.J., Zaanen, A.C.: Compactness of integral operators in Banach function spaces. Math. Ann. **149**, 150–180 (1963)
64. Malý, J., Pick, L.: An elementary proof of sharp Sobolev embeddings. Proc. Am. Math. Soc. **130**, 555–563 (2002)
65. Martín, J., Milman, M.: Symmetrization inequalities and Sobolev embeddings. Proc. Am. Math. Soc. **134**, 2335–2347 (2006)
66. Maz'ya, V.G.: Einbettungssätze für Sobolewsche Räume. Teil 1 (Embedding Theorems for Sobolev Spaces. Part 1). Teubner-Texte zur Mathematik, Leipzig (1979)
67. Maz'ya, V.G.: Sobolev Spaces. Springer, Berlin etc. (1985)
68. Milman, M., Pustylnik, E.: On sharp higher order Sobolev embeddings. Commun. Contemp. Math. **6**, (2004) 495–511
69. Muckenhoupt, B.: Hardy's inequality with weights. Studia Math. **44**, 31–38 (1972)
70. Neugebauer, C.J.: Weighted norm inequalities for averaging operators of monotone functions. Publ. Math. **35**, 429–447 (1991)
71. Oskolkov, K.I.: Approximation properties of summable functions on sets of full measure. Math. USSR Sb. **32**, 489–517 (1977)
72. Opic, B.: On boundedness of fractional maximal operators between classical Lorentz spaces. In: V. Mustonen and J. Rákosník (eds.) Function Spaces, Differential Operators and Nonlinear Analysis Vol. 4. Proceedings of the Spring School held in Syöte, June 1999, pp. 187-196. Mathematical Institute of the Academy of Sciences of the Czech Republic, Praha (2000)
73. Opic, B., Kufner, A.: Hardy type inequalities. Longman Sci., Harlow, (1991)
74. Pick, L.: Optimality of function spaces in Sobolev embeddings In: Maz'ya, V. (ed.) Sobolev Spaces in Mathematics I. Sobolev Type Inequalities, pp. 249–280. Springer and Tamara Rozhkovskaya Publisher, New York etc. (2009)
75. Pick, L.: Supremum operators and optimal Sobolev inequalities. In: V. Mustonen and J. Rákosník (eds.) Function Spaces, Differential Operators and Nonlinear Analysis Vol. 4. Proceedings of the Spring School held in Syöte, June 1999, pp. 207–219. Mathematical Institute of the Academy of Sciences of the Czech Republic, Praha (2000)
76. Pick, L.: Optimal Sobolev Embeddings. Rudolph-Lipschitz-Vorlesungsreihe no. 43, Rheinische Friedrich-Wilhelms-Universität Bonn (2002)
77. Pustylnik, E.: Optimal interpolation in spaces of Lorentz-Zygmund type. J. Anal. Math. **79**, 113–157 (1999)

78. Sawyer, E.: Weighted Lebesgue and Lorentz norm inequalities for the Hardy operator. Trans. Am. Math. Soc. **281**, 329–337 (1984)
79. Sawyer, E.: Boundedness of classical operators on classical Lorentz spaces. Studia Math. **96**, 145–158 (1990)
80. Sinnamon, G.: Weighted Hardy and Opial type inequalities. J. MathȦnalȦppl. **160**, 434–445 (1991)
81. Sinnamon, G., Stepanov, V.D.: The weighted Hardy inequality: new proofs and the case $p = 1$. J. London Math. Soc. **54**, 89–101 (1996)
82. Sparr, A.: On the conjugate space of the Lorentz space $L(\phi, q)$. Contemp. Math. **445**, 313–336 (2007)
83. Stepanov, V.D.: The weighted Hardy's inequality for nonincreasing functions. Trans. Am. Math. Soc. **338**, 173–186 (1993)
84. Stepanov, V.D.: Integral operators on the cone of monotone functions. J. London Math. Soc. **48**, 465–487 (1993)
85. Storozhenko, E.A.: Necessary and sufficient conditions for embeddings of certain classes of functions (Russian). Izv. AN SSSR Ser. Mat. **37**, 386–398 (1973)
86. Ul'yanov, P.L.: Embedding of certain function classes H_p^ω. Math. USSR Izv. **2**, 601–637 (1968)

Finite Rank Toeplitz Operators in the Bergman Space

Grigori Rozenblum

To Volodya Maz'ya,
an outstanding mathematician and person

Abstract We discuss resent developments in the problem of description of finite rank Toeplitz operators in different Bergman spaces and give some applications.

1 Introduction

Toeplitz operators arise in many fields of Analysis and have been an object of active study for many years. Quite a lot of questions can be asked about these operators, and these questions depend on the field where Toeplitz operators are applied.

The classical Toeplitz operator T_f in the Hardy space $H^2(S^1)$ is defined as

$$T_f u = Pfu, \tag{1.1}$$

for $u \in H^2(S^1)$, where f is a bounded function on S^1 (the weight function) and P is the Riesz projection, the orthogonal projection $P : L_2(S^1) \to H^2(S^1)$. Such operators are often called *Riesz–Toeplitz* or *Hardy–Toeplitz* operators (cf. [15], for more details). More generally, for a Hilbert space $\mathcal{H}$ of functions and a closed subspace $\mathcal{L} \subset \mathcal{H}$, the Toeplitz operator T_f in $\mathcal{L}$ acts as in (1.1), where P is the projection $P : \mathcal{H} \to \mathcal{L}$. In particular, in the case where $\mathcal{H}$ is the space $L_2(\Omega, \rho)$ for some domain $\Omega \subset \mathbb{C}^d$ and some measure ρ

Grigori Rozenblum
Department of Mathematics, Chalmers University of Technology and Department of Mathematics, University of Gothenburg, S-412 96, Gothenburg, Sweden
e-mail: grigori@math.chalmers.se

A. Laptev (ed.), *Around the Research of Vladimir Maz'ya III: Analysis and Applications*,
International Mathematical Series 13, DOI 10.1007/978-1-4419-1345-6_12,

and $\mathcal{L}$ is the Bergman space $\mathcal{B}^2 = \mathcal{B}^2(\Omega, \rho)$ of analytical functions in $\mathcal{H}$, such an operator is called *Bergman–Toeplitz*; we denote it by $\mathcal{T}_f$.

Among many interesting properties of Riesz–Toeplitz operators, we mention the following *cut-off* one. If f is a bounded function and the operator T_f is compact, then f should be zero. For many other classes of operators a similar cut-off on some level is also observed. The natural question arises, whether there is a kind of cut-off property for Bergman–Toeplitz operators. Quite long ago it became a common knowledge that at least direct analogy does not take place. In [13], the conditions were found on the function f in the unit disk $\Omega = D$ guaranteeing that the operator $\mathcal{T}_f$ in $\mathcal{B}^2(D, \lambda)$ with Lebesgue measure λ belongs to the Schatten class $\mathfrak{S}_p$. So, the natural question came up: probably, it is on the finite rank level that the cut-off takes place. In other words, if a Bergman–Toeplitz operator has finite rank it should be zero.

It was known long ago that the Schatten class behavior of $\mathcal{T}_f$ is determined by the rate of convergence to zero at the boundary of the function f. Therefore, the finite rank (FR) hypothesis deals with functions f with compact support not touching the boundary of Ω. In this setting, the FR hypothesis is equivalent to the one for Toeplitz operators on the Bargmann (Fock, Segal) space consisting of analytical functions in $\mathbb{C}$, square summable with a Gaussian weight. A proof of the FR hypothesis appeared in the same paper [13], about twenty lines long. Unfortunately, there was an unrepairable fault in the proof, so the FR remained unsettled.

It was only in 2007 that the proof of the FR hypothesis was finally found, even in a more general form. The Bergman projection $\mathbf{P} : L_2 \to \mathcal{B}$ can be extended to an operator from the space of distributions $\mathcal{D}'(\Omega)$ to $\mathcal{B}^2(\Omega, \lambda)$. Let μ be a regular complex Borel measure with compact support in Ω. With μ we associate the Toeplitz operator $\mathcal{T}_\mu : u \mapsto \mathbf{P}u\mu$ in $\mathcal{B}^2(\Omega, \lambda)$.

In [14], the following result was established.

Theorem 1.1. *Suppose that the Toeplitz operator $\mathcal{T}_\mu$ in $\mathcal{B}^2(\Omega, \lambda)$, $\Omega \subset \mathbb{C}$ has finite rank* $\mathbf{r}$. *Then the measure μ is the sum of* $\mathbf{r}$ *point masses,*

$$\mu = \sum_1^{\mathbf{r}} C_k \delta_{z_j}, \ z_j \in \Omega. \tag{1.2}$$

The publication of the proof of Theorem 1.1 induced an activity around it. In two years to follow several papers appeared, where the FR theorem was generalized in different directions, and interesting applications were found in Analysis and Mathematical Physics.

In this paper, we aim for collecting and systematizing the existing results on the finite rank problem and their applications. We also present several new theorems generalizing and extending these results.

In a more vague setting, the problem discussed in the paper can be understood in the following way: is it possible that the contribution of the positive part of a real measure μ and the contribution of the negative part of μ "eat up" each other, so that the resulting Toeplitz operator becomes "trivial." In this form, the relation arises with the results by Maz'ya and Verbitsky (cf., in particular, [16, 17], where the phenomenon of the mutual compensation of positive and negative parts of the weight for embedding of Sobolev spaces was studied in detail).

2 Problem Setting

Let Ω be a domain in $\mathbb{R}^d$ or $\mathbb{C}^d$. We suppose that a measure ρ is defined on Ω, jointly absolutely continuous with Lebesgue measure. Suppose that $\mathcal{L}$ is a closed subspace in $\mathcal{H} = L_2(\Omega, \rho)$, consisting of smooth functions, $\mathcal{L} \subset C^\infty(\Omega)$. In this case, the orthogonal projection $\mathbf{P} : \mathcal{L} \to \mathcal{H}$ is an integral operator with smooth kernel,

$$\mathbf{P}u(x) = \int P(x,y)u(y)d\rho(x). \tag{2.1}$$

We call $\mathbf{P}$ the *Bergman projection* and $P(x,y)$ the *Bergman kernel* (corresponding to the subspace $\mathcal{L}$).

Let F be a distribution, compactly supported in Ω, $F \in \mathcal{E}'(\Omega)$. We denote by $\langle F, \phi\rangle$ the action of the distribution F on the function $\phi \in \mathcal{E}$. Then one can define the Toeplitz operator in $\mathcal{L}$ with *weight* F:

$$(\mathcal{T}_F u)(x) = \langle F, P(x,\cdot)u(\cdot)\rangle. \tag{2.2}$$

Formula (2.2) can be also understood in the following way. The operator $\mathbf{P}$ considered as an operator $P : \mathcal{H} \to \mathcal{L}$ has an adjoint, $P' : \mathcal{L}' \to \mathcal{H}$, so PP' is the extension of $\mathbf{P}$ to the operator $\mathcal{L}' \to \mathcal{L}$, in particular, $\mathbf{P}$ extends as an operator from $\mathcal{E}'(\Omega)$ to $\mathcal{L}$. In this setting, $Fu \in \mathcal{E}'(\Omega)$ for $u \in \mathcal{E}(\Omega)$ and the Toeplitz operator has the form

$$\mathcal{T}_F u = \mathbf{P}Fu, \tag{2.3}$$

consistently with the traditional definition of Toeplitz operators.

It is more convenient to use the description of the Toeplitz operator by means of the sesquilinear form. For $u, v \in \mathcal{L}$, we have

$$(\mathcal{T}_F u, v) = (\mathbf{P}Fu, v) = \langle \sigma Fu, \overline{\mathbf{P}v}\rangle = \langle \sigma F, u\overline{v}\rangle, \tag{2.4}$$

where σ is the Radon–Nikodym derivative of ρ with respect to the Lebesgue measure. In particular, if F is a regular Borel complex measure $F = \mu$, the corresponding Toeplitz operator acts as

$$\mathcal{T}_\mu u(x) = \int_\Omega P(x,y)u(x)d\mu(x), \tag{2.5}$$

and the quadratic form is

$$(\mathcal{T}_F u, v) = \int_\Omega u\bar{v}\sigma d\mu(x).$$

Finally, when F is a bounded function, formula (2.4) takes the form

$$(\mathcal{T}_F u, v) = \int_\Omega u\bar{v}F(x)d\rho(x). \tag{2.6}$$

Classical examples of Bergman spaces and corresponding Toeplitz operators are produced by solutions of elliptic equations and systems.

Example 2.1. Let Ω be a bounded domain in $\mathbb{C}$, $\rho = \lambda$ the Lebesgue measure, and $\mathcal{L} = \mathcal{B}^2(\Omega)$ the space of L_2 functions analytical in Ω. This is the classical Bergman space.

Example 2.2. Let Ω be a bounded pseudoconvex domain in $\mathbb{C}^d$, $d > 1$, with Lebesgue measure ρ, and let the space $\mathcal{L}$ consist of L_2 functions analytical in Ω. This is also a classical Bergman space. Here, and in Example 2.1, measures different from the Lebesgue one are also considered, especially when Ω is a ball or a (poly)disk.

Example 2.3. For a bounded domain $\Omega \subset R^d$, we set $\mathcal{L}$ to be the space of L_2 solutions of the equation $Lu = 0$, where L is an elliptic differential operator with constant coefficients. In particular, if L is the Laplacian, the space $\mathcal{L}$ is called the *harmonic Bergman space.*

Example 2.4. If Ω is a bounded domain in $\mathbb{R}^d$ with *even* $d = 2\mathbf{m}$, and $\mathbb{R}^d$ is identified with $\mathbb{C}^{\mathbf{m}}$ with variables $z_j = (x_j, y_j)$, $j = 1, \dots, \mathbf{m}$, the Bergman space of functions which are harmonic with respect to each pair (x_j, y_j) is called $\mathbf{m}$-harmonic Bergman space; if on the other hand, the space of functions $u(z)$ such that $u_\zeta(\xi_1, \xi_2) = u(\zeta(\xi_1 + i\xi_2))$ is harmonic as a function of variables ξ_1, ξ_2 for any $\zeta \in \mathbb{C}^m \setminus \{0\}$, is called pluriharmonic Bergman space.

Example 2.5. Let Ω be the whole of $\mathbb{C}^{\mathbf{m}} = \mathbb{R}^d$, with the Gaussian measure $d\rho = \exp(-|z|^2/2)d\lambda$. The subspace $\mathcal{L} \subset L_2(\mathbb{C}^{\mathbf{m}}, \rho)$ of entire analytical functions in $\mathbb{C}^{\mathbf{m}}$ is called *Fock* or *Segal–Bargmann* space.

The study of Toeplitz operators in many cases is based upon the consideration of associated infinite matrices.

Let $\Sigma_1 = \{f_j(x), x \in \Omega\}$, $\Sigma_2 = \{g_j(x), x \in \Omega\}$ be two infinite systems of functions in $\mathcal{L}$. With these systems and a distribution $F \in \mathcal{E}'(\Omega)$ we associate the matrix

$$\mathcal{A} = \mathcal{A}(F) = \mathcal{A}(F, \Sigma_1, \Sigma_2, \Omega, \rho) = (\mathcal{T}_F f_j, g_k)_{j,k=1,\dots} = (\langle \sigma F, f_j \overline{g_k} \rangle). \quad (2.7)$$

So, the matrix $\mathcal{A}$ is the matrix of the sesquilinear form of the operator $\mathcal{T}_F$ on the systems Σ_1, Σ_2. We formulate the obvious but important statement.

Proposition 2.6. *Suppose that the Toeplitz operator $\mathcal{T}_F$ has finite rank* $\mathbf{r}$. *Then the matrix $\mathcal{A}$ also has finite rank; moreover,* $\operatorname{rank}(\mathcal{A}) \leqslant \mathbf{r}$.

The use of matrices of the form (2.7) enables one to perform important reductions. In particular, since the domain Ω does not enter explicitly into the matrix, the rank of this matrix does not depend on the domain Ω, as long as one can chose the systems Σ_1, Σ_2 dense simultaneously in the Bergman spaces in different domains. Thus, in particular, the FR problems for the analytical Bergman spaces in bounded domains and for the Fock space are equivalent (cf. the discussion in [21].)

3 Theorem of Luecking. Extensions in Dimension 1

In this section, we present the original proof given by Luecking in [14], and give extensions in several directions.

Theorem 3.1. *Let $\Omega \subset \mathbb{C}$ be a bounded domain, with Lebesgue measure. Suppose that for some regular complex Borel measure μ, absolutely continuous with respect to the Lebesgue measure, with compact support in Ω, the Toeplitz operator $\mathcal{T}_\mu$ in the Bergman space of analytical functions has finite rank* $\mathbf{r}$. *Then $\mu = 0$.*

We formulate and prove here Luecking's theorem only in the case of an absolutely continuous measure; the case of more singular measures will be taken care of later, as a part of the general distributional setting. In the proof, which follows [14], we separate a lemma that will be used further on.

Lemma 3.2. *Let ϕ be a linear functional on polynomials in $z, \overline{z}$. Denote by $\mathcal{A}(\phi)$ the matrix with elements $\phi(z^j \overline{z}^k)$. Then the following are equivalent:*

1) *the matrix $\mathcal{A}(\phi)$ has finite rank not greater than* $\mathbf{r}$,

2) *for any collections of nonnegative integers $J = \{j_0, \dots, j_{\mathbf{r}}\}$ and $K = \{k_0, \dots, k_{\mathbf{r}}\}$*

$$\phi^{\otimes N}\left(\textstyle\prod_{i \in (0,\mathbf{r})} z_i^{j_i} \det \overline{z_i}^{k_l}\right) = 0, \quad (3.1)$$

where $N = \mathbf{r} + 1$.

Proof. Since passing to linear combinations of rows and columns does not increase the rank of the matrix, it follows that for any polynomials $f_j(z), g_k(z)$, with $j, k = 0, \dots, \mathbf{r}$, the determinant $\operatorname{Det}(\phi(f_j \overline{g}_k))$ vanishes.

The determinant is linear in each column and ϕ is a linear functional, so we can write

$$\phi\left(f_0(z) \times \begin{vmatrix} \overline{g_0(z)} & \mu(f_1\overline{g}_0) & \dots & \phi(f_{\mathrm{r}}\overline{g}_0) \\ \overline{g_1(z)} & \phi(f_1\overline{g}_1) & \dots & \phi(f_{\mathrm{r}}\overline{g}_1) \\ \vdots & \vdots & \ddots & \vdots \\ \overline{g_{\mathrm{r}}(z)} & \phi(f_1\overline{g}_{\mathrm{r}}) & \dots & \phi(f_{\mathrm{r}}\overline{g}_{\mathrm{r}}) \end{vmatrix} \right) = 0.$$

We introduce the variable z_0 in place of z above and use ϕ_0 for ϕ acting in the variable z_0. We repeat this process in each column (using the variable z_j in column j and the notation ϕ_j for ϕ acting in z_j) to obtain

$$\phi_0\left(\phi_1\left(\dots\phi_{\mathrm{r}}\left(\textstyle\prod_{k=0}^{\mathrm{r}} f_k(z_k)\det\left(\overline{g_j(z_k)}\right)\right)\dots\right)\right) = 0. \tag{3.2}$$

We now specialize to the case where each $f_i = z^{j_i}$, $g_i = z^{k_i}$ and arrive at (3.1), thus proving the implication $1 \Rightarrow 2$. The converse implication follows by going along the above reasoning in the opposite direction. □

Proof of Theorem 3.1. We identify $\mathbb{C}$ and $\mathbb{R}^2$ with co-ordinates $z = x + iy$. Consider the functional $\phi(f) = \phi_\mu(f) = \int f(z)d\mu(z)$. Write Z for the N-tuple $(z_0, z_1, \dots, z_{\mathrm{r}})$ and $V_J(Z)$ for the determinant $\det\left(z_i^{k_j}\right)$. By Lemma 3.2,

$$\phi^{\otimes N}\left(Z^K\overline{V_J(Z)}\right) = 0. \tag{3.3}$$

Taking finite sums of equations (3.3), we get for any polynomial $P(Z)$ in N variables:

$$\phi^{\otimes N}\left(P(Z)\overline{V_J(Z)}\right) = 0. \tag{3.4}$$

By taking linear combinations of antisymmetric polynomials $V_J(Z)$ one can obtain any antisymmetric polynomial $Q(Z)$ (cf. [14]) for details). Thus,

$$\phi^{\otimes N}\left(P(Z)\overline{Q(Z)}\right) = 0 \tag{3.5}$$

for any polynomial $P(Z)$ and any antisymmetric polynomial $Q(Z)$. In its turn, the polynomial $Q(Z)$ is divisible by the lowest degree antisymmetric polynomial, the Vandermonde polynomial $V(Z) = \prod_{0\leqslant j\leqslant k\leqslant \mathrm{r}}(z_j - z_k)$, $Q(Z) = Q_1(Z)V(Z)$ with a symmetric polynomial $Q_1(Z)$. We write (3.5) for Q of this form and P having the form $P(Z) = P_1(Z)V(Z)$. So we arrive at

$$\phi^{\otimes N}\left(P_1(Z)\overline{Q_1(Z)}|V(Z)|^2\right) = 0 \quad \text{for all symmetric } P_1 \text{ and } Q_1. \tag{3.6}$$

It is clear that finite sums of products of the form $P_1(Z)\overline{Q_1(Z)}$ (with P_1 and Q_1 symmetric) form an algebra A of functions on $\mathbb{C}$ which contains the constants and is closed under conjugation. It does not separate points because each element is constant on sets of points that are permutations of one another. Therefore, we define an equivalence relation $\sim$ on $\mathbb{C}^N : Z_1 \sim Z_2$ if and only if $Z_2 = \pi(Z_1)$ for some permutation π. Let $Z = (z_0, \ldots, z_{\mathbf{r}})$, and let $W = (w_0, \ldots, w_{\mathbf{r}})$. If $Z \not\sim W$ then the polynomials $p(t) = \prod(t - z_j)$ and $q(t) = \prod(t - w_j)$ have different zeros (or the same zeros with different orders). This implies that the coefficient of some power of t in $p(t)$ differs from the corresponding coefficient in $q(t)$. Thus, there is an elementary symmetric function that differs at Z and W. Consequently, A separates equivalence classes.

We give the quotient space $\mathbb{C}^N/\sim$ the standard quotient space topology. If K is any compact set in $\mathbb{C}^N$ that is invariant with respect to $\sim$, then $K/\sim$ is compact and Hausdorff. Also, any symmetric continuous function on $\mathbb{C}^N$ induces a continuous function on $\mathbb{C}^N/\sim$ (and conversely). Thus, we can apply the Stone–Weierstrass theorem (on $K/\sim$) to conclude that A is dense in the space of continuous symmetric functions, in the topology of uniform convergence on any compact set. Therefore, for any continuous symmetric function $f(Z)$

$$\int_{\mathbb{C}^N} f(Z)|V(Z)|^2 \, d\mu^{\otimes N}(Z) = 0. \tag{3.7}$$

If f is an arbitrary continuous function, the above integral will be the same as the corresponding integral with the symmetrization of f replacing f. This is because the function $|V(Z)|^2$ and the product measure $\mu^{\otimes N}$ are both invariant under permutations of the coordinates. We conclude that this integral vanishes for *any continuous* f and so the measure $|V(Z)|^2 \, d\mu^{\otimes N}(Z)$ must be zero. Thus, $\mu^{\otimes N}$ is supported on the set where V vanishes, i.e., on the set of Lebesgue measure zero. Since $\mu^{\otimes N}$ is absolutely continuous, it must be zero. □

The initial setting of Theorem 3.1 dealt with arbitrary measures, as it is explained in the Introduction. A more advanced result was obtained in [2], where Luecking's theorem was carried over to distributions.

Theorem 3.3. *Suppose that $F \in \mathcal{E}'(\Omega)$ is a distribution with compact support in $\Omega \subset \mathbb{C}$ and the Toeplitz operator $\mathcal{T}_F$ has finite rank* $\mathbf{r}$. *Then the distribution F is a finite combination of δ-distributions at some points in Ω and their derivatives,*

$$F = \sum_{j \leqslant \mathbf{r}} L_j \delta(z - z_j), \tag{3.8}$$

L_j being differential operators.

We start with some observations about distributions in $\mathcal{E}'(\mathbb{C})$. For such a distribution we denote by $\operatorname{psupp} F$ the complement of the unbounded component of the complement of $\operatorname{supp} F$.

Lemma 3.4. *Let $F \in \mathcal{E}'(\mathbb{C})$. Then the following two statements are equivalent:*

a) *there exists a distribution* $G \in \mathcal{E}'(\mathbb{C})$ *such that* $\frac{\partial G}{\partial \overline{z}} = F$; *moreover,* $\operatorname{supp} G \subset \operatorname{psupp} F$,

b) F *is orthogonal to all polynomials of* z *variable, i.e.,* $\langle F, z^k \rangle = 0$ *for all* $k \in \mathbb{Z}_+$.

Proof. The implication $a) \Longrightarrow b)$ follows from the relation

$$\langle F, z^k \rangle = \left\langle \frac{\partial G}{\partial \overline{z}}, z^k \right\rangle = \left\langle G, \frac{\partial z^k}{\partial \overline{z}} \right\rangle = 0. \tag{3.9}$$

We prove that $b) \Longrightarrow a)$. Put $G := F * \frac{1}{\pi z} \in \mathcal{S}'(\mathbb{C})$, the convolution being well-defined because F has compact support. Since $\frac{1}{\pi z}$ is the fundamental solution of the Cauchy–Riemann operator $\frac{\partial}{\partial \overline{z}}$, we have $\frac{\partial G}{\partial \overline{z}} = F$ (cf., for example, [10, Theorem 1.2.2]). By the ellipticity of the Cauchy–Riemann operator, $\operatorname{singsupp} G \subset \operatorname{singsupp} F \subset \operatorname{supp} F$, in particular, this means that G is a smooth function outside $\operatorname{psupp} F$; moreover, G is analytic outside $\operatorname{psupp} F$ (by $\operatorname{singsupp} F$ we denote the singular support of the distribution F (cf., for example, [10], the largest open set where the distribution coincides with a smooth function). Additionally,

$$G(z) = \langle F, \frac{1}{\pi(z-w)} \rangle = \pi^{-1} \sum_{k=0}^{\infty} z^{-k-1} \langle F, w^k \rangle = 0$$

if $|z| > R$ and R is sufficiently large. By analyticity this implies $G(z) = 0$ for all z outside $\operatorname{psupp} F$. □

Proof of Theorem 3.3. The distribution F, as any distribution with compact support, is of finite order, therefore it belongs to some Sobolev space, $F \in H^s$ for certain $s \in \mathbb{R}^1$. If $s \geqslant 0$, F is a function and must be zero by Luecking's theorem. So, suppose that $s < 0$.

Consider the first $\mathbf{r} + 1$ columns in the matrix $\mathcal{A}(F)$, i.e.,

$$a_{kl} = (T_F z^k, z^l) = \langle \sigma F, z^k \overline{z}^l \rangle, \quad l = 0, \dots \mathbf{r}; \quad k = 0, \dots . \tag{3.10}$$

Since the rank of the matrix $\mathcal{A}(F)$ is not greater than $\mathbf{r}$, the columns are linearly dependent, in other words, there exist coefficients $c_0, \dots, c_{\mathbf{r}}$ such that $\sum_{l=0}^{\mathbf{r}} a_{kl} c_l = 0$ for any $k \geqslant 0$. This relation can be written as

$$\langle F, z^k h_1(\overline{z}) \rangle = \langle h_1(\overline{z}) F, z^k \rangle = 0, \quad h_1(\overline{z}) = \sum_{k=0}^{\mathbf{r}} c_l \overline{z}^l. \tag{3.11}$$

Therefore, the distribution $h_1(\overline{z}) F \in H^s$ satisfies the conditions of Lemma 3.4 and hence there exists a compactly supported distribution $F^{(1)}$ such that $\frac{\partial F^{(1)}}{\partial \overline{z}} = h_1 F$. By the ellipticity of the Cauchy–Riemann operator, the distribution $F^{(1)}$ is less singular than F, $F^{(1)} \in H^{s+1}$. At the same time,

$$\langle F^{(1)}, z^k \bar{z}^l \rangle = (l+1)^{-1} \Big\langle F^{(1)}, \frac{\partial z^k \bar{z}^{l+1}}{\partial \bar{z}} \Big\rangle$$
$$= (l+1)^{-1} \langle h_1(\bar{z}) F, z^k \bar{z}^l \rangle = (l+1)^{-1} \langle F, z^k \bar{z}^l h_1(\bar{z}) \rangle, \quad (3.12)$$

and therefore the rank of the matrix $\mathcal{A}(F^{(1)})$ does not exceed the rank of the matrix $\mathcal{A}(F)$.

We repeat this procedure sufficiently many (say, $n = [-s] + 1$) times and arrive at the distribution $F^{(n)}$ which is, in fact, a function in L_2, for which the corresponding matrix $\mathcal{A}(F^{(n)})$ has finite rank. By Luecking's theorem, this may happen only if $F^{(n)} = 0$.

Now, we go back to the initial distribution F. Since, by our construction,

$$\frac{\partial F^{(n)}}{\partial \bar{z}} = h_n(\bar{z}) F^{(n-1)},$$

we have $h_n(\bar{z}) F^{(n-1)} = 0$, and therefore supp $F^{(n-1)}$ is a subset of the set of zeroes of the polynomial $h_n(\bar{z})$. On the next step, since

$$\frac{\partial F^{(n-1)}}{\partial \bar{z}} = h_{n-1}(\bar{z}) F^{(n-2)},$$

we find that supp $F^{(n-2)}$ lies in the union of sets of zeroes of polynomials $h_{n-1}(\bar{z})$ and $h_n(\bar{z})$. After having gone all the way back to F, we find that its support is a finite set of points lying in the union of zero sets of polynomials h_j. A distribution with such support must be a linear combination of δ - distributions in these points and their derivatives, $F = \sum L_q \delta(z - z_q)$, where $L_q = L_q(D)$ are some differential operators. Finally, to show that the number of points z_q does not exceed $\mathbf{r}$, we construct for each of them the interpolating polynomial $f_q(z)$ such that $L_q(-D)|f_q|^2 \neq 0$ at the point z_q while at the points $z_{q'}$, $q' \neq q$, the polynomial f_q has zero of sufficiently high order, higher than the order of $L_{q'}$, so that $L_{q'}(f_q g)(z_{q'}) = 0$ for any smooth function g. With such a choice of polynomials, the matrix with entries $\langle F, f_q \overline{f_{q'}} \rangle$ is the diagonal matrix with nonzero entries on the diagonal, and therefore its size (that equals the number of the points z_q) cannot be greater than the rank of the whole matrix $\mathcal{A}(F)$, i.e., cannot be greater than $\mathbf{r}$. □

Remark 3.5. The attempt to extend directly the original proof of Theorem 3.1 to the distributional case would probably meet certain complications. The following property is crucial in this proof: the algebra generated by polynomials of the form $P_1(Z)\overline{Q_1(Z)}$ with symmetric P_1, Q_1 is dense (in the sense of the uniform convergence on compacts) in the space of symmetric continuous functions. This latter property is proved above by a reduction to the Stone–Weierstrass theorem.

Now, if F is a distribution that is not a measure, the analogy of reasoning in the proof would require a similar density property, however not in the sense of the uniform convergence on compacts, but in a stronger sense, the uniform convergence together with derivatives up to some fixed order (depending on the order of the distribution F.) The Stone–Weierstrass theorem seems not to help here since it deals with the uniform convergence only. Moreover, the required more general density statement itself is *wrong*, which follows from the construction below (cf. [2]).

Example 3.6. The algebra generated by the functions having the form $P_1(Z)\overline{Q_1(Z)}$, where P_1, Q_1 are symmetric polynomials of the variables $Z = (z_0, \dots, z_N)$ is not dense in the sense of the uniform C^l-convergence on compact sets in the space of C^l-differentiable symmetric functions, as long as $l \geqslant N(N-1)$. To show this, consider the differential operator $V(D) = \prod_{j<k}(D_j - D_k)$, $D_j = \frac{\partial}{\partial z}$. It is easy to check that $V(D)H$ is symmetric for any antisymmetric function $H(Z)$ and $V(D)H$ is antisymmetric for any symmetric function $H(Z)$. Further on, consider any function $H(Z)$ of the form $H(Z) = P_1(Z)\overline{Q_1(Z)}$ where $P_1(Z), Q_1(Z)$ are analytic polynomials. If at least one of them is symmetric, we have

$$(V(D)V(\overline{D}))H(0) = 0. \tag{3.13}$$

In fact,

$$V(D)V(\overline{D})P_1(Z)\overline{Q_1(Z)} = [(V(D)P_1(Z)][\overline{V(D)Q_1(Z)}].$$

In the last expression, for the symmetric polynomial P_1, the corresponding polynomial $V(D)P_1(Z)$ is antisymmetric, and therefore equals zero for $Z = 0$. Now, consider the symmetric function $|V(Z)|^2 = V(Z)\overline{V(Z)}$. We have

$$V(D)V(\overline{D})V(Z)\overline{V(Z)} = [V(D)V(Z)][V(\overline{D})\overline{V(Z)}].$$

Note that

$$V(Z) = \sum_{\varkappa} C_{\varkappa} \prod z_j^{\varkappa_j},$$

where the summing goes over multiindices $\varkappa = (\varkappa_1, \dots, \varkappa_N)$, $|\varkappa| = N$ and not all of real coefficients $C_{\varkappa}$ are zeros. Simultaneously,

$$V(D) = \sum_{\varkappa} C_{\varkappa} \prod D_j^{\varkappa_j}$$

with the same coefficients. We recall now that $\prod D_j^{\varkappa_j} \prod z_j^{\varkappa'_j} = 0$ if $|\varkappa| = |\varkappa'|$, $\varkappa \neq \varkappa'$ and it equals $\varkappa!$ if $\varkappa = \varkappa'$. Therefore,

$$V(D)V(Z) = \sum_{\varkappa} C_{\varkappa}^2 \varkappa!$$

is a positive constant. In this way, we have constructed the differential operator $V(D)V(\overline{D})$ of order $N(N-1)$ satisfying (3.13) for any function of the form $H(Z) = P_1(Z)\overline{Q_1(Z)}$ with symmetric P_1, Q_1, and not vanishing on some symmetric differentiable function $|V(Z)|^2$. Therefore, the function $|V(Z)|^2$ cannot be approximated by linear combinations of the functions $H(Z) = P_1(Z)\overline{Q_1(Z)}$ in the sense of the uniform $C^{N(N-1)}$ convergence on compacts.

Luecking's theorem was extended in a different direction by Le [11]. For the particular system of functions $f_k = z^k$ used above for the construction of the matrix $\mathcal{A}$, it turns out that its rank may even be infinite, but the assertion of the theorem still holds, as long as the range of the operator avoids sufficiently many analytical functions.

We say that the set of indices $\mathcal{J} = \{n_j\} \subset \mathbb{Z}_+ = \{n \in \mathbb{Z}, n \geqslant 0\}$ is *sparse* if the series $\sum_{n\in\mathcal{J}}(n+1)^{-1}$ converges.

Theorem 3.7 ([11]). *Suppose that μ is a regular complex Borel measure and $\mathcal{J} = \{n_j\} \subset \mathbb{Z}_+$ is sparse, $\mathcal{J}' = \mathbb{Z}_+ \setminus \mathcal{J}$. Consider the reduced matrix $\mathcal{A}^{\mathcal{J}}$ consisting of a_{jk}: $j, k \in \mathcal{J}'$. Suppose that the rank $\mathbf{r}$ of $\mathcal{A}^{\mathcal{J}}$ is finite. Then the support of μ consists of no more than $\mathbf{r}+1$ points.*

The original formulation of this theorem in [11] is given in the terms of the Toeplitz operator itself. Denote by M, N the space of polynomials spanned by monomials z^j with j, respectively, in $\mathcal{J}, \mathcal{J}'$. In Theorem 3.7, it is supposed that the operator $\mathcal{T}_\mu$, being restricted to $\overline{\mathrm{N}}$, has range in the linear span of $\overline{\mathrm{M}}$ and some finite-dimensional subspace in N.

In the next section, we establish a result that generalizes Theorem 3.7 in three directions: the multidimensional case will be considered, any distribution with compact support will replace the measure μ, and the condition of sparseness will be considerably relaxed.

4 The Multidimensional Case

In this section, we extend our main Theorem 3.3 to the case of Toeplitz operators in Bergman spaces of analytical functions of several variables. For the case of a measure acting as weight, there exist two ways of proving this result, in [5] and [21, 2]. The first approach generalizes the one used in [14] in proving Theorem 3.1, the other one uses the induction on dimension. As it follows from Remark 3.5, for the case of distribution the approach of [5] is likely to meet some complications. We present here the proof given in [2], with some modifications.

Theorem 4.1. *Let F be a distribution in $\mathcal{E}'(\mathbb{C}^d)$. Consider the matrix*

$$\mathcal{A}(F) = (a_{\alpha\beta})_{\alpha,\beta\in\mathbb{Z}_+^d};\ a_{\alpha\beta} = \langle F, \mathbf{z}^\alpha\overline{\mathbf{z}}^\beta\rangle,\ \mathbf{z} = (z_1, \ldots, z_d) \in \mathbb{C}^d. \qquad (4.1)$$

Suppose that the matrix $\mathcal{A}(F)$ *has finite rank* $\mathbf{r}$. *Then* $\operatorname{card}\operatorname{supp} F \leqslant \mathbf{r}$ *and* $F = \sum L_q \delta(\mathbf{z} - \mathbf{z}_q)$, *where* L_q *are differential operators and* $\mathbf{z}_q$, $1 \leqslant q \leqslant \mathbf{r}$, *are some points in* $\mathbb{C}^d$.

We notice first, following [21], that if the function g is analytical and bounded in some polydisk neighborhood of $\operatorname{supp} F$ and F_g is the distribution $|g|^2 F$, then $\operatorname{rank}\mathcal{A}(F_g) \leqslant \operatorname{rank}\mathcal{A}(F)$. To show this, we denote by M_g the bounded operator acting in $\mathcal{B}^2$ by multiplication by g. The adjoint operator M_g^* is, of course, bounded as well. Consider the quadratic form of the operator $\mathcal{T}_{F_g}$: for $u \in \mathcal{B}^2$,

$$(\mathcal{T}_{F_g}u, u) = \langle F, gu\overline{gu}\rangle = (\mathcal{T}_F(gu), gu) = (\mathcal{T}_F M_g u, M_g u) = (M_g^* \mathcal{T}_F M_g u, u).$$

So we see that the operator $u \mapsto \mathcal{T}_{F_g}u$ coincides with $M_g^*\mathcal{T}_F M_g$. The multiplication by bounded operators does not increase the rank of an operator, and the property follows.

Thus, it suffices to prove the statement that is, actually, only formally weaker than Theorem 4.1.

Theorem 4.2. *Suppose that for any function* $g(\mathbf{z})$, *analytic and bounded in a polydisk neighborhood of the support of the distribution* F, *the conditions of Theorem 4.1 are fulfilled with the distribution* F *replaced by* $|g(\mathbf{z})|^2 F \equiv F_g$. *Then* $\operatorname{card}\operatorname{supp} F \leqslant \mathbf{r}$ *and* $F = \sum_{1 \leqslant q \leqslant \mathbf{r}} L_q \delta(\mathbf{z} - \mathbf{z}_q)$, *where* L_q *are differential operators.*

Proof. We use the induction on d. For $d = 1$ the statement of Theorem 4.2 coincides with the one of Theorem 3.3 that was proved in Section 3. We suppose that we have established our statement in the complex dimension $d - 1$ and consider the d-dimensional case. We denote the variables as $\mathbf{z} = (z_1, \mathbf{z}')$, $\mathbf{z}' \in \mathbb{C}^{d-1}$.

For a fixed function $g(\mathbf{z})$ we denote by $G(g) = \pi_* F_g$ the distribution in $\mathcal{E}'(\mathbb{C}^{d-1})$ induced from F_g by the projection $\pi : \mathbf{z} \mapsto \mathbf{z}'$: for $u \in C^\infty(\mathbb{C}^{d-1})$,

$$\langle G(g), u\rangle = \langle F_g, 1_{\mathbb{C}^1} \otimes u\rangle. \tag{4.2}$$

Although the function g is defined only in a polydisk, the distribution (4.2) is well defined since this polydisk contains $\operatorname{supp} F$.

Consider the submatrix $\mathcal{A}'(F_g)$ in the matrix $\mathcal{A}(F_g)$ consisting only of those $a_{\alpha\beta} = \langle |g|^2 F, \mathbf{z}^\alpha \overline{\mathbf{z}}^\beta\rangle$ for which $\alpha_1 = \beta_1 = 0$. It follows from (4.2), that the matrix $\mathcal{A}'(F_g)$ coincides with the matrix $\mathcal{A}(G(g))$ constructed for the distribution $G(g)$ in dimension $d - 1$. Thus, the matrix $\mathcal{A}(G(g))$, being a submatrix of a finite rank matrix, has a finite rank itself; moreover, $\operatorname{rank}\mathcal{A}(G(g)) \leqslant \mathbf{r}$. By the inductive assumption, this implies that the distribution $G(g)$ has finite support consisting of $\mathbf{r}(g) \leqslant \mathbf{r}$ points $\zeta_1(g), \dots, \zeta_{\mathbf{r}(g)}$; $\zeta_q(g) \in \mathbb{C}^{d-1}$ (the notation reflects the fact that both the points and their quantity may depend on the function g). Among all functions g, we can find the one, $g = g_0$, for which $\mathbf{r}(g)$ attains its maximal value $\mathbf{r}_0 \leqslant \mathbf{r}$. Without losing in generality, we can assume that $g_0 = 1$.

Fix an $\epsilon > 0$, sufficiently small, so that 2ϵ-neighborhoods of $\zeta_q(1)$ are disjoint, and consider the functions $\varphi_q(\mathbf{z}') \in C^\infty(\mathbb{C}^{d-1})$, $q = 1, \dots, \mathbf{r}_0$ such that supp φ_q lies in the ϵ-neighborhood of the point $\zeta_q(1)$ and $\varphi_q(\mathbf{z}') = 1$ in the $\frac{\epsilon}{2}$-neighborhood of $\zeta_q(1)$. We fix an analytic function $g(\mathbf{z})$ and consider for any q the distribution $\Phi_q(t,g) \in \mathcal{E}'(\mathbb{C}^d)$, $\Phi_q(t,g) = |1 + tg(\mathbf{z})|^2\varphi_q(\mathbf{z}')F = \varphi_q(\mathbf{z}')F_{1+tg}$. For $t = 0$, $\Phi_q(t,g) = \varphi_q(\mathbf{z}')F$, the point $\zeta_q(1)$ belongs to the support of $\pi_*\Phi_q(0,g)$, and therefore for some function $u \in C^\infty(\mathbb{C}^{d-1})$, $\langle \pi_*\Phi_q(0,g), u\rangle \neq 0$. By continuity, for $|t|$ small enough, we still have $\langle \pi_*\Phi_q(t,g), u\rangle \neq 0$, which means that the ϵ-neighborhood of the point $\zeta_q(1)$ contains at least one point in the support of the distribution $G(1+tg)$. Altogether, we have not less than $\mathbf{r}_0$ points of the support of $G(1+tg)$ in the union of ϵ-neighborhoods of the points $\zeta_j(1)$. However, recall, the support of $G(1+tg)$ can never contain more than $\mathbf{r}_0$ points, so we deduce that for t small enough, there are no points of the support of $G(1+tg)$ outside the ϵ-neighborhoods of the points $\zeta_q(1)$, for $|t|$ small enough (depending on g.) Thus, the support of the distribution $G(1+tg)$ is contained in the ϵ-neighborhood of the set of points $\zeta_j(1)$ for any g.

We introduce a function $\psi \in C^\infty(\mathbb{C}^{d-1})$ that equals 1 outside 2ϵ-neighborhoods of the points $\zeta_q(1)$ and vanishes in ϵ-neighborhoods of these points. By the above reasoning, the distribution $\psi G(1+tg)$ equals zero for any g, for t small enough. In particular, applying this distribution to the function $u = 1$, we obtain

$$\langle \psi G(1+tg), 1\rangle = \langle \psi F, |1+tg|^2\rangle = \langle \psi F, 1 + 2t\operatorname{Re} g + t^2|g|^2\rangle = 0. \quad (4.3)$$

By the arbitrariness of t in a small interval, (4.3) implies that $\langle \psi F, |g|^2\rangle = 0$ for any g. By standard polarization, this implies that for any functions g_1, g_2 analytical in a polydisk neighborhood of supp F.

$$\langle \psi F, g_1\overline{g_2}\rangle = 0. \quad (4.4)$$

Any polynomial $p(\mathbf{z}, \overline{\mathbf{z}})$ can be represented as a linear combination of functions of the form $g_1\overline{g_2}$, so, (4.4) gives

$$\langle \psi F, p(\mathbf{z}, \overline{\mathbf{z}})\rangle = 0. \quad (4.5)$$

We take any function $f \in C^\infty(\mathbb{C}^d)$ supported in the neighborhood U of supp F such that $f = 0$ on the support of ψ. We can approximate f by polynomials of the form $p(\mathbf{z}, \overline{\mathbf{z}})$ uniformly on $\overline{U}$ in the sense of C^l, where l is the order of the distribution F. Passing to the limit in (4.5), we obtain $\langle \psi F, f\rangle = \langle F, f\rangle = 0$.

The latter relation shows that supp $F \subset \bigcup_q\{\mathbf{z} : |\mathbf{z}' - \zeta_q(1)| < 2\epsilon\}$. Since $\epsilon > 0$ is arbitrary, this implies that supp F lies in the union of affine subspaces $\mathbf{z}' = \zeta_j$, $j = 1, \dots, \mathbf{r}_0$ of complex dimension 1.

We repeat the same reasoning having chosen instead of $\mathbf{z} = (z_1, \mathbf{z}')$ another decomposition of the complex variable $\mathbf{z}$: $\mathbf{z} = (\mathbf{z}'', z_d)$. We find that for some points $\xi_k \in \mathbb{C}^{d-1}$, no more than $\mathbf{r}$ of them, the support of F lies in the union of subspaces $\mathbf{z}'' = \xi_k$. Taken together, this means that, actually, supp F lies in the intersection of these two systems of subspaces, which consists of no more than $\mathbf{r}^2$ points $\mathbf{z}_s$. The number of points is finally reduced to $\mathbf{r}_0 \leqslant \mathbf{r}$ in the same way as in Theorem 3.3, by choosing a special system of interpolation functions. □

The theorem just proved can be extended to the case of a sparse range, following the pattern of Theorem 3.7.

Definition 4.3. Let $\mathcal{J} \subset \mathbb{Z}_+^d$ be a set of multiindices, and let $\gamma \in \mathbb{Z}_+^d$ be a fixed multiindex. We say that the set $\mathcal{J}$ is *N-sparse* in the direction γ if for any $\alpha \in \mathbb{Z}_+^d$,

$$\limsup_{n\to\infty} n^{-1}\#\{(\alpha + ([0,n]\gamma)) \cap \mathcal{J}\} < N^{-1}. \tag{4.6}$$

In other words, the fact that the set $\mathcal{J}$ is N-sparse in direction γ means that along any half-line starting at some point of $\mathbb{Z}_+^d$ and going in the direction γ, the density of the points of $\mathcal{J}$ on this line is less than N^{-1}.

For a multiindex $\gamma = (k_1, \dots, k_d)$ we denote by $\mathbf{n}(\gamma)$ the set of indices j such that $k_j = 0$. Introduce the set $\mathcal{J}' = \mathbb{Z}_+^d \setminus \mathcal{J}$ and the reduced matrix $\mathcal{A}^{\mathcal{J}}(F)$ consisting of $a_{\alpha,\beta} : \alpha, \beta \in \mathcal{J}'$.

Theorem 4.4. *Suppose that for a distribution $F \in \mathcal{E}'(\mathbb{C}^d)$ the reduced matrix $\mathcal{A}^{\mathcal{J}}(F)$ has finite rank $\mathbf{r}$, and for some $\epsilon > 0$, the set $\mathcal{J}$ is $2\mathbf{r} + 2 + \epsilon$-sparse in some directions γ_l, $l = 1, \dots, \mathbf{l}$ such that*

$$\cup \mathbf{n}(\gamma_l) = \{1, \dots, d\}. \tag{4.7}$$

Then the distribution F has finite support consisting of no more than $\mathbf{r} + 1$ points. In the particular case where F is a function, the condition (4.7) can be dropped and if $\mathcal{J}$ is $2\mathbf{r} + 2 + \epsilon$-sparse just in one, arbitrary, direction, then $F = 0$.

Remark 4.5. For the case of measures, a version of Theorem 4.4 with a more restrictive notion of sparseness, was proved in [12].

Remark 4.6. The condition (4.7) is sharp in the sense that if it is violated, the support of F can be infinite. Consider the set $\mathcal{J}$ consisting of multiindices having zero in the first position. This set $\mathcal{J}$ is N-sparse for any N in any direction γ_l, with a non-zero in the first position. For such directions γ_l, we have $1 \notin \cup \mathbf{n}(\gamma_l)$. Let the distribution $F \in \mathcal{E}'(\mathbb{C}^d)$ be a measure supported in the subspace $\{z_1 = 0\}$. Then for any $\alpha, \beta \notin \mathcal{J}$, the functions $\mathbf{z}^\alpha, \overline{\mathbf{z}^\beta}$ vanish on $\{z_1 = 0\}$ and thus $\langle F, \mathbf{z}^\alpha \overline{\mathbf{z}^\beta}\rangle = 0$, so the statement on the finiteness of support becomes wrong. Of course, due to the second part of Theorem 4.4, such examples are impossible in the case where the distribution is, in fact, a function.

Proof. The proof follows the structure of the proof of Theorem 3.7 in [11], with two main ingredients replaced by their multidimensional analogies and with the extension to distributions and a more general notion of sparse sets.

First, similar to Lemma 3.2, the following two properties are equivalent:

1. the matrix $\mathcal{A}(F)$ has finite rank not greater than $\mathbf{r}$;
2. for any collections of $2N = 2\mathbf{r}+2$ multiindices $\alpha_0, \dots, \alpha_{\mathbf{r}}$, $\beta_0, \dots, \beta_{\mathbf{r}}$

$$\phi^{\otimes N}\left(\prod\nolimits_{i\in(0,\mathbf{r})} \mathbf{z}_i^{\alpha_i} \det \overline{\mathbf{z}_i}^{\beta_l}\right) = 0. \tag{4.8}$$

The proof of this fact is quite similar to the proof of Lemma 3.2 (one can also find details in [5, p.215]).

Now, we fix the multiindices $\alpha_0, \dots, \alpha_{\mathbf{r}}$, $\beta_0, \dots, \beta_{\mathbf{r}}$ and, supposing that the set $\mathcal{J}$ is $2\mathbf{r}+2+\epsilon$ - sparse in the direction γ, consider the set of multiindices

$$Y = \mathbb{Z}_+^d \setminus \left((\cup_{j=0}^{\mathbf{r}}(\mathcal{J}-\alpha_j))\bigcup(\cup_{j=0}^{\mathbf{r}}(\mathcal{J}-\beta_j))\right). \tag{4.9}$$

The set $\mathbb{Z}_+^d \setminus Y$ thus consists of $2N$ shifts of the set $\mathcal{J}$, therefore

$$\liminf_{n\to\infty} \{n^{-1}\#\{Y \cap \gamma[0,n]\}\} > \epsilon > 0.$$

In other words, the set of integers n such that $\alpha_j + n\gamma, \beta_j + n\gamma \notin \mathcal{J}$ for *all* j has positive density in $\mathbb{Z}_+$. In particular, this means that

$$\sum_{n:\gamma n\in Y} (n+1)^{-1} = \infty. \tag{4.10}$$

Consider the function of the complex variable w:

$$\begin{aligned}\Phi(w) &= \left\langle F^{\otimes N}, \prod \mathbf{z}_j^{\alpha_j+\gamma w}\overline{\det(\mathbf{z}_j^{\beta_k+\gamma w})}\right\rangle \\ &= \left\langle F^{\otimes N}, \prod \mathbf{z}_j^{\alpha_j}\overline{\det(\mathbf{z}_j^{\beta_k})}|\prod \mathbf{z}_j^{\gamma}|^{2Nw}\right\rangle.\end{aligned}$$

The function onto which the distribution $F^{\otimes N}$ acts, is not smooth for non-integer w, but we take care of this in the following way. The distribution F, having compact support, must have finite order, $\varkappa \geqslant 0$. Thus, F can be extended as a functional on $\varkappa$ times differentiable functions. The function $|\prod \mathbf{z}_j^{\gamma}|^{2Nw}$ belongs to $C^{\varkappa}$ for $2N \operatorname{Re} w \geqslant \varkappa$, so, in the half-plane $\mathbb{K} = \{\operatorname{Re} w > \frac{\varkappa}{2N}\}$ the function $\Phi(w)$ is well defined, analytical, and it is continuous in $\overline{\mathbb{K}}$. Therefore, if the support of F lies in the ball $|\mathbf{z}| < R$, the function $\Psi(w) = R^{-Nw}\Phi(w)$ is a bounded analytical function in $\mathbb{K}$. Since, by our construction, all $\alpha_j + \gamma n$, $\beta_n + \gamma n$ belong to $\mathbb{Z}_+ \setminus \mathcal{J}$ for all $n \in Y$, we have

$$\Phi(n) = \Psi(n) = 0, \quad n \in Y. \tag{4.11}$$

Let

$$H(\zeta) = \Psi\Big(\frac{1+(\zeta+\varkappa)}{1-(\zeta+\varkappa)}\Big).$$

Then H is a bounded analytical function on the unit disk. For any $n \in Y$, equation (4.11) implies that

$$H\Big(\frac{n-1-\varkappa}{n+1+\varkappa}\Big) = 0.$$

Now

$$\sum_{n\in Y}\Big(1 - \frac{n-1-\varkappa}{n+1+\varkappa}\Big) = \sum_{n\in Y}\frac{2\varkappa+2}{n+1+\varkappa} = \infty$$

by (4.10). The corollary to Theorem 15.23 in [22] shows that in this case H should be indentically zero on the unit disk, and therefore $\Phi(w) = 0, \operatorname{Re} w > k$. By continuity, $\Phi(\varkappa) = 0$, and this means, in particular that

$$(|\mathbf{z}^{\gamma}|^{2\varkappa}F)^{\otimes N}\Big(\prod_{j=0}^{\mathbf{r}} \mathbf{z}_j^{\alpha_j}\overline{\det(\mathbf{z}_j)^{\beta_k}}\Big) = 0.$$

In the reasoning above, the multiindices α_j, β_j are arbitrary, this means that (4.8) holds for the distribution $|\mathbf{z}^{\gamma}|^{2\varkappa}F$, and therefore, by Theorem 4.1, this distribution must have a support consisting of a finite number of points. Therefore, the support of F itself is contained in the union of the above points and the subset where $\mathbf{z}^{\gamma} = 0$. The latter subset is the union of the subspaces $z_m = 0$ for those m that do not belong to $\mathbf{n}(\gamma)$. In particular, if F is a function, this implies that $F = 0$. In the general case, we repeat the reasoning in the proof for any direction γ_l. Since, by the conditions of the theorem, $\cup\mathbf{n}(\gamma_l) = \{1, \dots, d\}$, the intersection of zero sets of $\mathbf{z}^{\gamma_l}$ consists only of the point 0, and this proves our statement. □

5 Applications

In this section, we give some applications of the results on the finite rank Toeplitz operators in analytical Bergman spaces.

5.1 Approximation

For a subset $Q \subset C(\Omega)$, we denote by $\mathcal{Z}(Q)$ the set of common zeros of functions in Q. Conversely, for a subset E of Ω, we denote by $J(E)$ the ideal in $C(\Omega)$ consisting of all functions vanishing on E. Given a subspace W in the Bergman space $\mathcal{B}(\Omega)$ of analytical functions, we denote by $\widehat{W}$ the closure in $C(\Omega)$ of the span of functions of the form $h(z) = f\overline{g}, f \in \mathcal{B}(\Omega), g \in W$ in the topology of uniform convergence on compacts in Ω.

Theorem 5.1 ([5]). *Let W be a subspace in $\mathcal{B}(\Omega)$ with finite codimension. Then $\mathcal{Z}(W)$ is a finite set and $\widehat{W} = J(\mathcal{Z}(W))$. In particular, if $\mathcal{Z}(W) = \varnothing$, then $\widehat{W} = C(\Omega)$.*

Proof. Endowed with the topology of uniform convergence on compact sets, the space $C(\Omega)$ is locally convex and its continuous linear functionals are identified with complex Borel measures supported on compact sets in Ω. Let Y be the space of measures orthogonal to $\widehat{W}$. If $Y \neq \varnothing$ there should exist a complex Borel measure $\mu \neq 0$ supported on a compact set in Ω such that

$$0 = \int_\Omega f\overline{g}d\mu = \int_\Omega (T_\mu f)\overline{g}d\lambda \tag{5.1}$$

for all $f \in \mathcal{B}, g \in W$. This shows that $T_\mu \mathcal{B}$ is contained in $W^\perp$, which is finite dimensional. By Theorem 3.3, μ must be supported on some finite set in Ω, say E_μ. It follows that $T_\mu \mathcal{B}$ is spanned by finitely many kernel functions $P(\cdot, w)$, $w \in E_\mu$. By (5.1), we have that these functions $P(\cdot, w)$ lie in $W^\perp$. Since these functions are also linearly independent, the union E of the sets E_μ, $\mu \in Y$, must be finite. By the reproducing property, $E \subset \mathcal{Z}(W)$. Moreover, we have $E = \mathcal{Z}(W)$, because point masses at the points of $\mathcal{Z}(W)$ belong to Y. Now, since each $\mu \in Y$ is supported in E, the ideal $J(E) = J(\mathcal{Z}(W))$ is annihilated by all $\mu \in Y$. Thus, $J(\mathcal{Z}(W)) \subset \widehat{W}$, and the converse inclusion is obvious. The case $Y = \varnothing$ is easily treated by the fact that $Y = \varnothing$ if and only if $\mathcal{Z}(W) = \varnothing$, which one may see from the proof above. □

Theorem 5.1 can be understood as saying that the linear combinations of the functions of the form $f\overline{g}$, $f \in W$, $g \in \mathcal{B}$, can approximate any continuous function uniformly on any compact not containing points from some finite set. By means of the more general Theorem 4.4, we can extend this approximation result in several directions.

Theorem 5.2. *Let $\Omega \subset \mathbb{C}^d$, and let $\mathcal{J} \subset \mathbb{Z}_+^d$ be some set of multiindices, $\mathcal{J}' = \mathbb{Z}_+^d \setminus \mathcal{J}$, satisfying the conditions of Theorem 4.4 with $\mathbf{r} < 2N+1$ for some N. Denote by $\mathcal{P} = \mathcal{P}(\mathcal{J})$ the space of polynomials of the form $\mathbf{p}(z) = \sum c_\alpha \mathbf{z}^\alpha, \alpha \in \mathcal{J}'$, Let U, V be linear subspaces in $\mathcal{P}(\mathcal{J})$ with codimension not greater than N. Then there are no more than $2N+1$ points $\mathbf{w}_\varkappa \in \Omega$ such that for any n, the space $\mathcal{R}$ of linear combinations of functions of the form $\mathbf{p}(\mathbf{z})\overline{\mathbf{q}(\mathbf{z})}$, $\mathbf{p}(\mathbf{z}) \in U, \mathbf{q}(\mathbf{z}) \in V$, is dense in $C^n(\Omega)$ in the sense of uniform convergence of all derivatives of order not higher than n, on any compact $K \subset \Omega$ not containing the points $\mathbf{w}_\varkappa$.*

Compared with Theorem 5.1, this theorem takes care of a more strong type of convergence, while the approximating set is considerably smaller.

Proof. Suppose that on some compact K the functions $\mathbf{p}(\mathbf{z})\overline{\mathbf{q}(\mathbf{z})}$ are not dense in the sense of uniform convergence on K with derivatives of order up to n. This means that there exists a distribution F with support in K such that $\langle F, \mathbf{p}(\mathbf{z})\overline{\mathbf{q}(\mathbf{z})}\rangle = 0$ for all $\mathbf{p}(\mathbf{z}) \in U, \mathbf{q}(\mathbf{z}) \in V$.

Since V has finite codimension in $\mathcal{P}(\mathcal{J})$, there exist no more than N polynomials $\phi_0, \dots, \phi_{N_0}$, $N_0 < N$, so that $\mathcal{P}(\mathcal{J}) = V + \operatorname{Span}(\phi_0, \dots, \phi_{N_0})$. Similarly, there exist no more than N polynomials $\psi_0, \dots, \psi_{N_1}$, $N_1 < N$, so that $\mathcal{P}(\mathcal{J}) = U + \operatorname{Span}(\psi_0, \dots, \psi_{N_1})$.

We choose some basis $\mathbf{p}_i(z)$ in U and some basis $\mathbf{q}_j(z)$ in V. Consider the infinite matrix $\mathcal{C}_0$ consisting of elements $b_{ij} = \langle F, \mathbf{p}_i(z)\overline{\mathbf{q}_j(z)}\rangle$, which are, of course, all zeros. Now, we append the matrix $\mathcal{C}_0$ by N_0 columns $b_{i,-s} = \langle F, \mathbf{p}_i(z)\overline{\phi_s(z)}\rangle$, $s = 0, \dots, N_0$ and then by N_1 horizontal rows $b_{-t,j} = \langle F, \psi_t(z)\overline{\mathbf{q}_j(z)}\rangle$, $b_{-t,-s} = \langle F, \psi_t(z)\overline{\phi_s(z)}\rangle$, $t = 0, \dots, N_1$, thus obtaining the matrix $\mathcal{C}$. Each of these two operations increases the rank of the matrix no more than by N, so $\operatorname{rank}(\mathcal{C}) \leqslant 2N$. Now, since any monomial $\mathbf{z}^k$, $k \in \mathcal{J}'$ is a finite linear combination of polynomials $\mathbf{p}_i, \psi_t$ and any monomial $\mathbf{z}^l$, $l \in \mathcal{J}'$ is a finite linear combination of polynomials $\mathbf{q}, \phi_s$, the matrix $\mathcal{A}^{\mathcal{J}}(F)$ consisting of $\mathbf{a}_{k,l} = \langle F, \mathbf{z}^k\overline{\mathbf{z}^l}\rangle$, $k, l \in \mathcal{J}'$, has rank not greater than $\operatorname{rank}(\mathcal{C})$, i.e., $\operatorname{rank}(\mathcal{A}^{\mathcal{J}}(F)) \leqslant 2N$.

Now, Theorem 4.2 implies that the distribution F has support consisting of no more than $2N + 1$ points; we denote this set $\mathcal{Z}(F)$. For some other distribution G, also vanishing on all functions of the form $\mathbf{p}(\mathbf{z})\overline{\mathbf{q}(\mathbf{z})}$, $\mathbf{p}(\mathbf{z}) \in U, \mathbf{q}(\mathbf{z}) \in V$, the support $\mathcal{Z}(G)$, by the same reasoning also consists of no more than $2N + 1$ points. By considering a linear combination of F and G, we see that still $\#\{\mathcal{Z}(F) \cup \mathcal{Z}(G)\} \leqslant 2N + 1$. So, the support of any distributions vanishing on $\mathbf{p}(\mathbf{z})\overline{\mathbf{q}(\mathbf{z})}$, $\mathbf{p}(\mathbf{z}) \in U, \mathbf{q}(\mathbf{z}) \in V$, is a part of some set $E \subset K$, that has no more than $2N + 1$ points. Therefore, any function in C^n can be approximated by the functions of the form $\mathbf{p}(\mathbf{z})\overline{\mathbf{q}(\mathbf{z})}$, $\mathbf{p}(\mathbf{z}) \in U, \mathbf{q}(\mathbf{z}) \in V$, in C^n uniformly on any compact $K' \subset K$, not containing these points. Finally, the compact K in our construction can be chosen arbitrarily large, while the set E can never have more than $2N + 1$ points and thus can be taken independently of K. □

We give an example of the application of Theorem 5.2. For the sake of simplicity, we take $\mathcal{J} = \varnothing$. For multiindices $\alpha, k \in \mathbb{Z}_+^d$ we write $\alpha \prec k$ if each component of α is not greater than the corresponding component of k, while at least one component is strictly less.

Example 5.3. Suppose that for any multiindex $k \in \mathbb{Z}_+^d$, $|k| > m$, two polynomials $\mathbf{p}_k(\mathbf{z})$ and $\mathbf{q}_k(\mathbf{z}), \mathbf{z} \in \mathbb{C}^d$ are given, of the form $\mathbf{p}_k(\mathbf{z}) = \mathbf{z}^k + \sum_{\alpha \prec k} c_{\alpha,k}\mathbf{z}^\alpha$, respectively, $\mathbf{q}_k(\mathbf{z}) = \mathbf{z}^k + \sum_{\beta \prec k} p_{\beta,k}\mathbf{z}^\beta$. These sets of polynomials have codimension not greater than $N = \binom{d+m}{d}$ in the space of all polynomials. Thus, Theorem 5.2 guarantees that there are no more than $2N+2$ points $\mathbf{w}_\varkappa$ such that any C^n- function can be approximated by linear combinations of $\mathbf{p}_k(\mathbf{z})\overline{\mathbf{q}_l(\mathbf{z})}$ on any compact not containing these points.

5.2 Products of Toeplitz operators

It was known since long ago that the product of two Toeplitz operators in the Hardy space on a circle can be zero only in the case one of them is zero (cf. [3]). This result was gradually extended to an arbitrary finite product of operators on a circle (cf. [1]) and to the multidimensional case, i.e., operators in the Hardy space on the torus, where the product of up to six Toeplitz operators is taken care of (cf. [8]).

Much less understandable is the situation with Toeplitz operators in the Bergman space; even for the case of a disk it is still not known if it is true that for $f, g \in L_\infty$, the relation $\mathcal{T}_f\mathcal{T}_g = 0$, or, more generally $\operatorname{rank}(\mathcal{T}_f\mathcal{T}_g) < \infty$, implies vanishing of g or f. Affirmative answers to this problem, as well as to its multidimensional versions, was obtained only in rather special cases, say, under the assumption that the functions f and g are harmonic or $\mathbf{m}$-harmonic (cf. [9, 6] where extensive references can also be found.)

We present here some very recent results on the finite rank product problem, essentially obtained by Le [11].

Theorem 5.4. *Let D be the unit disk in $\mathbb{C}^1$. Suppose that the function $f(z) \in L_2(D)$, $|z| \leqslant 1$ has in the expansion in polar coordinates the form*

$$f(re^{i\theta}) = \sum_{n=-\infty}^{M} f_n(r)e^{in\theta}, \tag{5.2}$$

and $\widehat{f_M}(l) = \int_0^1 f_M(r)r^l dr \neq 0$ for all l large enough, $l > l_0$. If for some distribution $G \in \mathcal{E}'(D)$, the product $\mathcal{T}_G\mathcal{T}_f$ has finite rank, the distribution G must have finite support. In particular, if G is a function, then $G = 0$.

Proof. The proof follows mostly the one in [11], with modifications allowed by more advanced finite rank theorems. First, recall that the Bergman space $\mathcal{B}^2$ on the disk has a natural orthonormal basis

$$\mathbf{e}_s(z) = \sqrt{s+1}z^s, \ s = 0, 1, \dots. \tag{5.3}$$

The matrix representation of the operator $\mathcal{T}_f$ in this basis has the form

$$(\mathcal{T}_f\mathbf{e}_k, \mathbf{e}_l) = C_{k,l}\int_0^1 f_{l-k}(r)r^{k+l+1}, \ k, l \geqslant 0,$$

$C_{kl} = 2\sqrt{(k+1)(l+1)}$. By our assumption about f, we have $(\mathcal{T}_f\mathbf{e}_k, \mathbf{e}_l) = 0$ whenever $l - k < M$. Thus, for $k \in \mathbb{Z}_+$, we can write

$$\mathcal{T}_f \mathbf{e}_k = \sum_{l=0}^{k+M} (\mathcal{T}_f \mathbf{e}_k, \mathbf{e}_l)\mathbf{e}_l = C_{k,k+M} \widehat{f}_M(2k+M+1)\mathbf{e}_{k+M}$$
$$+ \sum_{l=0}^{k+M+1} C_{k,l} \widehat{f}_{l-k}(k+l+1)\mathbf{e}_l.$$

This shows that when $k+M \geqslant 1$ and $2k+M+1 > l_0$, the function $\mathbf{e}_{k+M}$ can be expressed as a linear combination of $\mathcal{T}_f \mathbf{e}_k$ and $\mathbf{e}_l, l < M+k$. Suppose that $\mathcal{T}_G \mathcal{T}_f$ has finite rank $\mathbf{r}$ and let $\phi_1, \dots, \phi_{\mathbf{r}}$ be some basis in the range of $\mathcal{T}_G \mathcal{T}_f$. Then for any nonnegative integer k such that $k+M \geqslant 1$ and $2k+M+1 > l_0$, the function $\mathcal{T}_G \mathbf{e}_{k+M}$ is a linear combination of $\phi_1, \dots, \phi_{\mathbf{r}}$ and $\mathcal{T}_G \mathbf{e}_l$, $l \leqslant k+M$. We substitute consecutively this expression for $\mathcal{T}_G \mathbf{e}_{k+M}$ into the similar expression for $\mathcal{T}_G \mathbf{e}_{k'+M}$, for $k' > k$. Thus, all functions $\mathcal{T}_G \mathbf{e}_{k'+M}$, $k'+M > 1$, $2k'+M+1 > l_0$, is expressed as linear combinations of functions $\mathcal{T}_G \mathbf{e}_{k+M}$ with $2k+M+1 \leqslant l_0$ and the finite set of functions $\phi_1, \dots, \phi_{\mathbf{r}}$. This means that the matrix with entries $(\mathcal{T}_G \mathbf{e}_k, \mathbf{e}_l)$ has finite rank. We can apply Theorem 4.4 that grants the required properties for G. □

Since for any monomial $z^k \overline{z}^l$, $\widehat{r^{k+l}}(m)$ is never zero, the conditions of Theorem 5.4 are fulfilled for any f having the form $f(z) = p(z, \overline{z}) + \overline{h(z)}$ where p is a nonzero polynomial of $z, \overline{z}$ and h is a bounded analytical function.

Another type of results on finite rank products of Toeplitz operators in the analytical Bergman space in the unit disk or polydisk, established in [11, 12], covers the case where all Toeplitz weights, except one, are functions of a special form. We present here the formulation of the general theorem proved in [12], generalized to cover the case of distributional weights.

Theorem 5.5. *Let $f_1, \dots, f_{m_1+m_2}$ be bounded functions in the polydisk $D^d \subset \mathbb{C}^d$ such that each of them is radial, $f_j(z_1, \dots, z_d) = f_j(|z_1|, \dots, |z_d|)$ and none is identically zero. For a collection of multiindices $\alpha_j, \beta_j \in \mathbb{Z}_+^d$, $j = 1, \dots, m_1+m_2$ we set $g_j(z) = f_j(z)\mathbf{z}^{\alpha_j}\overline{\mathbf{z}}^{\beta_j}$. Suppose that F is a distribution with compact support in D^d and the operator*

$$A = \mathcal{T}_{g_1} \dots \mathcal{T}_{g_{m_1}} \mathcal{T}_F \mathcal{T}_{g_{m_1+1}} \dots \mathcal{T}_{g_{m_1+m_2}} \tag{5.4}$$

has finite rank. Then F has finite support. In particular, if F is a function, F is zero.

The proof is based upon the consideration of the kernel of the product of the operators $\mathcal{S}_1 = \mathcal{T}_{g_1} \dots \mathcal{T}_{g_{m_1}}$ and the range of $\mathcal{S}_2 = \mathcal{T}_{g_{m_1+1}} \dots \mathcal{T}_{g_{m_1+m_2}}$. The action of these operators is explicitly described in the natural basis in the Bergman space (and it is here the geometry of the polydisk is crucial.) The set of multiindices numbering the basis functions in the kernel of $\mathcal{S}_1$ and in the cokernel of $\mathcal{S}_2$ turns out to be sparse. Therefore, the finiteness of the rank of $\mathcal{S}_1 \mathcal{T}_F \mathcal{S}_2$ leads to the finiteness of the rank of the properly restricted operator $\mathcal{T}_F$. The reasoning concludes by the application of Theorem 4.4.

To demonstrate the idea, not going into complicated details, we present the proof, borrowed from [11, 12], for the most simple case, when $d = 1$, $m_1 = 1$, $m_2 = 0$, so the operator A in (5.4) has the form $A = T_g T_F$

Proof. As in the proof of Theorem 5.4, we consider the standard orthogonal basis $\mathbf{e}_s$ in the Bergman space, given by (5.3). In this base, the action of the operator $\mathcal{T}_g$ for $g(z) = f(|z|)z^\alpha \overline{z}^\beta$ is easily calculated,

$$\mathcal{T}_g \mathbf{e}_s = \begin{cases} 0, & s < \alpha - \beta, \\ C\widehat{f}(2s + 2\alpha + 1)\mathbf{e}_{s+\alpha-\beta}, & s \geqslant \alpha - \beta, \end{cases} \tag{5.5}$$

with some positive constants C, depending on all indices and exponents in (5.5).

Denote $\mathcal{J} = \{s : s < \alpha + \beta\} \bigcup \{s : \widehat{f}(2s + 2\alpha + 1) = 0\}$. Since the function f is nonzero, the set $\mathcal{J}$ is sparse by Müntz–Sász's theorem. For $s \notin \mathcal{J}$, we see from (5.5) that $\mathcal{T}_g \mathbf{e}_s \neq 0$ and $\mathbf{e}_{s+\alpha-\beta}$ is a multiple of $\mathcal{T}_g \mathbf{e}_s$. Suppose that $\varphi \in \mathcal{B}$ is a function such that $\mathcal{T}_g \varphi = 0$. Then

$$0 = \mathcal{T}_g \varphi = \mathcal{T}_g \Big(\sum_s (\varphi, \mathbf{e}_s)\mathbf{e}_s \Big) = \sum_s (\varphi, \mathbf{e}_s)\mathcal{T}_g \mathbf{e}_s.$$

By (5.5), this implies that $(\varphi, \mathbf{e}_s) = 0$ for all $s \notin \mathcal{J}$. Therefore, Ker $\mathcal{T}_g$ is contained in the closed span of $\{\mathbf{e}_s,\ s \in \mathcal{J}\}$. It follows that the rank of the matrix $(\mathcal{T}_F \mathbf{e}_s, \mathbf{e}_t)$, $s, t \notin \mathcal{J}$ is not greater than the rank of $\mathcal{T}_g \mathcal{T}_F$, thus it is finite. Finally, Theorem 4.4 applies. □

5.3 Sums of products of Toeplitz operators

Another interesting problem in the theory of Bergman spaces consists in determining the condition for some algebraic expression involving Toeplitz operators to be a Toeplitz operator again. The results existing by now concern only Toeplitz operators with weights of some special form.

In [7], this problem was considered in the following setting. Suppose that u_j, v_j, $j = 1, \ldots, n$, and w are $\mathbf{m}$–harmonic functions in the polydisk $D^{\mathbf{m}} \subset \mathbb{C}^{\mathbf{m}}$.

Theorem 5.6 ([7]). *The necessary and sufficient condition for the operator* $S = \mathcal{T}_w + \sum \mathcal{T}_{u_j} \mathcal{T}_{v_j}$ *to have finite rank,* $S = \sum_{l=1}^{\mathrm{r}} (\cdot, g_l) f_l$ *with some analytical functions* f_l, g_l *is*

$$\sum u_j v_j + w = \prod_{k=1}^{\mathbf{m}} (1 - |z_k|^2)^2 \sum f_l g_l$$

and

$$w + \sum \overline{\mathbf{P}u_j}\mathbf{P}v_j \text{ is } \mathbf{m}\text{-harmonic.}$$

The proof of this result, as well as other ones in [7], is based upon the finite rank theorems.

5.4 Landau Hamiltonian and Landau–Toeplitz operators

This topic was the source of the initial interest of the author in Bergman–Toeplitz operators. About the Landau Hamiltonian one can find a detailed information in [18, 19, 4] and the references therein. It is a second order differential operator H in $L_2(\mathbb{C}^1) = L_2(\mathbb{R}^2)$ that describes the dynamics of a quantum particle confined to a plane, under the action of the uniform magnetic field B acting orthogonal to the plane. The operator has spectrum consisting of eigenvalues $\Lambda_q = (2q+1)B$, $q = 0, 1 \ldots$, called *Landau levels*, with corresponding spectral subspaces X_q having infinite dimension. The subspace X_0 is closely related with the Fock space: X_0 consists of the functions $u(z) \in L_2$, $z \in \mathbb{C}^1$, having the form $u(z) = \exp(-\frac{B|z^2|}{4})f(z)$, where $f(z)$ is an entire analytical function.

The other spectral subspaces X_q are obtained from X_0 by the action of the so called *creation operator* $\overline{Q} = (2i)^{-1}(\partial + \frac{B}{2}(x_2 - ix_1))$:

$$X_q = \overline{Q}^q X_0. \tag{5.6}$$

Under the perturbation by the operator of multiplication by a real valued function $V(\mathbf{x})$, tending to zero at infinity, the spectrum, generally, splits into clusters around Landau levels, and a lot of interesting results were obtained, describing in detail the properties of these clusters. In particular, for the case of the perturbation V having compact support, the infiniteness of the clusters was proved only under the condition that V has constant sign (or it's minor generalizations). It remained unclear whether it is possible that some perturbation would not split some (or all) clusters.

The methods of the papers cited above relate this question with the following one: is it possible that for some function V with compact support, the *Landau–Toeplitz* operator $\mathbf{T}_q(V)$ in X_q has finite rank. Here $\mathbf{T}_q(V)$ is the operator $\mathbf{T}_q(V)u = P_q V u$ in X_q, where P_q is the orthogonal projection onto X_q. More exactly, the eigenvalues of $\mathbf{T}_q(V)$ give the main contribution to the spectrum of $H + V$ near Λ_q. If $\mathbf{T}_q(V)$ has finite rank, this does not immediately mean that the Landau level Λ_q does not split, but only that it may split by the interaction with $L_{q'}$, $q' \neq q$. In such a case, we say that there is no principal splitting.

The case $q = 0$ can be treated directly by means of Luecking's theorem. The subspace X_0 consists of analytical function multiplied by a Gaussian weight. Therefore, the infinite matrix $\mathcal{A}(V)$ constructed by means of the functions $f_j = \exp(-\frac{B|z^2|}{4})z^j$ is the same as the matrix $\mathcal{A}(F)$ for the function $F = \exp(-\frac{B|z^2|}{2})V$, constructed by means of monomials $g_j = z^j$; these two matrices have a finite rank simultaneously. Thus, by Theorem 3.1, the function F, and consequently the function V, should be zero. In the initial terms, this means that the lowest Landau level Λ_0 necessarily principally splits into an infinite cluster, as soon as the perturbation V is nonzero.

A more advanced technique is needed for higher Landau levels. The subspaces X_q in which the Toeplitz operators $\mathbf{T}_q(V)$ act, do not fit into the framework of Luecking's theorem. We can, however, use the relation (5.6) between different Landau subspaces.

The following fact was established in [4, Corollary 9.3].

Proposition 5.7. *Let V be a bounded function with compact support. Then for any q the Toeplitz operator $\mathbf{T}_q(V)$ is unitary equivalent to the operator $\mathbf{T}_0(W)$, where $W = \mathcal{D}_q(\delta)V$, $\mathcal{D}_q$ being a polynomial of degree q with positive coefficients.*

To be exact, the statement was proved in [4] for smooth functions V, but the proof extends automatically to the case of bounded functions V, and, of course, the expression $\mathcal{D}_q(\delta)V$ should be understood in the sense of distribution.

Now, suppose that for some q the operator $\mathbf{T}_q(V)$ has finite rank. By Proposition 5.7, the Toeplitz operator $\mathbf{T}_0(W)$ has finite rank as well, and we can apply Theorem 3.3. So, the distribution $W = \mathcal{D}_q(\delta)V$ must be a combination of a finite number of the δ-distributions and their derivatives. Therefore, the Fourier transform of W is a polynomial, the Fourier transform of V must be a rational function, and such V cannot be a function with compact support, by the analytic hypoellipticity.

We arrived at the following result.

Theorem 5.8. *Suppose that under the perturbation of the Landau Hamiltonian by a bounded function V with compact support there is no principal splitting for one of Landau levels. Then the perturbation is zero, and therefore there is no principal splitting on other Landau levels either.*

6 Other Bergman Spaces

The analytical Bergman spaces, considered above, have a vast advantage, the multiplicative structure that is used all the time. For other types of Bergman spaces, without the multiplicative structure, the results are therefore less extensive. An exception is constituted by the harmonic Bergman spaces in an even-dimensional space, due to their close relation to analytical functions.

6.1 Harmonic Bergman spaces

The aim of this section is to establish finite rank results for Toeplitz operators in Bergman spaces of harmonic functions. The results presented here generalize the ones in [2].

We start with the even-dimensional case, $d = 2\mathbf{m}$. Here the problem with harmonic spaces reduces easily to the analytical Bergman spaces.

For a distribution $F \in \mathcal{E}'(\mathbb{R}^d)$ we consider a matrix $H(F)$ consisting of elements $\langle F, f_j \overline{f_k} \rangle$, where f_j is some complete system of homogeneous harmonic polynomials in $\mathbb{R}^d = \mathbb{C}^{\mathbf{m}}$. It is convenient (but not obligatory) to suppose that real and imaginary parts of analytic monomials, $\mathrm{Re}(\mathbf{z}^\alpha), \mathrm{Im}(\mathbf{z}^\alpha)$, $\alpha \in \mathbb{Z}^d_+$, are among the polynomials f_j. For some subset $\mathcal{J} \subset \mathbb{Z}^d_+$, we denote by $H^{\mathcal{J}}(F)$ the matrix with entries $\langle F, f_j \overline{f_k} \rangle$, with $\mathrm{Re}(\mathbf{z}^\alpha), \mathrm{Im}(\mathbf{z}^\alpha)$, $\alpha \in \mathcal{J}$ removed.

Theorem 6.1. *Let $d = 2\mathbf{m}$ be an even integer. Suppose that for some N, the set $\mathcal{J}$ satisfies the conditions of Theorem 4.4, and for a distribution $F \in \mathcal{E}'(\mathbb{R}^n)$ the matrix $H^{\mathcal{J}}(F)$ has rank $\mathbf{r} \leqslant N$. Then the distribution F is a sum of $m \leqslant \mathbf{r}+1$ terms, each supported at one point: $F = \sum L_q \delta(\mathbf{x} - \mathbf{x}_q)$, $\mathbf{x}_q \in \mathbb{R}^d$, L_q are differential operators in $\mathbb{R}^d$.*

Proof. We identify the space $\mathbb{R}^d$ with the complex space $\mathbb{C}^{\mathbf{m}}$. Since the functions $\mathbf{z}^\alpha, \overline{\mathbf{z}}^\beta$ are harmonic, the matrix $\mathcal{A}^{\mathcal{J}}(F)$ (defined in Section 4) can be considered as a submatrix of $H^{\mathcal{J}}(F)$, and therefore it has rank not greater than $\mathbf{r}$. It remains to apply Theorem 4.4 to establish that the distribution F has the required form, with no more than $\mathbf{r}+1$ points $\mathbf{x}_q$. □

The same reasoning establishes similar properties for the Bergman spaces of pluriharmonic and $\mathbf{m}$-harmonic functions.

The odd-dimensional case requires considerably more work, and the results are less complete. We use again a kind of dimension reduction, as in Theorem 3.3, however, unlike the analytic case, we need projections of the distribution to one-dimensional subspaces. We have to restrict our considerations to distributions being regular complex Borel measures (so we use the notation μ instead of F) and from now on we will not consider the generalizations related with the removal of sparse subsets $\mathcal{J}$.

Let S denote the unit sphere in $\mathbb{R}^d$, $S = \{\zeta \in \mathbb{R}^d : |\zeta| = 1\}$, and let σ be the Lebesgue measure on S. For $\zeta \in S$, we denote by $\mathcal{L}_\zeta$ the one-dimensional subspace in $\mathbb{R}^d$ passing through ζ, $\mathcal{L}_\zeta = \zeta\mathbb{R}^1$. For a measure with compact support μ on $(\mathbb{R}^d)$ we define the measure μ_ζ on $(\mathbb{R}^1)$ by setting $\langle \mu_\zeta, \phi \rangle = \langle \mu, \phi_z \rangle$, where $\phi_z \in C^\infty(\mathbb{R}^d)$ is $\phi_z(\mathbf{x}) = \phi(\mathbf{x} \cdot z)$. The measure μ_ζ can be understood as result of projecting of μ to $\mathcal{L}_\zeta$ with further transplantation of the projection, $\pi_*^{\mathcal{L}_\zeta} \mu$, from the line $\mathcal{L}_\zeta$ to the standard line $\mathbb{R}^1$. The Fourier transform $\mathcal{F}\mu_\zeta$ of μ_ζ is closely related with $\mathcal{F}\mu$:

$$\mathcal{F}(\mu_\zeta)(t) = (\mathcal{F}\mu)(t\zeta). \tag{6.1}$$

The following fact in the harmonic analysis of measures was proved in [2].

Proposition 6.2. *For a finite complex Borel measure μ with compact support in $\mathbb{R}^d$ the following three statements are equivalent:*

a) *μ is discrete,*

b) *μ_ζ is discrete for all $\zeta \in S$,*

c) *μ_ζ is discrete for σ-almost all $\zeta \in S$.*

The proof of Proposition 6.2 can be found in [2, Corollary 5.3].

Now, we return to our finite rank problem.

Theorem 6.3. *Let $d \geqslant 3$ be an odd integer, $d = 2\mathbf{m} + 1$. Let μ be a finite complex Borel measure in $\mathbb{R}^d$ with compact support. Suppose that the matrix $H(\mu)$ has finite rank $\mathbf{r}$. Then* supp μ *consists of no more than* $\mathbf{r}$ *points.*

Proof. Fix some $\zeta \in S$ and choose some $d-1 = 2\mathbf{m}$-dimensional linear subspace $\mathcal{L} \subset \mathbb{R}^d$ containing $\mathcal{L}_\zeta$. We choose the co-ordinate system $\mathbf{x} = (x_1, \dots, x_d)$ in $\mathbb{R}^d$ so that the subspace $\mathcal{L}$ coincides with $\{\mathbf{x} : x_d = 0\}$. The even-dimensional real space $\mathcal{L}$ can be considered as the $\mathbf{m}$-dimensional complex space $\mathbb{C}^{\mathbf{m}}$ with co-ordinates $\mathbf{z} = (z_1, \dots, z_{\mathbf{m}})$, $z_j = x_{2j-1} + ix_{2j}$, $j = 1, \dots, \mathbf{m}$. The functions $(\mathbf{z}, x_d) \mapsto z^\alpha$, $(\mathbf{z}, x_d) \mapsto \overline{\mathbf{z}}^\beta$, $\alpha, \beta \in (\mathbb{Z}_+)^d$, are harmonic polynomials in $\mathbb{C}^d \times \mathbb{R}^1$. Moreover, by definition, $\langle \mu, \mathbf{z}^\alpha \overline{\mathbf{z}}^\beta \rangle = \langle \pi_*^{\mathbb{C}^n} \mu, \mathbf{z}^\alpha \overline{\mathbf{z}}^\beta \rangle$. Hence the matrix $\mathcal{A}(\pi_*^{\mathbb{C}^n} \mu)$ is a submatrix of the matrix $H(\mu)$, and the former has not greater rank than the latter, $\operatorname{rank}(\mathcal{A}(\pi_*^{\mathbb{C}^n} \mu)) \leqslant \mathbf{r}$. So we can apply Theorem 6.1 and find that the measure $\pi_*^{\mathbb{C}^{\mathbf{m}}} \mu$ is discrete and its support contains not more than $\mathbf{r}$ points. Now, we project the measure $\pi_*^{\mathbb{C}^n} \mu$ to the real one-dimensional linear subspace $\mathcal{L}_\zeta$ in $\mathcal{L}$. We obtain the same measure as if we had projected μ to $\mathcal{L}_\zeta$ from the very beginning, and not in two steps i.e., $\pi_*^{\mathcal{L}_\zeta} \mu$. As a projection of a discrete measure, $\pi_*^{\mathcal{L}_\zeta} \mu$ is discrete and has no more than $\mathbf{r}$ points in the support. By our definition of the measure μ_ζ as $\pi_*^{\mathcal{L}_\zeta} \mu$ transplanted to $\mathbb{R}^1$, this means that μ_ζ is discrete.

Due to the arbitrariness of the choice of $\zeta \in S$, we find that all measures μ_ζ are discrete. We can apply Proposition 6.2 which implies that the measure μ is discrete itself. Finally, in order to show that the number of points in supp μ does not exceed $\mathbf{r}$, we chose $\zeta \in S$ such that no two points in supp μ project to the same point in $\mathcal{L}_\zeta$. Then the point masses of μ cannot cancel each other under the projection, and thus $\operatorname{card}\operatorname{supp} \mu = \operatorname{card}\operatorname{supp} \mu_\zeta \leqslant \mathbf{r}$.

The number of points in the support of μ is estimated in the same way as in Theorem 3.3. □

The analysis of the reasoning in the proof shows that the only essential obstacle for extending Theorem 6.3 to the case of distributions is the limitation set by Proposition 6.2. If we were able to prove this proposition for distributions, all other steps in the proof of Theorem 6.3 would go through without essential changes. However, it turns out that not only the proof of Proposition 6.2 cannot be carried over to the distributional case, but, moreover, the corollary itself becomes wrong. The example, that can be found in

[2], does not disprove Theorem 6.3 for distributions, however it indicates that the proof, if exists, should involve some other ideas.

6.2 Helmholtz Bergman spaces

We consider now the Helmholtz equation

$$\Delta u + \mathbf{k}^2 u = 0 \tag{6.2}$$

in $\Omega \subset \mathbb{R}^d$, where $\mathbf{k} > 0$ (we set $\mathbf{k}^2 = 1$, without loosing in generality). Let Ω be a bounded domain, and let F be a distribution with compact support in Ω. Denote by $\mathrm{H} = \mathrm{H}(\Omega)$ the space of solutions of (6.2) in Ω belonging to $L_2(\Omega)$ (we consider the Lebesgue measure here). We call such solutions *Helmholtz functions.* For a distribution $F \in \mathcal{E}'(\Omega)$ we, as usual, define the Toeplitz operator $\mathcal{T}_F : \mathrm{H} \to \mathrm{H}$, by means of the quadratic form $(\mathcal{T}_f u, v) = \langle F, u\overline{v}\rangle$, $u, v \in \mathrm{H}$. For any two systems of linearly independent functions $\Sigma_1 = \{f_j\}, \Sigma_2 = \{g_k\} \subset \mathrm{H}$, we consider the matrix $\mathcal{A} = \mathcal{A}(F; \Sigma_1, \Sigma_2)$:

$$\mathrm{A}(F; \Sigma_1, \Sigma_2) = (\langle F, f_j(x)\overline{g_k(x)}\rangle), \quad f_j \in \Sigma_1,\ g_k \in \Sigma_2. \tag{6.3}$$

If the Toeplitz operator $\mathcal{T}_F$ has finite rank $\mathbf{r}$, the matrix (6.3) has rank not greater than $\mathbf{r}$ for any Σ_1, Σ_2. Moreover,

$$\operatorname{rank} \mathcal{T}_F = \max_{\Sigma_1, \Sigma_2} \operatorname{rank}(\mathcal{A}(F; \Sigma_1, \Sigma_2)).$$

Theorem 6.4. *Let $d \geqslant 3$, and let F be a function with compact support. Suppose that the Toeplitz operator $\mathcal{T}_F$ in the space of Helmholtz functions has finite rank. Then $F = 0$.*

Proof. Consider the systems Σ_1, Σ_2 consisting of functions having the form $f_j(x) = e^{-ix_1} h_j(x')$, $g_k(x) = e^{ix_1} h_k(x')$, where $h_j(x')$ is an arbitrary system of harmonic functions of the variable x' in the subspace $\mathcal{L} \subset \mathbb{R}^d : x_1 = 0$. Then the expression in (6.3) takes the form

$$\langle F, f_j(\mathbf{x})\overline{g_k(\mathbf{x})}\rangle = \int\int F(x_1, x')e^{-2ix_1} dx_1 h_j(x')\overline{h_k(x')}dx'. \tag{6.4}$$

This matrix has finite rank, not greater than $\mathbf{r}$. Now, we are in the conditions of Theorem 6.3 or Theorem 6.1, depending on whether d is even or odd, in dimension $d - 1 \geqslant 2$, applied to the function $\widetilde{F}(x') = \int F(x_1, x')e^{-2ix_1} dx_1$, i.e., the partial Fourier transform of F in x_1 variable, calculated in the point $\xi_1 = 2$. Since the matrix $\mathcal{A}(F; \Sigma_1, \Sigma_2)$ has finite rank for arbitrary system of harmonic functions $h_j(x')$, by the above finite rank theorems about harmonic Bergman spaces, the function $\widetilde{F}(x')$ must be zero. We make the Fourier transform of $\widetilde{F}$ in the remaining variables and find that the Fourier transform $\widehat{F}(\xi)$ of $F(\mathbf{x})$ equals zero for all ξ having the first component equal to 2.

Next we fix some $\varphi \in \mathbb{R}^d$, $|\varphi| = 1$ and consider the system $\Sigma_1 = \Sigma_2$ consisting of the functions having the form $f_j(\mathbf{x}) = e^{-i\varphi\mathbf{x}} h_j(x')$, $g_k(x) = e^{i\varphi\mathbf{x}} h_k(x')$ where x' is the variable in the subspace $\mathrm{L}(\varphi) \subset \mathbb{R}^d$, orthogonal to φ, and h_j are arbitrary harmonic functions. We repeat the reasoning above to obtain that $\widehat{F}(\xi) = 0$ for all ξ having the component in the direction of φ equal to 2. Note that for any $\xi \in \mathbb{R}^d$, $|\xi| \geqslant 2$, it is possible to find φ, $|\varphi| = 1$, such that ξ has φ-component equal to 2. Therefore, we find that $\widehat{F}(\xi) = 0$ for all $|\xi| \geqslant 2$. So, we conclude that $\widehat{F}$ has compact support. But, recall, F also has compact support. Therefore, F must be zero. □

6.3 An application: the Born approximation

In the quantum scattering theory, one of the main objects to consider is the scattering matrix; details can be found in many books on the scattering theory, for example, [23, 20]. We consider the Born approximation, which (up to a constant factor) is the integral operator $\mathbf{K}$ with kernel

$$K(\varphi, \varsigma) = \int_{\mathbb{R}^d} F(\mathbf{x}) e^{i\mathbf{x}(\varphi - \varsigma)} d\mathbf{x}, \tag{6.5}$$

where $|\varphi|^2 = |\varsigma|^2 = E > 0$, $F(\mathbf{x})$ is the potential (decaying at infinity sufficiently fast) and the operator acts on the sphere $S^{d-1} : |\varphi|^2 = E$. Further on, we suppose that $E = 1$.

The expression (6.5) coincides with the quadratic form of the Toeplitz operator $\mathcal{T}_F$ in the space of solutions of the Helmholtz equation, considered on the systems of functions $e^{i\varphi\mathbf{x}}, e^{i\varsigma\mathbf{x}}$. We consider the case where the operator $\mathbf{K}$ has finite rank. This implies that the matrix (6.3) has finite rank. So, applying Theorem 6.4, we obtain the following result.

Theorem 6.5. *Let $d \geqslant 3$, and let F be a function with compact support. Suppose that the Born approximation operator $\mathbf{K}$ has finite rank. Then $F = 0$.*

References

1. Aleman, A., Vukotić, D.: Zero products of Toeplitz operators. Duke Math. J. [To appear]
www.uam.es/personal_pdi/ciencias/dragan/respub/duke-toeplitz-pp.pdf
2. Alexandrov, A., Rozenblum G.: Finite rank Toeplitz operators. Some extensions of D. Luecking's theorem. J. Funct. Anal. **256**, 2291–2303 (2009)

3. Brown, A., Halmos, P.: Algebraic properties of Toeplitz operators. J. Reine Angew. Math. **213**, 89–102 (1963/1964)
4. Bruneau, V., Pushnitski, A., Raikov, G.: Spectral shift function in strong magnetic field. St. Petersburg Math. J. **16**, no. 1, 181–209 (2005)
5. Choe, B.R.: On higher dimensional Luecking's theorem. J. Math. Soc. Japan **61**, no. 1, 213–224 (2009)
6. Choe, B.R., Koo, H., Lee, Y.: Zero products of Toeplitz operators with n-harmonic symbols. Integral Equat. Operator Theory **57**, no. 1, 43–66 (2007)
7. Choe, B.R., Koo, H., Young, J.L.: Finite sums of Toeplitz products on the polydisk. http://math.korea.ac.kr/~choebr/papers/finitesum_polydisk.pdf
8. Ding, X.: Products of Toeplitz operators on the polydisk. Int. Equat. Operator Theory **45**, no. 4, 389–403 (2003)
9. Guo, K., Sun, S., Zheng, D.: Finite rank commutators and semicommutators of Toeplitz operators with harmonic symbols. Illinois J. Math. **51**, no. 2, 583–596 (2007)
10. Hörmander, L.: The Analysis of Linear Partial Differential Operators. V.1. Springer, 1983.
11. Le, T. : A refined Luecking's theorem and finite-rank products of Toeplitz operators. Complex Analysis and Operator Theory arXiv:0802.3925
12. Le, T.: Finite rank products of Toeplitz operators in several complex variables. Int. Equat. Operator Theory http://individual.utoronto.ca/trieule/Data/FiniteRankToeplitzProducts.pdf
13. Luecking, D.: Trace ideal criteria for Toeplitz operators. J. Funct. Anal. **73**, no. 2, 345–368 (1987)
14. Luecking, D.: Finite rank Toeplitz operators on the Bergman space. Proc. Am. Math. Soc. **136**, no. 5, 1717–1723 (2008)
15. Martines-Avendaño, R., Rosenthal, P.: An Introduction to Operators on the. Hardy-Hilbert Space. Springer (2007)
16. Maz'ya, V.G., Verbitsky, I.: The Schrödinger operator on the energy space: boundedness and compactness criteria. Acta Math. **188**, no. 2, 263–302 (2002)
17. Maz'ya, V.G., Verbitsky, I.: Form boundedness of the general second-order differential operator. Commun. Pure Appl. Math. **59**, no. 9, 1286–1329 (2006)
18. Melgaard, M., Rozenblum, G.: Eigenvalue asymptotics for weakly perturbed Dirac and Schrödinger operators with constant magnetic fields of full rank. Commun. Partial Differ. Equ. **28**, 697–736 (2003)
19. Raikov, G., Warzel, S.: Quasi-classical versus non-classical spectral asymptotics for magnetic Schroödinger operators with decreasing electric potentials. Rev. Math. Phys. **14**, 1051–1072 (2002)
20. Rodberg, L., Thaler, R.: Introduction to the Quantum Theory of Scattering. Academic Press (1965)
21. Rozenblum, G., Shirokov, N.: Finite rank Bergman–Toeplitz and Bargmann–Toeplitz operators in many dimensions. Complex Analysis and Operator Theory [To appear] arXiv:0802.0192.
22. Rudin, W.: Real and Complex Analysis. 3rd ed., McGraw-Hill Book Co., New York, 1987
23. Yafaev, D.: Scattering Theory; Some Old and New Problems. Lect. Notes Math. **1735**, Springer (2000)

Resolvent Estimates for Non-Selfadjoint Operators via Semigroups

Johannes Sjöstrand

Dedicated to V.G. Maz'ya

Abstract We consider a non-selfadjoint h-pseudodifferential operator P in the semiclassical limit ($h \to 0$). If p is the leading symbol, then under suitable assumptions about the behavior of p at infinity, we know that the resolvent $(z-P)^{-1}$ is uniformly bounded for z in any compact set not intersecting the closure of the range of p. Under a subellipticity condition, we show that the resolvent extends locally inside the range up to a distance $\mathcal{O}(1)((h \ln \frac{1}{h})^{k/(k+1)})$ from certain boundary points, where $k \in \{2, 4, \dots\}$. This is a slight improvement of a result by Dencker, Zworski, and the author, and it was recently obtained by W. Bordeaux Montrieux in a model situation where $k = 2$. The method of proof is different from the one of Dencker et al, and is based on estimates of an associated semigroup.

1 Introduction

In this paper, we are interested in bounds on the resolvent $(z-P)^{-1}$ of a non-selfadjoint h-pseudodifferential operator with leading symbol p as $h \to 0$, for z in a neighborhood of certain points on the boundary of the range of p. The interest in such questions arouse with that in pseudospectra of non-selfadjoint operators (cf. [22, 23]). Under reasonable hypotheses, we know that $(z-P)^{-1}$ is uniformly bounded for $h > 0$ small enough and for z in any fixed compact set in $\mathbf{C}$, disjoint from the closure of the range of p. On the other hand, by a quasimode construction of Davies [5], that was generalized by Zworski [24] by

Johannes Sjöstrand
IMB, Université de Bourgogne, 9, Av. A. Savary, BP 47870, FR-21078 Dijon Cédex and UMR 5584, CNRS, France
e-mail: johannes@u-bourgogne.fr

A. Laptev (ed.), *Around the Research of Vladimir Maz'ya III: Analysis and Applications*, International Mathematical Series 13, DOI 10.1007/978-1-4419-1345-6_13,

reduction to an old quasimode construction of Hörmander (cf. also [7] for a more direct approach), we also know that if $\mathbf{C} \ni z = p(\rho)$, where ρ is a point in phase space where $i^{-1}\{p, \overline{p}\} > 0$ and $\{\cdot, \cdot\cdot\}$ denotes the Poisson bracket, then we have quasimodes for $P - z$ in the sense that there exist $u = u_h \in C_0^\infty$, normalized in L^2, such that the L^2 norm of $(P - z)u_h$ is $\mathcal{O}(h^\infty)$, implying, somewhat roughly, that the norm of the resolvent (whenever it is defined) cannot be bounded by a negative power of h.

A natural question is then what happens when z is close to the boundary of the range of p. Boulton [1] and Davies [6] obtained some results about this in the case of the non-selfadjoint harmonic operator on the real line. As with the quasimode construction, this question is closely related to classical results in the general theory of linear PDE, and with Dencker and Zworski (cf. [7]) we were able to find quite general results closely related to the classical topic of subellipticity for pseudodifferential operators of principal type, studied by Egorov, Hörmander, and others (cf. [12]). This topic, in turn, is closely related to the oblique derivative problem and degenerate elliptic operators, where Maz'ya has made important contributions (cf. [16, 17]).

In [7], we obtained resolvent estimates at certain boundary points

(A) under a nontrapping condition, and

(B) under a stronger "subellipticity condition."

In case (A), we could apply quite general and simple arguments related to the propagation of regularity and, in case (B), we were able to adapt general Weyl–Hörmander calculus and Hörmander's treatment of subellipticity for operators of principal type (cf. [12]). In the first case, we found that the resolvent extends and has temperate growth in $1/h$ in discs of radius $\mathcal{O}(h \ln 1/h)$ centered at the appropriate boundary points, while in case (B), we got the corresponding extension up to distance $\mathcal{O}(h^{k/(k+1)})$, where the integer $k \geqslant 2$ is determined by a condition of "subellipticity type."

However, the situation near boundary points of type (B) is more special than the general subellipticity situations considered by Egorov and Hörmander, and the purpose of the present paper is to develop such an approach by studying an associated semigroup basically as a Fourier integral operator with complex phase in the spirit of Maslov [14], Kucherenko [13], and Melin–Sjöstrand [18] (cf. also the more recent works by Menikoff and Sjöstrand [20], Matte [15] extending the approach of [18] to inhomogeneous cases.) Finally, it turned out to be more convenient to use Bargmann–FBI transforms in the spirit of [21] and [9]. The semigroup method led to a strengthened result in case (B): The resolvent can be extended to a disc of radius $\mathcal{O}((h \ln 1/h)^{k/(k+1)})$ around the appropriate boundary points. This improvement was obtained recently by Bordeaux Montrieux [1] for the model operator $hD_x + g(x)$ when $g \in C^\infty(S^1)$ and the points of maximum or minimum are all nondegenerate. In that case, $k = 2$ and Bordeaux Montrieux also constructed quasimodes for values of the spectral parameter that are close to the boundary points.

We next state the results and outline the proof in case (B).

Let X be equal to $\mathbf{R}^n$ or a compact smooth manifold of dimension n. In the first case, let $m \in C^\infty(\mathbf{R}^{2n}; [1, +\infty[)$ be an order function (cf. [8] for more details about the pseudodifferential calculus) in the sense that for some $C_0, N_0 > 0$

$$m(\rho) \leqslant C_0 \langle \rho - \mu \rangle^{N_0} m(\mu), \quad \rho, \mu \in \mathbf{R}^{2n}, \tag{1.1}$$

where $\langle \rho - \mu \rangle = (1 + |\rho - \mu|^2)^{1/2}$. Let $P = P(x, \xi; h) \in S(m)$, meaning that P is smooth in x, ξ and satisfies

$$|\partial^\alpha_{x,\xi} P(x, \xi; h)| \leqslant C_\alpha m(x, \xi), \quad (x, \xi) \in \mathbf{R}^{2n}, \quad \alpha \in \mathbf{N}^{2n}, \tag{1.2}$$

where C_α is independent of h. We also assume that

$$P(x, \xi; h) \sim p_0(x, \xi) + h p_1(x, \xi) + \ldots \quad \text{in } S(m) \tag{1.3}$$

and write $p = p_0$ for the principal symbol. Impose the ellipticity assumption

$$\exists w \in \mathbf{C},\ C > 0 \ \text{ such that } \ |p(\rho) - w| \geqslant m(\rho)/C \quad \forall \rho \in \mathbf{R}^{2n}. \tag{1.4}$$

In this case, let

$$P = P^w(x, hD_x; h) = \operatorname{Op}(P(x, h\xi; h)) \tag{1.5}$$

be the Weyl quantization of the symbol $P(x, h\xi; h)$ that we can view as a closed unbounded operator on $L^2(\mathbf{R}^n)$.

In the second case where X is a compact manifold, let $P \in S^m_{1,0}(T^*X)$ (the classical Hörmander symbol space) of order $m > 0$, meaning that

$$|\partial^\alpha_x \partial^\beta_\xi P(x, \xi; h)| \leqslant C_{\alpha,\beta} \langle \xi \rangle^{m - |\beta|}, \quad (x, \xi) \in T^*X, \tag{1.6}$$

where $C_{\alpha,\beta}$ are independent of h. Assume that we have an expansion as in (1.3), now in the sense that

$$P(x, \xi; h) - \sum_0^{N-1} h^j p_j(x, \xi) \in h^N S^{m-N}_{1,0}(T^*X), \quad N = 1, 2, \ldots \tag{1.7}$$

and we quantize the symbol $P(x, h\xi; h)$ in the standard (nonunique) way, by doing it for various local coordinates and paste the quantizations together by means of a partition of unity. If $m > 0$, we impose the ellipticity condition

$$\exists C > 0 \ \text{ such that } \ |p(x, \xi)| \geqslant \frac{\langle \xi \rangle^m}{C}, \quad |\xi| \geqslant C. \tag{1.8}$$

Let $\Sigma(p) = \overline{p^*(T^*X)}$, and let $\Sigma_\infty(p)$ be the set of accumulation points of $p(\rho_j)$ for all sequences $\rho_j \in T^*X$, $j = 1, 2, 3, \ldots$ that tend to infinity. The following theorem is a partial improvement of the results in [7].

Theorem 1.1. *We adopt the general assumptions above. Let $z_0 \in \partial\Sigma(p) \setminus \Sigma_\infty(p)$, and let $dp \neq 0$ at every point of $p^{-1}(z_0)$. Then for every such a point ρ there exists $\theta \in \mathbf{R}$ (unique up to a multiple of π) such that $d(e^{-i\theta}(p - z_0))$ is real at ρ. We write $\theta = \theta(\rho)$. Consider the following two cases.*

(A) *For every $\rho \in p^{-1}(z_0)$ the maximal integral curve of $H_{\mathrm{Re}(e^{-i\theta(\rho)}p)}$ passing through the point ρ is not contained in $p^{-1}(z_0)$.*

(B) *There exists an integer $k \geqslant 1$ such that for every $\rho \in p^{-1}(z_0)$ there exists $j \in \{1, 2, \ldots, k\}$ such that $p^*(\exp tH_p(\rho)) = at^j + \mathcal{O}(t^{j+1})$, as $t \to 0$, where $a = a(\rho) \neq 0$. Here, p also denotes an almost holomorphic extension to a complex neighborhood of ρ, and $p^*(\mu) = \overline{p(\overline{\mu})}$. Equivalently, $H_p^j(\overline{p})(\rho)/(j!) = a \neq 0$.*

Then, in case (A), *there exists a constant $C_0 > 0$ such that for every constant $C_1 > 0$ there is a constant $C_2 > 0$ such that the resolvent $(z - P)^{-1}$ is well defined for $|z - z_0| < C_1 h \ln \frac{1}{h}$, $h < \frac{1}{C_2}$ and satisfies the estimate*

$$\|(z-P)^{-1}\| \leqslant \frac{C_0}{h} \exp\left(\frac{C_0}{h}|z - z_0|\right). \tag{1.9}$$

In case (B), *there exists a constant $C_0 > 0$ such that for every constant $C_1 > 0$ there is a constant $C_2 > 0$ such that the resolvent $(z - P)^{-1}$ is well defined for $|z - z_0| < C_1(h \ln \frac{1}{h})^{k/(k+1)}$, $h < \frac{1}{C_2}$ and satisfies the estimate*

$$\|(z-P)^{-1}\| \leqslant \frac{C_0}{h^{\frac{k}{k+1}}} \exp\left(\frac{C_0}{h}|z - z_0|^{\frac{k+1}{k}}\right). \tag{1.10}$$

In [7], we obtained (1.9) and (1.10) for $z = z_0$, implying that the resolvent exists and satisfies the same bound for $|z - z_0| \leqslant h^{k/(k+1)}/\mathcal{O}(1)$ in case (B) and with $k/(k+1)$ replaced by 1 in case (A). In case (A), we also showed that the resolvent exists with norm bounded by a negative power of h in any disc $D(z_0, C_1 h \ln(1/h))$. (The condition in case (B) was formulated a little differently in [7], but as we shall see later on, the two conditions lead to the same microlocal models and hence they are equivalent.) Actually, the proof in [7] also gives (1.9), so even if the methods of the present paper also most likely lead to that bound, we shall not elaborate the details in that case.

Let us now consider the special situation of potential interest for evolution equations, namely, the case where

$$z_0 \in i\mathbf{R}, \tag{1.11}$$

$$\mathrm{Re}\, p(\rho) \geqslant 0 \text{ in } \operatorname{neigh}(p^{-1}(z_0), T^*X). \tag{1.12}$$

Theorem 1.2. *We adopt the general assumptions above. Let $z_0 \in \partial\Sigma(p) \setminus \Sigma_\infty(p)$, and let (1.11) and (1.12) hold. Assume that $dp \neq 0$ on $p^{-1}(z_0)$, so that $d\,\mathrm{Im}\, p \neq 0$, $d\,\mathrm{Re}\, p = 0$ on that set. Consider the two cases of Theorem* 1.1:

(A) *For every $\rho \in p^{-1}(z_0)$ the maximal integral curve of $H_{\operatorname{Im} p}$ passing through the point ρ contains a point where $\operatorname{Re} p > 0$.*
(B) *There exists an integer $k \geqslant 1$ such that for every $\rho \in p^{-1}(z_0)$, we have $H^j_{\operatorname{Im} p} \operatorname{Re} p(\rho) \neq 0$ for some $j \in \{1, 2, \ldots, k\}$.*

Then, in case (A), *there exists a constant $C_0 > 0$ such that for every constant $C_1 > 0$ there is a constant $C_2 > 0$ such that the resolvent $(z - P)^{-1}$ is well defined for*

$$|\operatorname{Im}(z - z_0)| < \frac{1}{C_0}, \quad \frac{-1}{C_0} < \operatorname{Re} z < C_1 h \ln \frac{1}{h}, \quad h < \frac{1}{C_2}$$

and satisfies the estimate

$$\|(z-P)^{-1}\| \leqslant \begin{cases} \dfrac{C_0}{|\operatorname{Re} z|}, & \operatorname{Re} z \leqslant -h, \\ \dfrac{C_0}{h} \exp\big(\dfrac{C_0}{h} \operatorname{Re} z\big), & \operatorname{Re} z \geqslant -h. \end{cases} \tag{1.13}$$

In case (B), *there exists a constant $C_0 > 0$ such that for every constant $C_1 > 0$ there is a constant $C_2 > 0$ such that the resolvent $(z - P)^{-1}$ is well defined for*

$$|\operatorname{Im}(z - z_0)| < \frac{1}{C_0}, \quad \frac{-1}{C_0} < \operatorname{Re} z < C_1 \Big(h \ln \frac{1}{h}\Big)^{\frac{k}{k+1}}, \quad h < \frac{1}{C_2} \tag{1.14}$$

and satisfies the estimate

$$\|(z-P)^{-1}\| \leqslant \begin{cases} \dfrac{C_0}{|\operatorname{Re} z|}, & \operatorname{Re} z \leqslant -h^{\frac{k}{k+1}}, \\ \dfrac{C_0}{h^{\frac{k}{k+1}}} \exp\Big(\dfrac{C_0}{h} (\operatorname{Re} z)_+^{\frac{k}{k+1}}\Big), & \operatorname{Re} z \geqslant -h^{\frac{k+1}{k}}. \end{cases} \tag{1.15}$$

Case (A) in the theorems is practically identical with the corresponding results in [7] and can be obtained by inspection of the proof there, and from now on we concentrate on case (B). Away from the set $p^{-1}(z_0)$, we can use ellipticity, so the problem is to obtain microlocal estimates near a point $\rho \in p^{-1}(z_0)$. After a standard factorization of $P - z$ in such a region, we can further reduce the proof of the first theorem to that of the second one.

The main (quite standard) idea of the proof of Theorem 1.2 is to study $\exp(-tP/h)$ (microlocally) for $0 \leqslant t \ll 1$ and to show that, in this case,

$$\Big\| \exp -\frac{tP}{h} \Big\| \leqslant C \exp\Big(-\frac{t^{k+1}}{Ch}\Big) \tag{1.16}$$

for some constant $C > 0$. Noting that this implies $\| \exp -\frac{tP}{h} \| = \mathcal{O}(h^\infty)$ for $t \geqslant h^\delta$ when $\delta(k+1) < 1$ and using the formula

$$(z-P)^{-1} = -\frac{1}{h}\int_0^\infty \exp\Big(\frac{t(z-P)}{h}\Big)dt, \tag{1.17}$$

we obtain (1.15). (This has some relation to the works of Cialdea and Maz'ya [3, 4], where the L^p dissipativity of second order operators is characterized.)

The most direct way of studying $\exp(-tP/h)$, or rather a microlocal version of that operator, is to view it as a Fourier integral operator with complex phase (cf. [14, 13, 18, 15]) of the form

$$U(t)u(x) = \frac{1}{(2\pi h)^n}\iint e^{\frac{i}{h}(\phi(t,x,\eta)-y\cdot\eta)}a(t,x,\eta;h)u(y)dyd\eta, \tag{1.18}$$

where the phase ϕ should have a nonnegative imaginary part and satisfy the Hamilton–Jacobi equation

$$i\partial_t\phi + p(x,\partial_x\phi) = \mathcal{O}((\operatorname{Im}\phi)^\infty) \text{ locally uniformly} \tag{1.19}$$

with the initial condition

$$\phi(0,x,\eta) = x\cdot\eta. \tag{1.20}$$

The amplitude a will be bounded with all its derivatives and has an asymptotic expansion where the terms are determined by transport equations. This can indeed be carried out in a classical manner, for instance, by adapting the method of [18] to the case of inhomogeneous symbols following a reduction used in [20, 15]. It is based on making estimates on the function

$$S_\gamma(t) = \operatorname{Im}\Big(\int_0^t \xi(s)\cdot dx(s)\Big) - \operatorname{Re}\xi(t)\cdot\operatorname{Im}x(t) + \operatorname{Re}\xi(0)\cdot\operatorname{Im}x(0)$$

along the complex integral curves $\gamma : [0,T] \ni s \mapsto (x(s),\xi(s))$ of the Hamilton field of p. Note that here and already in (1.19), we need to take an almost holomorphic extension of p. Using property (B), one can show that $\operatorname{Im}\phi(t,x,\eta) \geqslant C^{-1}t^{k+1}$, and from that we can obtain (a microlocalized version of) (1.16) quite easily.

Finally, we preferred a variant that we shall now outline. Let

$$Tu(x) = Ch^{-\frac{3n}{4}}\int e^{\frac{i}{h}\phi(x,y)}u(y)dy$$

be an FBI – or (generalized) Bargmann–Segal transform that we treat in the spirit of Fourier integral operators with complex phase as in [21]. Here, ϕ is holomorphic in a neighborhood of $(x_0,y_0) \in \mathbf{C}^n \times \mathbf{R}^n$ and $-\phi'_y(x_0,y_0) = \eta_0 \in \mathbf{R}^n$, $\operatorname{Im}\phi''_{y,y}(x_0,y_0) > 0$, $\det \phi''_{x,y}(x_0,y_0) \neq 0$. Let $\varkappa_T$ be the associated canonical transformation. Then microlocally, T is bounded $L^2 \to H_{\Phi_0} := \operatorname{Hol}(\Omega) \cap L^2(\Omega, e^{-2\Phi_0/h}L(dx))$ and has (microlocally) a bounded inverse, where Ω is a small complex neighborhood of x_0 in $\mathbf{C}^n$. Here, the weight Φ_0 is smooth and strictly plurisubharmonic. If

$$\Lambda_{\Phi_0} := \left\{ \left(x, \frac{2}{i} \frac{\partial \Phi_0}{\partial x} \right) ; \, x \in \operatorname{neigh}(x_0) \right\},$$

then (in the sense of germs) $\Lambda_{\Phi_0} = \varkappa_T(T^*X)$.

The conjugate operator $\widetilde{P} = TPT^{-1}$ can be defined locally modulo $\mathcal{O}(h^\infty)$ (cf. also [11]) as a bounded operator from $H_\Phi \to H_\Phi$ provided that the weight Φ is smooth and satisfies $\Phi' - \Phi_0' = \mathcal{O}(h^\delta)$ for some $\delta > 0$. (Within the analytic framework, this condition can be relaxed.) Egorov's theorem applies in this situation, so the leading symbol $\widetilde{p}$ of $\widetilde{P}$ is given by $\widetilde{p} \circ \varkappa_T = p$. Thus (under the assumptions of Theorem 1.2), we have $\operatorname{Re} \widetilde{p}_{|_{\Lambda_{\Phi_0}}} \geqslant 0$, which, in turn, can be used to see that for $0 \leqslant t \leqslant h^\delta$ we have $e^{-t\widetilde{P}/h} = \mathcal{O}(1)\colon H_{\Phi_0} \to H_{\Phi_t}$, where $\Phi_t \leqslant \Phi_0$ is determined by the real Hamilton–Jacobi problem

$$\frac{\partial \Phi_t}{\partial t} + \operatorname{Re} \widetilde{p}\left(x, \frac{2}{i} \frac{\partial \Phi_t}{\partial x} \right) = 0, \quad \Phi_{t=0} = \Phi_0. \tag{1.21}$$

Now, the bound (1.16) follows from the estimate

$$\Phi_t \leqslant \Phi_0 - \frac{t^{k+1}}{C} \tag{1.22}$$

where $C > 0$. An easy proof of (1.22) is to represent the I-Lagrangian manifold Λ_{Φ_t} as the image under $\varkappa_T$ of the I-Lagrangian manifold $\Lambda_{G_t} = \{\rho + iH_{G_t}(\rho); \, \rho \in \operatorname{neigh}(\rho_0, T^*X)\}$, where H_{G_t} denotes the Hamilton field of G_t. It turns out that the G_t are given by the real Hamilton–Jacobi problem

$$\frac{\partial G_t}{\partial t} + \operatorname{Re}(p(\rho + iH_{G_t}(\rho))) = 0, \quad G_0 = 0, \tag{1.23}$$

and there is a simple minimax type formula expressing Φ_t in terms of G_t, so it suffices to show that

$$G_t \leqslant -t^{k+1}/C. \tag{1.24}$$

This estimate is quite simple to obtain: (1.23) first implies that $G_t \leqslant 0$, so $(\nabla G_t)^2 = \mathcal{O}(G_t)$. Then, if we Taylor expand (1.23), we get

$$\frac{\partial G_t}{\partial t} + H_{\operatorname{Im} p}(G_t) + \mathcal{O}(G_t) + \operatorname{Re} p(\rho) = 0$$

and we obtain (1.24) from a simple differential inequality and an estimate for certain integrals of $\operatorname{Re} p$.

The use of the representation with G_t is here very much taken from the joint work [9] with Helffer.

In Section 5, we discuss some examples.

2 IR-Manifolds Close to $\mathbf{R}^{2n}$ and Their FBI-Representations

Much of this section is just an adaptation of the discussion in [9] with the difference that we here use the simple FBI-transforms of generalized Bargmann type from [21], rather than the more complicated variant that was necessary to treat a neighborhood of infinity in the resonance theory of [9].

We shall work locally. Let $G(y,\eta) \in C^\infty(\mathrm{neigh}\,((y_0,\eta_0),\mathbf{R}^{2n}))$ be real-valued and small in the C^∞ topology. Then

$$\Lambda_G = \{(y,\eta) + iH_G(y,\eta);\ (y,\eta) \in \mathrm{neigh}\,((y_0,\eta_0))\},$$
$$H_G = \frac{\partial G}{\partial \eta}\frac{\partial}{\partial y} - \frac{\partial G}{\partial y}\frac{\partial}{\partial \eta}$$

is an I-Lagrangian manifold, i.e., a Lagrangian manifold for the real symplectic form $\mathrm{Im}\,\sigma$, where σ denotes the complex symplectic form $\sum_1^n d\widetilde{\eta}_j \wedge d\widetilde{y}_j$. Here, for notational reasons we reserve the notation (y,η) for the real cotangent variables and let the tilde indicate that we take the corresponding complexified variables.

We may also represent Λ_G by means of a nondegenerate phase function in the sense of Hörmander in the following way. Consider

$$\psi(\widetilde{y},\eta) = -\eta\cdot \mathrm{Im}\,\widetilde{y} + G(\mathrm{Re}\,\widetilde{y},\eta),$$

where $\widetilde{y}$ is complex and η is real according to the convention above. Then

$$\nabla_\eta \psi(\widetilde{y},\eta) = -\mathrm{Im}\,\widetilde{y} + \nabla_\eta G(\mathrm{Re}\,\widetilde{y},\eta),$$

and, since G is small, we see that $d\frac{\partial\psi}{\partial\eta_1},\dots,d\frac{\partial\psi}{\partial\eta_n}$ are linearly independent. So ψ is indeed a nondegenerate phase function if we drop the classical requirement of homogeneity in the η variables.

Let $C_\psi = \{(\widetilde{y},\eta) \in \mathrm{neigh}\,((y_0,\eta_0),\mathbf{C}^n\times\mathbf{R}^n);\ \nabla_\eta\psi = 0\}$. Consider the corresponding I-Lagrangian manifold

$$\Lambda_\psi = \Big\{\Big(\widetilde{y},\frac{2}{i}\frac{\partial\psi}{\partial\widetilde{y}}(\widetilde{y},\eta)\Big);\ (\widetilde{y},\eta)\in C_\psi\Big\}.$$

Here, we adopt the convention that $\dfrac{\partial}{\partial\widetilde{y}}$ denotes the holomorphic derivative since $\widetilde{y}$ are complex variables:

$$\frac{\partial}{\partial\widetilde{y}} = \frac{1}{2}\Big(\frac{\partial}{\mathrm{Re}\,\widetilde{y}} + \frac{1}{i}\frac{\partial}{\partial\,\mathrm{Im}\,\widetilde{y}}\Big).$$

Let us first check that Λ_ψ is I-Lagrangian, using only that ψ is a nondegenerate phase function: That Λ_ψ is a submanifold with the correct real dimension

$= 2n$ is classical since we can identify $\frac{2}{i}\frac{\partial\psi}{\partial\widetilde{y}}$ with $\nabla_{\mathrm{Re}\,\widetilde{y},\mathrm{Im}\,\widetilde{y}}\psi$. Further,

$$
\begin{aligned}
-\operatorname{Im}(\widetilde{\eta}\cdot d\widetilde{y})_{|\Lambda_\psi} \simeq -\operatorname{Im}\Big(\frac{2}{i}\frac{\partial\psi}{\partial\widetilde{y}}\cdot d\widetilde{y}\Big)_{|C_\psi} &= -\frac{1}{2i}\Big(\frac{2}{i}\frac{\partial\psi}{\partial\widetilde{y}}d\widetilde{y} + \frac{2}{i}\frac{\partial\psi}{\partial\overline{\widetilde{y}}}d\overline{\widetilde{y}}\Big)_{|C_\psi} \\
= \Big(\frac{\partial\psi}{\partial\widetilde{y}}d\widetilde{y} + \frac{\partial\psi}{\partial\overline{\widetilde{y}}}d\overline{\widetilde{y}}\Big)_{|C_\psi} &= d\psi_{|C_\psi}
\end{aligned}
$$

which is a closed form, and, using that $\operatorname{Im}\sigma = d\operatorname{Im}(\widetilde{\eta}\cdot d\widetilde{y})$, we get

$$-\operatorname{Im}\sigma_{|\Lambda_\psi} = 0.$$

We next check for our specific phase ψ that $\Lambda_\psi = \Lambda_G$: If $(\widetilde{y}, \frac{2}{i}\frac{\partial\psi}{\partial\widetilde{y}}(\widetilde{y},\eta))$ is a general point on Λ_ψ, then $\operatorname{Im}\widetilde{y} = \nabla_\eta G(\operatorname{Re}\widetilde{y},\eta)$ and

$$
\begin{aligned}
\frac{2}{i}\frac{\partial\psi}{\partial\widetilde{y}}(\widetilde{y},\eta) &= \frac{2}{i}\frac{1}{2}\Big(\frac{\partial}{\partial\operatorname{Re}\widetilde{y}} + \frac{1}{i}\frac{\partial}{\partial\operatorname{Im}\widetilde{y}}\Big)(-\eta\cdot\operatorname{Im}\widetilde{y} + G(\operatorname{Re}\widetilde{y},\eta)) \\
&= -\Big(\frac{\partial}{\partial\operatorname{Im}\widetilde{y}} + i\frac{\partial}{\partial\operatorname{Re}\widetilde{y}}\Big)(-\eta\cdot\operatorname{Im}\widetilde{y} + G(\operatorname{Re}\widetilde{y},\eta)) \\
&= \eta - i\nabla_y G(\operatorname{Re}\widetilde{y},\eta).
\end{aligned}
$$

Hence

$$\Big(\widetilde{y}, \frac{2}{i}\frac{\partial\psi}{\partial\widetilde{\eta}}\Big) = (y,\eta) + iH_G(y,\eta)$$

if we choose $y = \operatorname{Re}\widetilde{y}$. □

Now, consider an FBI (or generalized Bargmann–Segal) transform

$$Tu(x;h) = h^{-\frac{3n}{4}}\int e^{i\phi(x,y)/h}a(x,y;h)u(y)u(y)dy,$$

where ϕ is holomorphic near $(x_0,y_0)\in\mathbf{C}^n\times\mathbf{R}^n$, $\operatorname{Im}\phi''_{y,y}>0$, $\det\phi''_{x,y}\neq 0$, $-\frac{\partial\phi}{\partial y} = \eta_0\in\mathbf{R}^n$, and a is holomorphic in the same neighborhood with $a\sim a_0(x,y) + ha_1(x,y) + \dots$ in the space of such functions with $a_0\neq 0$. We can view T as a Fourier integral operator with complex phase, and the associated canonical transformation is

$$\varkappa = \varkappa_T : \Big(y, -\frac{\partial\phi}{\partial y}(x,y)\Big) \mapsto \Big(x, \frac{\partial\phi}{\partial x}(x,y)\Big)$$

from a complex neighborhood of (y_0,η_0) to a complex neighborhood of (x_0,ξ_0), where $\xi_0 = \frac{\partial\phi}{\partial x}(x_0,y_0)$. Complex canonical transformations preserve the class of I-Lagrangian manifolds and (locally)

$$\varkappa(\mathbf{R}^{2n}) = \Lambda_{\Phi_0} = \Big\{\Big(x, \frac{2}{i}\frac{\partial\Phi_0}{\partial x}(x)\Big);\ x\in\operatorname{neigh}(x_0,\mathbf{C}^n)\Big\},$$

where Φ_0 is smooth and strictly plurisubharmonic. Actually,

$$\Phi_0(x) = \sup_{y \in \mathbf{R}^n} -\operatorname{Im} \phi(x, y), \tag{2.1}$$

where the supremum is attained at the nondegenerate point of maximum $y_c(x)$ (cf. [21]).

Proposition 2.1. *We have* $\varkappa(\Lambda_G) = \Lambda_{\Phi_G}$, *where*

$$\Phi_G(x) = \mathrm{v.c.}_{\widetilde{y},\eta} - \operatorname{Im} \phi(x, \widetilde{y}) - \eta \cdot \operatorname{Im} \widetilde{y} + G(\operatorname{Re} \widetilde{y}, \eta), \tag{2.2}$$

and the critical value is attained at a nondegenerate critical point. Here, $\mathrm{v.c.}_{\widetilde{y},\eta}(\dots)$ *means "critical value with respect to* $\widetilde{y}, \eta$ *of ... . "*

Proof. At a critical point, we have

$$\begin{aligned} &\operatorname{Im} \widetilde{y} = \nabla_\eta G(\operatorname{Re} \widetilde{y}, \eta), \\ &\frac{\partial}{\partial \operatorname{Im} \widetilde{y}} \operatorname{Im} \phi(x, \widetilde{y}) + \eta = 0, \\ &-\frac{\partial}{\partial \operatorname{Re} \widetilde{y}} \operatorname{Im} \phi(x, \widetilde{y}) + (\nabla_y G)(\operatorname{Re} \widetilde{y}, \eta) = 0. \end{aligned}$$

If $f(z)$ is a holomorphic function, then

$$\frac{\partial}{\partial \operatorname{Im} z} \operatorname{Im} f = \operatorname{Re} \frac{\partial f}{\partial z}, \quad \frac{\partial}{\partial \operatorname{Re} z} \operatorname{Im} f = \operatorname{Im} \frac{\partial f}{\partial z}, \tag{2.3}$$

so the equations for our critical point become

$$\begin{aligned} &\operatorname{Im} \widetilde{y} = \nabla_\eta G(\operatorname{Re} \widetilde{y}, \eta), \\ &\eta = -\operatorname{Re} \frac{\partial \phi}{\partial \widetilde{y}}(x, \widetilde{y}), \\ &-\nabla_y G(\operatorname{Re} \widetilde{y}, \eta) = -\operatorname{Im} \frac{\partial \phi}{\partial \widetilde{y}} \end{aligned}$$

or, equivalently,

$$\left(\widetilde{y}, -\frac{\partial \phi}{\partial \widetilde{y}}(x, \widetilde{y})\right) = (\operatorname{Re} \widetilde{y}, \eta) + i H_G(\operatorname{Re} \widetilde{y}, \eta),$$

which says that the critical point $(\widetilde{y}, \eta)$ is determined by the condition that $\varkappa_T$ maps the point $(\widetilde{y}, \widetilde{\eta}) \in \Lambda_G$ to a point (x, ξ), situated over x. It is clear that the critical point is nondegenerate. We check it when $G = 0$. The Hessian matrix with respect to the variables $\operatorname{Re} y$, $\operatorname{Im} y$, η becomes

$$\begin{pmatrix} -\operatorname{Im}\phi''_{y,y} & B & 0 \\ {}^tB & C & -1 \\ 0 & -1 & 0 \end{pmatrix}$$

which is nondegenerate independently of B and C.

If $\Phi(x)$ denotes the critical value in (2.2), it remains to check that $\frac{2}{i}\frac{\partial \Phi}{\partial x} = \xi$ where $\xi = \frac{\partial \phi}{\partial x}(x, \widetilde{y})$, $(\widetilde{y}, \eta)$ denoting the critical point. However, since Φ is a critical value, we get

$$\frac{2}{i}\frac{\partial \Phi}{\partial x} = \frac{2}{i}\frac{\partial}{\partial x}(-\operatorname{Im}\phi(x, \widetilde{y})) = \frac{\partial \phi}{\partial x}(x, \widetilde{y}). \qquad \square$$

Note that, when $G = 0$, the formula (2.2) produces the same function as (2.1).

Write $\widetilde{y} = y + i\theta$ and consider the function

$$f(x; y, \eta; \theta) = -\operatorname{Im}\phi(x, y + i\theta) - \eta \cdot \theta, \tag{2.4}$$

which appears in (2.2).

Proposition 2.2. *f is a nondegenerate phase function with θ as fiber variables which generates a canonical transformation which can be identified with $\varkappa_T$.*

Proof. Since

$$\frac{\partial}{\partial \theta} f = -\operatorname{Re}\frac{\partial \phi}{\partial \widetilde{y}}(x, y + i\theta) - \eta,$$

f is nondegenerate. The canonical relation has the graph

$$\begin{aligned} &\{(x, \frac{\partial \phi}{\partial x}; y, \eta, \frac{\partial}{\partial y}\operatorname{Im}\phi(x, y + i\theta), \theta);\ \eta = -\operatorname{Re}\frac{\partial \phi}{\partial \widetilde{y}}(x, y + i\theta)\} \\ &= \{(x, \frac{\partial \phi}{\partial x}(x, y + i\theta); y, -\operatorname{Re}\frac{\partial \phi}{\partial \widetilde{y}}(x, y + i\theta), \operatorname{Im}\frac{\partial \phi}{\partial \widetilde{y}}(x, y + i\theta), \theta)\}, \end{aligned}$$

and up to reshuffling of the components on the preimage side and changes of sign, we recognize the graph of $\varkappa_T$. $\square$

Now, we have the following easily verified fact.

Proposition 2.3. *Let $f(x, y, \theta) \in C^\infty(\operatorname{neigh}(x_0, y_0, \theta_0), \mathbf{R}^n \times \mathbf{R}^n \times \mathbf{R}^N)$ be a nondegenerate phase function with $(x_0, y_0, \theta_0) \in C_\phi$ generating a canonical transformation which maps $(y_0, \eta_0) = (y_0, -\nabla_y f(x_0, y_0, \theta_0))$ to $(x_0, \nabla_x f(x_0, y_0, \theta_0))$. If $g(y)$ is smooth near y_0 with $\nabla g(y_0) = \eta_0$ and*

$$h(x) = \text{v.c.}_{y,\theta} f(x, y, \theta) + g(y)$$

is well defined with a nondegenerate critical point close to (y_0, θ_0) for x close to x_0, then we have the inversion formula

$$g(y) = \text{v.c.}_{x,\theta} - f(x, y, \theta) + h(x)$$

for $y \in \operatorname{neigh}(y_0)$, *where the critical point is nondegenerate and close to* (x_0, θ_0).

Combining the three propositions, we get

Proposition 2.4.

$$G(y,\eta) = \text{v.c.}_{x,\theta} \operatorname{Im} \phi(x, y+i\theta) + \eta \cdot \theta + \Phi_G(x). \tag{2.5}$$

If $(\widetilde{\Phi}, \widetilde{G})$ is a second pair of functions close to Φ_0, 0 and related through (2.2) and (2.5), then

$$G \leqslant \widetilde{G} \text{ if and only if } \Phi \leqslant \widetilde{\Phi}. \tag{2.6}$$

Indeed, if, for instance, $\Phi \leqslant \widetilde{\Phi}$, we introduce $\Phi_t = t\widetilde{\Phi} + (1-t)\Phi$ so that $\partial_t \Phi_t \geqslant 0$. If G_t is the corresponding critical value as in (2.5), then $\partial_t G_t = (\partial_t \Phi_t)(x_t) \geqslant 0$, where (x_t, θ_t) is the critical point.

3 Evolution Equations on the Transform Side

Let $\widetilde{P}(x,\xi;h)$ be a smooth symbol defined in $\operatorname{neigh}((x_0,\xi_0); \Lambda_{\Phi_0})$, with an asymptotic expansion

$$\widetilde{P}(x,\xi;h) \sim \widetilde{p}(x,\xi) + h\widetilde{p}_1(x,\xi) + \dots \quad \text{in} \quad C^\infty(\operatorname{neigh}((x_0,\xi_0), \Lambda_{\Phi_0})).$$

The same letter denotes an almost holomorphic extension to a complex neighborhood of (x_0, ξ_0):

$$\widetilde{P}(x,\xi;h) \sim \widetilde{p}(x,\xi) + h\widetilde{p}_1(x,\xi) + \dots \quad \text{in} \quad C^\infty(\operatorname{neigh}((x_0,\xi_0), \mathbf{C}^{2n}),$$

where $\widetilde{p}$, $\widetilde{p}_j$ are smooth extensions such that

$$\overline{\partial}\widetilde{p},\ \overline{\partial}\widetilde{p}_j = \mathcal{O}(\operatorname{dist}((x,\xi), \Lambda_{\Phi_0})^\infty).$$

Then, as developed in [11] and later in [19], if $u = u_h$ is holomorphic in a neighborhood V of x_0 and belongs to $H_{\Phi_0}(V)$ in the sense that $\|u\|_{L^2(V, e^{-2\Phi_0/h}L(dx))}$ is finite and of temperate growth in $1/h$ as h tends to zero, then we can define $\widetilde{P}u = \widetilde{P}(x, hD_x; h)u$ in any smaller neighborhood $W \Subset V$ by the formula

$$\widetilde{P}u(x) = \frac{1}{(2\pi h)^n} \iint_{\Gamma(x)} e^{\frac{i}{h}(x-y)\cdot\theta} \widetilde{P}\Big(\frac{x+y}{2}, \theta; h\Big) u(y) dy d\theta, \tag{3.1}$$

where $\Gamma(x)$ is a good contour (in the sense of [21]) of the form

$$\theta = \frac{2}{i}\frac{\partial \Phi_0}{\partial x}\Big(\frac{x+y}{2}\Big) + \frac{i}{C_1}\overline{(x-y)}, \quad |x-y| \leqslant 1/C_2, \quad C_1, C_2 > 0.$$

Then $\overline{\partial}\widetilde{P}$ is negligible $H_{\Phi_0}(V) \to L^2_{\Phi_0}(W)$, i.e., of norm $\mathcal{O}(h^\infty)$ and modulo such negligible operators, $\widetilde{P}$ is independent of the choice of a good contour. Solving a $\overline{\partial}$-problem (assuming, as we may, that our neighborhoods are pseudoconvex), we can always correct $\widetilde{P}$ with a negligible operator such that (after an arbitrarily small decrease of W)

$$\widetilde{P} = \mathcal{O}(1) : H_{\Phi_0}(V) \to H_{\Phi_0}(W).$$

Also, if $\Phi = \Phi_0 + \mathcal{O}(h \ln \frac{1}{h})$ in C^2, then it is clear that

$$\widetilde{P} = \mathcal{O}(h^{-N_0}) : H_\Phi(V) \to H_\Phi(W)$$

for some N_0. Using the Stokes formula, we can show that $\widetilde{P}$ will change only by a negligible term if we replace Φ_0 by Φ in the definition of $\Gamma(x)$. Then it follows that $\widetilde{P} = \mathcal{O}(1) : H_\Phi(V) \to H_\Phi(W)$.

Before discussing evolution equations, let us recall [19] that the identity operator $H_{\Phi_0}(V) \to H_{\Phi_0}(W)$ has up to a negligible operator the form

$$Iu(x) = h^{-n} \iint e^{\frac{2}{h}\Psi_0(x,\overline{y})} a(x,\overline{y};h) u(y) e^{-\frac{2}{h}\Phi_0(y)} dy d\overline{y}, \tag{3.2}$$

where $\Psi_0(x,y)$ and $a(x,y;h)$ are almost holomorphic on the antidiagonal $y = \overline{x}$ with $\Psi_0(x,\overline{x}) = \Phi_0(x)$, $a(x,y;h) \sim a_0(x,y) + ha_1(x,y) + \ldots$, and $a_0(x,\overline{x}) \neq 0$. More generally, a pseudodifferential operator like $\widetilde{P}$ takes the form

$$\widetilde{P}u(x) = h^{-n} \iint e^{\frac{2}{h}\Psi_0(x,\overline{y})} q(x,\overline{y};h) u(y) e^{-\frac{2}{h}\Phi_0(y)} dy d\overline{y} \tag{3.3}$$

$$q_0(x,\overline{x}) = \widetilde{p}\Big(x, \frac{2}{i}\frac{\partial \Phi_0}{\partial x}(x)\Big) a_0(x,\overline{x}),$$

where q_0 denotes the first term in the asymptotic expansion of the symbol q. In this discussion, Φ_0 can be replaced by any other smooth exponent Φ which is $\mathcal{O}(h^\delta)$ close to Φ_0 in C^∞, and we make the corresponding replacement of Ψ_0. Also recall that because of the strict plurisubharmonicity of Φ we have

$$2\,\mathrm{Re}\,\Psi(x,\overline{y}) - \Phi(x) - \Phi(y) \asymp -|x-y|^2, \tag{3.4}$$

so the uniform boundedness $H_\Phi \to H_\Phi$ follows from the domination of the modulus of the effective kernel by a Gaussian convolution kernel.

Next, consider the evolution problem

$$(h\partial_t + \widetilde{P})\widetilde{U}(t) = 0, \quad \widetilde{U}(0) = 1, \tag{3.5}$$

where t is restricted to the interval $[0, h^\delta]$ for some arbitrarily small, but fixed $\delta > 0$. We review how to solve this problem approximately by a geometrical optics construction. Look for $\widetilde{U}(t)$ of the form

$$\widetilde{U}(t)u(x) = h^{-n} \iint e^{\frac{2}{h}\Psi_t(x,\overline{y})} a_t(x,\overline{y};h)u(y)e^{-2\Phi_0(y)/h}dyd\overline{y}, \tag{3.6}$$

where Ψ_t and a_t depend smoothly on all the variables and $\Psi_{t=0} = \Psi_0$, $a_{t=0} = a_0$ in (3.3), so that $\widetilde{U}(0) = 1$ up to a negligible operator.

Note that, formally, $\widetilde{U}(t)$ is the Fourier integral operator

$$\widetilde{U}(t)u(x) = h^{-n} \iint e^{\frac{2}{h}(\Psi_t(x,\theta)-\Psi_0(y,\theta))} a_t(x,\theta;h)u(y)dyd\theta, \tag{3.7}$$

where we choose the integration contour $\theta = \overline{y}$. Writing $2\Psi_t(x,\theta) = i\phi_t(x,\theta)$ leads to more customary notation and we impose the eikonal equation

$$i\partial_t\phi + \widetilde{p}(x, \phi'_x(x,\theta)) = 0. \tag{3.8}$$

Of course, we are manipulating C^∞ functions in the complex domain, so we cannot hope to solve the eikonal equation exactly, but we can do so to infinite order at $t = 0$, $x = \overline{y} = \theta$. If we put

$$\Lambda_{\phi_t(\cdot,\theta)} = \{(x, \phi'_x(t,x,\theta))\}, \tag{3.9}$$

we have to ∞ order at $t = 0$, $\theta = x$:

$$\Lambda_{\phi_t(\cdot,\theta)} = \exp(t\widehat{H}_{\frac{1}{i}\widetilde{p}})(\Lambda_{\phi_0(\cdot,\theta)}). \tag{3.10}$$

with $\widehat{H}_{\widetilde{p}} = H_{\widetilde{p}} + \overline{H_{\widetilde{p}}}$ denoting the real vector field associated to the (1,0)-field H_p, and similarly for $\widehat{H}_{\frac{1}{i}\widetilde{p}}$. (We sometimes neglect the hat when integrating the Hamilton flows.) At a point where $\overline{\partial}\widetilde{p} = 0$, we have

$$\widehat{H}_{\widetilde{p}} = H^{\operatorname{Re}\sigma}_{\operatorname{Re}\widetilde{p}} = H^{\operatorname{Im}\sigma}_{\operatorname{Im}\widetilde{p}}, \quad \widehat{H}_{i\widetilde{p}} = -H^{\operatorname{Re}\sigma}_{\operatorname{Im}\widetilde{p}} = H^{\operatorname{Im}\sigma}_{\operatorname{Re}\widetilde{p}}, \tag{3.11}$$

where the other fields are the Hamilton fields of $\operatorname{Re}\widetilde{p}$ and $\operatorname{Im}\widetilde{p}$ with respect to the real symplectic forms $\operatorname{Re}\sigma$ and $\operatorname{Im}\sigma$ respectively (cf. [21, 19]). Thus, (3.10) can be written as

$$\Lambda_{\phi_t(\cdot,\theta)} = \exp(tH^{-\operatorname{Im}\sigma}_{\operatorname{Re}\widetilde{p}})(\Lambda_{\phi_0(\cdot,\theta)}). \tag{3.12}$$

A complex Lagrangian manifold is also an I-Lagrangian manifold (i.e., a Lagrangian manifold for $\operatorname{Im}\sigma$), so (3.12) can be viewed as a relation between I-Lagrangian manifolds and it defines the I-Lagrangian manifold $\Lambda_{\phi_t(\cdot,\theta)}$ in an unambiguous way, once we have fixed an almost holomorphic extension of $\widetilde{p}$ and especially the real part of that function. The general form of a smooth I-Lagrangian manifold Λ, for which the x-space projection $\Lambda \ni (x,\xi) \mapsto x \in \mathbf{C}^n$ is a local diffeomorphism, is locally $\Lambda = \Lambda_\Phi$ where Φ is real and smooth, and we define

$$\Lambda_\Phi = \left\{\left(x, \frac{2}{i}\frac{\partial\Phi}{\partial x}\right); x \in \Omega\right\}, \quad \Omega \subset \mathbf{C}^n \text{ open.}$$

With a slight abuse of notation, we can therefore identify the **C**-Lagrangian manifold Λ_{ϕ_0} with the I-Lagrangian manifold $\Lambda_{-\operatorname{Im}\phi_0}$ since for holomorphic functions (or, more generally, where $\overline{\partial}_x\phi_0 = 0$), we have

$$\frac{\partial\phi_0}{\partial x} = \frac{2}{i}\frac{\partial - \operatorname{Im}\phi_0}{\partial x}.$$

Formula (3.4) shows that

$$\Phi_0(x) + \Phi_0(\overline{\theta}) - (-\operatorname{Im}\phi_0(\cdot,\overline{\theta})) \asymp |x-\overline{\theta}|^2.$$

Thus, if we define

$$\Lambda_{\Phi_t} = \exp(tH_{\operatorname{Re}\widetilde{p}}^{-\operatorname{Im}\sigma})(\Lambda_{\Phi_0}) \tag{3.13}$$

and fix the t-dependent constant in this definition of Φ_t by imposing the real Hamilton–Jacobi equation,

$$\partial_t\Phi_t + \operatorname{Re}\widetilde{p}\Big(x, \frac{2}{i}\frac{\partial\Phi_t}{\partial x}\Big) = 0, \quad \Phi_{t=0} = \Phi_0, \tag{3.14}$$

and noticing that the real part of (3.8) is a similar equation for $-\operatorname{Im}\phi_t$,

$$\partial_t(-\operatorname{Im}\phi) + \operatorname{Re}\widetilde{p}\Big(x, \frac{2}{i}\frac{\partial}{\partial x}(-\operatorname{Im}\phi)\Big) = 0, \tag{3.15}$$

we get

$$\Phi_t(x) + \Phi_0(\overline{\theta}) - (-\operatorname{Im}\phi_t(x,\theta)) \asymp |x - x_t(\overline{\theta})|^2, \tag{3.16}$$

where

$$(x_t(\overline{\theta}), \xi_t(\overline{\theta})) := \exp(tH_{\operatorname{Re}\widetilde{p}}^{-\operatorname{Im}\sigma})\Big(\overline{\theta}, \frac{2}{i}\frac{\partial\Phi_0}{\partial x}(\overline{\theta})\Big).$$

Determining a_t by solving a sequence of transport equations, we arrive at the following result.

Proposition 3.1. *The operator $\widetilde{U}(t)$ constructed above is $\mathcal{O}(1) : H_{\Phi_0}(V) \to H_{\Phi_t}(W)$, ($W \Subset V$ being small pseudoconvex neighborhoods of a fixed point x_0) uniformly for $0 \leqslant t \leqslant h^\delta$, and it solves the problem* (3.5) *up to negligible terms. This local statement makes sense since, by (3.16), we have*

$$2\operatorname{Re}\Psi_t(x,\overline{y}) - \Phi_t(x) - \Phi_0(y) \asymp -|x - x_t(y)|^2. \tag{3.17}$$

Using standard arguments, we also obtain up to negligible errors

$$h\partial_t\widetilde{U}(t) + \widetilde{U}(t)\widetilde{P} = 0, \quad 0 \leqslant t \leqslant h^\delta. \tag{3.18}$$

Let us quickly outline an alternative approach leading to the same weights Φ_t (cf [19]).

Consider formally

$$(e^{-t\widetilde{P}/h}u|e^{-t\widetilde{P}/h}u)_{H_{\Phi_t}} = (u_t|u_t)_{H_{\Phi_t}}, \quad u \in H_{\Phi_0},$$

and try to choose Φ_t so that the time derivative of this expression vanishes to leading order. We get

$$0 \approx h\partial_t \int u_t \overline{u}_t e^{-2\Phi_t/h} L(dx)$$
$$= -\Big((\widetilde{P}u_t|u_t)_{H_{\Phi_t}} + (u_t|\widetilde{P}u_t)_{H_{\Phi_t}} + \int 2\frac{\partial \Phi_t}{\partial t}(x)|u|^2 e^{-2\Phi_t/h} L(dx) \Big).$$

Here,

$$(\widetilde{P}u_t|u_t)_{H_{\Phi_t}} = \int (\widetilde{p}_{|\Lambda_{\Phi_t}} + \mathcal{O}(h))|u_t|^2 e^{-2\Phi_t/h} L(dx),$$

and similarly for $(u_t|\widetilde{P}u_t)_{H_{\Phi_t}}$, so we would like to have

$$0 \approx \int \Big(2\frac{\partial \Phi_t}{\partial t} + 2\,\mathrm{Re}\,\widetilde{p}_{|\Lambda_{\Phi_t}} + \mathcal{O}(h)\Big)|u_t|^2 e^{-2\Phi_t/h} L(dx).$$

We choose Φ_t to be the solution of (3.14). Then the preceding discussion again shows that $e^{-t\widetilde{P}/h} = \mathcal{O}(1) : H_{\Phi_0} \to H_{\Phi_t}$.

Since $\mathrm{Re}\,\widetilde{p}$ is constant along the integral curves of $H_{\mathrm{Re}\,\widetilde{p}}^{-\mathrm{Im}\,\sigma}$, from (3.14) we see that the second term in (3.14) is $\geqslant 0$, so

$$\Phi_t \leqslant \Phi_0, \quad t \geqslant 0, \tag{3.19}$$

when

$$\mathrm{Re}\,\widetilde{p}_{|\Lambda_{\Phi_0}} \geqslant 0. \tag{3.20}$$

Recall that we limit our discussion to the interval $0 \leqslant t \leqslant h^\delta$.

The author found it simpler to get a detailed understanding by working with the corresponding functions G_t in the following way.

Let p be defined by $p = \widetilde{p} \circ \varkappa_T$. Define G_t up to a t-dependent constant by

$$\Lambda_{\Phi_t} = \varkappa_T(\Lambda_{G_t}).$$

Then we also have $\Lambda_{G_t} = \exp tH_p(\Lambda_0)$, where $\Lambda_0 = \mathbf{R}^{2n}$. In order to fix the t-dependent constant, we use one of the equivalent formulas (cf (2.2) and (2.5)):

$$\Phi_t(x) = \mathrm{v.c.}_{\widetilde{y},\eta}(-\mathrm{Im}\,\phi(x,\widetilde{y}) - \eta \cdot \mathrm{Im}\,\widetilde{y} + G_t(\mathrm{Re}\,\widetilde{y},\eta)), \tag{3.21}$$

$$G_t(y,\eta) = \mathrm{v.c.}_{x,\theta}(\mathrm{Im}\,\phi(x,y+i\theta) + \eta\cdot\theta + \Phi_t(x)). \tag{3.22}$$

If $(x(t,y,\eta),\theta(t,y,\eta))$ is the critical point in the last formula, we get

$$\frac{\partial G_t}{\partial t}(y,\eta) = \frac{\partial \Phi_t}{\partial t}(x(t,y,\eta)) = -\mathrm{Re}\,\widetilde{p}\Big(x, \frac{2}{i}\frac{\partial \Phi_t}{\partial x}\Big)_{|x=x(t,y,\eta)}. \tag{3.23}$$

As we have seen, the critical points in (3.21) and (3.22) are directly related to $\varkappa_T$, so (3.23) leads to

$$\frac{\partial G_t}{\partial t}(y,\eta) + \operatorname{Re} p((y,\eta) + iH_{G_t}(y,\eta)) = 0. \tag{3.24}$$

Note that $G_t \leqslant 0$ by (2.6) and (3.19).

Since we consider (3.24) only when G_t and its gradient are small, we can Taylor expand (3.24) and get

$$\frac{\partial G_t}{\partial t}(y,\eta) + \operatorname{Re} p(y,\eta) + \operatorname{Re}(iH_{G_t}p(y,\eta)) + \mathcal{O}((\nabla G_t)^2) = 0, \tag{3.25}$$

which simplifies to

$$\frac{\partial G_t}{\partial t}(y,\eta) + H_{\operatorname{Im} p}G_t + \mathcal{O}((\nabla G_t)^2) = -\operatorname{Re} p(y,\eta). \tag{3.26}$$

Now, $G_t \leqslant 0$, so $(\nabla G_t)^2 = \mathcal{O}(G_t)$ and we obtain

$$\Big(\frac{\partial}{\partial t} + H_{\operatorname{Im} p}\Big)G_t + \mathcal{O}(G_t) = -\operatorname{Re} p, \quad G_0 = 0. \tag{3.27}$$

Viewing this as a differential inequality along the integral curves of $H_{\operatorname{Im} p}$, we obtain

$$-G_t(\exp(tH_{\operatorname{Im} p})(\rho)) \asymp \int_0^t \operatorname{Re} p(\exp sH_{\operatorname{Im} p}(\rho))ds \tag{3.28}$$

for all $\rho = (y,\eta) \in \operatorname{neigh}(\rho_0, \mathbf{R}^{2n})$, $\rho_0 = (y_0,\eta_0)$.

Now, introduce the following assumption corresponding to case (B) in Theorem 1.2:

$$H^j_{\operatorname{Im} p}(\operatorname{Re} p)(\rho_0)\begin{cases} = 0, & j \leqslant k-1, \\ > 0, & j = k, \end{cases} \tag{3.29}$$

where k necessarily is even (since $\operatorname{Re} p \geqslant 0$). We will work in a sufficiently small neighborhood of ρ_0. Put

$$J(t,\rho) = \int_0^t \operatorname{Re} p(\exp sH_{\operatorname{Im} p}(\rho))ds, \tag{3.30}$$

so that

$$0 \leqslant J(t,\rho) \in C^\infty(\operatorname{neigh}(0,\rho_0), [0,+\infty[\times \mathbf{R}^{2n})$$

and

$$\partial_t^{j+1} J(0,\rho_0) = H^j_{\operatorname{Im} p}(\operatorname{Re} p)(\rho_0)\begin{cases} = 0, & j \leqslant k-1, \\ > 0, & j = k. \end{cases} \tag{3.31}$$

Proposition 3.2. *Under the above assumptions, there is a constant $C > 0$ such that*

$$J(t,\rho) \geqslant \frac{t^{k+1}}{C}, \quad (t,\rho) \in \operatorname{neigh}((0,\rho_0),]0,+\infty[\times \mathbf{R}^{2n}). \tag{3.32}$$

Proof. Assume that (3.32) does not hold. Then there is a sequence $(t_\nu, \rho_\nu) \in [0, +\infty[\times\mathbf{R}^{2n}$ converging to $(0, \rho_0)$ such that

$$\frac{J(t_\nu, \rho_\nu)}{t_\nu^{k+1}} \to 0.$$

Since $J(t, \rho)$ is an increasing function of t, we get

$$\sup_{0\leqslant t\leqslant t_\nu} \frac{J(t, \rho_\nu)}{t_\nu^{k+1}} \to 0.$$

Introduce the Taylor expansion

$$J(t, \rho_\nu) = a_\nu^{(0)} + a_\nu^{(1)}t + \ldots + a_\nu^{(k+1)}t^{k+1} + \mathcal{O}(t^{k+2})$$

and define

$$u_\nu(s) = \frac{J(t_\nu s, \rho_\nu)}{t_\nu^{k+1}}, \quad 0 \leqslant s \leqslant 1.$$

Then, on one hand,

$$\sup_{0\leqslant s\leqslant 1} u_\nu(s) \to 0, \quad \nu \to \infty,$$

and, on the other hand,

$$u_\nu(s) = \underbrace{\frac{a_\nu^{(0)}}{t_\nu^{k+1}} + \frac{a_\nu^{(1)}}{t_\nu^k}s + \ldots + a_\nu^{(k+1)}s^{k+1}}_{=:p_\nu(s)} + \mathcal{O}(t_\nu s^{k+2}),$$

so

$$\sup_{0\leqslant s\leqslant 1} p_\nu(s) \to 0, \quad \nu \to \infty.$$

The corresponding coefficients of p_ν have to tend to 0, and, in particular,

$$a_\nu^{(k+1)} = \frac{1}{(k+1)!}(\partial_t^{k+1} J(0, \rho_\nu) \to 0$$

which contradicts (3.31). □

Combining (3.28) and Proposition 3.2, we get the following assertion.

Proposition 3.3. *Under the assumption* (3.29), *there exists* $C > 0$ *such that*

$$G_t(\rho) \leqslant -\frac{t^{k+1}}{C}, \quad (t, \rho) \in \operatorname{neigh}((0, \rho_0), [0, \infty[\times\mathbf{R}^{2n}). \tag{3.33}$$

We can now return to the evolution equation for $\widetilde{P}$ and the t-dependent weight Φ_t in (3.14). From (3.33), (3.21), we get

Proposition 3.4. *Under the assumption (3.29), we have*

$$\Phi_t(x) \leqslant \Phi_0(x) - \frac{t^{k+1}}{C}, \quad (t,x) \in \operatorname{neigh}((0,x_0), [0,\infty[\times \mathbf{R}^{2n}). \tag{3.34}$$

4 The Resolvent Estimates

Let P be an h-pseudodifferential operator satisfying the general assumptions of the introduction.

Let $z_0 \in (\partial \Sigma(p)) \setminus \Sigma_\infty(p)$. We first treat the case of Theorem 1.2 so that,

$$z_0 \in i\mathbf{R}, \tag{4.1}$$

$$\operatorname{Re} p(\rho) \geqslant 0 \;\text{ in }\; \operatorname{neigh}(p^{-1}(z_0), T^*X), \tag{4.2}$$

$$\forall \rho \in p^{-1}(z_0),\ \exists j \leqslant k \;\text{ such that }\; H^j_{\operatorname{Im} p} \operatorname{Re} p(\rho) > 0. \tag{4.3}$$

Proposition 4.1. *There exists $C_0 > 0$ such that for all $C_1 > 0$ there is $C_2 > 0$ such that we have for z, h as in (1.14), $h < 1/C_2$, $u \in C_0^\infty(X)$*

$$|\operatorname{Re} z| \|u\| \leqslant C_0 \|(z-P)u\| \;\textit{when}\; \operatorname{Re} z \leqslant -h^{\frac{k}{k+1}}, \tag{4.4}$$
$$h^{\frac{k}{k+1}} \|u\| \leqslant C_0 \exp\left(\frac{C_0}{h} (\operatorname{Re} z)_+^{\frac{k+1}{k}}\right) \|(z-P)u\| \;\textit{when}\; \operatorname{Re} z \geqslant -h^{\frac{k}{k+1}}.$$

Proof. The required estimate is easy to obtain microlocally in the region where $P - z_0$ is elliptic, so we see that it suffices to show the following statement.

For every $\rho_0 \in p^{-1}(z_0)$ there exists $\chi \in C_0^\infty(T^*X)$ equal to 1 near ρ_0 and such that for z and h as in (1.14) and letting χ also denote the corresponding h-pseudodifferential operator, we have

$$|\operatorname{Re} z| \|\chi u\| \leqslant C_0 \|(z-P)u\| + C_N h^N \|u\| \;\textit{when}\; \operatorname{Re} z \leqslant -h^{\frac{k}{k+1}}, \tag{4.5}$$
$$h^{\frac{k}{k+1}} \|\chi u\| \leqslant C_0 \exp\left(\frac{C_0}{h} (\operatorname{Re} z)_+^{\frac{k+1}{k}}\right) \|(z-P)u\| + C_N h^N \|u\|, \tag{4.6}$$
$$\textit{when} \quad \operatorname{Re} z \geqslant -h^{\frac{k}{k+1}},$$

where $N \in \mathbf{N}$ can be chosen arbitrarily.

When $\operatorname{Re} z \leqslant -h^{k/(k+1)}$, this is an easy consequence of the semiclassical sharp Gårding inequality (cf., for instance, [8]), so from now on we assume that $\operatorname{Re} z \geqslant -h^{k/(k+1)}$.

If T is an FBI transform and $\widetilde{P}$ denotes the conjugate operator TPT^{-1}, it suffices to show that

$$\|u\|_{H_{\Phi_0}(V_1)} \leqslant h^{-\frac{k}{k+1}} C_0 \exp\left(\frac{C_0}{h} (\operatorname{Re} z)_+^{\frac{k+1}{k}}\right) \|(\widetilde{P} - z)u\|_{H_{\Phi_0}(V_2)}$$
$$+ \mathcal{O}(h^\infty) \|u\|_{H_{\Phi_0}(V_3)}, \tag{4.7}$$

$u \in H_{\Phi_0}(V_3)$, where $V_1 \Subset V_2 \Subset V_3$ are neighborhoods of x_0 given by $(x_0, \xi_0) = \varkappa_T(\rho_0) \in \Lambda_{\Phi_0}$.

From Proposition 3.4 and the fact that $\widetilde{U}(t) : H_{\Phi_0}(V_2) \to H_{\Phi_t}(V_1)$, we see that

$$\|\widetilde{U}(t)u\|_{H_{\Phi_0}(V_1)} \leqslant Ce^{-t^{k+1}/C}\|u\|_{H_{\Phi_0}(V_2)}. \tag{4.8}$$

Choose $\delta > 0$ small enough so that $\delta(k+1) < 1$ and put

$$\widetilde{R}(z) = \frac{1}{h}\int_0^{h^\delta} e^{\frac{tz}{h}}\widetilde{U}(t)dt. \tag{4.9}$$

We shall verify that $\widetilde{R}$ is an approximate left inverse to $\widetilde{P} - z$, but first we study the norm of this operator in H_{Φ_0}, starting with the estimate in $\mathcal{L}(H_{H_{\Phi_0}(V_2)}, H_{H_{\Phi_0}(V_1)})$:

$$\|e^{\frac{tz}{h}}\widetilde{U}(t)\| \leqslant C\exp\frac{1}{h}\Big(t\operatorname{Re} z - \frac{t^{k+1}}{C}\Big) \tag{4.10}$$

and note that the right-hand side is $\mathcal{O}(h^\infty)$ for $t = h^\delta$ since $\delta(k+1) < 1$ and $h^{-1}\operatorname{Re} z \leqslant \mathcal{O}(1)\ln\frac{1}{h}$.

We get

$$\|\widetilde{R}(z)\| \leqslant \frac{C}{h}\int_0^{+\infty}\exp\frac{1}{h}\Big(t\operatorname{Re} z - \frac{t^{k+1}}{C}\Big)dt = \frac{C^{\frac{k+2}{k+1}}}{h^{\frac{k}{k+1}}}I\Big(\frac{C^{\frac{1}{k+1}}}{h^{\frac{k}{k+1}}}\operatorname{Re} z\Big), \tag{4.11}$$

where

$$I(s) = \int_0^\infty e^{st - t^{k+1}}dt. \tag{4.12}$$

Lemma 4.2. *We have*

$$I(s) = \mathcal{O}(1) \text{ when } |s| \leqslant 1, \tag{4.13}$$

$$I(s) = \frac{\mathcal{O}(1)}{|s|} \text{ when } s \leqslant -1, \tag{4.14}$$

$$I(s) \leqslant \mathcal{O}(1)s^{-\frac{k-1}{2k}}\exp\left(\frac{k}{(k+1)^{\frac{k+1}{k}}}s^{\frac{k+1}{k}}\right) \text{ when } s \geqslant 1. \tag{4.15}$$

Proof. The first two estimates are straight forward, and we concentrate on the last one, where we may also assume that $s \gg 1$. A computation shows that the exponent $f_s(t) = st - t^{k+1}$ on $[0, +\infty[$ has a unique critical point $t = t(s) = (s/(k+1))^{1/k}$ which is a nondegenerate maximum,

$$f_s''(t(s)) = -k(k+1)^{\frac{1}{k}}s^{\frac{k-1}{k}},$$

with critical value

$$f_s(t(s)) = \frac{k}{(k+1)^{\frac{k+1}{k}}} s^{\frac{k+1}{k}}.$$

It follows that the upper bound in (4.15) is the one we would get by applying the formal stationary phase formula.

Now,

$$f_s''(t) = -(k+1)kt^{k-1} \lesssim f_s''(t(s)) \text{ for } \frac{t(s)}{2} \leqslant t < +\infty,$$

so

$$\int_{t(s)/2}^{\infty} e^{st-t^{k+1}} dt$$

satisfies the required upper bound.

On the other hand, we have

$$f_s(t(s)) - f_s(t) \geqslant \frac{s^{\frac{k+1}{k}}}{C} \text{ for } 0 \leqslant t \leqslant \frac{t(s)}{2}, \quad s \gg 1,$$

so

$$\int_0^{\frac{t(s)}{2}} e^{st-t^{k+1}} dt \leqslant \mathcal{O}(1) s^{\frac{1}{k}} \exp\Big(f_s(t(s)) - \frac{s^{\frac{k+1}{k}}}{C}\Big),$$

and (4.15) follows. □

Applying this to (4.11), we get the following assertion.

Proposition 4.3. *We have*

$$\|\widetilde{R}(z)\| \leqslant \frac{C}{h^{\frac{k}{k+1}}}, \quad |\operatorname{Re} z| \leqslant \mathcal{O}(1) h^{\frac{k}{k+1}}, \tag{4.16}$$

$$\|\widetilde{R}(z)\| \leqslant \frac{C}{|\operatorname{Re} z|}, \quad -1 \ll \operatorname{Re} z \leqslant -h^{\frac{k}{k+1}}, \tag{4.17}$$

$$\|\widetilde{R}(z)\| \leqslant \frac{C}{h^{\frac{k}{k+1}}} \exp\Big(C_k \frac{(\operatorname{Re} z)^{\frac{k+1}{k}}}{h}\Big), \quad h^{\frac{k}{k+1}} \leqslant \operatorname{Re} z \ll 1. \tag{4.18}$$

From the beginning of the proof of Lemma 4.2, or more directly from (4.10), we see that

$$\|e^{\frac{tz}{h}} \widetilde{U}(t)\| \leqslant C \exp \frac{C_k}{h} (\operatorname{Re} z)_+^{\frac{k+1}{k}},$$

which is bounded by some negative power of h since we have imposed the restriction

$$\operatorname{Re} z \leqslant \mathcal{O}(1)(h \ln \frac{1}{h})^{\frac{k}{k+1}}.$$

Working locally, we then see that modulo a negligible operator

$$\widetilde{R}(z)(\widetilde{P}-z) \equiv \frac{1}{h}\int_0^{h^\delta} e^{\frac{tz}{h}}(-h\partial_t - z)\widetilde{U}(t)dt \equiv 1,$$

where the last equivalence follows from an integration by parts and the fact that the integrand is negligible for $t = h^\delta$. Combining this with Proposition 4.3, we get (4.7), and this completes the proof of Proposition 4.1. □

We can now finish the proof of Theorem 1.2.

Proof of Theorem 1.2. Using standard pseudodifferential machinery (cf., for instance, [8]), we first note that P has discrete spectrum in a neighborhood of z_0 and $P - z$ is a Fredholm operator of index 0 from $\mathcal{D}(P)$ to L^2 when z varies in a small neighborhood of z_0. On the other hand, Proposition 4.1 implies that $P - z$ is injective and hence bijective for $\operatorname{Re} z \leqslant \mathcal{O}(k^{k/(k+1)})$, and we also get the corresponding bounds on the resolvent. □

Proof of Theorem 1.1. For the sake of simplicity, we can assume that $z_0 = 0$ and consider a point $\rho_0 \in p^{-1}(0)$. After conjugation with a microlocally defined unitary Fourier integral operator, we may assume that $\rho_0 = (0,0)$ and $dp(\rho_0) = d\xi_n$. Then from the Malgrange preparation theorem we get near $\rho = (0,0)$, $z = 0$

$$p(\rho) - z = q(x,\xi,z)(\xi_n + r(x,\xi',z)), \quad \xi' = (\xi_1,\dots,\xi_{n-1}), \tag{4.19}$$

where q, r are smooth and $q(0,0,0) \neq 0$, and as in [7], we note that either $\operatorname{Im} r(x,\xi',0) \geqslant 0$ in a neighborhood of $(0,0)$ or $\operatorname{Im} r(x,\xi',0) \leqslant 0$ in a neighborhood of $(0,0)$. Indeed; otherwise, there would exist sequences ρ_j^+, ρ_j^- in $\mathbf{R}^n \times \mathbf{R}^{n-1}$ converging to $(0,0)$ such that $\pm \operatorname{Im} r(\rho_j^\pm) > 0$. It is then easy to construct a simple closed curve γ_j in a small neighborhood of ρ_0, passing through the points $(\rho_j^\pm, 0)$, such that the image of γ_j under the map $(x,\xi) \mapsto \xi_n + r(x,\xi',0)$ is a simple closed curve in $\mathbf{C}\setminus\{0\}$, with winding number $\neq 0$. Then the same holds for the image of γ_j under p, and we see that $\mathcal{R}(p)$ contains a full neighborhood of 0, in contradiction with the assumption that $0 = z_0 \in \partial\Sigma(p)$.

In order to fix the ideas, let us assume that $\operatorname{Im} r \leqslant 0$ near ρ_0 when $z = 0$, so that $\operatorname{Re}(i(\xi_n + r(x,\xi',0))) \geqslant 0$. From (4.19) we get the pseudodifferential factorization

$$P(x,hD_x;h) - z = \frac{1}{i}Q(x,hD_x,z;h)\widehat{P}(x,hD_x,z;h) \tag{4.20}$$

microlocally near ρ_0 when z is close to 0. Here, Q and $\widehat{P}$ have the leading symbols $q(x,\xi,z)$ and $i(\xi_n + r(x,\xi',z))$ respectively.

We can now obtain a microlocal a priori estimate for $\widehat{P}$ as before. Let us first check that the assumption in (B) of Theorem 1.1 amounts to the statement that for $z = z_0 = 0$

$$H^j_{\mathrm{Re}\,\widehat{p}} \operatorname{Im} \widehat{p}(\rho_0) > 0 \tag{4.21}$$

for some $j \in \{1, 2, \dots, k\}$. In fact, the assumption in Theorem 1.1 (B) is obviously invariant under multiplication of p by nonvanishing smooth factors, so we drop the hats and assume from the start that $p = \widehat{p}$ and $\operatorname{Im} p \geqslant 0$. Put $\rho(t) = \exp tH_p(\rho_0)$ and $r(t) = \exp tH_{\mathrm{Re}\, p}(\rho_0)$. Let $j \geqslant 0$ be the order of vanishing of $\operatorname{Im} p(r(t))$ at $t = 0$. From $\dot{\rho}(t) = H_p(\rho(t))$ and $\dot{r}(t) = H_{\mathrm{Re}\, p}(r(t))$ we get

$$\frac{d}{dt}(\rho - r) = iH_{\operatorname{Im} p}(r) + \mathcal{O}(\rho - r),$$

so

$$\rho(t) - r(t) = \int_0^t \mathcal{O}(\nabla \operatorname{Im} p(r(s)))ds.$$

Here, if $p_2 = \frac{1}{2i}(p - p^*)$ is the almost holomorphic extension of $\operatorname{Im} p$, we get

$$\begin{aligned} p^*(\rho(t)) &= ip_2(\rho(t)) \\ &= ip_2(r(t)) + i\nabla p_2(r(t)) \cdot (\rho(t) - r(t)) + \mathcal{O}((\rho(t) - r(t))^2) \\ &= ip_2(r(t)) + i\nabla p_2(r(t)) \cdot \int_0^t \mathcal{O}(\nabla p_2(r(s)))ds \\ &\quad + \mathcal{O}(1)\Big(\int_0^t \mathcal{O}(\nabla p_2(r(s)))ds\Big)^2. \end{aligned}$$

Here, $\nabla p_2(r(t)) = \mathcal{O}(p_2(r(t))^{1/2}) = \mathcal{O}(t^{j/2})$, so $p^*(\rho(t)) = ia(\rho_0)t^j + \mathcal{O}(t^{j+1})$.

Then, if we conjugate with a Bargmann–FBI transform as above, we can construct an approximation $\widetilde{U}(t)$ of $\exp(-t\widetilde{\widehat{P}}/h)$ such that

$$\|\widetilde{U}(t)\| \leqslant C_0 e^{(C_0 t|z - z_0| - t^{k+1}/C_0)/h}$$

when $|z - z_0| = \mathcal{O}((h \ln \frac{1}{h})^{k/(k+1)})$.

From this we obtain a microlocal a priori estimate for $\widehat{P}$ analogous to the one for $P - z$ in Proposition 4.1, and the proof can be completed in the same way as for Theorem 1.2. □

5 Examples

Consider

$$P = -h^2 \Delta + iV(x), \quad V \in C^\infty(X; \mathbf{R}), \tag{5.1}$$

where either X is a smooth compact manifold of dimension n or $X = \mathbf{R}^n$. In the second case, we assume that $p = \xi^2 + iV(x)$ belongs to a symbol space $S(m)$, where $m \geqslant 1$ is an order function. It is easy to give quite general sufficient conditions for this to happen, let us just mention that if $V \in C_b^\infty(\mathbf{R}^2)$

then we can take $m = 1 + \xi^2$ and if $\partial^\alpha V(x) = \mathcal{O}((1+|x|)^2)$ for all $\alpha \in \mathbf{N}^n$ and satisfies the ellipticity condition $|V(x)| \geqslant C^{-1}|x|^2$ for $|x| \geqslant C$, for some constant $C > 0$, then we can take $m = 1 + \xi^2 + x^2$.

We have $\Sigma(p) = [0, \infty[+i\overline{V(X)}$. When X is compact, $\Sigma_\infty(p)$ is empty and, when $X = \mathbf{R}^n$, we have $\Sigma_\infty(p) = [0, \infty[+i\Sigma_\infty(V)$, where $\Sigma_\infty(V)$ is the set of accumulation points at infinity of V.

Let $z_0 = x_0 + iy_0 \in \partial\Sigma(p) \setminus \Sigma_\infty(p)$.

- In the case $x_0 = 0$, we see that Theorem 1.2 (B) is applicable with $k = 2$ provided that y_0 is not a critical value of V.
- Now assume that $x_0 > 0$ and y_0 is either the maximum or the minimum of V. In both cases, assume that $V^{-1}(y_0)$ is finite and each element of that set is a nondegenerate maximum or minimum. Then Theorem 1.2 (B) is applicable to $\pm iP$. By allowing a more complicated behavior of V near its extreme points, we can produce examples, where 1.2 (B) applies with $k > 2$.

Now, consider the non-selfadjoint harmonic oscillator

$$Q = -\frac{d^2}{dy^2} + iy^2 \tag{5.2}$$

on the real line, studied by Boulton [2] and Davies [6]. Consider a large spectral parameter $E = i\lambda + \mu$, where $\lambda \gg 1$ and $|\mu| \ll \lambda$. The change of variables $y = \sqrt{\lambda}x$ permits us to identify Q with $Q = \lambda P$, where $P = -h^2\frac{d^2}{dx^2} + ix^2$ and $h = 1/\lambda \to 0$. Hence $Q - E = \lambda(P - (1 + i\frac{\mu}{\lambda}))$ and Theorem 1.2 (B) is applicable with $k = 2$. We conclude that $(Q - E)^{-1}$ is well defined and of polynomial growth in λ (which can be specified further) respectively $\mathcal{O}(\lambda^{-\frac{1}{3}})$ when

$$\frac{\mu}{\lambda} \leqslant C_1(\lambda^{-1}\ln\lambda)^{\frac{2}{3}} \quad \text{and} \quad \frac{\mu}{\lambda} \leqslant C_1\lambda^{-\frac{2}{3}} \quad \text{respectively}$$

for any fixed $C_1 > 0$, i.e., when

$$\mu \leqslant C_1\lambda^{\frac{1}{3}}(\ln\lambda)^{\frac{2}{3}} \quad \text{and} \quad \mu \leqslant C_1\lambda^{\frac{1}{3}} \quad \text{respectively.} \tag{5.3}$$

We end by making a comment about the Kramers–Fokker–Planck operator

$$P = hy\cdot\partial_x - V'(x)\cdot h\partial_y + \frac{1}{2}(y - h\partial_y)\cdot(y + h\partial_y) \tag{5.4}$$

on $\mathbf{R}^{2n} = \mathbf{R}^n_x \times \mathbf{R}^n_y$, where V is smooth and real-valued. The associated semiclassical symbol is

$$p(x, y; \xi, \eta) = i(y\cdot\xi - V'(x)\cdot\eta) + \frac{1}{2}(y^2 + \eta^2)$$

on $\mathbf{R}^{4n}$, and we note that $\operatorname{Re} p_1 \geqslant 0$. Under the assumption that the Hessian $V''(x)$ is bounded with all its derivatives, $|V'(x)| \geqslant C^{-1}$ when $|x| \geqslant C$ for some $C > 0$, and that V is a Morse function, Hérau, Stolk, and the author [10] showed among other things that the spectrum in any given strip $i[\frac{1}{C_1}, C_1] + \mathbf{R}$ is contained in a half-strip

$$i\Big[\frac{1}{C_1}, C_1\Big] + \Big[\frac{h^{2/3}}{C_2}, \infty\Big[\tag{5.5}$$

for some $C_2 = C_2(C_1) > 0$ and that the resolvent is $\mathcal{O}(h^{-2/3})$ in the complementary half-strip. (We refrain from recalling more detailed statements about spectrum and absence of spectrum in the regions, where $|\operatorname{Im} z|$ is large and small respectively.)

The proof of this result employed exponentially weighted estimates based on the fact that $H_{p_2}^2 p_1 > 0$ when $p_2 \asymp 1$, $p_1 \ll 1$. This is of course reminiscent of Theorem 1.2 (B) with $k = 2$ or rather the corresponding result in [7], but actually more complicated since our operator is not elliptic near ∞ and we even have that $i\mathbf{R} \setminus \{0\}$ is not in the range of p, but only in $\Sigma_\infty(p)$. It seems likely that the estimates on the spectrum of the KFP-operator above can be improved so that we can replace h by $h \ln(1/h)$ in the confinement (5.5) of the spectrum of P in the strip $i[1/C_1, C_1] + \mathbf{R}$ and that there are similar improvements for large and small values of $|\operatorname{Im} z|$. This would be obtained by either a closer look at the proof in [10] or an adaptation of the proof above when $k = 2$.

Acknowledgment. The work was supported by l'Agence Nationale de la Recherche, ANR-08-BLAN-0228-01.

References

1. Bordeaux Montrieux, W.: Loi de Weyl presque sûre et résolvante pour des opérateurs différentiels non-autoadjoints. Thèse, CMLS, Ecole Polytechnique (2008)
2. Boulton, L.S.: Non-selfadjoint harmonic oscillator, compact semigroups and pseudospectra. J. Operator Theory **47(2)**, 413–429 (2002)
3. Cialdea, A., Maz'ya, V.: Criterion for the L^p-dissipativity of second order differential operators with complex coefficients. J. Math. Pures Appl. IX Sér. **84**, no. 8, 1067–1100 (2005)
4. Cialdea, A., Maz'ya, V.: Criteria for the L^p-dissipativity of systems of second order differential equations. Ric. Mat. **55**, no. 2, 233–265 (2006)
5. Davies, E.B.: Semi-classical states for non-selfadjoint Schrödinger operators. Commun. Math. Phys. **200**, no. 1, 35–41 (1999)
6. Davies, E.B.: Pseudospectra, the harmonic oscillator and complex resonances. Proc. Roy. Soc. London Ser. A **455**, 585–599 (1999)

7. Dencker, N., Sjöstrand, J., Zworski, M.: Pseudospectra of semiclassical (pseudo-) differential operators. Commun. Pure Appl. Math. **57**, no. 3, 384–415 (2004)
8. Dimassi, M., Sjöstrand, J.: Spectral asymptotics in the semiclassical limit. London Math. Soc. Lect. Note Ser. **268**. Cambridge Univ. Press, Cambridge (1999)
9. Helffer, B., Sjöstrand, J.: Résonances en limite semi-classique. Mém. Soc. Math. Fr., Nouv. Ser. 24/25 (1986)
10. Hérau, F., Sjöstrand, J., Stolk, C.: Semiclassical analysis for the Kramers–Fokker–Planck equation. Commun. Partial Differ. Equ. **30**, no. 5-6, 689–760 (2005)
11. Lascar, B., Sjöstrand, J.: Equation de Schrödinger et propagation des singularités pour des opérateurs pseudodifférentiels à caractéristiques réelles de multiplicité variable I. Astérisque **95**, 467–523 (1982)
12. Hörmander, L.: The Analysis of Linear Partial Differential Operators. I-IV. Springer, Berlin (1983, 1985)
13. Kucherenko, V.V.: Asymptotic solutions of equations with complex characteristics (Russian). Mat. Sb., N. Ser. **95(137)**, 163–213 (1974); English transl.: Math. USSR Sb. **24(1974)**, 159–207 (1976)
14. Maslov, V.P.: Operational Methods. Mir Publishers, Moscow (1976)
15. Matte, O.: Correlation asymptotics for non-translation invariant lattice spin systems. Math. Nachr. **281**, no. 5, 721–759 (2008)
16. Maz'ya, V.G.: The degenerate problem with an oblique derivative (Russian). Mat. Sb. **87(129)**, 417–454 (1972); English transl.: Math. USSR Sb. **16**, 429–469 (1972)
17. Maz'ya, V.G., Paneyah, B.: Degenerate elliptic pseudo-differential operators and the problem with oblique derivative (Russian). Tr. Moskov. Mat. O-va **31**, 237–295 (1974)
18. Melin, A., Sjöstrand, J.: Fourier integral operators with complex phase functions and parametrix for an interior boundary value problem. Commun. Partial Differ. Equ. **1**, no. 4, 313–400 (1976)
19. Melin, A., Sjöstrand, J.: Determinants of pseudodifferential operators and complex deformations of phase space. Methods Appl. Anal. **9**, no. 2, 177–238 (2002)
20. Menikoff, A.: Sjöstrand, J.: On the eigenvalues of a class of hypo-elliptic operators. Math. Ann. **235**, 55–85 (1978)
21. Sjöstrand, J.: Singularités analytiques microlocales. Astérisque **95** (1982)
22. Trefethen, L.N.: Pseudospectra of linear operators. SIAM Rev. **39**, no. 3, 383–406 (1997)
23. Trefethen, L.N., Embree, M.: Spectra and Pseudospectra. The Behavior of Nonnormal Matrices and Operators. Princeton Univ. Press, Princeton, NJ (2005)
24. Zworski, M.: A remark on a paper of E. B Davies [5]. Proc. Am. Math. Soc. **129**, no. 10, 2955–2957 (2001)

Index

References to Maz'ya's Publications Made in Volume III

Monographs

- Maz'ya, V.G.: Einbettungssätze für Sobolewsche Räume. Teil 1 Teubner-Texte zur Mathematik, Leipzig (1979)
  285, 329 Pick [66]
- Maz'ya, V.G.: Sobolev Spaces. Springer, Berlin etc. (1985)
  4, 24 Adams–Hrynkiv–Lenhart [25]
  109, 152 Frazier–Verbitsky [28]
  252, 258 Netrusov–Safarov [9]
  285, 329 Pick [67]
  cf. also Vols. I and II
- Maz'ya, V.G., Shaposhnikova, T.O.: Theory of Multipliers in Spaces of Differentiable Functions. Pitman, Boston etc. (1985) Russian edition: Leningrad. Univ. Press, Leningrad (1986)
  198, 200 D.Mitrea–M.Mitrea–S.Monniaux [17]
  cf. also Vols. I and II
- Kozlov, V.A., Maz'ya, V.G., Rossmann, J.: Elliptic Boundary Value Problems in Domains with Point Singularities. Am. Math. Soc., Providence, RI (1997)
  198, 200 D.Mitrea–M.Mitrea–S.Monniaux [12]
  cf. also Vols. I, II

Research papers

- Maz'ya, V., Sobolevskii. P.: On the generating operators of semigroups (Russian). Usp. Mat. Nauk **17**, 151–154 (1962)
  47, 75 Cialdea [26]
- Maz'ya, V.G.: The negative spectrum of the higher-dimensional Schrödinger operator (Russian). Dokl. Akad. Nauk SSSR **144**, 721–722 (1962); English transl.: Sov. Math. Dokl. **3**, 808–810 (1962)
  6, 24 Adams–Hrynkiv–Lenhart [22]
  53, 75 Cialdea [23]
  cf. also Vol. I
- Maz'ya, V.G.: On the theory of the higher-dimensional Schrödinger operator (Russian). Izv. Akad. Nauk SSSR. Ser. Mat. **28**, 1145–1172 (1964)
  6, 24 Adams–Hrynkiv–Lenhart [23]
  53, 75 Cialdea [24]
  cf. also Vol. I
- Maz'ya, V.G.: Solvability in $\dot{W}_2^2$ of the Dirichlet problem in a region with a smooth irregular boundary (Russian). Vestn. Leningr. Univ. **22**, no. 7, 87–95 (1967)
  198, 200 D.Mitrea–M.Mitrea–S.Monniaux [15]
- Maz'ya, V.G.: On Neumann's problem for domains with irregular boundaries (Russian). Sib. Mat. Zh. **9**, 1322-1350 (1968); English transl.: Sib. Math. J. **9**, 990–1012 (1968)
  248, 258 Netrusov–Safarov [8]
- Maz'ya, V.G.: Certain integral inequalities for functions of several variables (Russian). Probl. Mat. Anal. **3**, 33–68 (1972)
  6, 24 Adams–Hrynkiv–Lenhart [24]
- Maz'ya, V.G.: On Beurling's theorem on the minimum principle for positive harmonic functions (Russian). Zap. Nauch. Semin. LOMI **30**, 76–90 (1972)
  26, 36, 37, 43, 44, 46 Aikawa [27]

- Maz'ya, V.G.: The degenerate problem with an oblique derivative (Russian). Mat. Sb., N. Ser. **87(129)**, 417-454 (1972); English transl.: Math. USSR Sb. **16**, 429–469 (1972)
  260, 263, 277 Palagachev [17]
  360, 384 Sjostrand [16]
- Maz'ya, V.G.: The coercivity of the Dirichlet problem in a domain with irregular boundary (Russian). Izv. VUZ, Ser. Mat. no. 4, 64–76 (1973)
  198, 200 D.Mitrea–M.Mitrea–S.Monniaux [16]
- Maz'ya, V.G.: Paneyah, B.: Degenerate elliptic pseudo-differential operators and the problem with oblique derivative.Tr. Mosk. Mat. Ob. **31**, 237–295 (1974)
  260, 261, 277 Palagachev [18]
  360, 384 Sjostrand [17]
- Verzhbinskii, G.M., Maz'ya, V.G.: Closure in L_p of the Dirichlet-problem operator in a region with conical points. Sov. Math. (Izv. VUZ) **18**, 5–14 (1974)
  111, 152 Frazier–Verbitsky [38]
- Kresin, G., Maz'ya, V.: Criteria for validity of the maximum modulus principle for solutions of linear parabolic systems. Ark. Mat. **32**, 121–155 (1994)
  47, 49, 54, 61, 74 Cialdea [16]
- Kozlov, V., Maz'ya, V., Fomin, A.: The inverse problem of coupled thermoelasticity. Inverse Probl. **10**, 153–160 (1994)
  77, 104 Eskin [14]
- Hansson, K., Maz'ya, V.G., Verbitsky, I.E.: Criteria of solvability for multidimensional Riccati's equations. Ark. Mat. **37**, 87–120 (1999)
  106, 152 Frazier–Verbitsky [20]
- Langer, M., Maz'ya, V.: On L^p-contractivity of semigroups generated by linear partial differential operators. J. Funct. Anal. **164**, 73–109 (1999)
  47, 70–73, 75 Cialdea [18]
- Maz'ya, V.G., Verbitsky, I.E.: The Schrödinger operator on the energy space: boundedness and compactness criteria. Acta Math. **188**, 263–302 (2002)
  110, 152 Frazier–Verbitsky [29]
  333, 358 Rozenblum [16]
- Cialdea, A., Maz'ya, V.: Criterion for the L^p-dissipativity of second order differential operators with complex coefficients. J. Math. Pures Appl. **84**, 1067–1100 (2005)
  47, 49–51, 54–57, 74 Cialdea [4]
  364, 383 Sjostrand [3]
- Maz'ya, V., Shubin, M.: Discreteness of spectrum and positivity criteria for Schrödinger operators. Ann. Math. **162**, 1–24 (2005)
  206, 245 Molchanov–Vainberg [20]
- Maz'ya, V.G., Verbitsky, I.: Form boundedness of the general second-order differential operator. Commun. Pure Appl. Math. **59**, no. 9, 1286–1329 (2006)
  333, 358 Rozenblum [17]
- Cialdea, A., Maz'ya, V.: Criteria for the L^p-dissipativity of systems of second order differential equations. Ric. Mat. **55**, 233–265 (2006)
  47, 57, 63–70, 74 Cialdea [5]
  364, 383 Sjostrand [4]
- Maz'ya, V.: Analytic criteria in the qualitative spectral analysis of the Schrödinger operator. Proc. Sympos. Pure Math. **76**, no. 1, 257–288 (2007)
  53, 75 Cialdea [25]
  206, 245 Molchanov–Vainberg [19]
- Maz'ya, V., Mitrea, M., Shaposhnikova, T.: The Dirichlet Problem in Lipschitz Domains with Boundary Data in Besov Spaces for Higher Order Elliptic Systems with Rough Coefficients. Preprint (2008)
  198, 200 D.Mitrea–M.Mitrea–S.Monniaux [18]